ALLE ZEIT WACH
1842

Manfred Domrös

Peng Gongbing

The Climate of China

With 126 Figures

Springer-Verlag
Berlin Heidelberg New York
London Paris Tokyo

Prof. Dr. Manfred Domrös
Geographisches Institut
der Universität Mainz
D-6500 Mainz

Prof. Peng Gongbing
Institute of Geography
Academia Sinica
Beijing 10011
China

Scientific Adviser:
Dr. Wolf Tietze, Helmstedt, FRG

ISBN-13: 978-3-642-73335-2 e-ISBN-13: 978-3-642-73333-8
DOI: 10.1007/978-3-642-73333-8

Softcover reprint of the hardcover 1st edition 1988

Typesetting, printing, and binding: Konrad Triltsch, Graphischer Betrieb, Würzburg
2131/3130-543210

Foreword

Since the founding of the People's Republic of China in 1949 there has been a rapid advance in climatology in China. The number of climatological stations has increased from less than 100 to more than 2,000, and the research work of Chinese climatologists covers various fields. The climate of China is no longer just a description of the average weather for an area or locality, but covers many fields such as the monsoon climate, the fluctuation of climate, the spatial and temporal variations of the climatic elements and physical and dynamic climate. Four books on the climate of China, written in Chinese, have been published so far. There is, however, no excellent book written in English on the climate of China, although Volume 8 of the *World Survey of Climatology,* dealing with the climates of northern and eastern Asia, edited by H. Arakawa in 1969, contains a chapter on the climate of China and Korea written by I.E.M. Watts. The data sources for China are based mainly on observations from 1940–1952 and the climatological charts of China published by the Central Weather Bureau of China in 1953 and 1955.

This monograph on *The Climate of China* by Prof. Dr. M. Domrös and Prof. Peng Gongbing is the first comprehensive and advanced book in English on the climate of China. It is based on climate data for the recent 30-year period (1951–1980), published by the State Meteorological Administration of China and research papers by Chinese climatologists over the past 20 years. The purpose of this book is to give a concise description of the main characteristics of the climate of China. It is particularly intended for the climatologists and geographers of the Western countries who would like to know more about the basic weather and climatic conditions of China.

The book consists of three major parts. The first part deals with the climate-controlling factors with particular emphasis on the effect of the general circulation of the atmosphere. Part two discusses the spatial and temporal variations of the climatic elements such as temperature, precipitation, cloudiness, insolation and surface winds. The third part deals with the climate regionalization of China which includes classification criteria and characteristics of each climate zone. In the Appendix, there is a detailed set of climate data for 68 stations for the period 1951–1980, of which 22 stations are provincial capital cities; the tables are particularly useful as a reference source for both geographers and climatologists, but also for applied purposes.

In the preparation of this excellent book, a Workshop on the Climate of China was held at the Institute of Geography, Mainz University, West Germany, from August 25 to September 5, 1986 in which 20 eminent German and Chinese scientists discussed the advances of climatology in China, with particular attention given to the contents of this comprehensive textbook. The authors of the book have made every effort to make the contents accessible to geographers and climatologists.

I believe that this book will certainly serve the scientists of climatology, geography and other environmental and applied sciences in Western countries. The book will contribute to a better understanding of the climatic characteristics of China.

Professor TAO SHI-YAN

President, Chinese Meteorological Society
and Member of Academia Sinica,
Geoscience Division, Beijing, China

Preface

The Climate of China is a joint venture carried out by the two authors who worked in close scientific collaboration during a more than 2-year research visit of Prof. Peng Gongbing, Institute of Geography, Chinese Academy of Sciences, Beijing, to the Institute of Geography, Mainz University, West Germany, by invitation from its director, Prof. Dr. Manfred Domrös. The research fellowship of Prof. Peng to Germany was generously granted by the Max-Planck Society, Munich, and covered the period from October 1984 to February 1987. This grant is greatly appreciated.

The common interest of the two authors to the climate of China developed from a mere idea to a mature plan for a textbook. The challenging plan, in fact, was facilitated by a unique and voluminous data set of climatic observations for China, provided by the State Meteorological Administration of China (SMA). The data available refer to a grand total of 279 stations, with a fair distribution in most parts of China. Some gaps in the climatic data exist in the western and northern Qinghai-Xizang (Tibetan) Plateau and the surrounding high mountains. The observation period covers the recent meteorological standard period from 1951–1980, either completely or partly for a number of years with available data. Thus, a fair degree of homogeneity over space and time could be obtained. We would like to express our gratitude to the SMA for its assistance in providing the climatic data on which this book is based.

Part of the observation data was contributed through the cooperation with the Deutscher Wetterdienst, Seewetteramt Hamburg, which is gratefully acknowledged.

The climatic data were computerized in the Computer Division of the Institute of Geography at Mainz University, carried out by Mr. Jürgen Wenzel under the supervision of Mr. Peter Spehs. Their outstanding help, both in the data input and processing as well as in the development of computer programs has accompanied this work to completion. Their collaboration is greatly appreciated. Part of the maps and diagrams presented in the book were also drafted and finally printed through specially developed computer programs. Once more, we express our warmest thanks to Mr. Jürgen Wenzel who also readily computerized the climate tables which are given in the Appendix.

Cartographic works for the book were carried out by Miss Susanne Kurz and Mr. Horst Engelhardt. With their support by the compilation of the manuscript, Miss Wiebke Krause and Mr. Hans Joachim Fuchs, both at the Institute of Geography, Mainz University, and Mr. Chen Deliang, at the Institute of Geography, Academy of Sciences of China, readily assisted the two authors. To all of them we are sincerely thankful for the close cooperation leading to the successful completion of the present book.

Thanks are also due to Miss Anette Kirchmajer, Institute of Geography, Mainz University, who patiently and efficiently carried out the tremendous burden of typing the manuscript, from the beginning to the final version. Typing on discs was done in the Central Computer Division of Mainz University (Prof. Dr. Schmutzler, Director), whose staff, particularly instructor Mrs. Doris Müller, we wish to thank for their generous cooperation.

We are greatly indebted to the President of the University of Mainz, Prof. Dr. P. Beyermann, and to the "Freunde der Universität Mainz", for financial grants given in order to carry out the tremendous bulk of the computer and cartographic works. Professor Domrös also wishes to express his thanks to the Hon. Minister of Cultural Affairs of the State of Rhineland-Palatinate for granting a sabbatical during the summer term 1986 in order to study the climate of China, including three visits to China in September/October and November 1986 as well as May and September 1987, which were further supported by the Academy of Sciences of China, the South China Academy of Tropical Crops and the Max-Planck Society, Munich. The visits to China included climatological field observations in different parts of the country in connection with productive scientific exchange on the climate of China.

During the progress of this work, we were gratefully granted scientific assistance and guidance by distinguished Chinese scientists: Prof. Tao Shi-yan, President of the Meteorological Society of China and Member of the Academy of Sciences of China, Beijing; Prof. Huang Bing-wei, Honorary Director, Institute of Geography, Chinese Academy of Sciences, Beijing; and Prof. Jiang Ai-liang, climatologist in the Commission of Integrated Survey of Natural Resources, Chinese Academy of Sciences, Beijing. The three eminent scientists were invited as chief guests to a German-Chinese Workshop on the Climate of China which took place at the Institute of Geography, Mainz University, from August 25 to September 5, 1986, organized by Prof. Dr. Manfred Domrös. The Workshop as such and the invitation to the three chief guests from China were possible due to the generous support given by the German Academic Exchange Service (DAAD), Bonn, the Ministry of Research and Technology of the Federal Republic of Germany as well as by several other funding agencies. To all of them we wish to express our gratitude.

In particular, Prof. Tao Shi-yan stimulated and inspired the work on this book. He also read the final draft. We are greatly indebted to him for his encouring comments, suggestions and personal communications. We also sincerely acknowledge Prof. Tao's cooperation in our work by presenting his foreword to the book. Dr. Wolf Tietze, Helmstedt, has acted as a scientific adviser to the book; his firm interest and support are very much appreciated.

Last, but not least, we are deeply obliged to the Springer Verlag and particularly to Dr. Dieter Czeschlik, for their willing acceptance of our concept and ideas concerning the book's contents and presentation.

With great pleasure and thanks, we wish to dedicate this book to our wifes, Mrs. Gisela Domrös and Mrs. Wang Zhi-min. Their patience was a continuous encouragement for the progress of this work.

Mainz and Beijing, June, 1988

Prof. Dr. MANFRED DOMRÖS
Prof. PENG GONGBING

Contents

1	**Introduction**	1
1.1	Aims and Concept of the Study	1
1.2	Climate Data	2
1.3	Review of Climate Studies on China	14
2	**Controlling Factors of the Climate**	20
2.1	Latitude, Longitude and Location	20
2.2	Topography and Landforms	23
2.3	Distribution of Land and Sea and Nature of the Underlying Ground	26
2.4	Seasons	28
3	**Circulation**	30
3.1	Seasonal Pressure Distribution at Sea Level	30
3.2	Seasonally Prevailing Winds and Air Masses	34
3.3	Winter and Summer Monsoon	39
3.3.1	Characteristics of the Monsoon in General	39
3.3.2	Onset and Duration of the Winter Monsoon	43
3.3.3	Periods of Active and Weak Winter Monsoon	45
3.3.4	Damage Due to Strong Cold Outbreaks of Winter Monsoon	50
3.3.5	Onset and Duration of the Summer Monsoon	52
3.3.6	Some Characteristics of the Summer Monsoon	57
3.4	Frontology	59
3.4.1	Mean Front Position in January and July	59
3.4.2	The Stationary Fronts in February and March as well as in the Pre-Typhoon Season in South China	59
3.4.3	Some Characteristics of the Mei-Yu Front	62
3.5	The Transient Disturbances	65
3.5.1	The Upper Westerly Troughs in the Westerlies	66
3.5.2	Extratropical Cyclones and Anticyclones	68
3.5.3	Typhoons	71
4	**Temperature**	77
4.1	Mean Annual Air Temperature Distribution	77
4.2	Mean Seasonal Temperature Distribution	80

4.3 Annual Range and Annual Variation of Temperature 86
4.4 Onset and End of Certain Limited Temperatures and Their Duration 101
4.4.1 Mean Daily Air Temperature ≤0 °C 103
4.4.2 Mean Daily Air Temperature ≥10 °C 107
4.4.3 Maximum Daily Air Temperature ≥35 °C 110
4.4.4 Other Extreme Limited Temperatures 112
4.5 Vertical Distribution of Temperature 114
4.6 Comparison of Temperature at the Same Latitude . 116
4.7 Diurnal Range of Temperature 119
4.8 Interannual Variability of Temperature 122
4.8.1 Variability of Annual Mean Temperature 123
4.8.2 Variability of Monthly Mean Temperatures 124
4.9 Historical-Climatic Change of Temperature During the last 5,000, 500 and 100 Years 130

5 Precipitation 139

5.1 Mean Annual Precipitation Distribution 139
5.2 Mean Seasonal Precipitation Distribution 141
5.3 Annual Variation of Precipitation 155
5.3.1 Specific Precipitation Types and Their Distribution 155
5.3.2 Variation of Wet and Dry Months over Space and Time 160
5.3.3 Summer Precipitation 168
5.4 Interannual Precipitation Variability 168
5.4.1 Variability of Annual Precipitation 169
5.4.2 Variability of Monthly Precipitation 174
5.4.3 Variability of Annual and Monthly Precipitation at Beijing 174
5.5 Precipitation Frequency Expressed in Rainy Days . 182
5.6 Precipitation Intensity 183
5.7 Rainstorms and Certain Events of Heavy Rainfall . 186
5.8 Diurnal Variation of Precipitation 190
5.9 Influence of Topography and Elevation on Precipitation 191
5.9.1 Influence of the Exposition of Slopes on Precipitation 191
5.9.2 Effect of Elevation on Precipitation 193
5.10 Historical Change of Precipitation 195
5.11 Snow 201
5.11.1 Mean Length of Snow Cover Period 201
5.11.2 Number of Snowfall Days 205
5.11.3 Maximum Depth of Snow 206
5.11.4 Altitude of the Snow Line 207

6	**Cloudiness and Sunshine**	210
6.1	Mean Annual Cloudiness and January and July Amount	210
6.2	Sunshine	212
6.2.1	Annual Sunshine Duration	213
6.2.2	Sunshine Duration in January and July and Annual Variation	216
6.3	Global Radiation	218
6.4	Fog	220
7	**Surface Wind**	222
7.1	Mean and Extreme Wind Velocities	222
7.2	Local Wind Systems	224
7.2.1	Mountain and Valley Breezes	224
7.2.2	Land and Sea Breezes, Lake Breeze	227
7.2.3	Plateau Monsoon	227
7.2.4	Local Dry and Hot Winds	229
8	**Climate Classification and Division of China**	230
8.1	General Objectives and Fundamentals of Climate Regionalization	230
8.2	China Within Global Climate Classifications	232
8.3	National Climate Classifications of China	240
8.4	Climate Division of China According to HUANG Bing-wei (1986)	253
9	**Climate Zones of China**	258
9.1	Cold Temperate Zone (I)	259
9.2	Middle Temperate Zone (II)	260
9.3	Warm Temperate Zone (III)	263
9.4	Northern Subtropical Zone (IV)	266
9.5	Middle Subtropical Zone (V)	268
9.6	Southern Subtropical Zone (VI)	270
9.7	Peripheral Tropical Zone (VII)	272
9.8	Middle Tropical Zone (VIII)	274
9.9	Southern Tropical Zone (IX)	275
9.10	Alpine Plateau Zone (H0)	276
9.11	Subalpine Plateau Zone (HI)	277
9.12	Temperate Plateau Zone (HII)	278
Appendix: Climate Tables		281
References		351
Subject Index		358

1 Introduction

1.1 Aims and Concept of the Study

Climate is by nature a rather complex theme, because of the manifold earth-atmosphere interaction which considerably varies over space and time and finally creates a specific type of climate at a particular location. The amalgamation of the individual climate types at various locations results in a distinct climate type for a whole region. Considering the climate as a scientific subject, its description depends upon the specific reason and purpose of application. In the widest and most general sense, climate can be understood as a particular composition of the atmospheric conditions on earth which produce a distinct climatic environment in a biospheric context. Any climate description necessarily depends upon available instrumental readings which have to be processed and interpreted.

To deal with the climate of China is a challenging and, at the same time, sensitive task because China is a large country with a vast territory and extremely complex landforms so that, almost at first glance, a considerably large variation of climate must be expected. Evidence of this has already been given from the early approach towards a climate classification of the earth as represented in the first global regionalization scheme of the climate by Köppen (since 1884). He has shown that the global variation of climate meets the climate variation in China alone. This was confirmed by nearly all climate classifications compiled after Köppen, although different criteria have been applied (cf. Chap. 8).

It is obvious from climate classifications that the main feature of the climate of China is therefore its diversity and contrariety which together may lead to the existence of a great number of climate types. On this background, a review of the climate of China represents a particularly sensitive task, depending upon the method of investigation chosen and the aim and intention of the study presented. For this survey, a synthetic approach of the climate has been strived for. Its major concept includes the following:

1. First, the principles of circulation over China and the East Asia-Pacific region are described in order to give basic information for a genetic understanding of the climate elements and their characteristics over space and time.
2. Second, the climate elements are presented *individually,* with particular attention given to the two main elements, temperature and precipitation, and under consideration of the most common variables over time (e.g. the annual total, seasonal or monthly total resp. average). The climate data used are normally mean values, preferably for the recent meteorological standard period 1951–1980, occasionally added by recorded extreme values.
   The climate data given are mostly surface (screen) records, and their interpretation includes the effects caused by both the orographical structure and

landforms of China as well as the nature of the underlying ground on the prevailing air masses and circulation systems. Characterized as a climate-geographical approach, this study mostly aims at a presentation of climate as a result of interaction between various factors of the physical spheres. According to the data sources, the study mainly considers the large-scale climatic characteristics on the surface.

3. Finally, the study aims at a contribution towards a regionalization of the climate of China, by considering climate types and climate regions according to their homogeneity over space and time. Therefore, the latest version of the climate classification of China after HUANG Bing-wei (1986) was chosen [1]. The macro-scale climate regions are each briefly viewed.
4. A comprehensive set of climate tables for 68 stations, containing monthly and annual means of 28 variables as far as available, is given in the Appendix for the reader's specific interest and interpretation.

Since this work is an account on the spatial and temporal variations of the climate of China, the major research discipline is Regional Climatology. Although concentrated on the climate of China, the book may also be a useful source of comparative studies on Regional Climatology. From this viewpoint, climate statistics are an indispensable source of information.

This book is intended to meet a variety of needs and readers. Students and scientists in climatology, meteorology, geography, botany, agriculture and other disciplines concerned with the impact of climate upon human and natural affairs, may thus gain useful information.

Although this study is limited in its content due to the climate data available, it nevertheless aims to contribute towards a comprehensive account of the climatic characteristics and variations over space (both in a horizontal and vertical direction) and over time. Results and findings are of a poorer content in certain parts of China, due to the paucity of climatological records.

1.2 Climate Data

This study is based on records from a total of 279 stations which, on average, means one station for every 34,000 km^2. The distribution of stations in practice, however, is uneven and shows considerable gaps in western China where in major parts of the Qinghai-Xizang (Tibetan) Plateau, including the surrounding high mountains, stations are completely lacking. In all of eastern China, a relatively dense distribution of climate stations exists. Therefore, the network of stations and data available can be regarded to a fair extent as a sound source of climatological information.

[1] Presented by Prof. HUANG Bing-wei at the German-Chinese Workshop on *The Climate of China*, which took place in the Institute of Geography, University of Mainz, FRG, August 25–September 5, 1986; see also Chap. 8.

Fig. 1.1. Dispersion map of the climate stations, giving their distribution and reference numbers (cf. Table 1.1). Stations indicated by a square are climatological observatories, while stations with a cross are temperature and rainfall recording stations only

For the purpose of quick reference to the climate stations, the reference number of each station (1–279) is added to the station name (for example: Beijing/61).

The distribution of all 279 reference stations is given in Fig. 1.1. The names and reference numbers of all stations (1–279) are explained in Table 1.1, together with their coordinates and altitude above sea level. Since China shows an extremely large altitudinal extent and major parts are occupied by mountains, particular attention has to be paid to the mountain climate and thus to climate stations in the mountains. This aim could be reasonably fulfilled up to 4,500 m only, by a fair number of stations, particularly in the southeastern Qinghai-Xizang (Tibetan) Plateau and in the mountains of Qinghai and Qamdo. In total, 82 of all 279 stations are located above 1,000 m above sea level, the highest

(continued p. 13)

Table 1.1. List of climate stations in numerical order including reference number (1–279), latitude and longitude, elevation above sea level (m)

No.	Name	Latitude	Longitude	Elevation
1	Mohe	53° 28' N	122° 22' E	296.0 m
2	Huma	51° 43' N	126° 39' E	177.4 m
3	Sunwu	49° 26' N	127° 21' E	234.5 m
4	Nenjiang	49° 10' N	125° 13' E	222.3 m
5	Bugt	48° 46' N	121° 55' E	738.7 m
6	Hailar	49° 13' N	119° 45' E	612.9 m
7	Arxan	47° 10' N	119° 57' E	1026.5 m
8	Qiqihar	47° 23' N	123° 55' E	145.9 m
9	Keshan	48° 03' N	125° 53' E	236.9 m
10	Hailun	47° 26' N	126° 58' E	239.4 m
11	Yichun	47° 43' N	128° 54' E	231.3 m
12	Fujin	47° 14' N	131° 59' E	64.2 m
13	Tonghe	45° 58' N	128° 44' E	108.6 m
14	Anda	46° 23' N	125° 19' E	149.3 m
15	Harbin	45° 41' N	126° 37' E	171.7 m
16	Shangzhi	45° 13' N	127° 58' E	189.7 m
17	Jixi	45° 17' N	130° 57' E	233.1 m
18	Suifenhe	44° 23' N	131° 09' E	496.7 m
19	Mudanjiang	44° 34' N	129° 36' E	241.4 m
20	Qian Gorlos	45° 07' N	124° 50' E	134.7 m
21	Dong Ujmqin Qi	45° 31' N	116° 58' E	839.1 m
22	Xi Ujimqin Qi	44° 35' N	117° 36' E	995.9 m
23	Bairin Zuoqi	44° 00' N	119° 12' E	483.4 m
24	Jarud Qi	44° 34' N	120° 54' E	265.0 m
25	Tongliao	43° 40' N	122° 15' E	179.5 m
26	Changling	44° 15' N	123° 58' E	191.9 m
27	Changchun	43° 54' N	125° 13' E	236.8 m
28	Siping	43° 11' N	124° 20' E	164.2 m
29	Huadian	42° 59' N	126° 45' E	263.3 m
30	Dunhua	43° 22' N	128° 12' E	523.7 m
31	Yanji	42° 53' N	129° 28' E	176.8 m

(contd.)

Table 1.1 continued

No.	Name	Latitude	Longitude	Elevation
32	Linjiang	41° 43' N	126° 55' E	332.5 m
33	Jian	41° 06' N	126° 09' E	177.7 m
34	Kuandian	40° 43' N	124° 47' E	260.1 m
35	Dandong	40° 03' N	124° 20' E	15.1 m
36	Yingkou	40° 40' N	122° 16' E	3.3 m
37	Benxi	40° 53' N	123° 54' E	233.1 m
38	Shenyang	41° 46' N	123° 26' E	41.6 m
39	Zhangwu	42° 25' N	122° 32' E	79.4 m
40	Linxi	43° 30' N	118° 03' E	806.6 m
41	Abagnar Qi	43° 57' N	116° 04' E	989.5 m
42	Abag Qi	43° 41' N	114° 29' E	1126.1 m
43	Erenhot	43° 39' N	112° 00' E	964.8 m
44	Sonid Youqi	42° 37' N	112° 50' E	1150.5 m
45	Huade	41° 54' N	114° 00' E	1482.5 m
46	Duolun	42° 11' N	116° 28' E	1245.4 m
47	Weichang	41° 56' N	117° 45' E	842.3 m
48	Chifeng	42° 16' N	118° 55' E	571.1 m
49	Chaoyang	41° 33' N	120° 27' E	168.7 m
50	Jinzhou	41° 08' N	121° 07' E	168.7 m
51	Chengde	40° 58' N	117° 50' E	375.2 m
52	Fengning	41° 12' N	116° 32' E	659.7 m
53	Huailai	40° 24' N	115° 30' E	536.8 m
54	Zhangjiakou	40° 47' N	114° 53' E	723.9 m
55	Jining	40° 58' N	113° 03' E	1416.5 m
56	Bailingmiao	41° 42' N	110° 26' E	1375.9 m
57	Haliut	41° 40' N	108° 48' E	1288.2 m
58	Hohhot	40° 49' N	111° 41' E	1063.0 m
59	Datong	40° 6' N	113° 20' E	1067.6 m
60	Yuxian	39° 50' N	114° 34' E	909.5 m
61	Beijing	39° 48' N	116° 28' E	31.2 m
62	Tianjin	39° 6' N	117° 10' E	3.3 m

(contd.)

Table 1.1 continued

No.	Name	Latitude	Longitude	Elevation
63	Leting	39° 25' N	118° 54' E	10.5 m
64	Dalian	38° 54' N	121° 38' E	93.5 m
65	Cangzhou	38° 20' N	116° 55' E	11.4 m
66	Baoding	38° 50' N	115° 34' E	17.2 m
67	Yuanping	38° 45' N	112° 42' E	836.7 m
68	Yulin	38´ 14' N	109° 42' E	1057.5 m
69	Taiyuan	37° 47' N	112° 33' E	777.9 m
70	Shijiazhuang	38° 04' N	114° 26' E	81.8 m
71	Dezhou	37° 28' N	116° 13' E	21.2 m
72	Huimin	37° 30' N	117° 32' E	11.3 m
73	Weifang	36° 42' N	119° 05' E	44.1 m
74	Rongcheng	37° 24' N	122° 41' E	47.7 m
75	Qingdao	36° 04' N	120° 20' E	76.0 m
76	Yiyuan	36° 11' N	118° 09' E	304.5 m
77	Jinan	36° 41' N	116° 59' E	51.6 m
78	Tai Shan	36° 15' N	117° 06' E	1533.7 m
79	Xingtai	37° 04' N	114° 30' E	76.8 m
80	Jiexiu	37° 03' N	111° 56' E	748.8 m
81	Yan'an	36° 36' N	109° 30' E	957.6 m
82	Anyang	36° 07' N	114° 22' E	75.5 m
83	Heze	35° 15' N	115° 26' E	49.7 m
84	Yanzhou	35° 34' N	116° 51' E	51.6 m
85	Linyi	35° 03' N	118° 21' E	87.9 m
86	Sheyang	33° 46' N	120° 15' E	2.0 m
87	Xuzhou	34° 17' N	117° 18' E	43.0 m
88	Bo Xian	33° 53' N	115° 46' E	37.7 m
89	Xihua	33° 47' N	114° 31' E	52.6 m
90	Zhengzhou	34° 43' N	113° 39' E	110.4 m
91	Luoyang	34° 40' N	112° 25' E	154.5 m
92	Lushi	34° 00' N	111° 01' E	568.8 m
93	Hua Shan	34° 29' N	110° 05' E	2064.9 m

(contd.)

Table 1.1 continued

No.	Name	Latitude	Longitude	Elevation
94	Yuncheng	35° 2' N	111° 00' E	367.8 m
95	Xi'an	34° 18' N	108° 56' E	396.9 m
96	Baoji	34° 21' N	107° 08' E	612.4 m
97	Tianshui	34° 35' N	105° 45' E	1131.7 m
98	Xifengzhen	35° 44' N	107° 38' E	1421.9 m
99	Pingliang	35° 33' N	106° 40' E	1346.6 m
100	Tongwei	35° 23' N	105° 00' E	2450.6 m
101	Wudu	33° 23' N	104° 41' E	1079.1 m
102	Songpan	32° 39' N	103° 34' E	2827.7 m
103	Pingwu	32° 25' N	104° 31' E	876.5 m
104	Hanzhong	33° 04' N	107° 02' E	508.4 m
105	Wanyuan	32° 04' N	108° 02' E	674.0 m
106	Ankang	32° 43' N	109° 02' E	290.8 m
107	Guanghua	32° 25' N	111° 40' E	91.1 m
108	Nanyang	33° 02' N	112° 35' E	129.8 m
109	Xinyang	32° 7' N	114° 05' E	75.9 m
110	Zhumadian	32 58' N	114° 03' E	83.7 m
111	Fuyang	32° 56' N	115° 50' E	30.6 m
112	Gushi	32° 10' N	115° 40' E	56.9 m
113	Huoshan	31° 24' N	116° 20' E	68.1 m
114	Hefei	31° 51' N	117° 17' E	23.6 m
115	Bengbu	32° 57' N	117° 22' E	21.0 m
116	Nanjing	32° 00' N	118° 48' E	8.9 m
117	Dongtai	32° 51' N	120° 18' E	6.3 m
118	Zhenjiang	33° 36' N	119° 02' E	17.5 m
119	Shanghai	31° 10' N	121° 26' E	4.5 m
120	Dinghai	30° 02' N	122° 07' E	35.7 m
121	Shipu	29° 12' N	121° 57' E	128.4 m
122	Sheng Xian	29° 36' N	120° 49' E	104.3 m
123	Hangzhou	30° 14' N	120° 10' E	41.7 m
124	Huangshan	30° 08' N	118° 09' E	1840.4 m

(contd.)

Table 1.1 continued

No.	Name	Latitude	Longitude	Elevation
125	Anqing	30° 31' N	117° 2' E	44.0 m
126	Wuhan	30° 38' N	114° 04' E	23.3 m
127	Jiangling	30° 24' N	112° 05' E	32.7 m
128	Zhongxiang	31° 10' N	112° 34' E	65.8 m
129	Yichang	30° 42' N	111° 05' E	131.1 m
130	Enshi	30° 17' N	109° 28' E	437.2 m
131	Nanchong	30° 48' N	106° 05' E	297.7 m
132	Mianyang	31° 28' N	104° 41' E	470.8 m
133	Barkan	31° 54' N	102° 14' E	2664.4 m
134	Chengdu	30° 40' N	104° 01' E	505.9 m
135	Ya'an	29° 59' N	103° 00' E	627.6 m
136	Emei Shan	29° 31' N	103° 20' E	304.7 m
137	Jiulong	28° 59' N	101° 33' E	2987.3 m
138	Xichang	27° 53' N	102° 18' E	1590.7 m
139	Yibin	28° 49' N	104° 32' E	340.9 m
140	Neijiang	29° 35' N	105° 03' E	352.3 m
141	Luzhou	28° 53' N	105° 26' E	334.8 m
142	Chongqing	29° 35' N	106° 28' E	260.6 m
143	Nunchuan	28° 57' N	107° 08' E	1905.9 m
144	Youyang	28° 50' N	108° 46' E	663.7 m
145	Yuanling	28° 28' N	110° 24' E	143.2 m
146	Changde	28° 55' N	111° 33' E	36.7 m
147	Changsha	28° 12' N	113° 04' E	44.5 m
148	Yueyang	29° 23' N	113° 05' E	51.6 m
149	Xiushui	29° 02' N	114° 34' E	146.8 m
150	Nanchang	28° 36' N	115° 55' E	46.7 m
151	Jingdezhen	29° 10' N	117° 15' E	46.3 m
152	Qu Xian	28° 58' N	118° 52' E	66.1 m
153	Lishui	28° 27' N	119° 55' E	60.8 m
154	Wenzhou	28° 01' N	120° 40' E	6.0 m
155	Fuding	27° 20' N	120° 12' E	36.2 m

(contd.)

Table 1.1 continued

No.	Name	Latitude	Longitude	Elevation
156	Nanping	26° 39' N	118° 10' E	127.2 m
157	Pucheng	27° 55' N	118° 32' E	283.3 m
158	Nancheng	27° 33' N	116° 36' E	80.9 m
159	Guangchang	26° 51' N	116° 20' E	143.8 m
160	Jian	27° 07' N	114° 58' E	76.4 m
161	Yichun	27° 48' N	114° 23' E	128.5 m
162	Shaoyang	27° 14' N	111° 28' E	249.8 m
163	Wugang	26° 44' N	110° 38' E	340.2 m
164	Zhijiang	27° 27' N	109° 41' E	272.2 m
165	Sinan	27° 57' N	108° 15' E	416.3 m
166	Zunyi	27° 42' N	106° 53' E	843.9 m
167	Bijie	27° 18' N	105° 14' E	1510.6 m
168	Weining	26° 52' N	104° 17' E	2239.5 m
169	Zhaotang	27° 20' N	103° 45' E	1949.5 m
170	Huili	26° 39' N	102° 15' E	1787.1 m
171	Zhanyi	25° 35' N	103° 50' E	1898.7 m
172	Xingren	25° 26' N	105° 11' E	1378.5 m
173	Guiyang	26° 35' N	106° 43' E	1071.2 m
174	Luodian	25° 26' N	106° 46' E	440.3 m
175	Dushan	25° 50' N	107° 33' E	972.2 m
176	Rongjiang	25° 58' N	108° 32' E	285.7 m
177	Guilin	25° 20' N	110° 18' E	166.7 m
178	Quanzhou	25° 48' N	111° 36' E	191.7 m
179	Lingling	26° 14' N	111° 36' E	174.5 m
180	Ganzhou	25° 50' N	114° 50' E	123.8 m
181	Changting	25° 51' N	116° 22' E	317.5 m
182	Yong'an	25° 58' N	117° 21' E	206.0 m
183	Fuzhou	26° 05' N	119° 17' E	84.0 m
184	Pingtan	25° 31' N	119° 47' E	32.5 m
185	Taibei	25° 02' N	121° 31' E	9.0 m
186	Hualian	24° 01' N	121° 37' E	14.0 m

(contd.)

Table 1.1 continued

No.	Name	Latitude	Longitude	Elevation
187	Hengchun	22° 00' N	120° 45' E	24.0 m
188	Tainan	23° 00' N	120° 13' E	14.0 m
189	Magong	23° 31' N	119° 34' E	22.0 m
190	Taichong	24° 09' N	120° 41' E	78.0 m
191	Dehua	25° 43' N	118° 06' E	1650.0 m
192	Xiamen	24° 27' N	118° 04' E	63.2 m
193	Zhangzhou	24° 30' N	117° 39' E	30.0 m
194	Shantou	23° 24' N	116° 41' E	1.2 m
195	Mei Xian	24° 18' N	116° 07' E	77.5 m
196	Haifeng	22° 47' N	115° 22' E	5.8 m
197	Heyuan	23° 44' N	114° 41' E	41.1 m
198	Shenzen	22° 33' N	114° 06' E	18.2 m
199	Guangzhou	23° 8' N	113° 19' E	6.3 m
200	Shaoguan	24° 48' N	113° 35' E	69.3 m
201	Lian Xian	24° 47' N	112° 23' E	97.6 m
202	Mengshan	24° 12' N	110° 31' E	144.0 m
203	Guiping	23° 24' N	110° 05' E	42.2 m
204	Wuzhou	23° 29' N	111° 18' E	119.2 m
205	Yangjiang	21° 52' N	111° 56' E	23.3 m
206	Zhanjiang	21° 13' N	110° 24' E	25.3 m
207	Haikou	20° 02' N	110° 21' E	14.1 m
208	Qianghai	19° 14' N	110° 28' E	23.5 m
209	Xisha	16° 50' N	112° 20' E	4.7 m
210	Dongfang	19° 6' N	108° 37' E	8.4 m
211	Dan Xian	19° 31' N	109° 35' E	168.7 m
212	Beihai	21° 29' N	109° 06' E	14.6 m
213	Qinzhou	21° 57' N	108° 36' E	4.0 m
214	Longzhou	22° 22' N	106° 45' E	128.3 m
215	Nanning	22° 49' N	108° 21' E	72.2 m
216	Hechi	24° 42' N	108° 03' E	213.9 m
217	Bose	23° 55' N	106° 32' E	173.2 m

(contd.)

Table 1.1 continued

No.	Name	Latitude	Longitude	Elevation
218	Guangnan	24° 02' N	105° 02' E	1250.5 m
219	Mengzi	23° 23' N	103° 23' E	1300.7 m
220	Yuanjiang	23° 34' N	102° 09' E	396.6 m
221	Jinghong	21° 52' N	101° 04' E	552.7 m
222	Lancang	22° 34' N	99° 06' E	1054.0 m
223	Simao	22° 40' N	101° 24' E	1302.1 m
224	Lincang	23° 57' N	100° 13' E	1463.5 m
225	Kunming	25° 1' N	102° 41' E	1891.4 m
226	Chuxiang	25° 01' N	101° 32' E	1772.0 m
227	Tengchong	25° 07' N	98° 29' E	1647.8 m
228	Dali	25° 43' N	100° 11' E	1990.5 m
229	Lijiang	26° 52' N	100° 26' E	2393.2 m
230	Deqen	28° 39' N	99° 10' E	3588.6 m
231	Litang	30° 00' N	100° 16' E	3948.9 m
232	Batang	30° 00' N	99° 06' E	2589.2 m
233	Garze	31° 38' N	99° 59' E	3393.5 m
234	Dege	31° 44' N	98° 34' E	3201.2 m
235	Qamdo	31° 11' N	96° 59' E	3240.7 m
236	Nyingchi	29° 34' N	94° 28' E	3000.0 m
237	Yadong	27° 44' N	89° 05' E	4300.0 m
238	Xigaze	29° 13' N	88° 55' E	3836.0 m
239	Lhasa	29° 42' N	91° 8' E	3658.0 m
240	Nagqu	31° 29' N	92° 03' E	4507.0 m
241	Sog Xian	31° 53' N	93° 47' E	4022.8 m
242	Dengqen	31° 9' N	95° 36' E	3893.1 m
243	Yushu	33° 6' N	96° 45' E	3702.6 m
244	Darlag	33° 45' N	99° 39' E	3967.9 m
245	Madoi	34° 55' N	98° 13' E	4272.3 m
246	Tongde	35° 3' N	100° 20' E	3289.4 m
247	Lanzhou	36° 3' N	103° 53' E	1517.2 m
248	Zhongning	37° 29' N	105° 40' E	1183.3 m

(contd.)

Table 1.1 continued

No.	Name	Latitude	Longitude	Elevation
249	Yanchi	37° 47' N	107° 24' E	1347.8 m
250	Yinchuan	38° 29' N	106° 13' E	1111.5 m
251	Jaitai	39° 47' N	105° 45' E	1031.8 m
252	Bayan Mod	40° 45' N	104° 30' E	1328.1 m
253	Zhangye	38° 56' N	100° 26' E	1482.7 m
254	Minqin	38° 38' N	103° 05' E	1367.0 m
255	Tianzhu	37° 12' N	102° 52' E	3045.1 m
256	Xining	36° 35' N	101° 55' E	2261.2 m
257	Dulan	36° 18' N	98° 06' E	3191.1 m
258	Golmud	36° 12' N	94° 38' E	2807.7 m
259	Da Qaidam	37° 50' N	95° 17' E	3173.2 m
260	Lenghu	38° 50' N	93° 23' E	2733.0 m
261	Jiuquan	39° 46' N	98° 31' E	1477.2 m
262	Yumenzhen	40° 16' N	97° 02' E	1526.0 m
263	Dunhuang	40° 09' N	94° 41' E	1138.7 m
264	Ruoqiang	39° 2' N	88° 10' E	888.3 m
265	Hami	42° 49' N	93° 31' E	737.9 m
266	Qijiaojing	43° 29' N	91° 38' E	873.2 m
267	Turpan	42° 56' N	89° 12' E	34.5 m
268	Qitai	44° 01' N	89° 34' E	793.6 m
269	Ürümqi	43° 54' N	87° 28' E	653.5 m
270	Altay	47° 44' N	88° 5' E	735.1 m
271	Hoboksar	46° 47' N	85° 43' E	1291.6 m
272	Tacheng	46° 44' N	83° 00' E	548.0 m
273	Tinghe	44° 37' N	82° 54' E	320.1 m
274	Yining	43° 57' N	81° 20' E	662.5 m
275	Kuqa	41° 43' N	82° 57' E	1099.0 m
276	Hotan	37° 8' N	79° 56' E	1374.6 m
277	Bachu	39° 48' N	78° 34' E	1116.5 m
278	Shache	38° 26' N	77° 16' E	1231.2 m
279	Kashi	39° 28' N	75° 59' E	1288.9 m

mountain station is at 4,507 m above sea level (at Nagqu/240); according to their altitude, the mountain stations can be divided as follows:

1,000–1,999 m: 52 stations
2,000–2,999 m: 11 stations
3,000–3,999 m: 15 stations
≥ 4,000 m: 4 stations.

A dispersion diagram, showing all stations above 1,000 m, demonstrates the vertical distribution of the mountain stations (Fig. 1.2); 200-m classes were used (1,000–1,200, . . . , 4,400–4,600 m). Although to a limited extent, reliable climatological information can nevertheless also be given for the lower mountain regions of China.

Since all 279 climate stations are under the authority of the State Meteorological Administration of China, the readings also confirm reliability and homogeneity in terms of their recording techniques and data processing methods. Without any adjustment and correction, the readings of all stations are comparable.

The climate stations used are of two types:
(1) 100 stations which represent fully equipped climatological observatories of which the following 28 climatological variables are considered for this study; data were taken as monthly means and annual totals resp. means, occasionally added to by recorded extreme values:

1. Atmospheric pressure
2. Mean temperature
3. Highest recorded maximum temperature
4. Lowest recorded minimum temperature
5. Mean maximum temperature
6. Mean minimum temperature
7. Number of days $<0°C$
8. Number of days $<-10°C$
9. First and last date showing a maximum temperature $\geq 0°C$

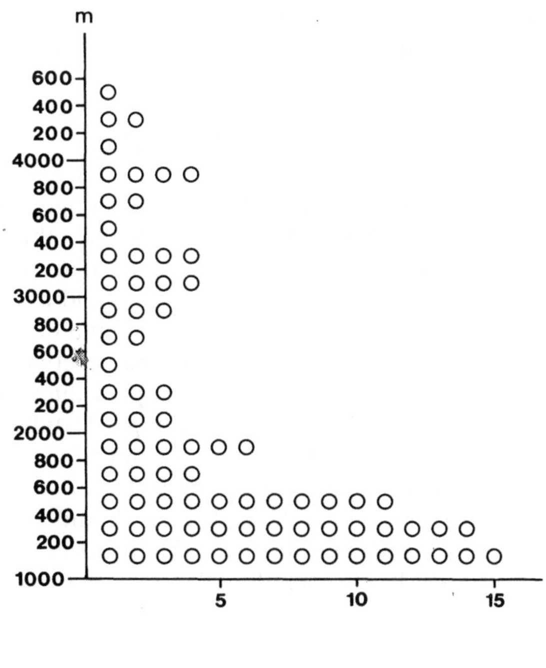

Fig. 1.2. Altitudinal dispersion diagram of mountain climate stations (≥ 1,000 m a.s.l.). Each circle expresses one station

10. First and last date showing a maximum temperature $\geq 10°C$
11. Number of frost days
12. Relative humidity
13. Cloudiness
14. Precipitation amount
15. Number of days >0.1 mm
16. Number of days >5 mm
17. Number of days >10 mm
18. Number of days >25 mm
19. Evaporation
20. Snow depth
21. Number of snowfall days
22. Length of the snow cover period
23. Sunshine duration (total)
24. Sunshine duration (percentage of the duration maximum)
25. Wind velocity
26. Prevailing wind direction
27. Percentage of prevailing wind direction
28. Percentage of calms

For 68 stations, a climate table is given in the Appendix, containing monthly and annual means as far as available for all variables over the 30-year standard period 1951–1980 or part of it. The number of reference years is added to each variable and station under consideration.

(2) 179 stations with mean monthly and annual records of temperature and precipitation only. The observations refer to the period 1951–1980.

As exceptional cases, the observation period from 1951–1980 was extended for monthly and annual precipitation totals, at Beijing/61 for a 257-year period from 1724–1980, and at Shanghai/119 for a 108-year period from 1873–1980. However, it has to be considered that the data for Beijing are partly heterogeneous over time since the location of the observatory has changed several times, while it remained at the same location in the case of Shanghai. Records at Shanghai (and Hongkong) represent the longest, homogeneous data source in all of China.

All climate information used in this study is represented by original data, according to the official readings from the State Meteorological Administration of China. Unless otherwise mentioned, maps and diagrams in this study represent original drafts, compiled for this study from original records.

1.3 Review of Climate Studies on China

Climatology in China gained rapid progress only after the foundation of the People's Republic of China in 1949 when the number of meteorological stations increased from less than 100 to more than 2,000 and research work was stepwise extended to a large number of basic and applied fields as well. With the availability of sufficient observation data and qualified scientific personnel, climatology has advanced to an international level only over the last 10 years. Since then, international links of scientific cooperation and exchange were successfully established. As a result, recent climatological studies and research on China were so far mostly carried out by scientists from China.

In all, scientists from three different categories of institutions in China are engaged with climatological research and studies:

1. The State Meteorological Administration of China (SMA) with its branches at different levels, including the state as a whole, her provinces and counties. The

state and provinces are also in charge of the weather forecast services and issue the daily weather report; they administer the meteorological stations and their readings and observations. Each county has at least one station, giving daily reports to the SMA in Beijing where the data are processed according to the standards of the World Meteorological Organization [2].
2. Institutes of Atmospheric Physics and of Geography under the Academy of Sciences of China.
3. Institutes of Meteorology and Geography in the Universities.

Under (2) and (3), climatological research and studies are normally carried out by special sections and departments, depending on the staff members and their research topics as well as on certain research programs and projects, directed to an individual scientist or group. Research programs are mostly carried out by order of the government, at a state, province or county level.

Climatological research in China, in general and concerned with China, is mainly directed to three major subjects: (1) geographical climatology, (2) synoptic climatology and (3) physical climatology.

Climatological studies in China can be summarized into six topics:

1. Processing of Climate Data, Climate Analysis and Climate Mapping

According to the guidelines of the State Meteorological Administration of China (SMA), data are not available to the public and restricted to official use only. They are available only with a special permit from the SMA. Besides the daily weather report, no yearbooks or other periodicals have been published so far. Recently, the SMA has entered into a closer cooperation with several national weather agencies, opening a computerized exchange of weather observations.

Strong emphasis was placed on climate mapping which resulted in various atlas publications containing greatly desirable climate information. Noteworthy is the first statewide Climate Atlas of China, published in 1966 which is based, however, on a 10-year data period only (1951–1960). The atlas contains a large number of maps showing the distribution of all important climate variables, in addition to climate diagrams and a map of climate division of China. Based on 20-year observations (1949–1969), another Climate Atlas of China [3] was published by the Central Weather Bureau of China [4] in 1978, containing maps of mostly the same climate variables as the earlier atlas, but referring to a 20-year observation period and thus expressing a more reliable and precise source of climate information. The scale of the maps varies between 1:14, 1:20 and 1:28 million. The atlas, containing 226 large-sized pages, includes all climate elements and numerous variables, which are shown in their spatial variation on

[2] China became a member of the World Meteorological Organization in 1973.
[3] The full title reads *Climate Atlas of the People's Republic of China* (Beijing 1978).
[4] The name was changed to State Meteorological Administration of China in 1979.

a monthly and annual basis. The atlas contains the most useful regional-climatological information available on China so far [5].

Particularly noteworthy is a climate atlas of China (published by the State Meteorological Administration of China in 1981) which shows the spatial distribution of dryness and wetness by annual maps for a 510-year (!) period from 1470 to 1979; the scale of the maps is 1:36 million. This atlas is a unique source of information on the climate change in historical times.

2. Climate Classification of China

Chinese climatologists have given much effort in establishing a valid climate regionalization scheme of China and thus numerous attempts of a climate division have been made in the past. The most important work has been carried out by the Working Committee of Natural Division of China, established already in 1956 under Prof. Chu Co-ching (Zhu Kezhen) as Chairman and Prof. Huang Bing-wei and Prof. Tu Changwang as Vice Chairmen; the Climatology Section of the Working Committee, headed by Prof. Zhang Paokun, already presented in 1959 a concise climate classification of China which thereafter had been only slightly revised in two directions: (1) in a more meteorological one (most recently by CHEN et al. 1981) and (2) a more geographical one (by HUANG Bing-wei, at last 1986). The revision resulted from the increasing climatological and geographical knowledge of China; the meteorological or geographical orientation of the revision was according to the interest of the leading scientists concerned and the practical application expected from the classification. Although some threshold values vary between the two valid versions of the climate division of China, the hierarchial system of climate regionalization in both cases refers to three levels of specific thermal, hygric and orographical (natural) conditions. The differences between the two versions can be regarded as of minor importance. For a climato-geographical consideration of China, the division of HUANG Bing-wei (1986) appears even more suitable, particularly because of the more detailed emphasis on the physio-geographical setting as a whole. An overview of the climate division schemes will be given in Chapter 8.

3. Regional Climatology of Different Areas in China

Studies in Regional Climatology, in China commonly called Geographical Climatology, are mostly carried out by individual scientists and refer preferably to regions and areas established by the natural physical setting and typical units by

[5] Worth mentioning is also a set of climate maps, contained in the *National Physiogeographic Atlas of the People's Republic of China* (Beijing 1965), issued by the Institute of Geography, Academy of Sciences of China. A new edition is in preparation for publication in 1991. Climate maps can also be found in the forthcoming atlases, issued by the same Institute: *The Qinghai-Xizang Plateau Atlas, The Atlas of Endemic Diseases and Their Environments in the People's Republic of China* and *The Atlas of Population of the People's Republic of China*. All atlases will be published by the Science Press Beijing, the latter also in an English version jointly with Oxford University Press, Hongkong et al.

certain landforms, such as river basins, plains and mountains. These studies are particularly concerned with practical application, such as land use systems and their planning.

4. Climatic Change in China

Based on rich information on certain climate events in the past which are contained in historical writings and annals, substantial efforts have been involved in reconstructing the climate of China in the past. Due to the nature of the climate sources, most studies are concerned with the change in precipitation and temperature over the past centuries. In exceptional cases, the climatic change was even studied over a 5,000-year period (cf. Chap. 3).

5. Formation of the Climate of China

The different factors which are involved in the formation of climate have been studied, either individually or in combination. Studies carried out include the general circulation of the atmosphere and its effects upon climate (cf. Chap. 3). Other studies were concerned with astronomical factors (such as solar radiation, solar activity and the motion of the planets) or with sea flows and ice-snow cover and geographical factors (for example, the pole displacement and the variation of the earth's rotation). These studies particularly promoted the development of dynamic climatology.

6. Applied Climatology

Although most climatological studies specifically aim at application, strong scientific attention was given to research on various branches of applied climatology, for example in agricultural climatology, urban and city climate, aviation climatology, etc.

In an overview of studies and research on the climate of China, some *monographs* on this subject also need particular attention. These works can be considered as substantial reference books and textbooks written at a high scientific level and which present genuine research findings of Chinese scientists. These monographs are issued in Chinese, without an English summary. In some cases, maps may, however, be readable.

The first textbook on the climate of China was written by LU Wo more than 35 years ago[6] (1952). Despite scarce observation data at that time, the book nevertheless represented for the first time a profitable source of reference on the climate of China. In 1957, CHEN Shixun issued another comprehensive mono-

[6] LU Wo (1952) *General comments of the Climate of China,* Beijing.

graph on the same subject which mainly served as a valuable textbook to geography students [7]. In 1962, ZHU Binghai presented another monograph on the same subject which, however, contained much more in-depth observation, which was also due to the greater amount of observation data available [8]. Showing the successful development of research in climatology, TAO Shiyan (1984) edited a collective monograph on the climate of China which was published in a series of a comprehensive textbook on *The Physical Geography of China* [9], which represents a source of sound information on the subject, clearly expressing the rapid advance in climatology due to the many Chinese scientists.

Only recently (1985), ZHANG and LIN have published an extensive monograph on *The Climate of China,* containing 589 pages and expressing precise scientific accuracy on almost all climate elements in their variations over space and time; also aspects of climate regionalization and classification are included, as well as the application of climate, its changes in historical times and climate hazards. Presented at a high scientific level, this book may serve as a standard textbook and a rich reference source, offering profound information on the climate of China. It is based on substantial climate data, which mostly refer to the 30-year period from 1950–1979. Statistical charts and tables are also added for quick reference purposes. This book was taken as the main source of information for the present study.

Very recently (1986), SHENG Cheng-yu et al. have issued another rich climate monograph on China, titled *General Comments on the Climate of China.* It contains 538 pages on which the authors present a concise description of the climate of China; the major topics are the formation of the climate, the various climate elements and their variations over space and time, climate regionalization and the different climate types and regions, climate hazards and climate change as well as relationships between climate and economy. The latter two monographs, voluminous and profound in their scientific findings, are considered as authoritative sources on the climate of China.

As far as Chinese *periodicals* are concerned, three scientific journals place particular emphasis on the climate of China; these are *Acta Meteorologica Sinica* (edited by the Chinese Meteorological Society), *Acta Atmospherica Sinica* (edited by the Institute of Atmospheric Physics, Beijing) and *Acta Geographica Sinica* (edited by the Chinese Geographical Society). The latter one had already published up to its 50th anniversary in 1984 (see Vol. 39) [10] a total of 166 research papers on the climate of China.

Also noteworthy in view of the scientific research published on the climate of China are the following periodicals: *Journal of Geographical Research* (published by the Institute of Geography, Academia Sinica, Beijing); *Tropical Geography*

---

[7] CHEN Shixun (1957) *The Climate of China,* Beijing.

[8] ZHU Binghai (1962) *The Climate of China,* Beijing.

[9] TAO Shiyan (1984) (ed) *The Climate of China.* Part of *Physical Geography of China,* Beijing, Chinese Geographical Society.

[10] XU Shuying and ZHENG Sizhong (1979) Thirty years of climatology in China. *Acta Geographica Sinica* 4: 293–304. The paper can also be considered as an overview of climatological research in China after 1949.

(published by the Guangzhou Institute of Geography); *Arid Land Geography* (published by the Xinjiang Institute of Geography, Academia Sinica); *Geographical Science* (published by the Changchun Institute of Geography). At the present, the articles also contain an English abstract.

So far, the *foreign* contributions to the climate of China are very scarce. As a general textbook, the English presentation of I.E.M. WATTS is commonly regarded; it is a part of Vol. 8 "The Climates of Northern and Eastern Asia" (edited by H. ARAKAWA) of the World Survey of Climatology (1969, ed. H. E. LANDSBERG; Amsterdam: Elsevier). The climates of China (which also include Korea) are described on 117 pages and mainly include a presentation of the climate elements, in addition to regional classifications of the climate of China; the contents are based on older references and climate data sources, which have to be up-dated due to the heavily increased research activity in China since the 1960s and particularly over the last 10 years.

Expressing the most recent research findings, two books (written in English) need particular attention:

1. D. YE, C. FU, J. CHAO, M. M. YOSHINO (1987) (eds.) *The Climate of China and Global Climate;* Beijing: China Ocean Press, Berlin et al.: Springer; pp 357. The book represents the Proceedings of the Beijing International Symposium on Climate, 1984, and contains a total of 67 papers on five major aspects: Climate history in the past 2,000 years, air-sea interaction in the short-term climate variation, land surface processes related to climate variation, impact of human activity and some natural factors on climate, prediction methods of monthly and seasonal climate variation.
2. C. L. BAO (1987) *Synoptic meteorology in China;* Berlin et al.: Springer. The book deals with weather systems and weather processes valid for China. It contains the following 10 chapters: The background of the atmospheric circulation, cold outbreak and winter weather, cyclones and spring weather, rainy seasons in China, the subtropical high, weather systems on the Tibetan Plateau, typhoon, other tropical weather systems, autumn weather.

The so far widely neglected international scientific research on the climate of China is also expressed by a small number of papers and articles only, written by foreign researchers on the climate of China (see References).

2 Controlling Factors of the Climate

Since climate represents the characteristic biospheric conditions at a location or area, it chiefly results from the varying solar and atmospheric moisture and circulation conditions over space and time. Climate at any space or time level is, therefore, represented by certain expressions of the various atmospheric elements, which are concerned with radiation, temperature, sunshine, precipitation and others. The superior impact of the atmosphere upon climate is governed by various factors which control the climate; these climate-controlling factors are, in the widest sense, of a topographical nature (see Sects. 2.1–2.3). Additionally, seasons will be considered as a climate-controlling factor, thus referring to the major time scale of climate impact (see Sect. 2.4).

2.1 Latitude, Longitude and Location

Covering an area of 9,598,022 km², China (including Taiwan[11], Hongkong[12] and Macau[13]) is the third largest country on earth; only the Soviet Union (22.4 million km²) and Canada (9.8 million km²) are larger over space. Expressed as a percentage of the global area, excluding oceans, China comprises 6.4%, however, the Soviet Union 15.0% and Canada 6.7% respectively.

China covers an area of a subcontinental dimension if the extremely far distances in a horizontal direction are considered (Fig. 2.1). The farthest air distances measure around 4,500 km in a west to east direction and between 4,200 and 5,600 km in a north to south direction, depending on the southernmost locality referred to, whether it is South Hainan and thus referring to the major distribution of lands in China, or the southernmost group of Nan Sha Islands in the South China Sea which would reflect the official Chinese borderline. These distances form a relatively compact shape of China .

In addition to the huge territory as such and the far, extreme air distances, the latitudinal and longitudinal distances also represent an important climate-controlling factor. With regard to latitude, the northernmost point of China lies at 54° N, on the Amur River, bordering Heilungkiang Province against the Soviet Union. The pendant in the extreme south is situated either at 18° N (South Hainan) or at 3° N (southern Nan Sha Islands). At a maximum, China therefore extends between 36–51° in latitude. For this study, the southernmost climate station available is located at 17° N. Although not of such a huge latitudinal

[11] Taiwan 35,981 km².

[12] Hongkong (British Colony) 1,045 km².

[13] Macau (under Portuguese administration) 16 km².

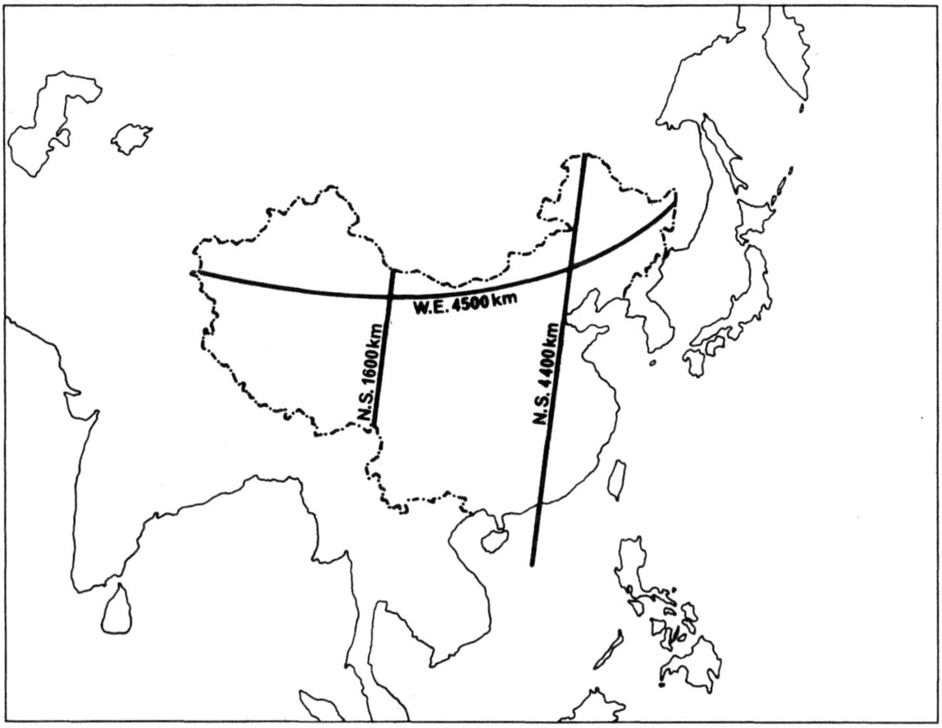

Fig. 2.1. Location and dimension of China over space within the East Asia Pacific region

dimension, nevertheless the northwestern parts of China, which belong to Xinjiang Autonomous Region, also reach as far north to 49° N, while the southernmost mountains of Yunnan Province even reach 21° N. The large latitudinal extent of China can also be shown by the shortest north to south air distance which still comes to over 1,600 km, between 42° N (in southwestern Inner Mongolia) and 20° N (in the southeastern Tibetan Plateau); see also Fig. 2.1.

From its latitudinal location, China mostly belongs to the mid-latitudes and only a small part to the low latitudes. In a broad and generalized approach, therefore, climate for the main parts of China may be described as temperate, and for a small part only as subtropical.

Although of a lesser climate-controlling impact, the longitudinal distance of China also reaches a large dimension. The westernmost point lies 71° E, on the borderline between Xinjiang Autonomous Region and the Soviet Union, while the pendant in the east is situated at 135° E, on the Ussuri River bordering Heilungkiang Province against the Soviet Union. At a maximum, China extends over 64° in longitude. As far as the climate is concerned, the large longitudinal extent of the country may contribute towards a modification of the major latitudinal climate zones.

With regard to the climate of China, the impact of space cannot only be applied to China itself, but also to the whole Asia-Pacific region. Continental Asia, on the one hand, and the Pacific Ocean, on the other, create two major

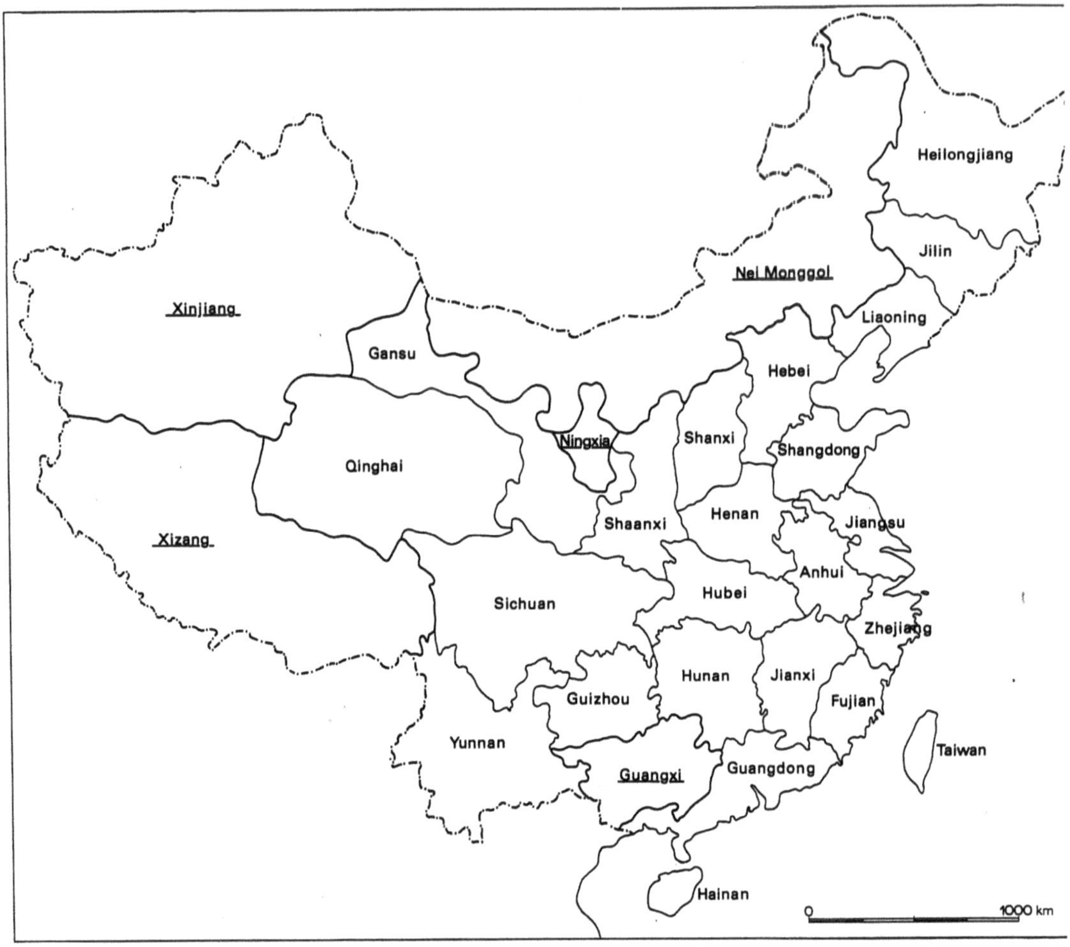

Fig. 2.2 Division of China into provinces and autonomous regions. Names of autonomous regions are underlined

surfaces, but neither of them shows topographical and structural uniformity. The Pacific Ocean, although characterized by a vast extent of water, is interspersed with thousands of islands, particularly on its western margin (Japan, Philippines, Indonesia). Continental Asia, in contrast, which encompasses China on three sides, represents a sharply contrasting and highly diversified surface configuration, whether it is the Siberian mountains, or the South Russian Steppes and desert belt, or the South and Southeast Asian subcontinents.

Although in geographical terms commonly regarded to be a part of East Asia, China can more plausibly be classified with both Inner or Central Asia, on the one side, and East or Pacific Asia, on the other. This structural bipartition already sketches the basic features of the climate division.

In this study, localization and regionalization of climate observations are often attempted by the administrative division of China into provinces and autonomous regions (see Fig. 2.2).

2.2 Topography and Landforms

Considering the macro-scale location, China flanks the Eurasian continent towards the Pacific Ocean, thus sharing the two major types of landforms for eastern continental Asia: China participates in both the coastal lowland corridor around Asia and the mountainous core of inner Asia which represents the earth's compactest dimension. A considerable part of it obviously falls into China, of which more than half belongs to mountainous inner Asia. Among the earth's highest mountains, the Himalayas, Pamir, Tian Shan, Altai and others form distinct wall-like physical boundaries, dividing China into distinct regions, which are either sheltered from or exposed to air masses. Only smaller parts of China can be considered as comparably open in terms of their physical and structural setting, and are thus easily invaded by air masses from outside, for example the Gobi desert, the Great Plain, the Mandzuria and, to a fair extent, also the South China Hills and the Yunnan mountains. The coast towards the China Seas acts not only as a natural, but also an open boundary, which enables an unhindered penetration of oceanic air masses.

As the most important structural feature of China in general and, at the same time, the most effective climatic control in particular, China represents a predominantly mountainous country with a very distinct structural pattern. The mountainous structure is expressed by a great number of mountain ranges as well as highlands and plateaux, strikingly marked and even approaching the highest peak on earth, Mt. Qomolangma (Everest) 8,848 m a.s.l. In contrast, lowlands and downlands occupy only a small part of China.

China thus creates a complex orographical structure which obviously leads to different patterns of physical regionalization, depending upon the scale of investigation. As the most commonly applied macro-scale structural regionalization of China, the characteristic steplike relief is referred to. On the basis of altitude, three steps descend from west to east:

1. The highest step is formed by the gigantic Qinghai-Xizang (Tibetan) Plateau, known as the "roof of the earth", located at more than 4,000 m elevation and surrounded by high mountain ranges; among them are the Himalayas, Pamir, Kunlun Shan and Altun Shan which, at the same time, rank among the highest mountains on earth.
2. In a northward and eastward direction follows the second, medium step in altitude, mainly formed by vast plateaux and enormous intra-montane basins. The larger plateaux are the Inner Mongolian Plateau, the Huangtu (Loess) Plateau and the Yungui Plateau. Basins include the Junggar, Tarim, Turpan and Sichuan Basins.
3. Eastward of a line from the Da Hinggan, Taihang and Wu Mountains to the eastern periphery of the Yungui Plateau lies the third and lowest step, at a mean elevation at or below 500 m a.s.l. Plains and downlands represent the dominant landforms. The Northeast China Plain, North China Plain, Middle and Low Changjiang (Jangtze) Plains are among the large plains. Generally, all these plains include undulating downlands and occasional tall hills as well. Part of the eastern step of China also includes the over 5,000 islands, the largest of

them are Taiwan and Hainan Islands, which all lie off-shore the 18,000-km-long coastline.

Each of the three structural steps shows, however, complex landforms which include different physical units, such as mountain ranges and hills, plateaux and plains, and lowlands as well. Nevertheless, the three steps represent the main physical division of China into three parts, i.e. upland, midland and lowland China.

The boundary between lowland and midland China extends in a NNE to SSW direction, from Da Hinggan Ling via Taihang Shan, Fangton Shan and Talon Shan to the South Yunnan mountains. The boundary between midland and upland China extends approximately along 105° E longitude from Ningxia Autonomous Region to Yunnan Province. For the climate of China, the 105° E longitude boundary is commonly considered as one of the most significant dividing lines.

The longitudinally descending landforms can be expressed by a series of west to east cross-sections, extending over a distance of 5° latitude each, between 25° and 45° N (Fig. 2.3, after Hsieh 1973).

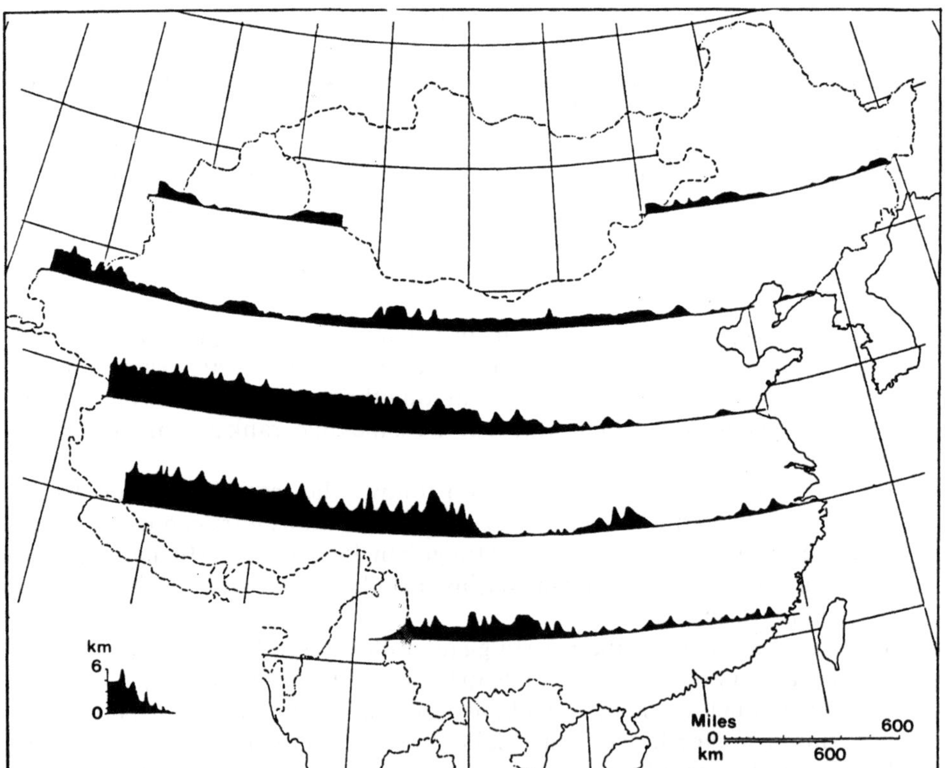

Fig. 2.3. West to East cross sections through China, along 25, 30, 35, 40 and 45° N. (After Hsieh 1973)

The three main physical parts of China, which correspond to the steplike relief, are generally divided into smaller physical regions, according to the prevailing structural conditions. As a whole, the complex and mountainous landforms of China are expressed by the percentages of various categories of landforms against the total land area of China:

- Mountains 33%
- Plateaux 26%
- Basins 19%
- Plains 12%
- Downlands 10%

The mountainous nature of China is also underlined by the percentages of various altitudinal belts (according to HSIEH 1973):

< 500 m	14%
500–1,000 m	18%
1,000–2,000 m	35%
2,000–5,000 m	17%
> 5,000 m	16%

As a structural characteristic, all major mountain ranges extend in two principal directions only, i.e. west to east and north to south. To the first category, for example, belong the Himalayas, the Kunlun Shan, Tian Shan and Altai, which are among the backbonelike high mountains of Inner Asia and which rank among the great dividing mountains of western China; their highest elevations, more than 4,000 to 5,000 m a.s.l., are even glaciered. Outside the Inner-Asian mountains, other distinct west to east extending mountain ranges are, for example, Chin Ling Shan, Tapa Shan and Nan Ling, which are of a minor dimension, over both length and height, but with far-reaching climatic importance. The Chin Ling Shan is even considered to represent the major climate barrier which, in climatological respects, separates North from South China.

The second category of mountain ranges, extending in a predominantly north to south direction, is explicitedly represented, for example, by the Da Hinggan Ling, Siao Hinggan Ling, Chang Kwansai Ling, Taihang Shan and a large number of ranges in western Sichuan and Yunnan, like Chiungshia Shan, Tasueh Shan, Yun Ling Shan and Nu Shan.

Not following the two principal directions (west to east, north to south) is the irregular distribution of the South China Hills which are widespread over the entire southeastern part of the country. Most strikingly, however, this part of China also represents a mountainous area, although the altitude is lower and does not exceed 2,000 m, the only exception being Taiwan (maximum elevation 3,998 m, Tung Shan).

The extremely varied landforms of China may affect the climate conditions in various aspects. The climatic impact of orography may influence temperature and precipitation in particular. With regards to temperature, climatic control in mountainous areas is dependent on several factors, e.g., the altitude above sea level, slope exposition and sheltering effects from air masses and airflows from outside so that in particular intramontane basins may produce a specific thermal climate. It can be expected that a direct relationship exists between the height of mountains and their temperature control. As to precipitation, mountains effectively control the flow and passage of air masses, which may either be blocked, diverted or channelled. Mountain ranges act as effective barriers which block the further passage of air masses, thus causing a strongly developed foehn effect, associated with a different pattern of precipitation on the windward and leeward side of the mountain if sufficient moisture is present. Also, with regards to precipitation, climatic control by mountains is greater, the more the elevation of a mountain range increases. The most favourable conditions exist when the air masses meet a mountain range rectangular and containing a high percentage of air moisture.

The role of mountain ranges as a climate division has to be extended also to other climate elements, especially wind, where they produce local wind systems, e.g. foehn winds and slope and valley winds.

2.3 Distribution of Land and Sea and Nature of the Underlying Ground

Due to the location of China on the southeastern sector of the Eurasian continent towards the Pacific Ocean, air masses of either continental or maritime origin affect the climate of China; the circulation pattern and air masses will be dealt with in detail in Chapter 3. Resulting from the large latitudinal distances of China, the air masses over China will not only be of a continental or maritime type, but they also exhibit different characteristics for each type with regards to temperature and moisture, depending upon the latitudinal origin, and also vary according to different seasons.

More than 18,000 km of coastline has been recorded in China between the Yalukiang mouth (Korea Bay) and the Bay of Tongking. Its considerable length results mostly from the crumpled coastline type. The coast is fringed throughout by islands of widely contrasting sizes, the largest of them are Taiwan and Hainan Islands. The dissolved coastline is distinctly shaped by the two major peninsulas, Liaoning and Shantung. In addition, two major river estuaries, formed by the Yangtze and Si Kiang, also markedly subdivide the disrupted coastal configuration.

The land and water surface of and partly around China leads to a considerably varying heat budget between a static and opaque land surface and a mobile and fluid water body. This creates different climate conditions in regions which are governed by either continental or maritime influences. The distribution of land and water over space is, therefore, a major factor that controls the climate conditions of a whole region or individual locality. In the case of China, climatic control through land and water must be strongly considered, due to the vast continental extent of China itself, on the one hand, and due to the huge extent of both continental and maritime surroundings as well, on the other. Towards the west and north, China is surrounded by the Russian and Siberian continental land masses, to the east and southeast, however, by the huge Pacific Ocean, which includes the Yellow Sea (Hwang Hai), East China Sea (Tung Hai) and South China Sea (Nan Hai).

The Indian Ocean, although for some parts of western China nearer than the Pacific Ocean, may only marginally affect the climate of China, since the Himalayas represent a marked barrier to the influences from the Indian Ocean.

The huge extent of continental landmasses, on the one hand, and of the Pacific Ocean water bodies, on the other, leads logically to a strong continental and oceanic impact upon climate. Its degree, principally caused by the different heat budgets between land and water, directly influences temperature conditions, and through them, wind and precipitation conditions are also affected. As the principal rule for the continental landmasses, all energy absorbed is concentrated in a shallow layer, producing a larger temperature increase from incoming radiation, and, conversely, a more rapid temperature decrease from radiation loss. A deeper water body, however, spreads the incoming energy and warming effect over a much thicker layer so that surface temperatures tend to fluctuate less and slower in comparison with the land surface.

Continental and oceanic effects on the climate of China may vary gradually, from strong oceanic influences over the sea and coastal areas to strong continental influences in the most inland and remote areas. A wide transitional region showing continental and oceanic effects, overlapping and active at the same time, is thus expected in the coastal hinterlands. Their extent over space, however, may be increased in the case of downlands and plains (e.g. the Great Plains), but restricted if mountains occupy the coastal hinterland as in parts of the South China Hills.

Distinct land-sea effects, such as a well-established system of land and sea breezes, must be particularly considered along the coastlines and on the islands as well which fringe the China Seas. Climatic impact of this type is, however, restricted to a meso-scale.

The striking climatic continentality to be expected in major parts of China can be expressed by the distance from the open sea. For western China, the distance from the Pacific Ocean varies between 1,500 and 4,000 km. However, besides the great distance from the open sea, the orographical setting may also hinder far-reaching oceanic influences because the mountains, extending north to south and arranged in several parallel formations, effectively block inflowing oceanic air masses from the central and western parts of China.

However, climatic effects of a small scale only must also be taken into consideration from inland lakes, a greater number of which are spread over the lower Yangtze lowlands (e.g. the Tung Ting Hu, Poyang Hu, Tai Hu and Kaoyu Hu), but also over other parts of China, in particular on the Tibetan Plateau where Tengri Nor (lake) ranks first. The climatic effects, caused by inland lakes, are mainly concerned with temperature, leading to a more even annual and daily distribution of temperature.

Particular attention must be paid to the climatic influence exerted by ocean and coastal currents. They tend to equalize horizontal differences in temperature, thus reducing the temperature extremes which might otherwise occur. The degree of temperature influence of a current depends upon its temperature conditions in relation to the ones of the surrounding ocean and over land.

In summer, huge quantities of warm water are transported off East Asia by two major currents:

1. A southwesterly current which flows from the Java Sea through the South China and East China Seas even beyond Korea; the latter part is called *Tsushima*.
2. The *Kuroshio* which is part of the Northern Equatorial Current; this crosses the Pacific Ocean in the passates belt and is diverted northeastwards off Taiwan in order to flow east of Japan.

In winter, a cold current flows southwest from the Japan Sea through the China Seas, thus affecting the coastal waters along the coast. Further off the coast, the warm Kuroshio continues to flow northeastwards east of Japan, while another diversion of the warm Northern Equatorial Current turns in a southwestern direction and flows through the South China Sea.

Depending upon the coastal current, sea surface temperatures along the China coasts vary considerably. In summer, figures range from 29° C around Hainan to 20° C in the Yellow Sea. Air temperatures along the coast are 1° to 2° C higher. In winter, sea surface temperatures along the China coasts fall rapidly and are between 20° C around Hainan and 10° C in the Yellow Sea. However, the temperatures of surface waters along the China coasts are still a few degrees higher than the air temperature. In later winter and early spring, the relatively cold coastal current often leads to the so-called *crachin* fogs along the South China coast. These typical fogs are formed of warm moist air that passes westwards over the northern Philippines and which is cooled when crossing the cool waters of the South China Sea.

2.4 Seasons

The four common seasons, winter, spring, summer and autumn, are normally also distinguished in the weather calendar of China. Winter and summer are, however, generally regarded as the dominating seasons, while spring and autumn are only transitional seasons which experience weather conditions often closely related to those in winter and summer. Beyond any doubt, winter and summer mark the amplitude of climatic events which may occur throughout the year. Due to the enormous dimensions of China over space, the length of the seasons and their exact dates are controversial. It is noteworthy that from a meteorological and climatological viewpoint, the four common seasons are normally applied in a statistically simplified way that divides the year into four periods of 3-month duration each:

Winter: December – February
Spring: March – May
Summer: June – August
Autumn: September – November.

Furthermore, since the discussion about seasons is a matter of controversy, whether from an academic or practical viewpoint, it therefore mostly depends upon the intended purpose or application of seasons.

In view of seasons, the term *monsoon* is commonly applied, both for winter and summer, using synonymously the terms winter and winter monsoon, on the

Table 2.1. Duration (in months) of the seasons. (After CHANG, Pao Kun 1934)

Region	Spring	Summer	Autumn	Winter	Spring and autumn
Xinjiang	2 –3	2	2	5 –6	–
Northeast China	2 –3	1.5–2.5	2	5.5–7	–
Inner Mongolia	2 –3	1 –3	1.5–2.5	5.5–6.5	–
North China	1.5–3	2 –4.5	1.5–2	4.5–6	–
Central China and Yangtze	2 –3	3 –5	2 –3	2.5–4.5	–
Yunnan-Guizhou Plateau	–	–	–	2 –3	9–10
South China	–	6 –8	–	–	4–6

one hand, and summer and summer monsoon, on the other. In terms of the monsoon, China experiences two rather than four seasons.

Among various classifications of the seasons, CHANG Paokun (1934) has particularly studied the regionally varying duration of the four seasons by using 5-day mean temperature limits of 10° C and 22° C to define spring and autumn respectively. CHANG determined remarkably large differences in the duration of the seasons in different parts of China (Table 2.1). The most significant observations were found for the length of summer and winter, providing evidence that from north to south, the duration of winter is gradually shortened, from 6 months (in Xinjiang and North and Northeast China) to 3 months (on the Yunnan-Guizhou Plateau), whereas in South China it is completely missing. In the opposite direction, the length of summer decreases gradually from south to north. According to CHANG, spring and autumn are of a major significance only for the Yunnan-Guizhou Plateau and South China.

The four determined seasons in China were sometimes modified by two additional seasons which are distinguished by particular weather conditions:

1. The *"mei-yu" season,* confined to the southern provinces of China, represents the late spring and early summer rains of April, May and June, associated with small depressions extending seawards along a trough which lies in an east-west direction between the Yangtze and the southern coast. This season is named after the so-called mei-yu depressions, but the popular term "plum season" or "plum rains" (see Sect. 3.4.3) is often used.
2. Exhibiting also a sharp limitation over time and space, the *"crachin season",* which falls in late February, March and early April, is characterized by small anticyclones which sometimes move from the Asian continent to the China Seas. Here, water temperatures are fairly high and thus the air is warm and moist, unlike the cold and dry continental air. The collision of maritime and continental air masses leads to the typical *"crachin"* formation of fog or low stratus along the coast of the China Seas (see Sect. 6.4).

3 Circulation

3.1 Seasonal Pressure Distribution at Sea Level

Pressure distribution at sea level, expressed for winter, spring, summer and autumn, shows a distinctly marked seasonal variation over the Asia-Pacific region. Most clearly to be seen, the largest differences in atmospheric pressure occur between winter and summer, whereby January and July can be considered as representative months (Figs. 3.1 and 3.2). For a close understanding of the pressure distribution at sea level in China, particular attention is given to the Asia-Pacific region.

Winter

As can be seen from the mean sea level pressure in January (Fig. 3.1), the Asia-Pacific region is divided into an intensely developed anticyclone over mid-Siberia and Mongolia and a strongly established cyclone over the nothwestern Pacific Ocean. The continental anticyclone which at the same time represents the largest one in the northern hemisphere, is centred around the Baikal Sea and Mongolia, while the centre of the NW-Pacific cyclone lies over the Bering Sea and the Aleutean Islands. Since both pressure systems practically lie in the same latitude of 50° to 55° N, a steep pressure gradient occurs which produces strong and persistent northerlies. They carry dry continental and cold polar air masses over the whole East Asia region.

Atmospheric pressure at the centre of the continental anticyclone reaches high values at sea level, and varies only slightly over China. According to mean values at sea level, pressure in January reaches a maximum of 1040 mb in the centre of the anticyclone and decreases to 1020 mb only over southern China.

For South Asia, Southeast Asia and the Philippines, pressure at sea level continues to decrease to about 1015 mb and even less (see Fig. 3.1).

As a typical feature, the Mongolian and mid-Siberian anticyclone is only a rather shallow pressure system. At the 700 mb surface, it turns into a ridge of high pressure.

As a counterpart to the continental anticyclone, also the Aleutean Low is strongly developed. In its centre, pressure at sea level drops below 1000 mb. The cyclonic pressure field is comparably less established over the Northwest Pacific Ocean only, while the anticyclonic pressure field occupies the whole of continental East Asia, including China.

As a third pressure system in winter, the equatorial Low over Australia and New Guinea has to be considered, although it is limited to southeastern China.

The strong anticyclonic circulation during winter over China is valid not only over space, but also over time. Proof of the stable pressure pattern in winter is given by the monthly pressure means from November to February at any of the climate reference stations. At Shanghai/119, for example, monthly means of pressure at sea level, from November to February, amount to 1023, 1026, 1026 and 1024 mb. The corresponding values for Tianjin/62 are 1028, 1029, 1027 and 1025 mb, while Guangzhou/199 records 1017, 1020, 1020 and 1018 mb monthwise from November to February. During winter as a whole, January experiences the strongest anticyclonic pressure field which reaches a maximum of 1040 mb, while the maximum values amount to 1030 mb in November, and 1035 mb each in December and February. As clear proof of the strong anticyclonic pressure pattern over China during winter, the pressure gradient is similarly weakly developed from November to February.

Spring

The intense continental anticyclone over China still remains rather stable through March, showing only a slightly weaker pressure in the centre by about 5 mb in comparison with January. But in April the pressure gradient weakens rapidly, resulting in a modest value in May when the mean pressure at sea level varies from 1007.5 to 1012.5 mb only, without showing a significant regional variation. The strong impact of the continental anticyclone ultimately disappears in May, at the same time indicating the end of continental air masses over China. With regards to the Aleutean cyclone, low pressure continues through March, but weakens in April and May. Similarly, the equatorial Low also weakens as compared to winter conditions.

Summer

The pressure pattern at sea level during summer differs completely from winter conditions. As can be seen from the mean pressure field in the Asia-Pacific region in July (Fig. 3.2), a strong cyclone is located over the northwestern Indian-Pakistan subcontinent.

For the whole region, however, the pressure field shows a relatively weak gradient so that for China only a moderate variation of pressure is experienced. As for July (Fig. 3.2), mean pressure at sea level drops to 995 mb in the centre of the Low, while it rises to a maximum of 1025 mb over the northern Pacific Ocean. Over most parts of China, pressure at sea level in July mostly occurs at an intermediate level between 1000 and 1005 mb, thus showing a weak pressure gradient which means a less pronounced variation in pressure over space.

A relatively homogeneous pressure field is valid for each of the three summer months June, July and August. Proof is given by the mean monthly pressure at sea level at selected climate stations. At Shanghai/119, for example, mean pressure values are 1006, 1004 and 1006 mb for June, July and August respectively. The corresponding values for Tianjin/62 are 1005, 1002 and 1007 mb, for Guangzhou/199, 1005, 1004 and 1004 mb. Mean monthly values for these three stations clearly express the only weak pressure gradient over time, respectively summer, and space. Even more distinctly than in July (see Fig. 3.2), a weak pressure gradient over space is true for June and August for which the monthly

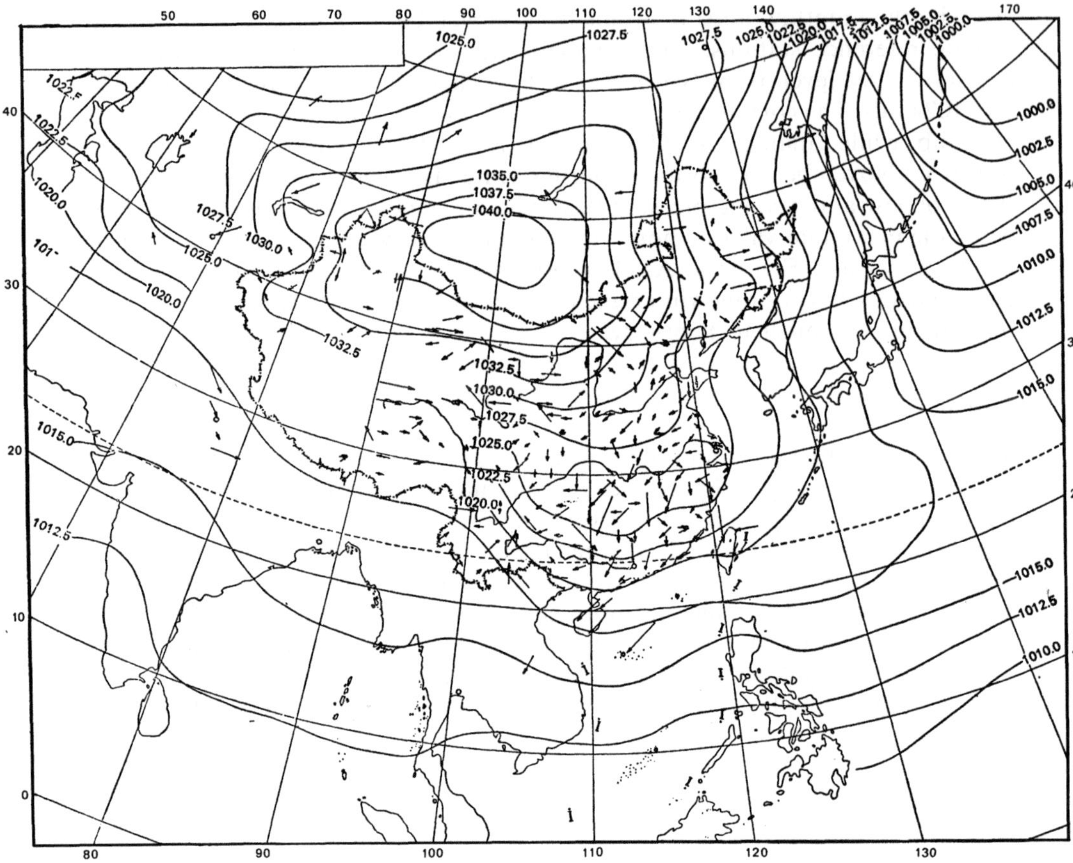

Fig. 3.1. Mean sea level isobars in January. (After ZHANG Jiacheng and LIN Zhi-quang 1985)

pressure means merely vary between 1002.5 and 1005 mb and 1000 and 1002.5 mb respectively, over most parts of China.

The weak cyclonic pressure gradient over China in summer is also associated with a weak, but anticyclonic pressure gradient over the northern Pacific Ocean. Thus, the resulting surface winds are of a moderate nature, but they exhibit a persistent flow from a mainly southern direction.

In summer, the equatorial Low is significantly weakened, without showing a major occurrence.

Autumn

Through September, the intense cyclone over the northwestern Indian-Pakistan subcontinent quickly disappears as pressure in the centre rises to an average value of 1005 mb. While pressure throughout China increases in comparison with summer conditions, a continental anticyclone is already weakly established, having its centre over Mongolia. During October and November, the continental

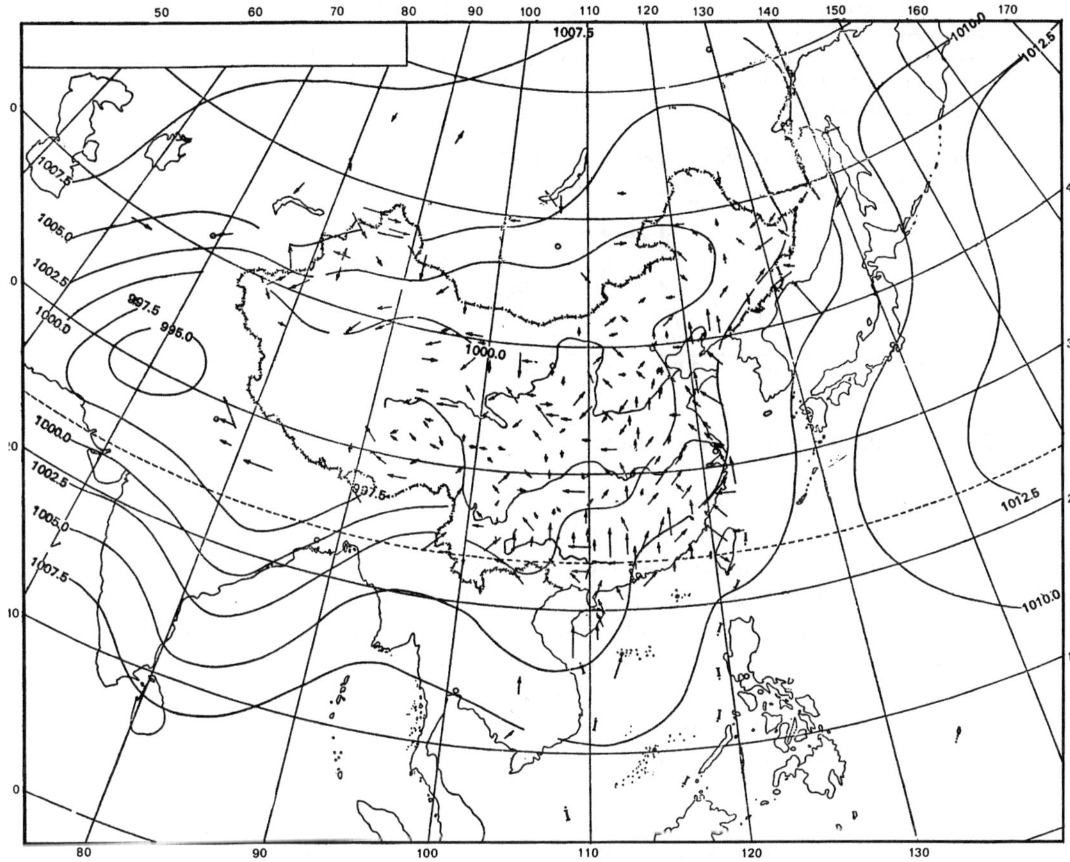

Fig. 3.2. Mean sea level isobars in July. (After ZHANG Jiacheng and LIN Zhi-quang 1985)

anticyclone becomes strongly established. In October, the centre of the anticyclone shows a mean pressure of 1025 mb at sea level, in November 1030 mb.

In correspondence to the gradually established anticyclone the Aleutean Low also progressively develops during October and November. Accordingly, the equatorial low pressure field is strengthened. As a whole, the typical pressure distribution of winter is already fully developed in November, consisting of the continental anticyclone over Mongolia and mid-Siberia, the Aleutean Low and the equatorial Low.

Pressure distribution over China, as a whole, is of a transitional type during spring and autumn, showing in spring a strong development towards cyclonic, and in autumn a distinct development towards an anticyclonic pressure field. Therefore, an anticyclonic field over China prevails from about September/October to April/May and thus covers about 7 to 8 months, while a cyclonic pressure pattern persists from May/June until September and thus covers 4 to 5 months.

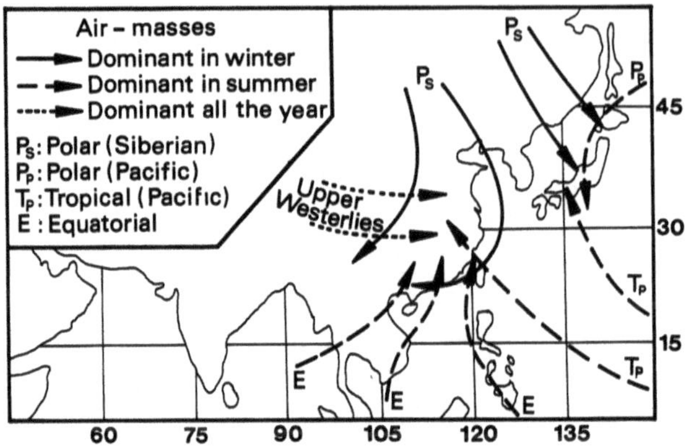

Fig. 3.3. Dominant wind directions and air masses in the Asia-Pacific region

3.2 Seasonally Prevailing Winds and Air Masses

As a result of the largely differing pressure distribution over space and time (Sect. 3.1), the prevailing winds over China are also expected with a great seasonal variation. According to the major differences in pressure between winter and summer, also in the case of prevailing winds, the differences between winter and summer reliably express a large annual variation. Observations, therefore, mainly refer to the conditions in winter and summer which occur on ground, both in terms of direction and velocity of surface winds.

Basic statistical information on wind is contained in the climate tables (see Appendix) where four wind variables are considered: prevailing wind direction and its percentage, calms and velocity. Figures are given as monthly means from 1951–1980.

The prevailing winds over China alternate in their direction from winter to summer. In a simplified way, prevailing surface winds in winter can be described as northerlies, associated with dry and cold, polar air masses by origin. They may be distinguished into polar-Siberian and polar-Pacific air masses. In summer, however, prevailing winds on ground are mainly from the south, representing tropical-Pacific and equatorial air masses from a southeastern and southwestern direction respectively; correspondingly, both are warm and moisture-laden by origin. Northerlies and southerlies are shallow and are overlaid by westerlies, above 2,000 m, which are dry and are apparently almost constant winds of great velocity (Fig. 3.3). Due to the general relief conditions of China, the western highland parts are governed by persistent westerlies, while the eastern parts, including Northeast and Northwest China, mainly record seasonally opposing northerlies and southerlies.

For a more detailed presentation of surface wind directions and their variation between winter and summer, observations are given for 13 stations, representing 30-year averages (1951–1980) for January and July; data under consideration are percentage figures, applied to the commonly valid eight directions N, NE, . . . , NW, in addition to calms. The names of the stations under consideration are given in Table 3.1. For part of the stations, data are also presented as wind roses (see Fig. 3.4).

Table 3.1. Surface wind directions for January and July at selected stations, given as percentage figures from 1951–1980

		N	NE	E	SE	S	SW	W	NW	Calms
Harbin	1	6	2.5	2	5.5	15.5	21	18.5	13	7
	7	5	6.5	14	15	25	16.5	7.5	4.5	8
Abagqi	1	6	5.5	2	2	2	9	26.5	10	39
	7	8.5	13.5	12	8.5	9	9	14	8.5	17
Zhangjiakou	1	25.5	1	2	3	2	2.5	10.5	36.5	18
	7	10	1	11	24.5	6	4	5.5	12	25
Shijiazhung	1	15.5	6	4.5	9.5	6	2.5	10	15	31
	7	12	7.5	5.5	16	8	1.5	3.5	9	36
Zhengzhou	1	5	17.5	5.5	5.5	11.5	7	17	14	16
	7	5	16	11	14	20.5	6.5	6	7	15
Wuhan	1	22.5	24.5	11	8	4.5	4	4	9.5	12
	7	8	10	10	15.5	17.5	16	5	6	12
Changsha	1	18	2.5	2	5	4	2	3	37.5	21
	7	5.5	3.5	4	17	33.5	8.5	2.5	12.5	13
Guangzhou	1	36.5	7	5	6	3.5	1.5	2	11.5	29
	7	3	7	16	24.5	16	3.5	3	2	26
Hangzhou	1	25	8.5	5	4.5	9	7.5	3	20.5	19
	7	7.5	7.5	10	9	26	20.5	2	8.5	10
Chengdu	1	17	17	3.5	3.5	5	3.5	2	3.5	45
	7	14.5	12.5	2.5	2.5	8	7	5	7	41
Qamdo	1	7.5	3.5	2	3.5	7.5	5.5	6	10.5	56
	7	13	4.5	2	6	11.5	4.5	6.5	15.5	37
Lhasa	1	4.5	6	25	13	5	8	10	5.5	24
	7	4	3.5	17	13.5	2	4.5	15	9.5	30
Ürümqi	1	9	12	4	10	18	10	2	7	30
	7	16	9.5	2	8.5	16.5	12.5	5	22	11
Haikou	1	12.5	45.5	18.5	7	1.5	0	0	0.5	12
	7	4.5	7	7	26	26.5	6.5	5.5	6.5	10

1 = January, 7 = July.

Winter

Prevailing winds over most parts of China are from N, NW and NE which result from the intensely developed anticyclonic wind field over the region. Winds turn to the east over the Qinghai-Xizang Plateau and to the south over western Xinjiang. Exceptional wind directions occur over the northeastern parts of China, i.e. Heilungkiang and Kirin Provinces, where westerlies and even southerlies are predominantly developed.

As for January (Table 3.1; Fig. 3.4), surface winds at all stations under investigation are mainly from a northern direction; maximum direction varies from NW to NE. The total percentage of northern surface winds, as summed by winds from

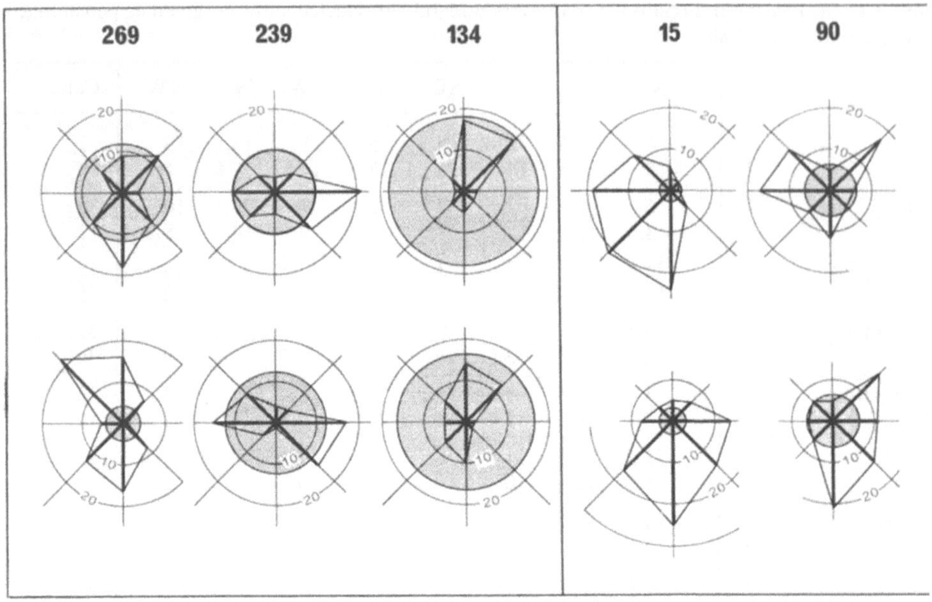

Fig. 3.4. Wind roses showing surface conditions for January and July at selected stations. Data are from Table 3.1. Top row shows wind roses for January, bottom row for July. Screened circle gives percentages of calms. 269 – Ürümqi, 239 – Lhasa, 134 – Chengdu, 15 – Harbin, 90 – Zhengzhou, 126 – Wuhan, 147 – Changsha, 199 – Guangzhou, 207 – Haikou

NW, N and NE, increases to even two-thirds of all directions (for example at Changsha/147: 63%, Haikou/207: 58.5%), but it may also fall to one-third only or even less (at Chengdu/134: 37.5%, Zhengzhou/90: 36.5%). These figures confirm typical northerlies in winter which are derived from the continental anticyclonic pressure field.

As also proved from statistical figures, different wind conditions are established over some marginal parts of China. Harbin/15, which records 50% of all directions from SE, S and SW, underlines atypical wind conditions which exist over northeastern China. The same is true for Lhasa/239 which experiences 44% of all directions from NE, E and SE, and Ürümqi/269 which records 38% of all winds with a southern component. These three stations demonstrate rare wind conditions which do not coincide with the typical northern winds in winter.

Wind velocity on ground, expressed as monthly means, shows rather moderate values in winter over most parts of China. Monthly means of wind velocity vary mostly between 2 and 3 $m\ s^{-1}$ which represent comparably low values (see Table 3.2). Despite the moderate values, winter is normally considered to be the period of maximum wind velocity for China.

Particular attention must be given to the extremely large percentage of calms (Table 3.1) which are a distinct criterion in winter. In January, calms were recorded on almost every second occasion (see Chengdu/134: 45%). Although generally of a smaller value, nevertheless, the figures for calms at all stations under investi-

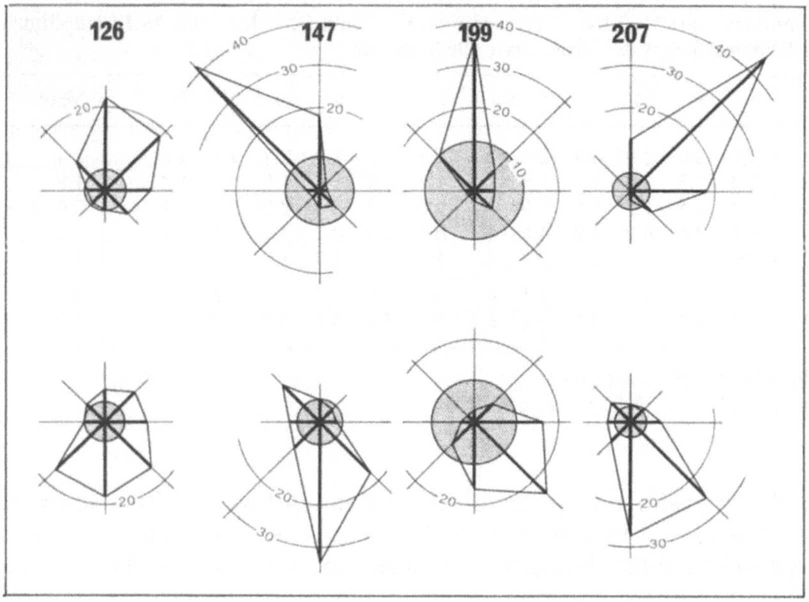

gation are remarkably high and amount to between 15 and 30% in most cases. If calms were excluded from the statistical surface wind composition as shown in Table 3.1, the strong prevalence of winds from a northern direction would be even more intensively expressed; they then would amount to even 50–75% (at Changsha/147, for example, even 80%). For the high percentage of calms, two reasons seem to be important: (1) the relief conditions, in general, and mountains, in particular, which block the passage of air masses; (2) the shallow thickness of the northerlies which amounts to less than 2,000 m.

The air masses which mostly affect China during winter are of a persistent dry and cold, polar origin, although they can be warmed and also partly moistened when they move southward. As a typical character of the constantly invading continental air, which is commonly called *winter monsoon* (see Sect. 3.3), cold waves frequently travel from Xinjiang and Mongolia even up to the southernmost parts of China (see Sect. 3.3). Associated with a sharp drop of temperature and frequent rains, the effects of cold waves are weakened with increasing distance from their origin. Continental air of a dry and cold origin prevails from October through March and even falls into April.

Summer

As a result of the strong heat low pressure over India and Pakistan, surface winds over South, Central and East China show a persistent southern direction which slightly varies from southwesterlies to southeasterlies if no major orographically induced diversions occur. Only over northeastern China, Mongolia and Xinjiang

Table 3.2. Monthly means of surface wind velocities at selected stations of China and the India-Pakistan subcontinent (m s^{-1}, 30-year average figures). (Data from MÜLLER, M.J.: Handbuch ausgewählter Klimastationen der Erde, Trier 1983, 3rd. ed.)

Station	J	F	M	A	M	J	J	A	S	O	N	D	Year
Beijing	2.3	2.6	2.9	3.3	2.9	2.3	1.7	1.6	1.9	2.2	2.3	2.5	2.4
Xian	1.4	1.3	2.0	1.7	1.8	2.1	1.8	1.9	1.5	1.5	1.4	1.4	1.8
Ürümqi	1.1	1.5	1.6	2.6	2.9	2.3	2.0	2.3	2.3	2.2	1.6	1.6	2.0
Shanghai	4.6	4.6	4.9	4.9	4.6	4.4	4.9	4.7	4.1	3.9	4.2	4.5	4.5
Guangzhou	2.1	2.0	1.8	1.8	1.7	1.6	1.8	1.7	1.7	1.9	1.9	1.9	1.8
Karachi [a]	3.3	3.6	4.3	5.1	6.1	6.7	6.7	6.2	5.2	3.5	2.8	3.1	4.7
Hydarabad [b]	2.2	2.5	2.7	3.0	3.4	6.6	6.1	5.1	3.5	2.5	2.2	2.0	3.5

[a] Karachi, 24°48′N/66°59′E, 4 m a.s.l.
[b] Hydarabad, 17°26′N/78°27′E, 542 m a.s.l.

do winds with a prevailing northern direction occur. All winds with a southern component, after having passed over the China Seas, are moist and warm and thus cause plentiful rainfall during the "summer monsoon" season. The prevailing winds from a southern direction are commonly called *summer monsoon* (see Sect. 3.3).

The summer monsoon occurs from June through August, but most typically in July (Table 3.1; Fig. 3.4). As can be seen from the statistics and wind roses for selected stations, in most cases the maximum direction varies from SE to SW. The percentage of all southern winds, represented by the total of winds from SW, S and SE, amounts to 40–60% (see Table 3.1). Thus, the valid winds from a southern quarter, as can be derived from the pressure field over China in summer, are confirmed.

Wind velocity on ground in summer is generally regarded as light and is normally smaller than in winter, for which comparably moderate values have already been shown (see Table 3.2).

Some exceptional observations should also be noted. Occasionally, higher wind velocities may occur on the coasts, for example at Shanghai/119 (see Table 3.2). In contrast, relatively low wind velocity values are recorded in inland areas, particularly in intramontane basins, valleys and wind-sheltered locations. This has been confirmed at Chengdu/134, Yibin/139, Guanghua/107, Yichang/129 and Lancang/222, where the wind velocity falls to the lowest values, ranging between 0.5 and 1.3 m s^{-1}. It may also occur that fairly even values are recorded monthwise throughout the year, so that the annual course is rather homogeneous, as, for example, at Chengdu/134 which records monthly values of wind velocity at around 1.5 m s^{-1} only. For some stations in China, the winter maximum of surface wind velocity is shifted towards April and May. In contrast to the common annual course, a winter minimum and summer maximum occur at some locations, for example at Ürümqi/269.

In accordance with the observations in winter, calms are also recorded in summer with a remarkable frequency. In July, for example, at most climate stations under consideration, calms represent between 15 and 30% of all cases, at

Chengdu/134 even 41%. Calms in summer seem to occur at a slightly smaller percentage, which still remains high on the whole.

Exceptionally high wind velocities in summer may occur if tropical cyclones occur which range from weak tropical depressions to mature typhoons. Wind velocity increases accordingly. Tropical cyclones of a different stage hit the coastal regions and lowlands most heavily and then quickly weaken over land. A detailed description of typhoons is given in Section 3.5.3.

The air masses invading China in summer, are of a moist, warm and unstable nature. Their climatic effectiveness weakens with decreasing distance from the coasts and lowlands of the China Seas.

During spring and autumn, the circulation as well as the nature of air masses show a transitional type, either from winter to summer or vice versa. Wind direction and velocity correspond well with the conditions in winter and summer; the same is true for the nature of the prevailing seasons in terms of winds and air masses.

3.3 Winter and Summer Monsoon

3.3.1 Characteristics of the Monsoon in General

The climate of China is generally considered as part of the monsoon climate of East Asia. The monsoons represent the overwhelming climate and weather regime for China which governs the climate conditions more effectively than any other regime. As a distinct characteristic of the monsoon system, two different components are commonly distinguished; they are the *winter monsoon* and *summer monsoon*, both synonymously used for the winter and summer seasons. Since spring and autumn show transitional conditions between winter and summer and vice versa, the climate events throughout the year are decisively determined by the winter and summer monsoon.

The study of monsoon climate in China progressed greatly in the last 10 years (Tao Shiyan 1984). Particular attention was given to the atmospheric circulation of the monsoon, on the one hand, and the climate characteristics, on the other.

As far as the atmospheric circulation of the monsoon is concerned, a revision of the original explanation of the monsoon was made in that the *summer monsoon* in East Asia and the Indian summer monsoon were no longer accepted as a single monsoon system only. In-depth scientific studies (Tao Shiyan 1984) have clearly shown the following major components for the monsoon in East Asia (Fig. 3.5) which, however, significantly differ from the monsoon in India: the anticyclone over Australia, the cross-equatorial current from the southern hemisphere (near 105° E), the monsoon trough or Innertropical Convergence Zone (ITCZ) in the South China Sea and western Pacific, the tropical easterly jet, the subtropical West Pacific High, the mei-yu front and the mid-latitude disturbances.

Like the summer monsoon in China, the atmospheric circulation of the *winter monsoon* was also verified by exhaustive studies during the recent past. Particularly, the thermal and dynamic influence of Qinghai-Xizang (Tibetan) Plateau on

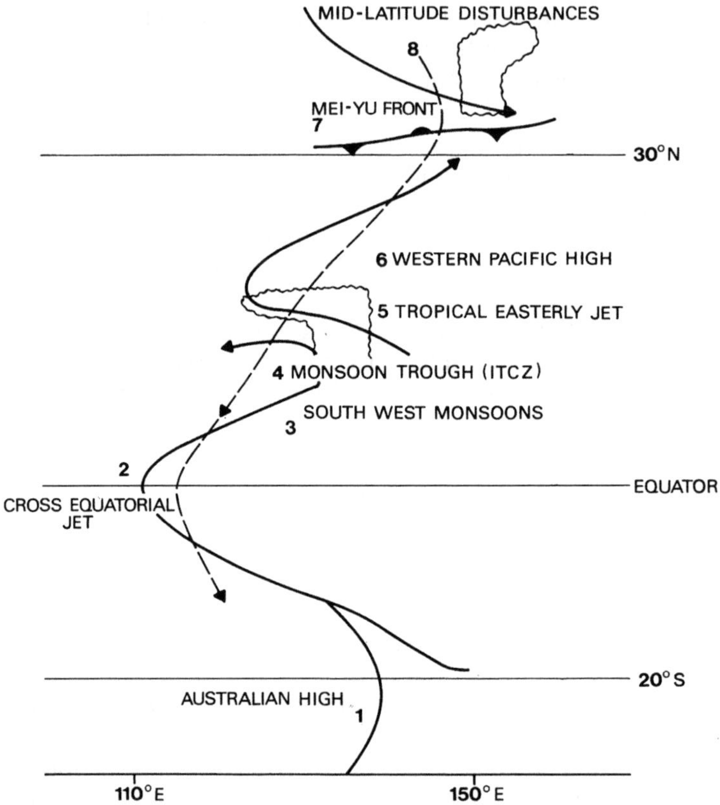

Fig. 3.5. A schematic diagram of the East Asia summer monsoon components over 110°–150° E. (After TAO Shiyan and CHEN Longxun 1987)

the south and north jet streams near the plateau and the regime of the winter monsoon over East Asia have been revealed more thoroughly (TAO Shiyan 1984). For example, it has been found that the effect of the cold source of the plateau at heights of 4 to 5 km intensifies the winter monsoon over China. As a sequence of the dynamic effect of the plateau and the thermal difference between the continent and ocean, a deep upper trough over the coastal area of East Asia can be observed very often and the winter monsoon is closely related to the trough.

With particular scientific emphasis and effort, research on the climate characteristics of the monsoons was concerned with both the winter and summer monsoon. Among all climate elements affected by the monsoon, precipitation was found to be the main element. As far as the summer monsoon is concerned, oscillation types of the monsoon system were found, showing periods of 4 to 5 days, 14 days and 40 to 50 days (CHEN Longxun et al. 1982). The summer monsoon was also shown as the major source of precipitation in China, thus providing a close correlation between the rainy season and the period of an active monsoon which roughly corresponds to the total length of the period between onset and withdrawal of the monsoon. However, the total amount of monsoon rainfall and the intramonsoonal variation of precipitation may vary considerably, and the interannual rainfall variability of the summer monsoon may also be extensive (see Chap. 5).

The nature of winter and summer monsoon in China is basically determined by an established system of winds with almost opposing directions during winter and summer. The major direction of the prevailing winds is from a northern sector in winter and from a southern sector in summer. With reference to the most frequent direction of surface winds, the winter monsoon is also sometimes called "northeast monsoon" in contrast to the "southeast monsoon" during summer, due to prevailing southeasterlies.

Although a matter of controversy, the definition of monsoon given by ALISSOW (1954), CHROMOW (1950, 1957) and FLOHN (1950) is internationally widely accepted. Since the monsoon is regarded as part of the circulation system of the atmosphere, the authors categorically described the monsoon as alternating winds between winter and summer, the direction of which varies more than 120°. Conversely, the authors also regarded all seasonally alternating wind systems as monsoon if the critical "monsoon angle" is fulfilled. As for China (and East Asia as a whole), a seasonally alternating wind system is justified by prevailing winds from a northern sector in winter and major winds from a southern sector in summer. As a part of the general circulation system of the atmosphere, FLOHN (and others) defined the East Asia monsoon by the seasonally alternating belt of westerlies in winter and trade winds in summer; the East Asia monsoon is called "extra-tropical", in comparison to the "inner-tropical" monsoon system in South and Southeast Asia. However, the original wind directions of the two major wind belts of the atmosphere are remarkably changed over China, due to the wide-ranging effects from the land–ocean contrast to the pressure field and circulation pattern in the East Asia-Pacific region.

According to Chinese meteorologists, the monsoon was defined as an alternation of two kinds of airflows with different properties: (1) prevailing winds differ largely in winter and summer; (2) since winter and summer monsoons originate in different regions, there are substantial differences in their air-mass properties; (3) they are accompanied by various weather phenomena, thus bringing a great diversity of seasons, such as rainy and dry summer and winter seasons respectively (ZHANG Jiacheng 1983).

The alternating wind direction between winter and summer monsoon was also taken as a major criterion for calculating a so-called *monsoon index* (after ZHANG Jiacheng 1983) which expresses the strength of the alternating wind directions. The monsoon index I was calculated as follows:

$$I=(P_1-P_7)+(P_7'-P_1'),$$

where P_1 represents the frequency of the prevailing wind direction (%) in January among all eight rhumbs and P_7 the frequency of that direction in July. P_7' denotes the frequency of the prevailing wind direction in July and P_1' the frequency of that wind direction in January. Calculation of the monsoon index I was carried out for different elevations: at sea level, at 500, 1,000, 1,500, 2,000 and 3,000 m. The results show that the monsoon is strongest between 500 and 1,000 m a.s.l., while

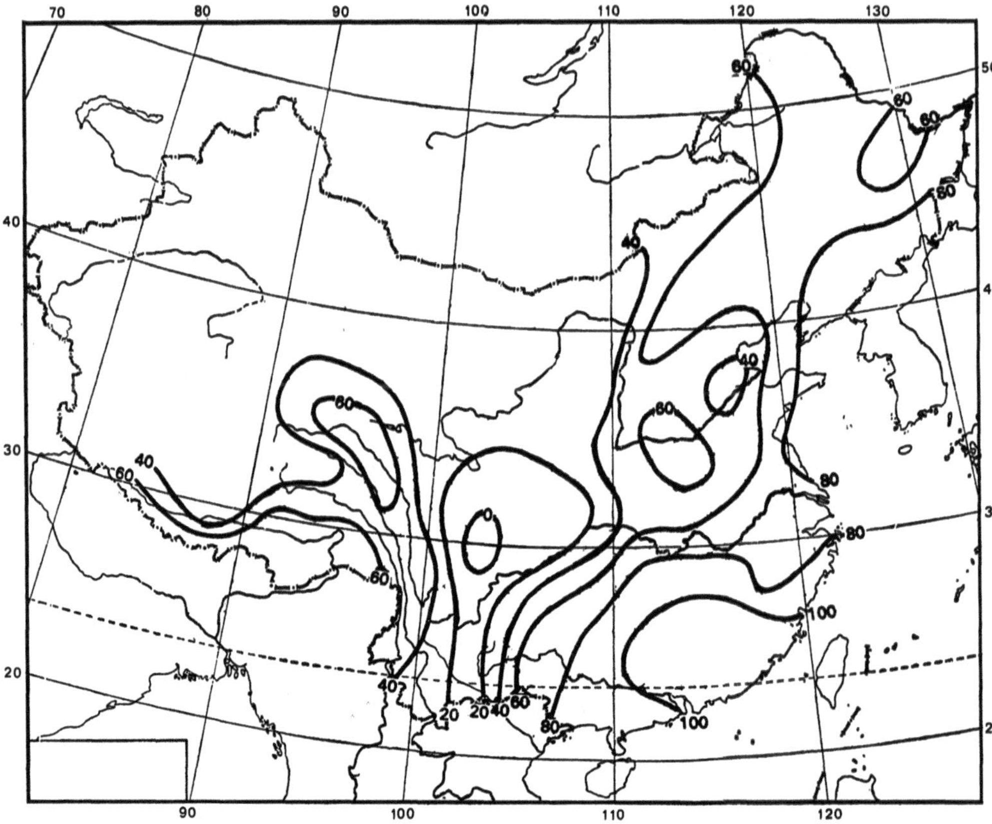

Fig. 3.6. The "monsoon indices" for surface winds. Large values show the increasing strength of alternating winds between January and July. (After ZHANG Jiacheng and LIN Zhi-guang 1983)

it is less marked on the ground and between 1,000 and 2,000 m. Above 2,000 m a.s.l. the monsoon is not clearly established. For the monsoon on the ground (Fig. 3.6), maximum values of the monsoon index occur in the area of Nanelking Mountain, Fujian and Guangdong Provinces, Hengduan Mountain and southern Tibet. The absolutely highest monsoon indices are valid in the area of Sichuan and Yunnan Provinces which represent a boundary area between two monsoons. The parts to the east belong to the East Asia monsoon, where both winter and summer monsoon are distinctly developed, although winter monsoon is markedly stronger; monsoon precipitation is associated with the polar front. The parts west of the boundary region are influenced by the Indian monsoon, showing a strongly developed summer monsoon which is also responsible for most of the precipitation.

A common feature of most definitions is that the monsoon is basically regarded as a system of alternating winds between winter and summer which are associated with air masses of different origin and thus a distinct nature. The climatic impact of the monsoon is, therefore, strongly linked with precipitation and temperature. Due to its origin from mid-Siberia and Mongolia, the winter monsoon can be characterized by cold and dry air masses, the primary character of which is gradually moderated with increasing distance from their origin. This means that the characteristics of the winter monsoon are weakened from North to South China, and that over the southern parts warmer and moister air masses of an oceanic origin may even govern the climatic conditions in winter. In an abrupt contrast to the continental winter monsoonal conditions, warm and moist air masses of a tropical and equatorial origin prevail during summer. They invade China although their original nature is weakened with increasing distance from their origin. The effects of the summer monsoon are heavily weakened over West and Northwest China where geographical and topographical conditions prevent the invasion of the moist and warm summer-monsoonal air.

The different hygric nature between winter and summer monsoon air masses also leads to a clear seasonal difference in precipitation (see Chap. 5). As a general rule, winter represents a dry, summer a wet period. However, when monsoonal air masses are moderate and weak, then the hygric features are also modified, and thus the alteration between a wet and dry season is less pronounced. The length of the wet and dry season logically varies, depending mostly upon the distance of the region concerned from the origin of the monsoonal air masses, but it is also modified by the effects of the relief on the air masses.

Although the climate of China is commonly described as a monsoon climate, it is nevertheless often a matter of controversy whether or not China as a whole experiences a monsoon climate. If the term "monsoon" is strictly applied to refer only to a seasonally alternating system of winds associated with a distinct wet and dry season, non-monsoonal conditions occur in the following parts of China: Xinjiang, central and western part of North Qaidam Basin, western part of North Qinghai-Xizang (Tibetan) Plateau, north part of Inner Mongolia (Helanshan and Yinshan Mountains).

Taking the monsoon climate as a whole, the onset of either the winter and summer monsoon and the duration of the monsoon periods are normally regarded as the most significant events which, therefore, shall be regarded with particular attention. In contrast, the retreat of the monsoon is given only little attention.

3.3.2 Onset and Duration of the Winter Monsoon

Winter monsoon, which is commonly considered as persistently invading continental, cold and dry air masses, associated with winds from a mostly northern quadrant, is quickly established over Northeast and North China during the first 10-day period of September when the summer monsoon has retreated to the Yangtze valley. During the second and third 10-day periods of September, winter monsoon occupies nearly all eastern parts of China, east of 105° E longitude. At the same time, the summer monsoon quickly retreats to the coastal areas of South and Southeast China. During the first 10-day period of October, the coastal

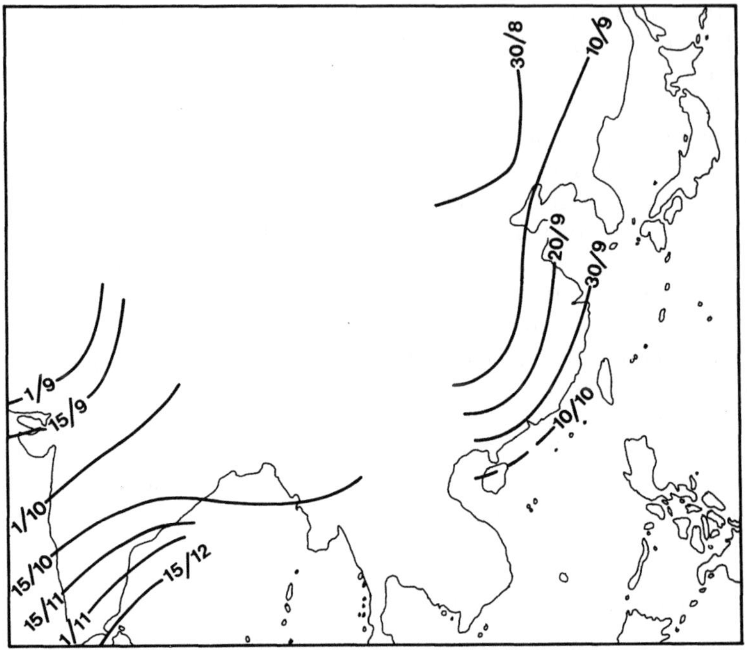

Fig. 3.7. Mean onset date of the winter monsoon over East and South Asia. (After TAO Shiyan and CHEN Longxun 1987)

regions of South and Southeast China are also penetrated by the winter monsoon, and after October 10, the penetration of the China Seas as well as Hainan and Taiwan Islands follows (see Fig. 3.7).

It is interesting to note that the onset of the winter monsoon takes place more rapidly over China than over the India-Pakistan subcontinent, where the winter monsoon penetrates between September 1 and December 15, thus covering a period of 3.5 months, instead of only 1.5 months for China (see Fig. 3.7).

The distribution of pressure and circulation over China is thus abruptly reverted in September, from summer monsoon conditions with winds of a southern direction to winter monsoon conditions with winds of a northern direction. The reason for the rapid onset of the winter monsoon in September is, in particular, the initiating effect from the earth's surface which leads to a swift reversion of the pressure and wind patterns. The sudden change of the atmospheric circulation in September represents a very specific criterion for the climate of China. In addition, the dynamic effect of the Tibetan Plateau contributes greatly to a quick invasion of the winter monsoon into southern China.

The abrupt onset of the winter monsoon in September is due to the pressure field over the Asia Pacific region. Most typically, the heat Low over the northwestern India-Pakistan subcontinent quickly weakens in September, while the pressure throughout China slightly increases, thus indicating a continental anticyclone centred over Mongolia. Northerly winds become prevailing. In October and November, the pressure pattern becomes markedly pronounced: while the India-

Pakistan heat Low completely dissolves, the continental High over Mongolia and mid-Siberia is strongly established. Cold and dry northerly winds dominate the fully settled winter monsoon circulation. As a typical feature of the well-established winter monsoon from October onwards, cold waves invade China.

Winter monsoon prevails continuously from October to March, as long as the strong anticyclone over Mongolia and mid-Siberia remains stably established. As far as the pressure pattern is concerned, January shows the maximum development of the winter monsoon because the pressure pattern at sea level experiences the highest values. In the centre of the anticyclone, pressure on ground even exceeds 1040 mb which, in association with the well-established Aleutean Low, leads to an intense pressure gradient and thus to the strongly established winter monsoon winds from the northern quadrant.

Following the changing pressure pattern during winter, the months October through December show a gradually increasing winter monsoon, from February through March a decreasing winter monsoon.

During winter monsoon, two major currents of cold air invade China on certain tracks from Siberia and Mongolia, one current in an eastern or southeastern direction to the Aleutean Low, the other in a southern direction through southern China and finally flowing into the equatorial Low.

Winter monsoon is fully settled until the end of March. The transition from winter to summer monsoon takes place in April and May when air masses of an either southern or northern origin and direction compete to dominate over China. This causes a swinging shift northward or southward of the monsoon front (see Sect. 3.3.5).

3.3.3 Periods of Active and Weak Winter Monsoon

The common pressure field over the Asia-Pacific region, which induces the winter monsoon over China, can be changed by anomalies of the Siberian anticyclone and the Aleutean cyclone. This leads to a stronger or weaker pressure gradient, causing an either stronger or weaker circulation and resulting in a comparably cold or warm air advection of the monsoonal air masses in winter. These events are called "active" and "weak" winter monsoon respectively.

The anomalous pressure field which typically expresses the active or weak winter monsoon is shown by the mean pressure field at sea level and its deviation for 30 cold pentads and 34 warm pentads respectively, using the data of 1951–1960 (Zheng 1965; see also Figs. 3.8 and 3.9). In the case of a cold "active" winter monsoon, the Siberian anticyclone and the Aleutean Low are strongly developed. The positive anomaly of the anticyclone reaches 8 mb, the negative one 4 mb. Thus, the pressure gradient is very intensive. Under these conditions, a strong outbreak of the winter monsoon and extremely cold air masses over China can be observed. As a result of the active winter monsoon, temperatures in the regions south of the Yangtze and Huaihe reach very low values.

In the case of a "weak" winter monsoon (see Fig. 3.9) when the Siberian anticyclone is comparably weak and only a small high pressure cell, while the

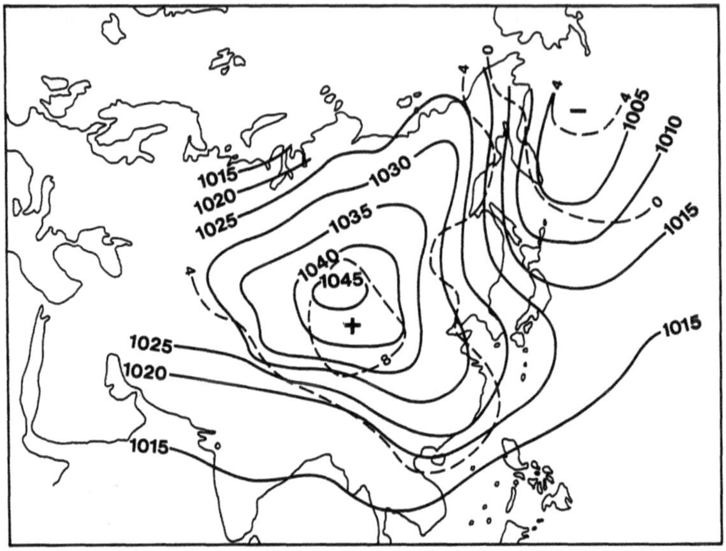

Fig. 3.8. Mean pressure field at sea level (*solid lines*) and its deviation (*dotted line*) for 30 cold pentads between 1951–1960. (After Chinese Geographical Society 1984)

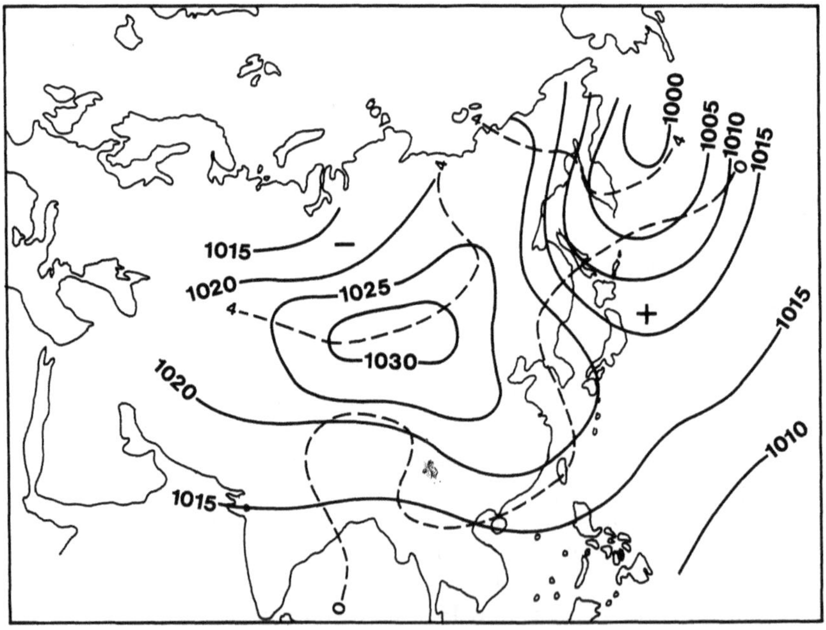

Fig. 3.9. Mean pressure field at sea level and its deviation for 34 warm pentads between 1951–1960 (After Chinese Geographical Society 1984)

Aleutean Low is still strongly developed, the pressure gradient between land and sea in the Asia-Pacific region decreases. Subsequently, relatively warm air masses invade China which cause fairly high winter temperatures.

The mean height of the 500-mb level and its anomalous height field for active (cold) and weak (warm) monsoon conditions in winter are shown in Figs. 3.10 and 3.11 (after Chinese Geographical Society 1984). The height fields for anomalous monsoon conditions are established in an opposing pattern. In an active (cold) winter monsoon a meridional circulation pattern prevails over the Asia-Pacific region. In the area of the Ural Mountains a large ridge of high pressure appears. The huge trough over East Asia is extremely well established. Over the northwestern part of East Asia a positive height anomaly occurs, while it is negative over the southeastern part. Opposite conditions appear in a weak (warm) winter monsoon when a zonal circulation pattern prevails over East Asia. A negative height anomaly of the 500-mb level can be seen for the northwestern part, but a positive one for the southeastern part. During weak winter monsoon conditions, cold air of polar origin cannot penetrate southward, due to the strong zonal circulation of westerlies.

Active and weak winter monsoon conditions can also be expressed by the mean pressure difference at sea level, shown for different latitudes between 110° and 160° E (Table 3.3); figures are taken from the period 1951–1960 (after ZHENG 1965). From the figures it can easily be deduced that the pressure difference is remarkably higher under active (cold) than weak (warm) winter monsoon conditions. The difference is about two-fold greater under active than under weak conditions. The absolutely largest values for the pressure difference thus express the strongest active or weak winter monsoon which occurs at 40° and 50° N (Table 3.3), indicating the most markedly established active and weak winter monsoon over North and Northeast China.

A significant characteristic of the winter monsoon is represented by successive outbursts of cold air, commonly called *"cold waves"* or "surges of the winter monsoon". Defined by a sharp fall in air temperature by over 10° C within 24–48 h and originating from the Arctic seas, cold waves may travel through China southward even to the southernmost parts of China. The progress of cold waves is, however, strongly affected by orography so that mountains and hills may block the passage of cold air, while it may be channelled by valleys and basins. As a general rule, the climatic effectiveness of cold waves weakens with increasing distance from their origin. The climatic effects of cold waves are predominantly concerned with a sharp drop of temperature to even 20° C, but frequently they are also associated with rain and snow. In this respect, however,

Table 3.3. Mean pressure difference at sea-level (mb) under active and weak winter monsoon conditions at different latitudes between 110° and 160° E for 1951–1960. (After ZHENG 1965)

Latitude (°N)	20	30	40	50	60	70
Active monsoon	4	10	31	41	22	1
Weak monsoon	2	4	16	20	14	0

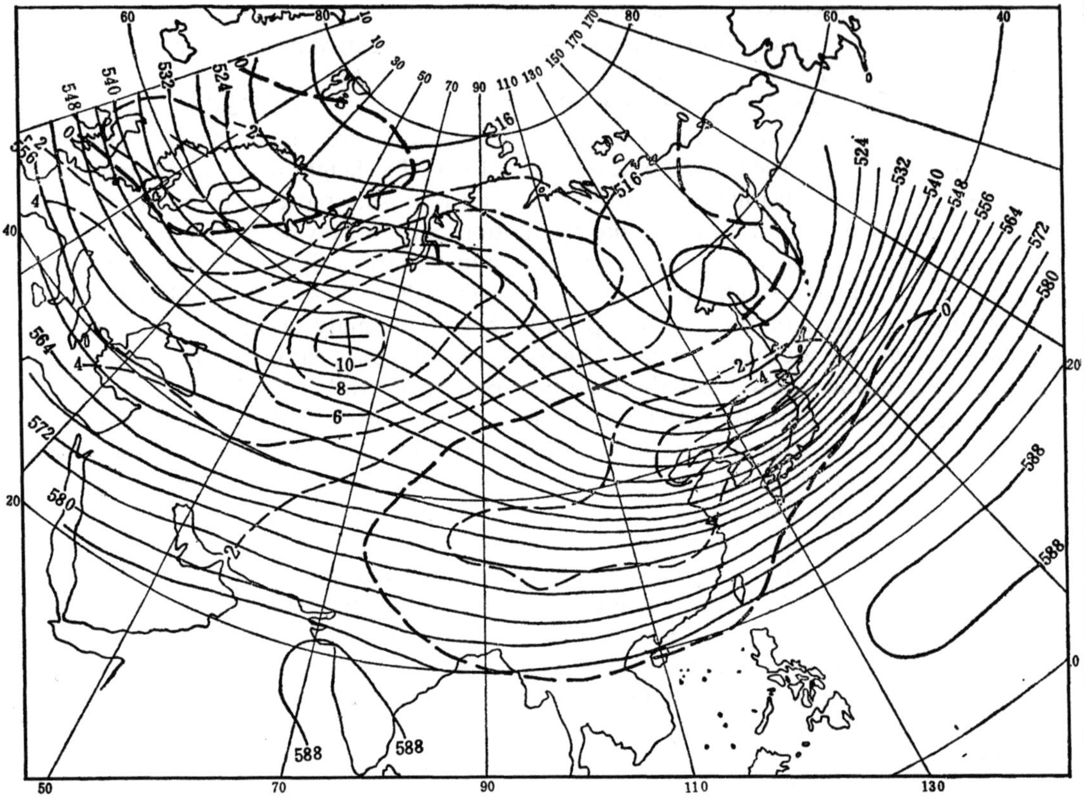

Fig. 3.10. Mean height of the 500-mb level (*solid lines*) and its anomalies (*dotted lines*) for "active" (cold) winter monsoon. (After Chinese Geographical Society 1984)

Xinjiang represents an exceptional case because very vigorous outbursts of cold air, channelled through Tian Shan, Altai and Dzungaria, may produce exceptionally high winter precipitation, particularly in wind-exposed parts (Lu 1937).

The frequency of occurrence of cold waves (Table 3.4) comes to 5.2 $year^{-1}$ which represents an average value for 1951–1976. Of a total of 135 cold waves during the period of observation, a monthly mean maximum of 29 (or 21.5% of all cases) occurred in November, followed by March (27 cases; 20.0%) and February (22/16.2%). The period of cold wave risk is restricted from November to April into which a total of 97.1% of all cases fell (see Table 3.4).

The frequency of cold waves shows a large interannual variation. Over the observation period 1951/52–1975/76 (see Table 3.4), the maximum number of cold waves was ten (for 1965/66 and 1968/69), while the minimum number was one only (1974/75).

Cold waves over China originate from the Arctic seas and the Russian and Siberian lowlands; they then travel far southward and turn towards the east, most heavily affecting the northern- and northeasternmost parts of China (Fig. 3.12).

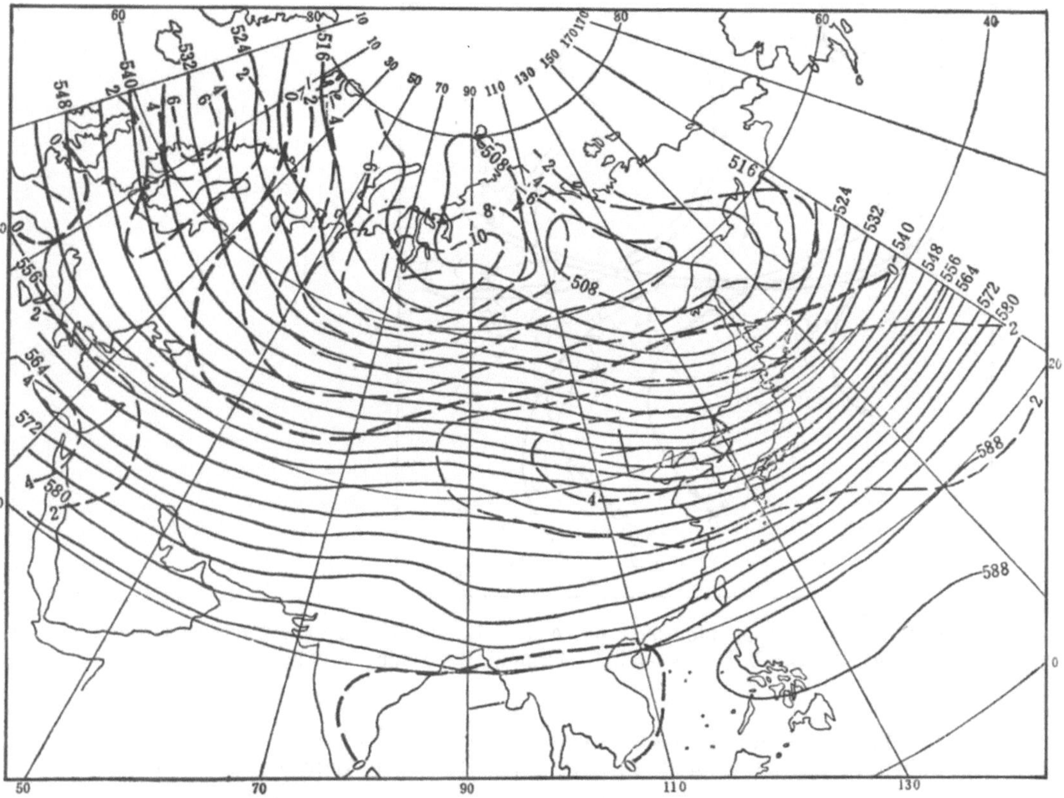

Fig. 3.11. Mean height of the 500-mb level and its anomalies for "weak" (warm) winter monsoon. (After Chinese Geographical Society 1984)

From the three typical main streams of cold waves over North Asia and Siberia (Nos. I, II, III; Fig. 3.12), different branch streams turn to the south, thus affecting also all other parts of China, although to a lesser extent than the north and northeast. Cold fronts, migrating southward, extend to a maximum of 2,000 m above ground level so that their progress is largely dependent upon the relief conditions. Mountains may delay the advance of cold fronts, while basins and valleys permit the passage of cold waves.

Table 3.4. Absolute and mean number of cold waves, 1951–1976

Month	Oct	Nov	Dec	Jan	Feb	Mar	Apr	May	Year
Total No.	3	29	16	17	22	27	20	1	135
Mean No.	0.12	1.12	0.73	0.78	0.85	1.04	0.79	0.04	5.2
% of annual No.	2.2	21.5	11.9	12.6	16.2	20.0	14.8	0.7	100

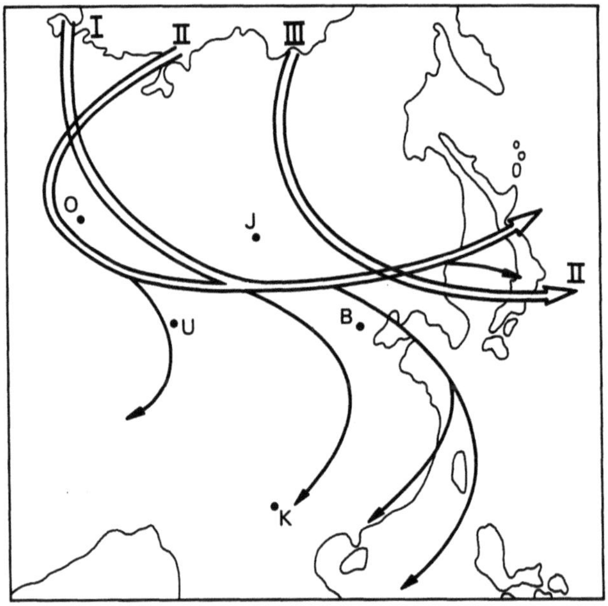

Fig. 3.12. Typical paths of cold waves over Asia and China for the 500-mb surface. (After TAO Shiyan 1984). O – Omsk, I – Irkutsk, U – Ürümqi, B – Beijing, K – Kunming

The typical pressure field at sea level over Asia, valid for a cold wave, is shown for February 22, 1974, 08.00 h, (Fig. 3.13). It clearly demonstrates an intense High over Siberia which has a maximum air pressure of 1070 mb and a very distinct and strong anticyclonic pressure field throughout Asia. Only over the northwestern Pacific and southern China is a low pressure field well established.

Representing a typical phenomenon of the winter monsoon, cold waves migrate far southward throughout China and finally even invade Hainan Island. A cold wave is defined by a temperature drop of $\geq 10°C$ which leads to a daily minimum temperature below 5°C for a few consecutive days. The period of cold air risk in Hainan Island corresponds to the winter monsoon season from the first 10-day period of October to the first 10-day period of February which was defined accordingly by the strong prevalence of winds from an ENE direction.

Particular weather phenomena of active winter monsoon periods also include occasionally occurring monsoon gales of cold air which may last up to 1 week and which affect the coasts and coastal lowlands more heavily than the inland areas where they quickly weaken. During the winter monsoon, outbursts of cold air are sometimes also associated with violent snowstorms which particularly strike the northern lowlands.

3.3.4 Damage Due to Strong Cold Outbreaks of Winter Monsoon

Winter monsoon in China is characterized by frequent, sudden drops of temperature during strong cold outbreaks from Siberia. The advance of cold waves may

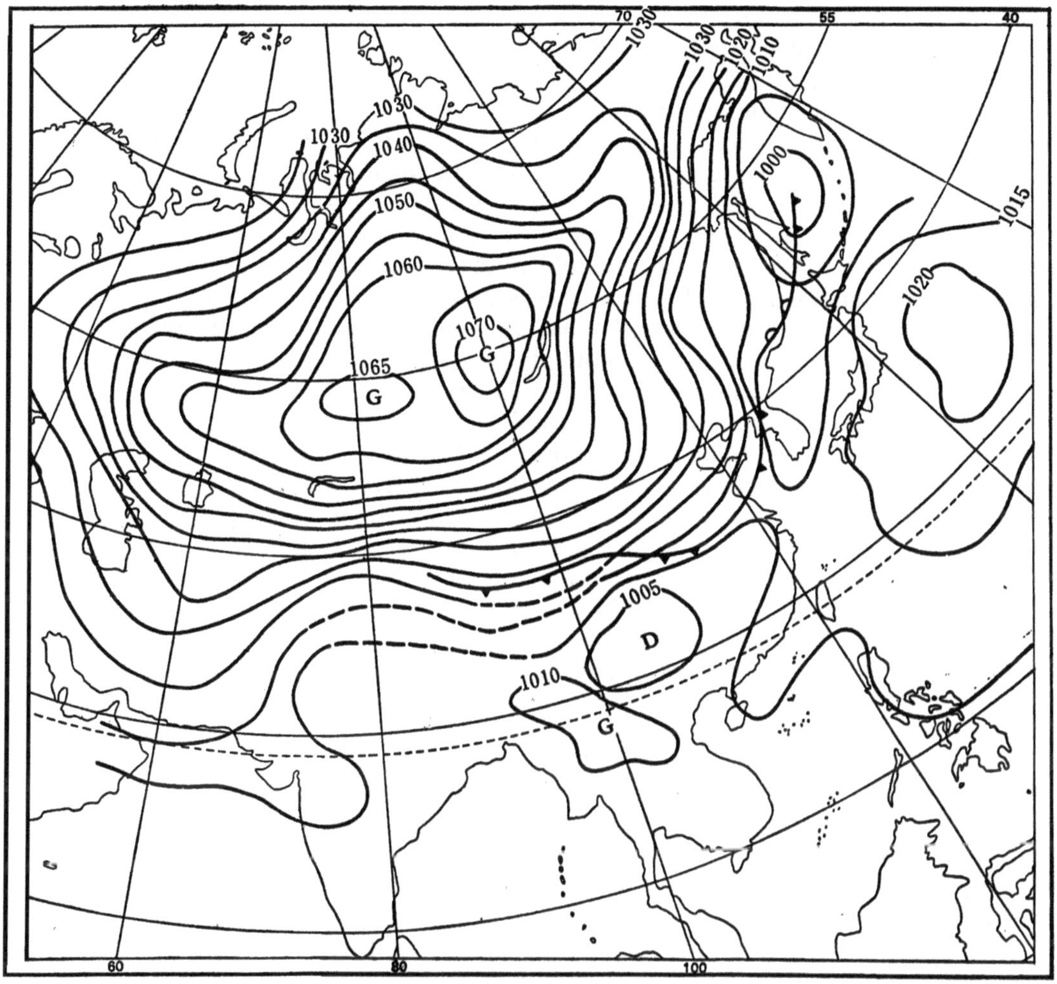

Fig. 3.13. The sea-level pressure pattern on February 22, 1974, 08.00 h during an intense cold wave. (After ZHANG Jiacheng and LIN Zhi-quang 1985)

migrate as far as southern China. Cold air advection, being associated with a sudden temperature drop of more than 10° C for some consecutive days, may even cause frost, which represents the most harmful hazard, particularly to sensitive subtropical and tropical plants (see Sect. 4.4.1). Normally, strong cold outbreaks last between 4 and 6 days; their agroclimatic implications depend upon the extent of temperature drop, the plants affected and the locality of their cultivation which is responsible for the topo-climate.

JIANG Ailiang (1984) has studied the effects of frost damage and chilling injury of subtropical and tropical plants in southern China, both types of cold injury depending upon the seriousness of cold outbreaks. According to JIANG, frost damage and chilling injury may correspondingly be caused by a radiation as well as an advection type of temperature drop. Both types are accompanied by the incidence of cold outbreaks; thus, advectional cold injury occurs during the advance stage of a cold wave, whereas radiational cold injury occurs in the later stage of intrusion of a cold wave, during quick radiation loss on a starry, cloudless night and early morning bright sunshine.

Among subtropical plants, citrus (in particular *Citrus unshiu*) is reported to be affected by light, moderate or heavy frost damage when the minimum temperature drops below –7°, –9° or –11° C respectively (JIANG Ailiang 1984). As for tropical plants, rubber trees (which are widely cultivated in parts of southern China) are particularly sensitive to cold outbreaks and can be badly injured. Radiational chilling injury of rubber trees occurs on cloudless nights. Such occasions were reported for Xishuangbanna, Yunnan Province and for subtropical Guanxi Province (with 11 cases of cold injury to rubber trees from 1951–1980), but even for tropical Hainan Island (18°–20° N).

During the 30-year observation period 1950–1979, severe cold injuries to rubber trees in Hainan Inland were recorded in 1955, 1963, 1967, 1974 and 1976. In January 1955, the severest cold damage to rubber trees was observed when 10 to 30% of young (2 to 3 years old) trees suffered chilling injury; minimum air temperature dropped to 0.4° C at Danxian, on January 12, 1955. In the 1963 severe cold outbreak, air temperature in northern Hainan even dropped to a minimum of –0.6° C. Especially at lower elevation, serious cold injuries to rubber trees could be observed in the past. However, the cold-resistant cultivar RRIM 600 showed no significant damage by a cold wave. In principal, the extent of cold injury to rubber trees depends widely upon the rubber cultivars and the topo-climatological conditions (JIANG Ailiang 1984). However, the occurrence of serious cold injury indicates the rather unusual, but severe occurrence of cold waves as a winter monsoon hazard, with far-reaching agroclimatic effects to southern China and affecting rubber trees and other thermophilic tropical and subtropical plants.

A radiational subtype of chilling injury to rubber trees, called "rotting foot" by the local inhabitants of some parts of Yunnan Province, was reported by JIANG Ailiang (1982). This type of injury to rubber trees occurs when sunny and cloudless weather with daily minimum temperatures $<5°$ C (but $>0°$ C) prevails for 1 or even 2 weeks in winter. The persistently low temperatures injure the bark of the base of rubber trees.

Cold damage is, however, not restricted to subtropical and tropical plants only, but may also occur to temperate crops, such as winter wheat which may suffer freeze injury or even death in the case of an unusual occurrence of extremely low temperatures.

3.3.5 Onset and Duration of the Summer Monsoon

While winter monsoon quickly ceased through April, the onset of the summer monsoon from South to Northeast China starts in the first 10-day period of May and finally terminates in the last 10-day period of July (see Fig. 3.14). Onset of the summer monsoon is usually defined by an abrupt increase of rainfall. The mean dates of the summer monsoon are therefore derived from long-term rainfall records.

The regions initially influenced by the summer monsoon in the beginning of May, are Hainan Island and the southern part of Taiwan Island. Thereafter, the monsoon moves stepwise northward. The monsoon advances to the coast of southern China on May 10, then advances slowly northward. During the last 10 days of May, the monsoon front stagnates in southern China, then moves northward rapidly in early June. Summer monsoon arrives in the middle and low reaches of Yangtze in mid-June when the mei-yu (or plum) rains begin. After a stagnation of the mei-yu front until the first 10-day period in July, the monsoon then advances quickly northward again, arriving in northern and northeastern China in late July. It is, however, typical that the northward advance of the summer monsoon takes place in alternating phases of quick and slow development.

The dates given for the onset and advance of the summer monsoon over China (Fig. 3.14) are average values; in absolute terms, they can vary up to 4 weeks, and accordingly the advance of the monsoon can also be accelerated or weakened so that phases of a rapid advance may alternate with those of a slow advance. On average, the summer monsoon advances comparably slowly northward during the last 10-day period of May and June as well, while a quick advance to the north is observed for the second 10 days of both June and July. In individual years, even phases of a stagnant or retreating monsoon may occur, for example during the

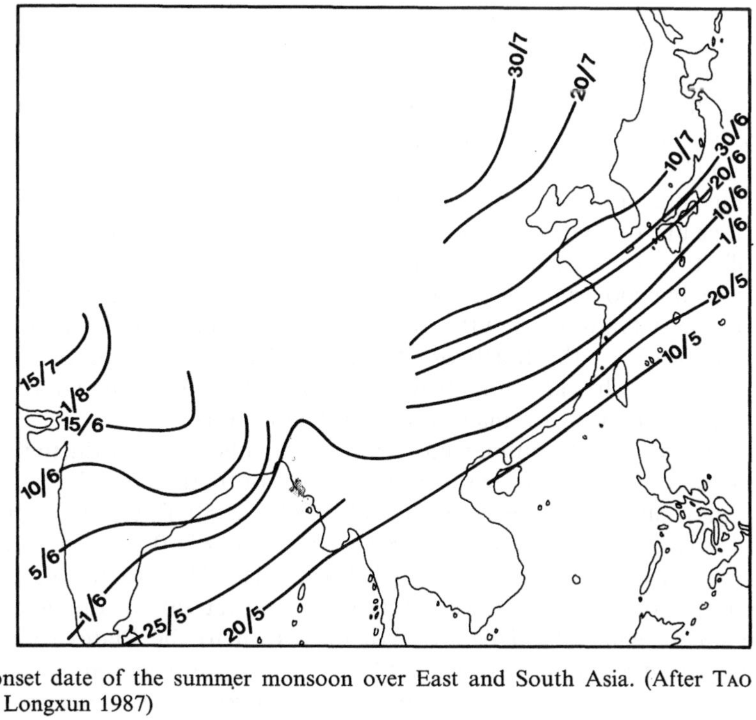

Fig. 3.14. Mean onset date of the summer monsoon over East and South Asia. (After TAO Shiyan and CHEN Longxun 1987)

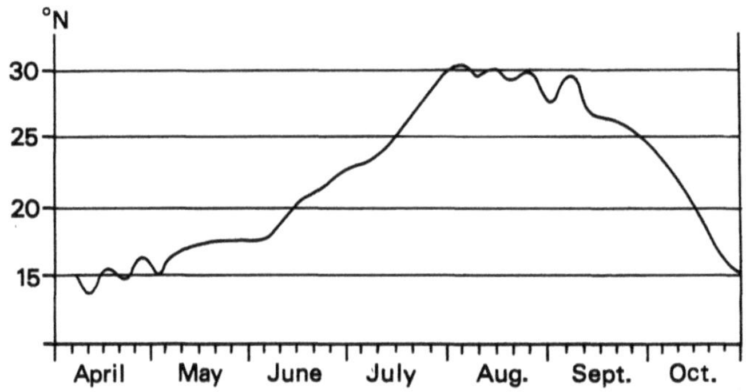

Fig. 3.15. The pentad mean position of the ridge line of the subtropical anticyclone in the West Pacific at 500-mb for 1971–1980. (After LIN Chunyu 1982)

last 10-day period of June. The interannual variation of the onset and advance of the summer monsoon is therefore an aspect of particular concern.

In comparison with the India-Pakistan subcontinent the summer monsoon over China starts half a month earlier and terminates about half a month later so that the total duration of the summer monsoon over China is about 1 month longer. This longer period of duration roughly coincides with the greater south to north distance of China than India. On average, the period of active duration of the summer monsoon over China varies between 5 months, in the southernmost parts, and 1 month only, in the extreme northeast. Thus, the duration of the summer monsoon gradually decreases in China from south to north.

The advance of the summer monsoon has a close connection with the northward migration of the subtropical anticyclone in the western Pacific (TAO Shiyan and CHENG Longxun 1986). Based on studies on the mean pentad position of the ridge line of the subtropical anticyclone at 500 mb for 1971–1980 (Fig. 3.15), the ridge line oscillates around 15° N in April. It moves to 18° N in the second pentad of May at which time the summer monsoon advances to southern China. The ridge line moves to the north of 20° N in the second pentad of June, during which the mei-yu rains start in the Yangtze valley. In the second pentad of July, the ridge line moves to the north of 25° N, and the mei-yu rains in the Yangtze valley terminate. The ridge line then moves to 39° N in the fifth pentad of July when the summer monsoon rains in northern China start.

Onset of the summer monsoon represents a period during which the whole circulation pattern over the northern and southern hemisphere in the Asia-Pacific region rapidly changes (ZHAO et al. 1984). At 80° E, easterlies and westerlies are clearly separated from each other to the south of 20° N with a strong westerly jet on the Tibetan Plateau which is fully established before the onset of the monsoon. During onset, the westerly jet weakens and retreats to the north of the Tibetan Plateau, while the Tibetan High moves on to the Plateau and the southwest monsoon pushes northward. At 105° E, the westerly jet remained quasi-stationary at 35° N. During onset, the boundary between easterlies and westerlies moves northward in response to the northward movement of the ridge line of the subtropical anticyclone to 20° N. Examination of cross-sections along 30° N shows that the meridional wind is primarily longitudinally dependent before onset and vertically dependent after onset (ZHAO et al. 1984). Before onset of the summer monsoon, transient waves are found propagating eastward in the westerlies of the southern hemisphere. During onset, the long waves become stationary with troughs at 60° and 120° E.

The formation, nature and characteristics of the summer monsoon over China were intensively investigated by Chinese meteorologists. In the course of the studies, the summer monsoon was found to represent an independent wind and pressure system (CHENG Longxun et al. 1980), which shows different sources of origin in comparison to the Indian summer monsoon. The summer monsoon over China includes the following major components: the anticyclone over Australia, the cross-equatorial flow from the southern hemisphere, the Innertropical Convergence Zone, the tropical easterly jet, the subtropical anticyclone in the West Pacific, the mei-yu frontal zones and the mid-latitude disturbances, see Fig. 3.5, p. 40 (see Sect. 3.3.1; TAO Shiyan and CHEN Longxun 1987). It is important to note that the circulation pattern in the southern hemisphere has a close connection with the formation and advance of the summer monsoon over China (TAO Shiyan and HE Dakang 1983). The dynamic processes of the summer monsoon over China are ruled by a heat source which is centred in the South China Sea and a heat sink with its centre in Australia (CHEN and LI 1981).

Summer monsoon in China has two sources of air masses, one from the subtropical anticyclone in the Northwest Pacific, another from the cross-equatorial flow from the southern hemisphere. These two sources establish two different branches of the Chinese summer monsoon, a southeast and a southwest branch. The southeast monsoon component is much stronger, it extends further and is also maintained longer. This leads to the frequently used term "southeast monsoon" for the summer monsoon.

The retreat of the summer monsoon in China is a straightforward, rather quick process which only lasts around 1 month. The ridge line of the subtropical anticyclone starts to retreat southward in early September. It arrives at 25° N in late September and retreats to 17° N until mid-October. In the middle of September the low-level circulation pattern in eastern Asia changes suddenly from a cyclonic to an anticyclonic pressure field. Thus, cold and dry anticyclonic air masses frequently invade from the north, resulting in fine autumn weather in China. The average position of the polar front progresses continuously southward and arrives in southernmost China in mid-October.

The retreat of the summer monsoon in Eastern Asia is closely connected with the southward migration of the subtropical anticyclone (Fig. 3.15). The ridge line of the subtropical anticyclone starts to retreat southward in early September, quickly moving southward and arriving at 25°N in late September, thereafter only slowly retreating to 17° N by late October.

Early September, when the summer monsoon in the lower troposphere quickly retreats southward, and subsequently the winter monsoon invades China, at the upper levels, the summer circulation conditions are still maintained. South of the Yangtze valley, the upper easterly flow prevails which overlays the northerly winds on ground. In South China and the mid-lower sections of the Yangtze valley, fine mid-autumn weather prevails.

In the second 10-day period of October, the ridge of the West Pacific High retreats to the South China Sea. Westerlies therefore prevail over South China and lead to the cessation of autumn weather.

Compared to the above mentioned areas, conditions are opposite in western China in September and October. Winter monsoon in the lower troposphere advances to the areas of Sichuan and Guizhou Provinces, but the heat Low above the Qinghai-Xizang Plateau still acts forcefully. During this period, the plateau summer monsoon, in particular the southwest monsoon over the eastern parts of the plateau, is still strongly established. A southwest current occurs above the low-level winter monsoon and causes autumn rainfall over western China. In the second 10-day period of October, this process comes to an end, and subsequently the autumn rainfall in those areas is also terminated.

The duration of both the summer and winter monsoon varies considerably for different parts of China. In South China, both monsoons last an average of 5 months each. In central China, the duration of the winter monsoon is longer than the summer monsoon. In North and Northeast China, winter monsoon lasts between 9 and 10 months in comparison to only 1 month for the summer monsoon. In addition, both the winter and summer monsoons show a remarkably large interannual variation in terms of their onset and duration.

The prevalence of southerlies during the summer monsoon is clearly shown by the monthly frequency of the prevailing wind direction, recorded at a height of 500 m at four selected stations (Table 3.5); figures are given as percent values of a 360° wind field, 0° represents the northern direction. As can be seen, at Guangzhou/199, from March to August, prevailing winds from the south occur, corresponding to the summer monsoon. Southerlies prevail in Changsha/147 from April to August and in Wuhan/126 from June to July. The withdrawal of the summer monsoon takes place rapidly from north to south of China. The summer frequency of southerlies in Beijing/61, which is the border area of the summer monsoon, is relatively low in comparison to the other three stations. The higher

Table 3.5. Monthly frequency of the prevailing wind direction at 500 m height at selected stations (figures are in %, related to a 360° wind field)

Station	Wind direction	J	F	M	A	M	J	J	A	S	O	N	D
Guangzhou	10– 49	38	27	18	11	7	6	2	6	18	32	37	37
	150–189	13	16	32	41	38	33	41	28	18	11	6	8
Changsha	350– 29	37	33	24	24	21	14	8	20	44	40	45	32
	170–209	7	16	21	28	27	33	49	27	8	5	9	14
Wuhan	10– 49	31	30	22	23	22	11	10	20	35	39	33	33
	190–229	9	8	13	16	14	19	41	18	4	4	7	10
Beijing	310–349	41	28	19	15	13	10	8	10	13	18	25	37
	190–229	12	19	21	31	27	23	23	24	27	27	17	11

value of southerlies in spring and autumn is caused by cyclonic and anticyclonic activities, but not by the summer monsoon.

3.3.6 Some Characteristics of the Summer Monsoon

Besides the generally valid, enormous heat Low over India and Pakistan and air masses of both tropical and equatorial origin, which characterize the summer monsoon regime over China, characteristics of summer monsoon have been studied with particular attention by Chinese meteorologists (see the review by TAO Shiyan 1984). Accordingly, summer monsoon has been shown to be closely related to precipitation in China. As a general rule, the rainy season coincides with the period of the summer monsoon; its onset and withdrawal represent, at the same time, the beginning and termination of the rainy season. The exact time period of the rainy season varies according to the delay or prematurity of the summer monsoon.

The correlation between the intensity of the summer monsoon, expressed by wind velocity, and rainfall has been studied by ZHU Kezhen (1954). He calculated for July the correlation coefficients between both variables for Shanghai and Beijing. It was shown that there is a negative correlation at Shanghai, a positive one at Beijing. This means that less rainfall occurs at Shanghai and higher rainfall at Beijing, if the summer monsoon is strongly developed; in contrast, rainfall increases at Shanghai and decreases at Beijing, if the summer monsoon is weakly developed.

For a closer understanding of the relationship between the intensity of summer monsoon and rainfall in July in eastern China, GUO Qiyun (after TAO Shiyan 1984) has calculated the correlation coefficient between "summer monsoon intensity" and rainfall for 88 stations evenly distributed over eastern China:

$$\text{Summer Monsoon Index (SMI)} = \frac{\sum_i \sum_j (\Delta p \leqq -5\,\text{mb})}{M},$$

where $i = 1, 2, 3, \ldots, 12$ for months, $j = 10\,°\text{N}, 20\,°\text{N}, \ldots 50\,°\text{N}$ for latitudes, $\Delta p = p_{110°\text{E}} - p_{160°\text{E}}$, the difference in the sea-level pressure and M is calculated according to the formula:

$$M = \sum_{1951}^{1980} [\sum_i \sum_j (\Delta p \leqq -5\,\text{mb})]/30\,.$$

According to the correlation indices calculated, eastern China can be divided into three regions (Fig. 3.16). There are positive correlations in North and South China respectively, while there is a negative correlation in the Yangtze valley. This means that in the case of a strong summer there are above normal rainfalls in North and South China, but below normal rainfall in the Yangtze valley.

Onset and advance of summer monsoon in China have been found as events of a remarkably great variation over time. Significant inter-annual variations of the dates of onset and withdrawal of summer monsoon were observed in China, and similarly the strength of the monsoon varies from year to year. The advance-

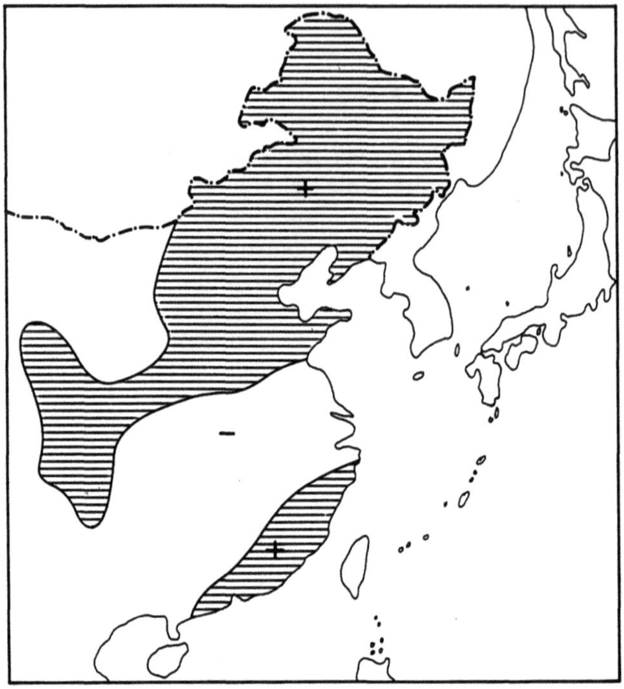

Fig. 3.16. The correlations between precipitation and "summer monsoon index" in July in eastern China. (After GUO Qiyun, in TAO Shiyan 1984)

ment of the summer monsoon northward consists of individual phases of rapid and slow migration. A quick advancement, for example, occurs during the second 10-day period of both June and July, while only a moderate migration or even a stagnation is normally found for the last 10 days of both May and June.

A remarkably large intramonsoonal variation of rainfall also represents a characteristic feature of summer monsoon in China. The front of the advancing equatorial air masses provides most of the monsoonal precipitation, while the interior air masses lead to less rainfall and to spells of fine weather which last a few consecutive days. The northward advance of the front of equatorial air masses leads in the first half of June in the middle and lower reaches of the Yangtze to ample rainfall called "mei-yu" or "plum rains", which is associated with very hot and damp air, massive low cloud and depressing weather. The northward advances of the front lead to a maximum rainfall during May and June in southern and southeastern China.

During summer monsoon, heavy precipitation also originates from thunderstorms in shallow, thermal depressions, lee-side depressions, orographical lifting on wind-exposed slopes and typhoons.

3.4 Frontology

3.4.1 Mean Front Position in January and July

Figures 3.17 and 3.18 represent the mean location of climatological fronts in January and July respectively. The thin isolines express the mean number of fronts in an area of 2.5 latitude and longitude grid points each. The thick lines indicate the axis of the maximum number of fronts.

Due to the influence of the Tibetan Plateau, the frequency of climatological fronts and cyclones in the area between Huang Ho and Yangtze is very low. Two main belts of fronts occur over North and South China respectively, which show significant seasonal variations.

In winter, the northern belt of the fronts is located over Xinjiang and northeastern China, but the southern belt is located in Yunnan and over the southern provinces. South of the Yangtze, including southwestern and southern China, quasi-stationary fronts are established and formed between transformed polar, continental air masses and a warm southwest air current. They play a very important role for weather and climate in these areas.

In spring, tropical air masses and the polar front belt gradually migrate to the north, followed in the same direction by the cloud-rainfall belt. This process leads to the typical "plum rains" in the Yangtze basin.

In summer, southerly air currents reach the Inner Mongolia Plateau. In this case, the northern front belt represents the polar front between polar and tropical air masses. The southern front belt is also shifted to the north and maximum frequency of these fronts occurs over the lower regions of the Yangtze. Due to the influence of the subtropical High, southeast China belongs to the frontolytical sector.

In autumn, the front belts shift rapidly back to the south, resulting in less frontal activity in central and South China.

3.4.2 The Stationary Fronts in February and March as well as in the Pre-Typhoon Season in South China

Being of far-reaching importance to the weather and climate in southern China, the frequent occurrence of the stationary fronts represents a noteworthy phenomenon in February and March as well as in the pre-typhoon season.

In winter, when strong cold waves migrate from north to south, the cold front often moves to South China and finally to the South China Sea. If, however, the cold air mass is not so strong the front moves slowly or can even be stopped south of Yangtze and becomes here a stationary or quasi-stationary front. This process can be observed both in winter and spring. These types of fronts cause over larger regions a heavy cloud cover and abundant rainfall. In South China, stationary fronts are formed not only because of cold air outbreaks from the north, but also from local frontogenesis processes. Figure 3.19 shows the mean location of the quasi-stationary front and the associated cloud and rainfall area for different seasons in South China.

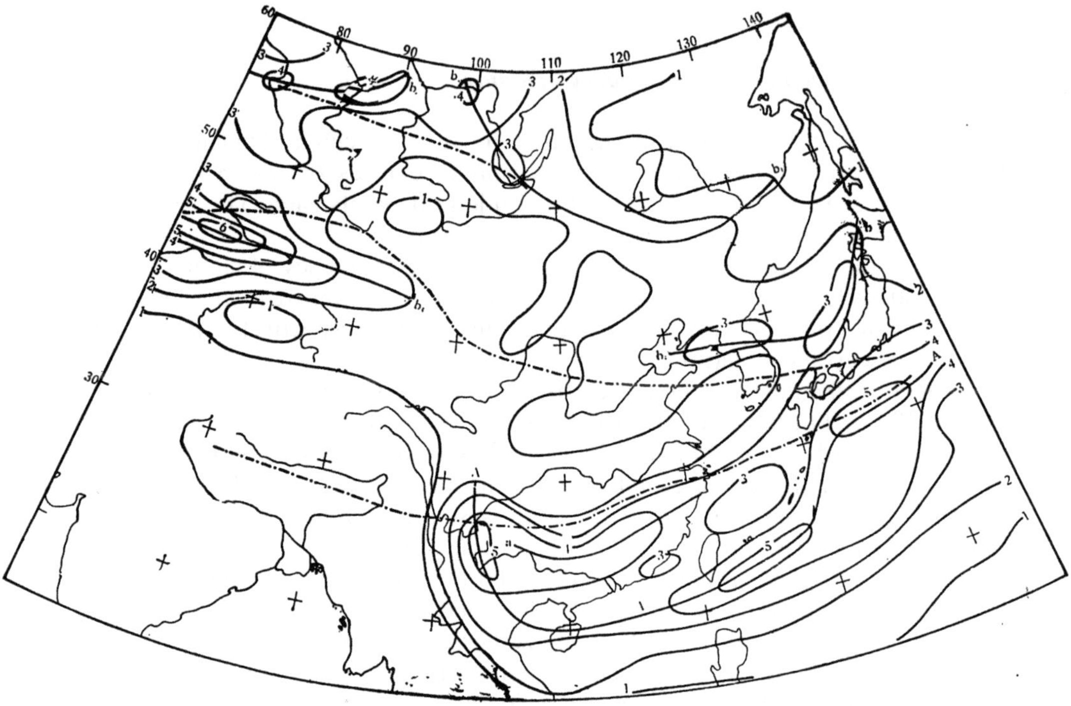
60
80
90
100
110
120
130
140
50
40
30
A
b₁
b₂
b₃
b₄

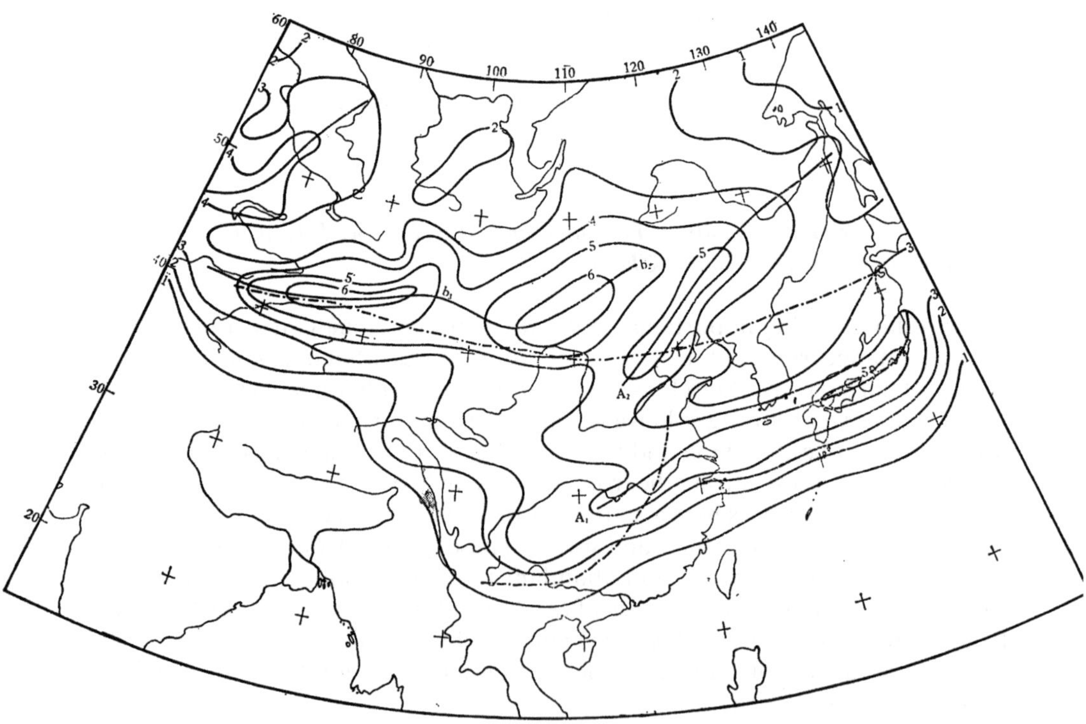
60
80
90
100
110
120
130
140
50
40
30
20
A₁
A₂
b₁
b₂

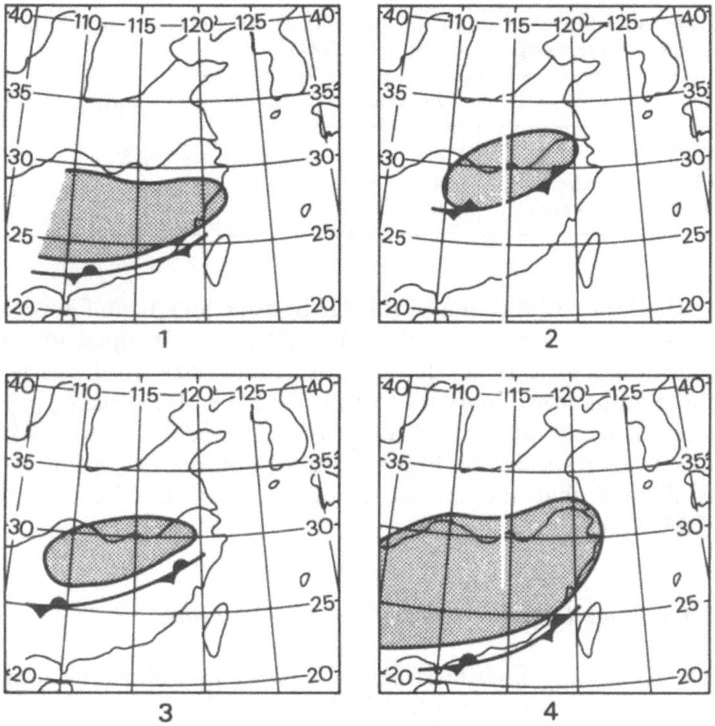

Fig. 3.19. Mean position of the quasi-stationary front and the associated cloud and rainfall area for different seasons in South China. *1* Spring, *2* Summer, *3* Autumn, *4* Winter

Table 3.6 expresses the mean as well as the maximum and minimum number of stationary fronts in South China from February to June for 1958–1975, observed in the area between Hengyang (28° N) to Beihai (21° N). For these observations, the stationary front has been taken as the speed of movement which is less than 1° latitude per day. It can be seen that from February to June, on average, between two to three stationary fronts were experienced. The absolute maximum value amounted to between four and six, the minimum zero to two stationary fronts. On average, the largest number of fronts occurred in May, while the absolute maximum (six fronts) was recorded in June.

◄

Fig. 3.17. Mean frequency and location of climatological fronts in January. (After SHENG Cheng-yu et al. 1986). *A* Polarfront (A1, A2 etc), *a* Nanling front, *b1* Northeast China front, *b2* Menggu front, *b3* Tianshan front, b4 Bohai and Japan Sea front; *1* axis of jet-stream (south subtropic and north temperate jet-stream)

Fig. 3.18. Mean frequency and location of climatological fronts in July. Abbreviations see legend of Fig. 3.17. (After SHENG Cheng-yu et al. 1986)

Table 3.6. Number of stationary fronts in South China, from February to June, 1958–1975. (After Chinese Geographical Society, 1984)

	February	March	April	May	June
Mean	2.3	2.7	2.8	3.2	2.8
Absolute maximum	4	5	5	5	6
Absolute minimum	0	1	1	2	1

Duration of stationary fronts in South China from February to June may vary from 1 to 3 days only up to even 20 days. Grouped into different classes of duration, a 4- to 6-day duration of a stationary front clearly prevails in each of the months from February to June (Table 3.7). Fronts of a 4- to 6-day duration amount to about 50% of all cases, followed by 7- to 10-day and 1- to 3-day durations respectively. From the total number of stationary fronts, it has been confirmed that the largest number in South China occurred in May (see Table 3.7).

3.4.3 Some Characteristics of the Mei-Yu Front

The commonly described *"mei-yu"* season is confined to South China and represents the quasi-stationary belt of heavy rainfall in pre- and early summer in April, May and June. It is associated with small depressions passing seawards along a quasi-stationary front which lies in an east-west direction between the Yangtze and the southern coast. This season is named after the so-called mei-yu front, but it is also often given the popular name "plum season" or "plum rains".

The pressure field and circulation pattern of mei-yu, together with the associated rainfall conditions, have been particularly studied by TAO Shiyan (1986, lecture given at a German-Chinese Workshop on "The Climate of China", at the Institute of Geography, Mainz University, FRG). Some of the major findings are cited in this section.

A belt of heavy rainfall, which migrates northward from the subtropics, occurs in the transition period from winter to summer monsoon. It generally lasts for about 1 month or less over a locality. In southern China, heavy rain occurs mainly in May, in Taiwan from May 15 to June 15 (CHEN and CHI 1981), and in

Table 3.7. Number of stationary fronts in South China for different durations, from February to June, 1958–1975. (After Chinese Geographical Society, 1984)

Duration of front	1–3	4–6	7–10 days	11–15	16–20	Total
February	8	19	10	4	0	41
March	13	26	10	0	0	49
April	7	27	12	4	1	51
May	9	30	13	3	3	58
June	16	27	4	3	1	51

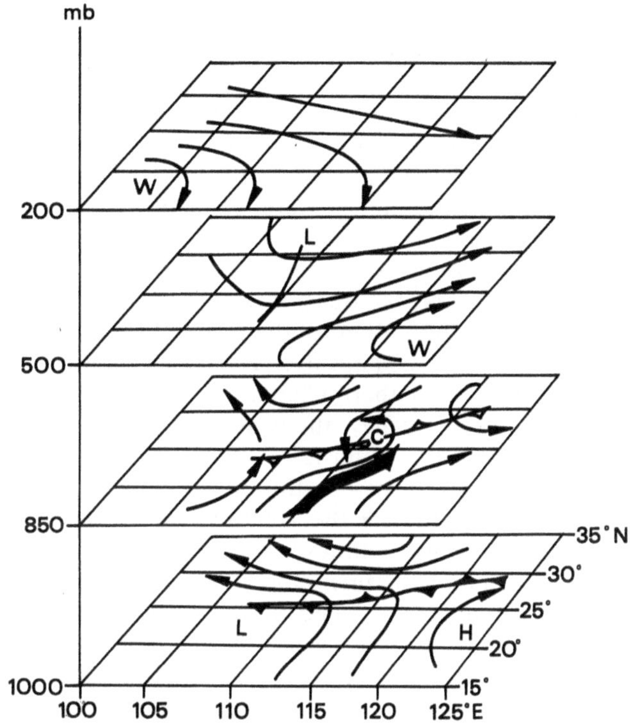

Fig. 3.20. Schematic flow patterns during heavy rainfall in South China in May and June; thick arrow shows the jet stream. (After TAO Shiyan and CHEN Longxun 1987)

the Yangtze valley from about June 10 to July 10 when the term mei-yu is officially used. This rain belt is associated with a quasi-stationary front, which extends from northeast to southwest, and rainfall develops along this front.

In southern China, the quasi-stationary front and its associated rainfall belt from May to early June represent the first of the two heavy rain periods. The other occurs mainly in mid-summer, although sometimes it may shift to late summer or autumn. The May to early June rain period is caused by the interaction between the summer monsoon from the south and air flow of mid-latitudinal origin from the north. The mid-summer rain period, however, is caused by the influence of typhoons. It has already been shown (Sect. 3.3) that before the onset of the summer monsoon in the South China Sea the ridge line of the subtropical Western Pacific anticyclone is located to the south of 16° N. During the first 10-day period of May, the ridge line advances northward to 18° N, and the heat Low on land becomes established, the cross-equatorial flow at 105° E intensifies, the upper level tropical easterlies with speeds greater than 20 ms^{-1} appear at 200 mb over the northern part of the South China Sea and the southwest monsoon with a low level jet prevails throughout the South China Sea. At this time rainfall in southern China rapidly increases.

Rainfall develops along the quasi-stationary front with a major portion falling on the warm side of it. Huge, stationary and persistent thunderstorms are initiated and sustained by upper tropospheric divergence and lower tropospheric convergence in the southwest monsoon. Figure 3.20 is a schematic diagram showing the flow pattern at different levels. The surface front is located near 25° N. To the south of the front, there is a low-level jet in the southwest monsoon, where the stratification is unstable. The low-level jet is an important factor for the formation of heavy rainfall. Thirty-eight of 40 rainstorm cases in 1971–1978 occurred when low-level jet streams were present in the southwest monsoon. In southern China, most of the non-typhoon rainstorms occur in late May and early June (Tropical Oceanic and Meteorological Research Institute, 1984).

In the middle of June, the ridge line of the subtropical High advances to latitude 20°–25° N, the quasi-stationary front in southern China moves northward and heavy rainfall in southern China terminates rapidly. The appearance of stable easterlies at 200 mb over Hongkong is a criterion for the termination of heavy rainfall in southern China.

In the Yangtze valley, the heavy rainfall period from mid-June to mid-July, called "mei-yu" or "plum rain", occurs when the stationary front has arrived. During the onset of mei-yu, the Indian summer monsoon is established and the westerly jet stream over the Tibetan Plateau retreats to the north. At the same time, the subtropical anticyclone in the western Pacific moves northward to 20°–25° N, and the summer monsoon over China advances to 30° N. At that time, the rain belt appears to the north of the Yellow River which signifies the beginning of a heavy rainfall period in northern China and the end of mei-yu.

During the mei-yu season, a stable flow pattern exists over the middle and high latitudes of Asia, with a blocking High located at 50°–70° N and a zonal flow prevailing in 35°–50° N. Transient disturbances move eastward within this zonal flow, advecting cold air southward to intersect the monsoon air along the Yangtze valley, thus forming the stationary mei-yu front.

Unlike the polar front, the temperature contrast across the mei-yu front is small, but there is a marked moist tongue near the frontal zone (Tao Shiyan 1980; Akiyama 1973a).

In the mei-yu front itself, periodic cyclonic vortices of a scale of 500–1,000 km, propagating from western China, move eastward along the front and bring about heavy rainfall. These vortices are not distinct in the surface pressure field, but are quite pronounced in the wind field. In the southeastern quadrant of the vortex, convection is very strong and heavy rainfall occurs (Tao Shiyan 1980).

Severe rainstorms in the mei-yu period can cause damaging floods in the Yangtze valley. For example, the persistent heavy rainfall in the mei-yu periods of 1931 and 1954 caused devastating floods in the middle and lower Yangtze valley. The rainstorm in the upper Yangtze valley on July 3–8, 1935, produced more than 1,200 mm of rainfall in 5 days.

Significant interannual variations are observed both in the length of the mei-yu period and in the total amount of rainfall. Table 3.8 shows the beginning and ending date of the mei-yu period in the Yangtze valley and the percentage of departure from the normal rainfall amount in June–July for each year from 1954–1983. The earliest date of onset was June 1 (in 1954), the latest date July 7 (in 1982). In a few years (such as 1958 and 1978), the mei-yu period occurred for a few days only, giving evidence to the so-called "empty mei-yu" years in China.

Table 3.8. Date of onset and termination of mei-yu in the Yangtze Valley and the percentage of departure from the normal rainfall amount in June–July for 1954–1983

Year	Date of onset	Date of termination	Length of mei-yu period (No. of days)	Percentage of departure in rainfall in June–July
1954	June 1	August 2	63	+119
1955	17	July 8	22	+ 22
1956	5	19	45	− 12
1957	20	12	23	− 12
1958	27	June 29	Empty mei-yu	− 41
1959	28	July 7	10	− 11
1960	18	June 29	12	− 16
1961	6	16	11	− 20
1962	17	July 7	21	+ 16
1963	22	8	17	− 35
1964	23	June 28	6	+ 1
1965	23	July 6	14	− 17
1966	24	12	19	− 2
1967	24	9	16	− 22
1968	23	11	19	− 16
1969	24	16	23	+ 53
1970	18	18	31	+ 24
1971	9	June 26	17	− 18
1972	20	July 3	14	− 29
1973	16	June 29	14	+ 12
1974	10	July 18	39	+ 19
1975	17	16	30	+ 12
1976	16	16	31	− 6
1977	17	1	15	+ 2
1978	23	June 26	Empty mei-yu	− 48
1979	18	July 24	37	+ 9
1980	9	21	43	+ 19
1981	22	3	12	− 26
1982	July 7	27	21	+ 21
1983	June 20	18	29	+ 37
Average	June 18	July 18	23	

There will be an empty mei-yu if the summer monsoon is strong. In such cases, the rain belt moves quickly into northern China and the Yangtze valley suffers from drought. In contrast, in years with a weak monsoon, the mei-yu front will stagnate in the Yangtze valley for a long time, causing a large amount of rainfall. In 1982 and 1983, for example, a severe drought was experienced in northern China, while in the Yangtze valley rainfall was above normal.

3.5 The Transient Disturbances

The main transient atmospheric circulation systems over eastern Asia are the upper westerly troughs in the westerlies, some of which are accompanied by cyclones and anticyclones in the lower troposphere, the subtropical anticyclone in the northwestern Pacific, fronts, the Innertropical Convergence Zone (ITCZ), typhoons and the easterly waves in the tropics. In this section, mainly the upper westerly troughs in the westerlies, the extratropical cyclones and anticyclones and the typhoons are discussed.

3.5.1 The Upper Westerly Troughs in the Westerlies

In late autumn, winter and spring, China in its entirety comes under the regime of westerlies. In summer, the area north of 40° N is also located under this wind belt. Because of the influence of the Tibetan Plateau, the westerlies are diverted into two stream flows, a northern and a southern branch called the "northern westerlies" and "southern westerlies" respectively. In the upper flow of this belt, transient disturbances with different amplitudes often appear which are transported from west to east. The waves occur typically with troughs and ridges of pressure systems. Some of them are accompanied by cyclones and anticyclones on the earth surface.

The majority of the troughs which migrate in the northern branch of the westerlies moves from west to east through Xinjiang and Inner Mongolia into Northeast China, then into the North China Sea. While progressing to the east, the southern part of the troughs has influence on Gansu, Inner Mongolia, Ningxia and North China and in some cases even on Qinghai and North Tibet.

Table 3.9 gives the frequency of westerly troughs for January, April, July and October, the figures represent the mean number and, in parentheses, the minimum and maximum number for 1968–1972. It can be deduced that the total number of troughs reaches its maximum in April (27 cases) and January (25), while July and October record remarkably lower figures (18 and 19 respectively). Thus, the frequency of transient disturbances is high throughout the year, but highest in spring and winter.

The speed of migration of upper westerly troughs is rather large. In the case of northwestern troughs, they move towards the east with a speed of 10° to 15°

Table 3.9. Mean frequency of upper westerly troughs for selected months, 1968–1972.

	Direction From west	From northwest	From southwest
January	12 (10–17)	5 (0–8)	8 (5–11)
April	14 (11–17)	3 (1–8)	10 (6–14)
July	13 (7–16)	5 (2–7)	0
October	15 (11–19)	2 (1–4)	2 (0– 5)

Fig. 3.21. Mean height of the 500-mb level in January, fields of pressure (*solid lines*) and temperature (*dotted lines*). (After SHENG Cheng-yu et. al. 1986)

Fig. 3.22. Mean height of the 500-mb level in July, fields of pressure (*solid lines*) and temperatures (*dotted lines*). (After SHENG Cheng-yu et. al. 1986)

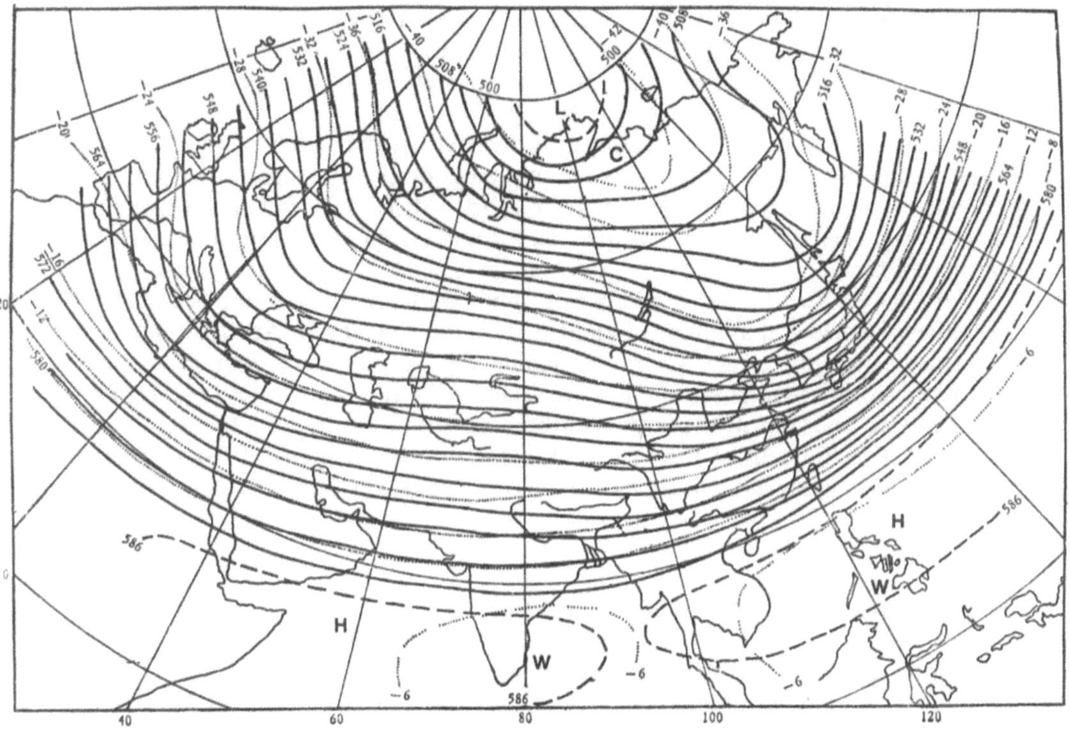
L
C
H
W
H
W
586
500
508
516
524
532
540
548
556
564
572
580
20
0
40
60
80
100
120

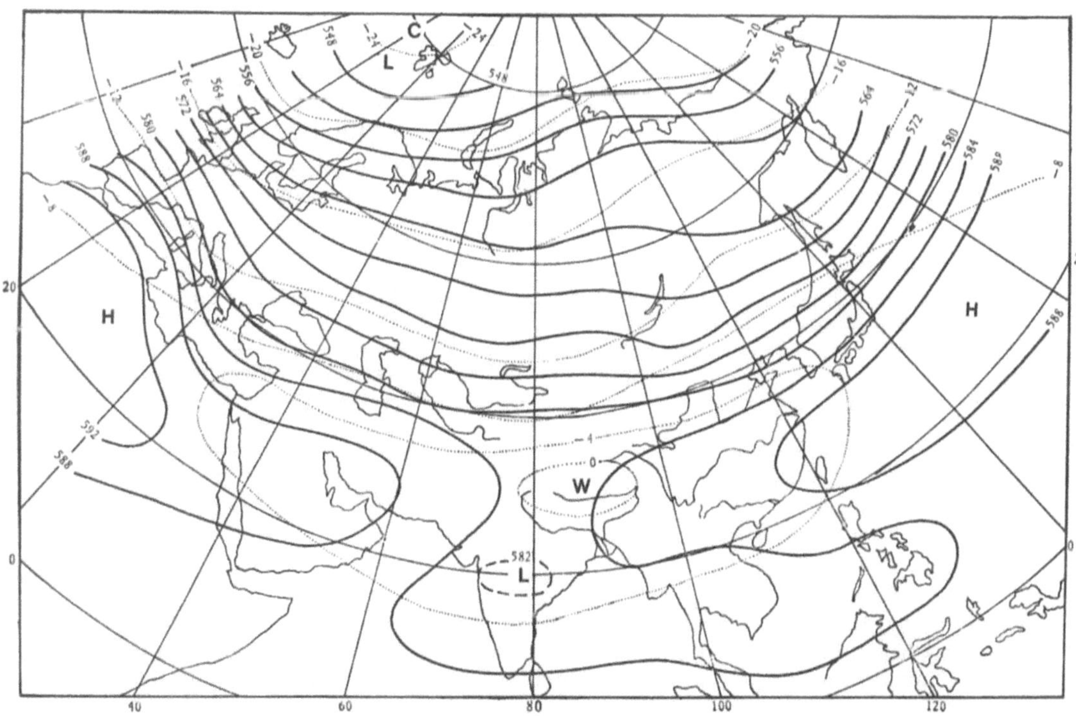
C
L
H
H
W
L
548
556
564
572
580
584
588
592
582
20
0
40
60
80
100
120

longitude per day. Thus, a trough which has arrived in Xinjiang needs 4 days only to approach Northeast China. Most of the upper troughs, passing China, cannot move directly towards the east, but are transformed partly into smaller troughs and some even dissipate completely.

The second branch of the westerlies, diverted to the south of the Tibetan Plateau enters southern China. This branch, including its transient disturbances, originates from the Mediterranean Sea and North Africa. Upper westerly troughs of that origin enter China via North India and Burma and bring moist air to southern China.

For 1968–1972, the mean number of upper westerly troughs in the southern westerlies reached a maximum of eight in January and ten in April so that the frequency of upper troughs was much smaller than in the case of troughs for the northern westerlies stream flow. However, a large interannual variation must also be taken into consideration for the southern stream flow. In July, the westerlies shift towards the north of the Tibetan Plateau and the troughs of the southern branch of the westerlies disappear. They appear again in the second half of October.

3.5.2 Extratropical Cyclones and Anticyclones

With the exception of winter when the strong and huge anticyclone centred over mid-Siberia and Inner Mongolia dominates China (see Sect. 3.1), all other seasons are characterized by a high frequency and often an alteration of cyclones and anticyclones.

In China most of the cyclones belong to the extratropical cyclones. Commonly, four types of cyclones are distinguished in China: the cyclones of (1) Northeast China, (2) Huanghe, (3) Yangtze-Huaihe and (4) East China Sea. As can be seen from the number of cyclones for the different seasons (Table 3.10), remarkably large seasonal differences appear, not only in the total number of cyclones, but also with respect to the different types. The maximum number of cyclones occurs

Table 3.10. Mean seasonal frequencies of anticyclones and cyclones of different types. (After ZHANG Jiacheng and LIN Zhi-quang 1985)

Type		Spring	Summer	Autumn	Winter	Year
Anticyclones		26	21	24	20	91
Cyclones[a]:	Northeast China	16	10	16	8	50
	Huanghe	1	4	2	2	9
	Yangtze and Huaihe	10	12	3	4	29
	East China Sea	9	1	6	8	24
	Total	36	27	27	22	112

[a] SHENG Cheng-yu et al. (1986) distinguished the following eight types of cyclones: Northeast China (I and II), Huanghe or Yellow River (I and II), Chanjiang or Yangtze (I and II), Huaihe, East China Sea (see Fig. 3.23).

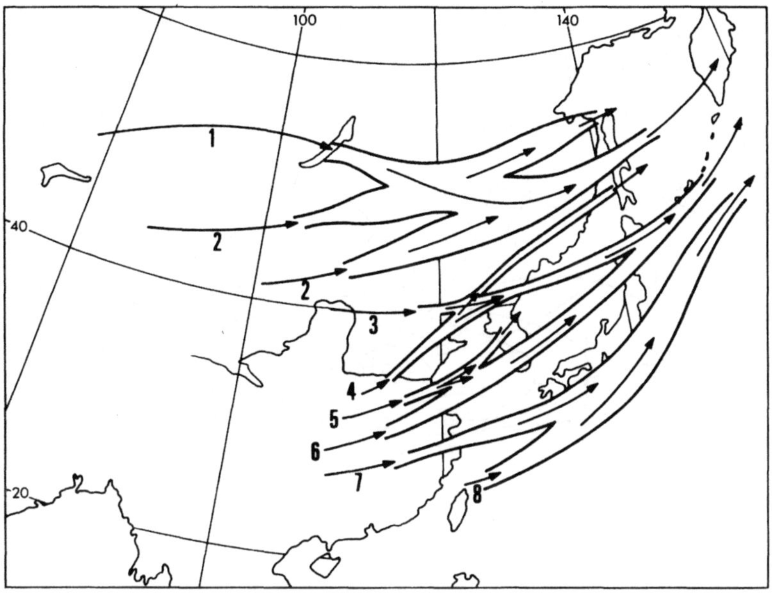

Fig. 3.23. Characteristic paths of cyclones in China. *1* Northeast type I, *2* Northeast type II, *3* Huanghe type I, *4* Huanghe type II, *5* Huaihe type, *6* Changjiang type I, *7* Changjiang type II, *8* East China Sea type. (After SHENG Cheng-yu et al. 1986)

in spring, the minimum number in winter. In spring, moving pressure systems and front cyclones often pass through China. Summer and autumn show a medium frequency of 27 cyclones each from June to August and September to November respectively. In summer, cyclones mainly result from the mei-yu front, due to which during the time of maximum development of the mai-yu front (which is June) also the greatest number of cyclones are observed. In autumn, a greater number of moving pressure systems leads to a medium number of cyclones as stated above (see Table 3.10).

The different types of cyclones vary considerably, both for the year as a whole and for each season individually (see Table 3.10). The largest number of cyclones are of the Northeast China type, amounting to 50 of 112 cases in all (45%). This type is also the leading one in spring and autumn as well as winter, representing 44, 59 and 36% of all cyclones respectively. In summer only, the Northeast China type of cyclones ranks second after the Yangtze-Huaihe type (37 resp. 44%). For the annual total and a seasonal breakdown, the Huanghe type of cyclones shows the lowest number, while the remaining Yangtze-Huaihe and East China Sea types are of medium frequencies (see Table 3.10).

The different types of cyclones vary in their paths in China (Fig. 3.23). The cyclones of the Northeast China type, which are related to the upper westerly troughs in the northern jet stream, move from the Inner Mongolia to Northeast China and then to Sachalin Island. Some of them are formed in the Inner Mongolia. The cyclones of the Huanghe types affect, on the one side, He Tao and the

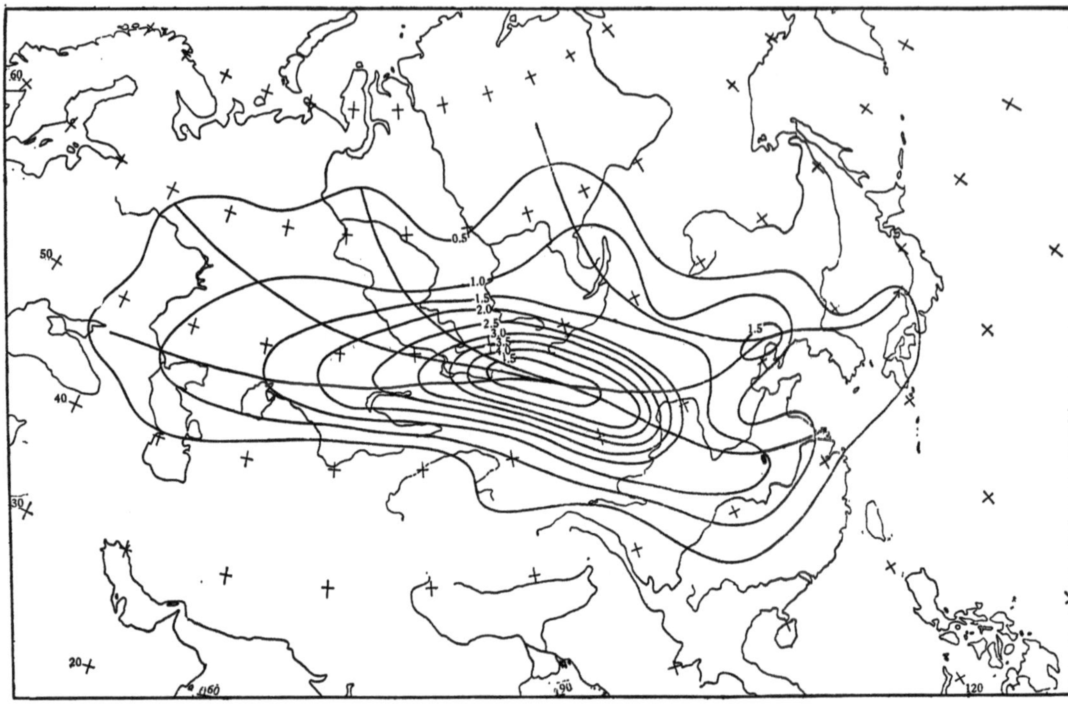

Fig. 3.24. Characteristic paths of anticyclones in China in January (mean of 1951–1976)

North China Plain before moving to North Korea while they occur, on the other side, in the lower reaches of the Yellow River before proceeding either to Northeast China and then to Sachalin Island or to North Korea and the Japan Sea. The cyclones of the Huaihe type move to Korea and the Japan Sea. Most of the Yangtze and Huaihe cyclones occur in the lower reaches of Yangtze, and a small part of them only appears in middle reaches of Zhejiang Province. Some of the cyclones belong to the waves of the stationary fronts. They move to the East China Sea and Japan. The cyclones of the East China Sea type have the same formation conditions and movement paths as the Yangtze-Huaihe cyclones.

The frequency of anticyclones in China is smaller than that of cyclones, showing for the whole year 91 compared to 112 cases (less 19%), which are, however, relatively evenly distributed over the different seasons (Table 3.10). The source regions of anticyclones in China are Siberia and the Mongolian Plateau. Only a small part of them are formed in North China. The paths of anticyclones show a characteristic pattern (Fig. 3.24); they approach the Mongolian Plateau from the west, northwest and north and usually move to China in an east- and southeastward direction. Only a small part of the anticyclones move in a southern direction.

The extratropical cyclones and anticyclones represent prominent and wide-ranging weather systems in China which may cause changeable weather, varying between rainfall and snow, on the one hand, and between warm, cloudy and cold, clear weather, on the other.

Table 3.11. Frequency of southwest cyclonic vortices from April to September for 1970–1974

	April	May	June	July	Aug.	Sept.
1970	10	10	4	5	1	7
1971	7	12	11	7	3	8
1972	9	14	11	3	0	9
1973	8	12	15	11	2	9
1974	10	8	9	9	10	0
Total	44	56	50	35	16	33
Mean	8.8	11.2	10.0	7.0	3.2	8.2

There are cyclonic vortices of a scale of 500 to 1,000 km which originate in southwestern China and move eastward to the south of 35° N. Their scale is less than that of extratropical cyclones. These disturbances are called southwest cyclonic vortices. Their frequency, given as monthly figures from April to September for 1970–1974 (Table 3.11), shows a large monthly as well as an interannual variation. The maximum frequency is observed in May, followed by June, while the minimum occurs in August. These vortices have a great climatological implication since they lead to heavy rainfalls.

3.5.3 Typhoons

Typhoons represent an important weather system in China which mainly appears during summer. They are associated with strong winds and heavy rainfall in South, East and North China and they strike the coastal lowlands most seriously, but do not persist further than 500 km inland from the coast.

A typhoon is a distinct type of an intense tropical cyclone formed in the Northwest Pacific Ocean. A *tropical cyclone* with a wind speed between 17.2 and 32.6 ms^{-1} in its centre is defined as a *typhoon*. When the wind speed is over 32.6 ms^{-1}, a so-called *strong typhoon* occurs, even peak gusts of 50 to 80 ms^{-1} may occur. In these cases of extremely high wind speeds, *storm surges* are mostly responsible for fatalities and devastations. The sea-level pressure in the centre of a typhoon is extraordinarily low; in the centre or "eye" of a typhoon, pressure is generally below 950 mb.

The number of typhoons comes to 36% of all tropical cyclones in the world. The frequency of typhoons in the Northwest Pacific Ocean (Table 3.12) shows a relatively large annual figure, on the one hand, and a remarkably large variation throughout the year, on the other. The typhoon season is the period from June to November, with the maximum frequency from July through October, showing three to four cases of typhoons a month if the observation period 1880–1940 is considered. For 1949–1969, however, the average number of typhoons from July to October even rose to 4–6 cases a month.

According to 70-year observations for 1884–1953 (Watts 1969), on average, 21.9 typhoons occurred per year over the Northwest Pacific Ocean and the China Seas; observations were based within 50° N to 30° N and 105° E. Considering the

Table 3.12. Average number of typhoons in the Northwest Pacific Ocean

Month	J	F	M	A	M	J	J	A	S	O	N	D	Year
Mean (1880–1940)	0.7	0.3	0.4	0.5	0.9	1.1	3.3	3.6	4.1	3.2	1.5	0.9	20.4
Mean (1949–1969)	0.43	0.33	0.48	0.81	1.05	1.81	4.10	6.14	5.42	3.90	2.76	1.48	28.71
Maximum	5	4	2	3	4	5	8	10	10	7	5	5	
Minimum	0	0	0	0	0	0	0	0	1	0	0	0	

only 18-year period 1947–1964, the annual total of typhoons, on average, amounted to 31 occasions with a maximum of 45 in 1961 and a minimum of 20 in 1951 (WATTS 1969). The reason for the rapidly increased number of typhoons in recent years may be due to the improvement of observational techniques which lead to the detection of more typhoons than previously.

The number of typhoons observed in China comes to a much smaller number than the frequency of typhoons in the Northwest Pacific Ocean (cf. Tables 3.12 and 3.13). On average, for the observation period 1949–1969, the maximum figure of typhoons in China amounted to 2.0 for September (of 5.42 typhoons in the Northwest Pacific Ocean). For the whole year, an average of 6.6 typhoons struck China; the typhoon season is restricted to May through November, while December to April are free of typhoons.

According to observations for 1949–1969 (cf. Tables 3.12 and 3.13), only 32.4% of all typhoons in the Northwest Pacific Ocean were also observed in mainland China. According to observations for 1884–1955 (KAO and TSANG 1957), of a total number of 438 typhoons, 256 occasions, i.e. 58% of all typhoons, were also recorded on the southern and southeastern coasts, from Hainan Island to Wenchow.

In Table 3.14, the annual frequencies of tropical cyclones (D) as well as typhoons (T) and strong typhoons (T_s) are given for the observation period 1949–1976, both for the Northwest Pacific Ocean and for the South China Sea.

It is obvious that all three types of tropical cyclones and typhoons respectively, show a much greater frequency in the Northwest Pacific Ocean; subsequently, both totals are much higher than in the South China Sea. Considering the whole observation period 1949–1976, it is interesting to note that in the case of the

Table 3.13. Average number of typhoons observed in China, 1949–1969

Months	A	M	J	J	A	S	O	N	D	Year
Typhoons	0	0.2	0.5	1.6	1.6	2.0	0.4	0.3	0	6.6
Tropical cyclones	0	0.1	0.2	0.7	1.1	0.7	0.1	0.1	0	3.0

Table 3.14. Number of tropical cyclones and typhoons in the Northwest Pacific Ocean and in the South China Sea

No. Year	Northwest Pacific Ocean					South China Sea				
	D	T	T_s	$T+T_s$	$D+T+T_s$	D	T	T_s	$T+T_s$	$D+T+T_s$
1949	5	10	15	25	30	1	2	0	2	3
1950	6	15	15	30	36	4	2	0	2	6
1951	7	4	15	19	26	3	1	0	1	4
1952	4	8	20	28	32	1	2	1	3	4
1953	5	5	17	22	27	3	4	0	4	7
1954	6	4	15	19	25	4	4	0	4	8
1955	5	7	18	25	30	4	1	2	3	7
1956	5	3	18	21	26	9	2	0	2	11
1957	3	3	17	20	23	2	1	1	2	4
1958	1	6	21	27	28	2	5	1	6	8
1959	3	5	17	22	25	5	2	0	2	7
1960	4	6	20	26	30	6	2	2	4	10
1961	8	8	18	26	34	2	5	2	7	9
1962	2	5	23	28	30	3	3	1	1	7
1963	3	3	20	23	26	5	2	0	2	7
1964	4	10	20	30	34	9	3	3	6	6
1965	5	8	18	26	31	4	5	1	6	10
1966	6	10	20	30	36	3	1	4	5	8
1967	8	15	24	39	47	5	0	1	1	6
1968	7	4	22	26	33	2	2	1	3	5
1969	7	6	14	20	27	3	2	0	2	5
1970	15	10	11	21	36	6	5	1	6	12
1971	10	9	25	34	44	5	2	0	2	7
1972	5	6	18	24	29	1	2	5	7	8
1973	2	6	9	15	17	3	5	4	9	12
1974	4	16	16	32	36	0	5	0	5	5
1975	5	6	14	20	25	2	2	1	3	5
1976	3	8	14	22	25	1	0	3	3	4
Total	148	206	494	700	848	89	72	34	106	195
Mean	5.3	7.3	17.7	25.0	30.3	3.2	2.6	1.2	3.8	7.0

Northwest Pacific Ocean, $D<T<T_s$, but in the South China Sea $D>T>T_s$. The Pacific Ocean is, therefore, the most important source region of typhoons affecting China. Moreover, the typhoons from the Bay of Bengal also influence some areas of the Tibetan Plateau.

The region of typhoon formation is the Pacific Ocean east of Taiwan and the Philippines and in the South China Sea where the warm North Equatorial Current between the Marianas and Luzon Island represents relatively high sea surface temperatures. All typhoons typically travel towards the west, northwest and north, but may already dissipate or weaken over the China Seas, without occurring in mainland China. If arriving in China, typhoons usually quickly weaken as they pass inland.

The air flows at the 500-mb and 200-mb levels have an obvious influence on the paths of typhoons. Therefore, the paths are related to the pressure field at

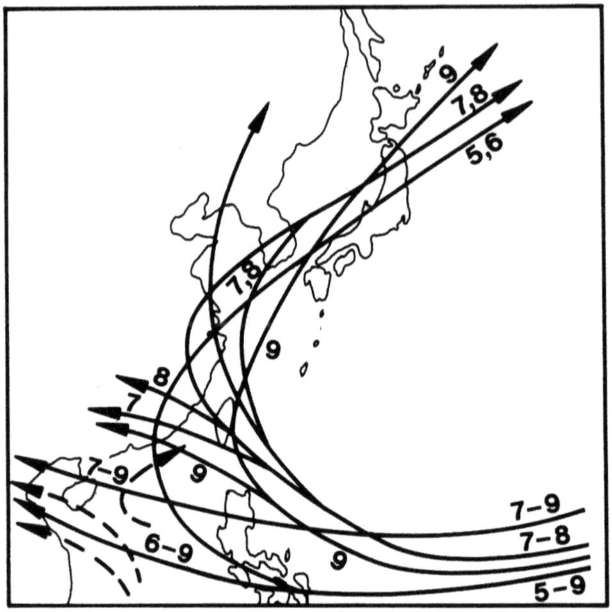

Fig. 3.25. Paths of typhoons from May to September; 5 . . . 9 = May . . . September. (After Chinese Geographical Society 1984)

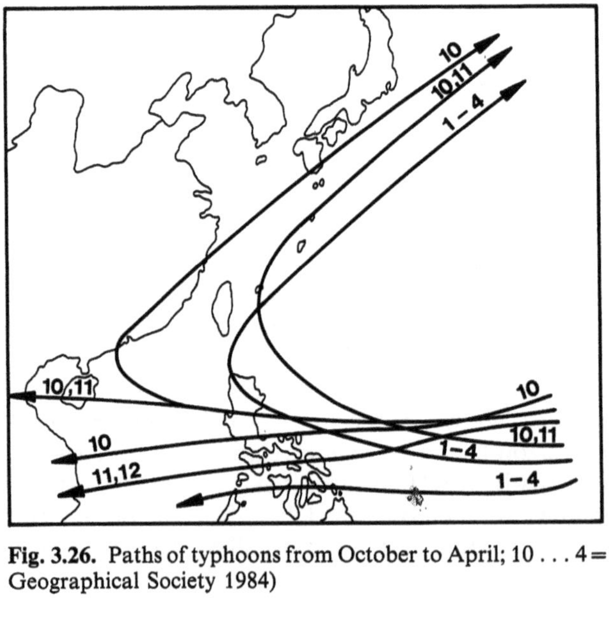

Fig. 3.26. Paths of typhoons from October to April; 10 . . . 4 = October . . . April. (After Chinese Geographical Society 1984)

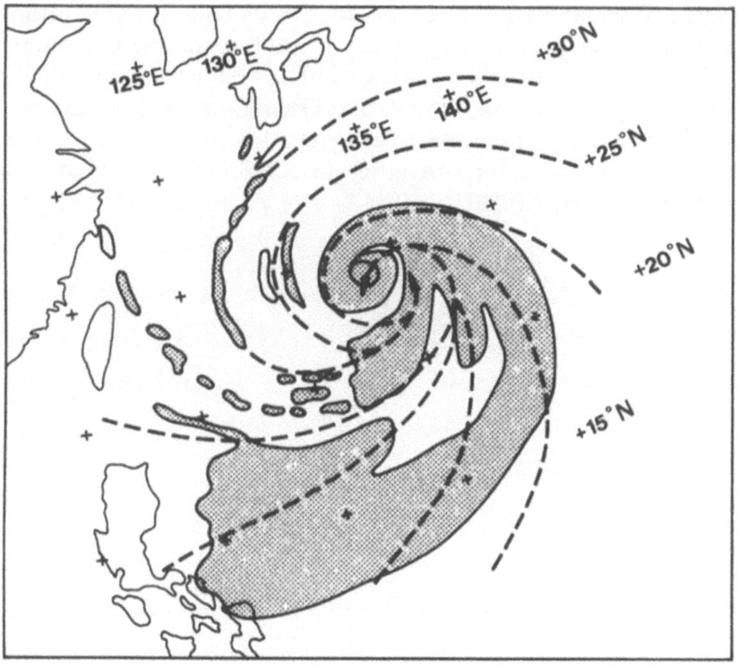

Fig. 3.27. Satellite cloud imagy and winds on sea level during a typhoon at 08.00 h on July 17, 1972; wind observations were made by usual instruments. (After Chinese Geographical Society 1984)

these levels. The paths of typhoons can be divided into two types: one moves from Carolines Islands directly towards the west; the other crosses Carolines Islands and then turns west to northwest and finally even northeast, thus showing a typical parabolic course. The two types of paths are valid for any months, as can be seen for the paths of typhoons from May to September (Fig. 3.25) and from October to April (Fig. 3.26).

Typhoons are formed over the oceans where the sea surface temperature must be higher than 27.5° C. Such conditions are found for summer and early autumn in the Northwest Pacific Ocean. Another condition for the formation of a typhoon is the convergence of air currents, ideally fulfilled in case of the Intertropical Convergence Zone and waves in the easterlies. These two systems occur south of the Subtropical High in the belt of easterlies. Therefore, the position of the Northwest Pacific Subtropical High plays a tremendous role in the formation of typhoons in China. When the High moves towards a northern position, the area of easterly wind extends and the typhoon then has ideal conditions for development, and vice versa. The ridge line of the Subtropical High represents the northern boundary of typhoon activity.

Figure 3.27 gives an example of winds and cloud distributions during a typhoon, recorded on July 17, 1972. Cloud distributions were derived from satellite imagery. Wind at sea level was recorded at 08.00 h by conventional observations. The dotted lines express the flow lines and the oblique lines show the cloud belt.

Typhoon rainfall is very important for South China. For example, in the coastal areas of Zhejiang, Fujian and Guangdong Provinces, typhoon rainfall in July, August and September represents 50% or more of the annual amount. It also has importance for other regions of South China, but the percentage is obviously less than 50%.

4 Temperature

4.1 Mean Annual Air Temperature Distribution

The mean annual temperature distribution demonstrates gigantic thermal differences in China (Fig. 4.1). Most significantly, temperatures vary over space and produce a rather confusing picture of temperature distribution, giving evidence of a considerably large variation, both in latitudinal and longitudinal respects. Considering the distribution of mean annual isotherms (Fig. 4.1), China may be broadly divided into a western and eastern part, each of them showing a particular temperature pattern.

The *western* parts of China, mostly occupying Xinjiang and the Qinghai-Xizang (Tibetan) Plateau, including Gansu and Qinghai Provinces, show remarkable temperature differences, although due to the lack of climate stations and records, the western and northern parts of the Qinghai-Xizang Plateau could not be mapped out (Fig. 4.1).

The large temperature differences in western China are expressed by the maximum and minimum isotherms which amount to 16° and −4° C respectively. However, due to lacking data >4,500 m above sea level, temperature conditions in high mountains could not be investigated.

As shown by climate records, the huge intramontane basins, like the Dsungaria, Tarim, Qaidam, and the eastern and southern parts of the Qinghai-Xizang Plateau are characterized by comparably high mean annual temperatures, evidence of which is given by accurate temperature figures for selected stations, for example:

Ürümqi/269 (653.5 m a.s.l.):	5.7° C
Turpan/267 (34.5 m):	13.9° C
Kuqa/275 (1,099 m):	11.4° C
Bachu/277 (1,116.5 m):	11.7° C
Ruoqiang/264 (888.3 m):	11.5° C
Hotan/276 (1,374.6 m):	12.2° C

Despite their high elevation, even the eastern and southern parts of the Qinghai-Xizang Plateau are characterized by a relatively large heat surplus, as expressed by the mean annual temperature, for example:

Lhasa/239 (3,658 m):	7.5° C
Xigaze/238 (3,836 m):	6.3° C
Nyingchi/236 (3,000 m):	8.5° C

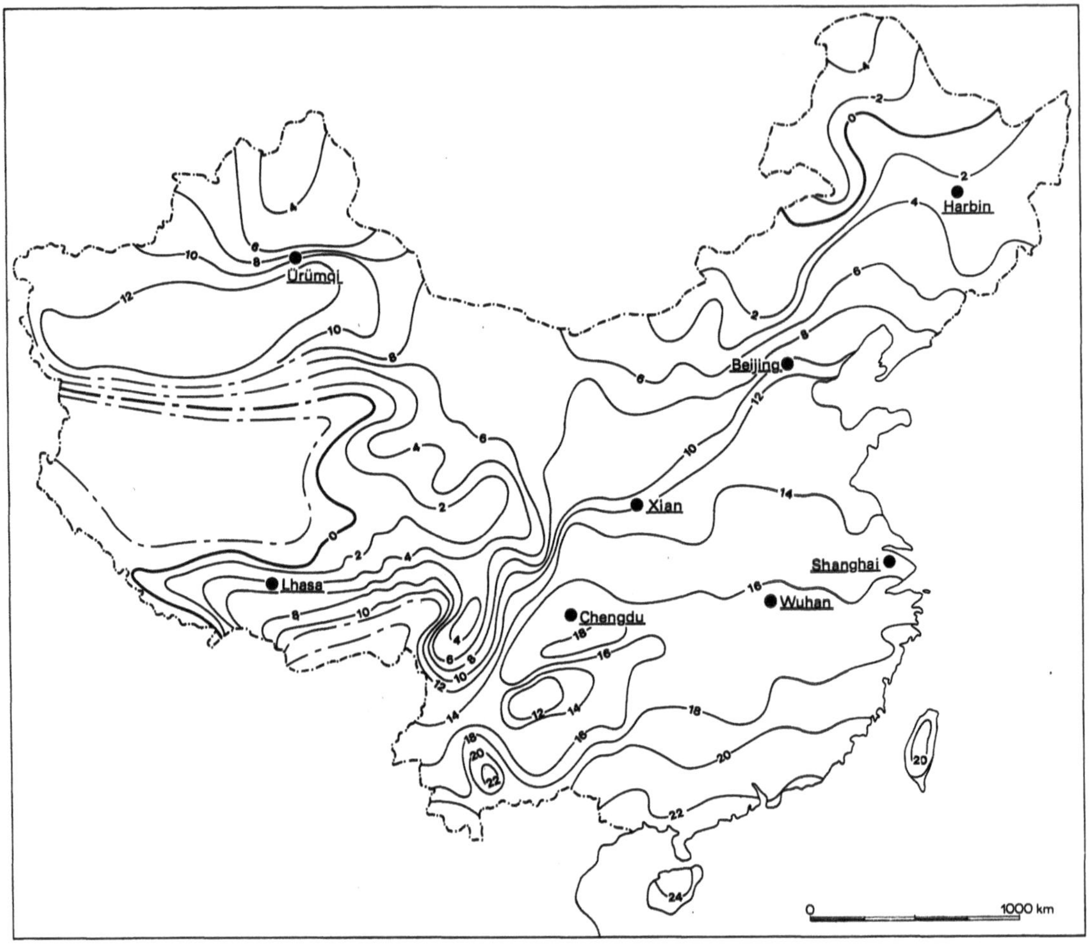

Fig. 4.1. Mean annual air temperature distribution

Although, in total, only sparse data were available, the heat surplus of western China is demonstrated by a comparison between stations in a corresponding geographical latitude in western and eastern China (Table 4.1).

It can be seen that despite the 1,000 to 1,500 m higher elevation of the stations in western China, their mean annual temperature is in the same order as that for the stations in eastern China. Only in the case of the Qinghai-Xizang (Tibetan) Plateau (see stations Lhasa/239, Nyingchi/236, Xigaze/238), the mean annual temperature, in absolute terms, drops considerably below the figures at stations in eastern, lowland China. However, in relative terms, expressed by the mean temperature gradient, the Tibetan Plateau is characterized as a remarkably large heat source, proved by the fact that a 3,000 to 3,800 m decrease in altitude is connected with a comparable meagre temperature drop of 9° to 10° C only. This would result in a mean vertical temperature gradient of only 0.2° to 0.4° C 100 m^{-1} (cf. Sect. 4.5).

Table 4.1. Mean annual temperature at selected stations at the same latitude in western (left) and eastern China (right)

Latitude (°N)	Station No.	Elevation (m)	Temp. (°C)	Station No.	Elevation (m)	Temp. (°C)	Temperature gradient °C 100 m^{-1}
47	Hoboksar/271	1,292	3.0	Fujin/12	64	2.5	±0
44	Qitai/268 Ürümqi/269	794 654	4.7 5.7	Changchun/27 Tongliao/25	237 180	4.9 6.0	±0
42	Kuqa/275	1,099	11.4	Shenyang/38	42	7.8	+0.5
40	Dunhuang/263 Yumenzhen/262 Bachu/277	1,139 1,526 1,117	9.3 6.9 11.7	Beijing/61	31	11.5	±0/−0.3
39	Kashi/279	1,289	11.7	Tianjin/62 Leting/63	3 11	12.2 10.1	±0
38	Shache/278	1,231	11.4	Huimin/72	11	12.2	−0.1
37	Hotan/276	1,375	12.2	Weifang/73	44	12.3	±0
30	Lhasa/239 Nyingchi/236	3,658 3,000	7.5 8.5	Dinghai/120	36	16.3	−0.2/−0.3
29	Xigaze/238	3,836	6.3	Shipu/121 Jingdezhen/151 Changde/146	128 46 37	16.2 17.0 16.8	−0.4

This exceptionally small vertical temperature gradient, together with the non-existing gradient in Xinjiang, give strong evidence of a heat source from the huge intramontane basins, plateaux and highlands in western China. As far as the Qinghai-Xizang (Tibetan) Plateau and its specific temperature conditions are concerned, FLOHN (1968) has shown that, during summer, the Plateau represents a large-scale, elevated heating surface (about 2×10^6 km^2 with an average elevation of 4,500 m). According to FLOHN, the Tibetan Plateau therefore acts in summer as a heat engine with a giant chimney in its southeastern corner where heat is continuously carried upward into the high troposphere. This results in a relatively stationary warm anticyclone in summer which is centred in the middle and upper troposphere above the southeastern Tibetan Plateau.

The comparably high mean annual temperatures in the intramontane basins and plateaux of western China are therefore controlled by a well-pronounced heating effect of a large-scale, elevated surface in summer which weakens or even completely neutralizes the usually effective vertical lapse rate of temperature.

In the mountain ranges of western China, mean annual temperatures drop to much lower values than in the plateaux and depressions, as can be seen from the data available; unfortunately, records >4,500 m a.s.l. were missing. The following high mountain stations have recorded the lowest mean annual temperatures:

Tianzhu/255	3,045 m	−0.2°C
Darlag/244	3,968 m	−1.3°C
Madoi/245	4,272 m	−4.1°C
Yadong/237	4,300 m	−0.2°C
Nagqu/240	4,507 m	−2.1°C

All other mountain stations for which data were available recorded a positive mean annual temperature.

In sharp contrast to the disrupted distribution of temperature in western China, a comparably clear and schematic temperature distribution can be seen in the *eastern* parts (Fig. 4.1). With a few exceptions only, the isotherms prefer a zonal direction, showing gradually increasing temperature values from North to South China. The northernmost station recorded −4.8°C (Mohe/1), while in the southernmost Hainan Island the mean annual temperature reached 24.5°C (Dongfang/210). Thus, the range of temperature over space in eastern China amounted to 29.3°C.

The strong latitudinal change of temperature underlines the fact that temperature conditions are governed by sun insolation; therefore, a different latitudinal temperature pattern between summer and winter can be expected (see Sect. 4.2).

Significant deviations from the strong latitudinal direction of the isotherms occur in the Yunnan Plateau and in Guizhou Province where a cellular type dominates, indicating a more striking effect of elevation on temperature. In northeastern China, the isotherms show a remarkable distribution towards a southwest-northeast direction which indicates higher temperatures towards the Yellow Sea side and lower temperatures inland. Reasons for this are the oceanic effects of warming and the cold monsoonal northeasterlies from the Mongolian anticyclone during winter.

4.2 Mean Seasonal Temperature Distribution

In order to obtain a broad description of the seasonal temperature distribution, the mean temperature field for January, April, July and October was regarded as representative.

January

As can be seen from the mean temperature distribution (Fig. 4.2), January shows an extremely complex and, at the same time, extraordinarily variable pattern of temperature. The large variations of temperature over space for January can even be regarded as the most striking feature of the temperature conditions and also one of the major characteristics of the climate of China in total.

Despite the great complexity and irregularity of the temperature distribution in January, a distinct zonation nevertheless can be worked out in two major directions: (1) temperature rapidly decreases with increasing latitude, i.e. from South to North China; (2) for the same latitude, temperature also decreases with increasing altitude above sea level, i.e. from the eastern lowlands to the western highlands, intramontane basins and mountain ranges.

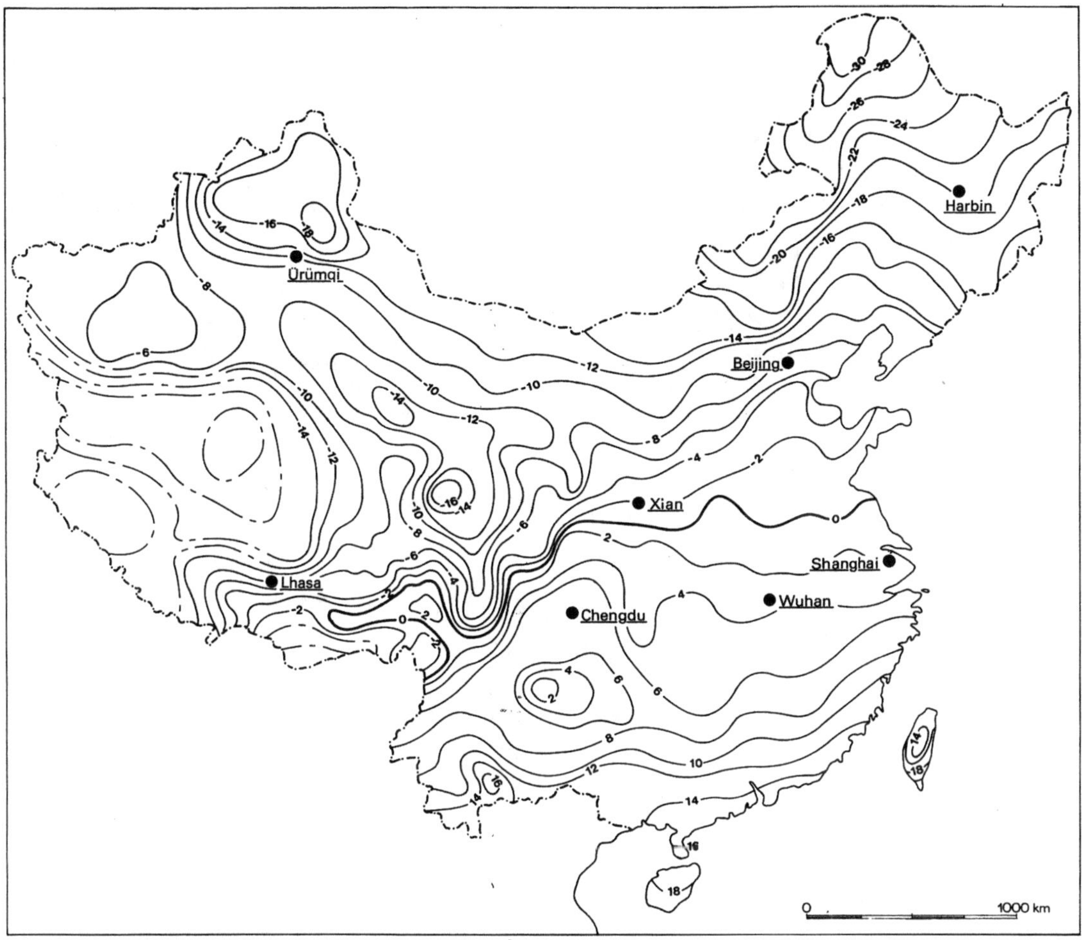

Fig. 4.2. Mean air temperature distribution in January

Underlining the strong latitudinal effects upon temperature, isotherms preferably run in a zonal direction. However, according to a distinct north to south direction of the mountains in central China, also a longitudinal direction of isotherms appears in the transitional regions from eastern lowland to western upland China (Fig. 4.2). In the whole eastern part, a rather schematic and gradual sequence of isotherms can be noted, commencing with the 18°C isotherm in southern China (Hainan Island) and terminating with the −30°C isotherm in North Heilungkiang Province. In northeastern China, however, temperatures decrease more rapidly than in southern China. In total, for January a gigantic temperature range of as much as 48°C is experienced in eastern China.

The extremely low winter temperatures, experienced in Northeast China, in particular Heilungkiang Province are, however, not found for western upland

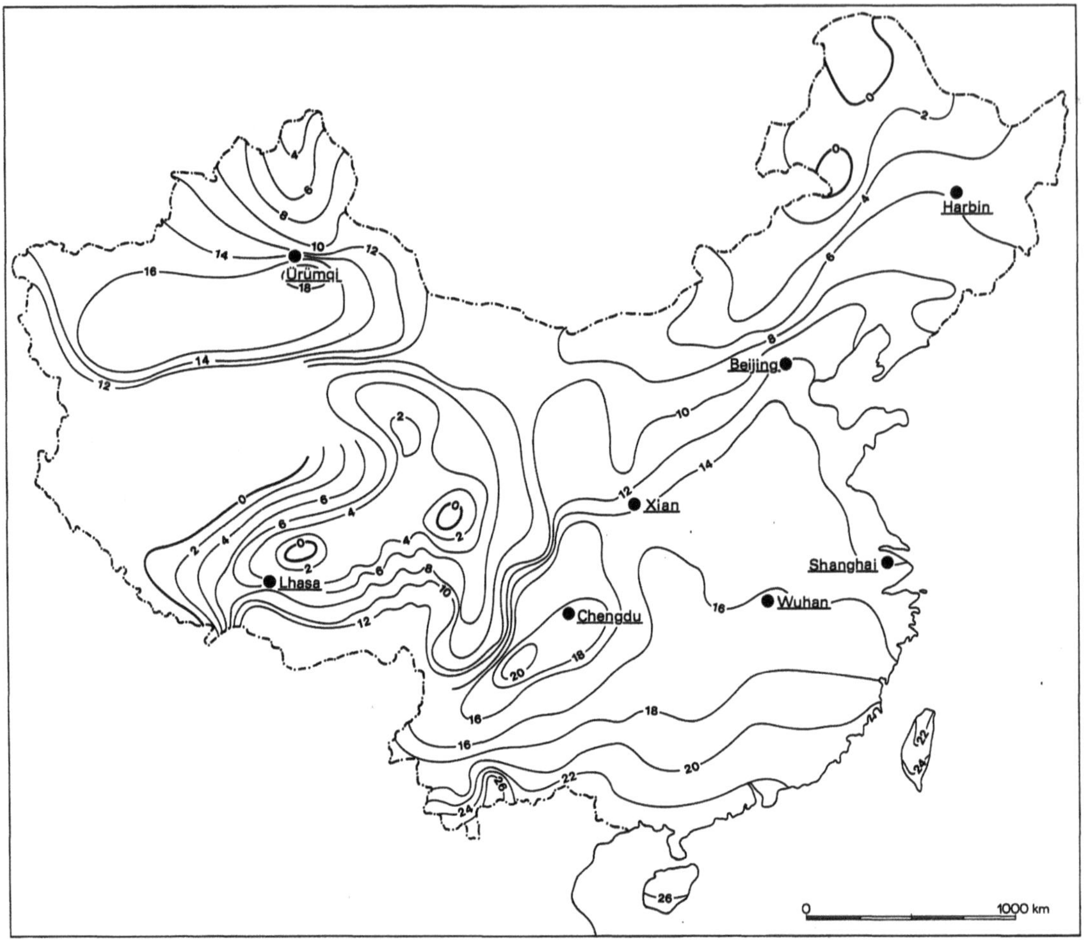

Fig. 4.3. Mean air temperature distribution in April

China as far as accurate records show; for Tian Shan and Kunlun Shan a sharp drop of temperature certainly has to be taken into consideration, although exact data were not available. The lowest recorded mean January temperature amounted to −18.3° C (at Qitai/268), defining a "cell of coldness" in the Dsungaria. Temperatures in the Tarim Basin do not drop as low as in the Dsungaria, whereas the Qinghai-Xizang (Tibetan) Plateau recorded fairly mild temperatures in January (Lhasa −2.3° C, Xigaze −4.1° C). Temperatures for the central and western parts of the Tibetan Plateau were unfortunately lacking.

Effected by the vertical lapse rate, it is by no means surprising that temperatures in the mountains of Qinghai and western Sichuan considerably dropped (–16.8° C at Madoi/245), establishing a pronounced cell of coldness in the mountain "corridor" of China.

Evidence of the serious winter temperatures in China can also be given by the 0° C isotherm in Januar which is located as far south as 33° N (Fig. 4.2). It runs from the Yellow Sea, Kiangsu Province, directly in an east to west direction until it reaches the high mountain corridor (Tasurkai Shan, Min Shan), then it sharply turns in a southern direction, roughly proceeding along the upper Yangtze towards the easternmost parts of the Tibetan Plateau. From a global comparison of temperature distribution in January, it is worth noting that in China the 0° C isotherm is shifted further south and thus nearer to the equator than elsewhere in the northern hemisphere. Therefore, the by far major part of China experiences in January a mean temperature below zero, while the comparably small southern and southeastern sector only records positive temperatures. From a global comparison, China experiences unusually harsh January and thus winter temperatures.

April

In all of China, temperatures from January to April increase considerably. Corresponding to the temperature distribution in January, in April temperatures also decrease in a predominantly latitudinal direction in eastern China, while they show a rather irregular and partly well-established cellular distribution in western China. A cell of coldness occupies the western and northern part of the Qinghai-Xizang (Tibetan) Plateau, while a heat cell is located in the Tarim Basin.

The mean temperature in April (Fig. 4.3) exceeds the 26° C-isotherm, with an absolute maximum of 27.1° C at Xizha/209, but it drops, at a minimum, below 0° C, both in the northernmost part of Heilungkiang Province (−1.2° C at Mohe/1) and in the northern and western Tibetan Plateau (−2.9° C at Madoi/245), although exact observation data here are scarce. In China as a whole, the range of temperature in April amounts to nearly 30° C which still underlines a relatively large temperature gradient over space.

Although not shown by accurate records, a drop of temperature in the mountains must also be taken into consideration, due to the vertical lapse rate of temperature (cf. Sect. 4.5).

Comparing the mean temperatures for January and April, it is obvious that in all of China the temperature remarkably increases, although to a spatially different extent. From January through April, temperature increases by 10° C in southern China, but up to 25° C in northeastern China; a similarly large temperature increase is also valid in most parts of western China.

July

The temperature distribution for July (Fig. 4.4) shows a very disrupted picture in western China, while in major parts of eastern, particularly southeastern China, a comparably even temperature field prevails.

In western China, mean temperatures in July vary remarkably although, due to the lack of records, temperature conditions can only be described incompletely; no data were available for central and western Xizang and for the mountain ranges. It can nevertheless undoubtedly be seen that extremely high summer temperatures occur in Xinjiang where, in July, temperatures gradually increase to the intramontane centres. At a maximum, the recorded mean temperature in July here even climbed to 32.7° C (Turpan/267). Even at locations at a higher elevation,

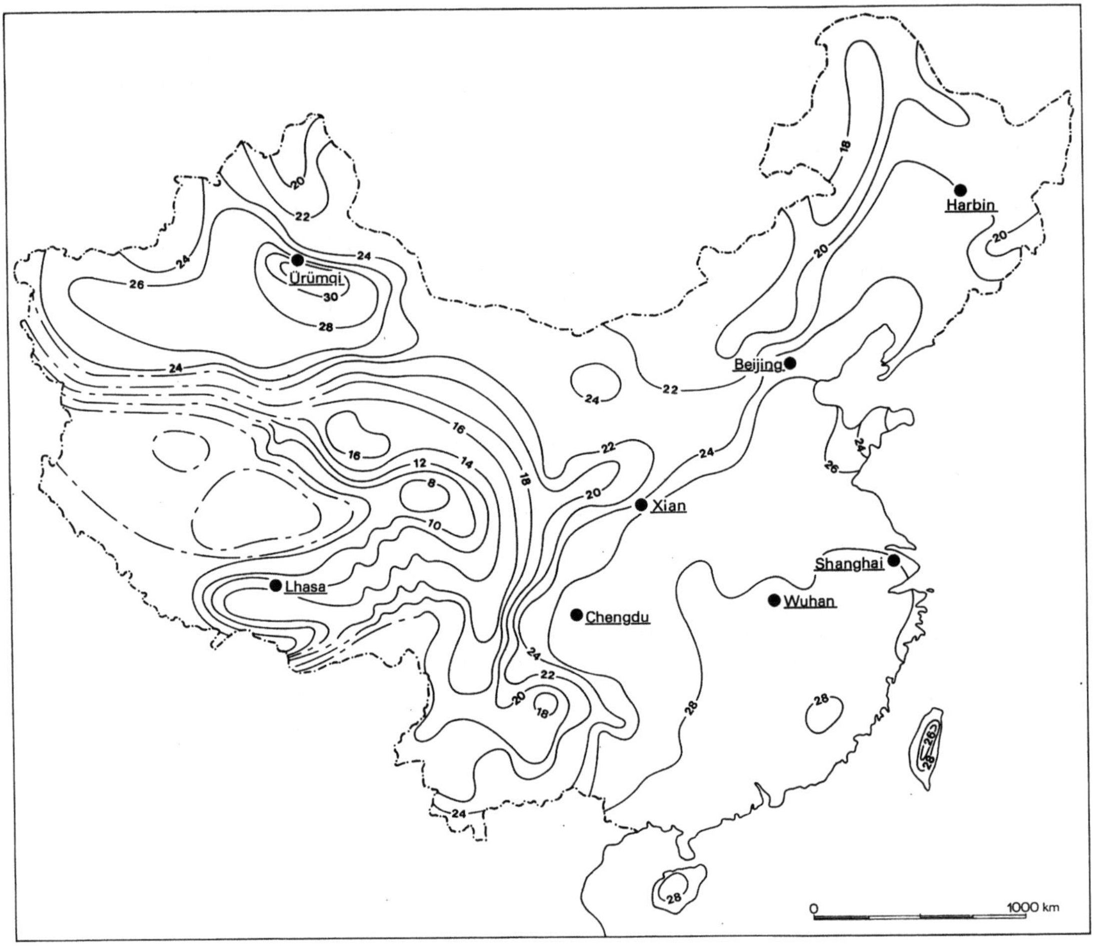

Fig. 4.4. Mean air temperature distribution in July

the temperature still remains rather high as can be seen at the following stations: Kuqa/275 (1,099 m): 25.9° C; Kashi/279 (1,289 m): 25.9° C; Hotan/276 (1,375 m); 25.5° C. Although much lower than in Xinjiang, still relatively high temperatures, with regards to the great elevation, occur in the eastern parts of the Tibetan Plateau (Lhasa/239 14.9° C, Xigaze/238 14.2° C) and intermediate high temperatures in the central and northern Yunnan mountains (between 16° and 20° C).

Unlike the highest mean temperatures in July, given for the intramontane and thus warm or even hot regions, no records were available for the lowest mean temperatures in July which are to be expected in the high-altitude mountains of western China.

In eastern China, the temperature distribution is comparably even over space. It shows fairly equal temperatures, clearly underlined by the fact that from the North China Plain up to the southern coast, which extends over an air distance of some 4,000 km, temperature in July only varies between 26° and 29° C. In northeastern China, however, a remarkable temperature variation occurs, showing gradually decreasing figures up to $<18°$C.

In China as a whole, July shows a further decreasing range of temperature over space in comparison to January and April (see Figs. 4.2 and 4.3). According to available records, the spatial range of temperature amounts to 20° C (January 36° C, April 30° C), if high mountain regions above 4,500 m a.s.l. are excluded.

Temperature distribution in July (Fig. 4.4) obviously expresses the effects of different factors on temperature. The effect of latitude on insolation is shown by the rather zonal sequence of isotherms in eastern China, while the lower temperature in the Tibetan Plateau is principally due to the relatively high altitude. In summer, however, the Tibetan Plateau acts as a large-scale, elevated heating surface which leads to comparably high summer (July) temperatures in the Plateau (cf. Sect. 4.1). The strong impact of thermal continentality is clearly seen for the intramontane basins, such as Tarim Basin, Dzungaria and Qaidam; they all represent well-established heat cells on a large scale. The strong effect of continentality is strikingly expressed in the case of the Turpan depression, located at the northeastern edge of Tarim Basin, where mean July temperature was even recorded at 32.7° C (at Turpan/267). This value is even higher than the highest mean July temperature in southernmost China (Dongfang/210, 29.1° C). Therefore, the effect of thermal continentality, strongly established in Xinjiang, even exceeds the effect of tropicality in southern China which logically follows from the low geographical latitude.

October

Corresponding to the already described temperature distribution in January, April and July, the mean temperature in October (Fig. 4.5) also shows a rather complex and variable picture over space. In comparison with July the temperature in all of China drops considerably, to a greater extent, however, in western, northern and northeastern China than in southern and southeastern China. From July to October, recorded mean temperatures drop at the most by 18° C in northern Heilungkiang Province, but only by 4° C in Hainan Island. Large temperature drops between July and October, as they are valid in Northeast China, also occur in western China, approaching 16° to 18° in Xinjiang and 6° to 12° in the Tibetan Plateau.

Similar to the preceding three seasons (as shown for January, April and July respectively), temperature distribution in October also divides China into two parts with different temperature fields: obviously, a zonal distribution of isotherms prevails in eastern China, which shows gradually increasing temperatures from Northeast to South China, thus underlining prevailing latitudinal and thus insolation effects on temperature. In contrast, western China shows a predominantly cellular type of isotherms which expresses prevailing effects of both continentality and elevation on temperature. However, it must be pointed out that the Tibetan Plateau may still be considered as a heating surface, although to a smaller extent than in summer.

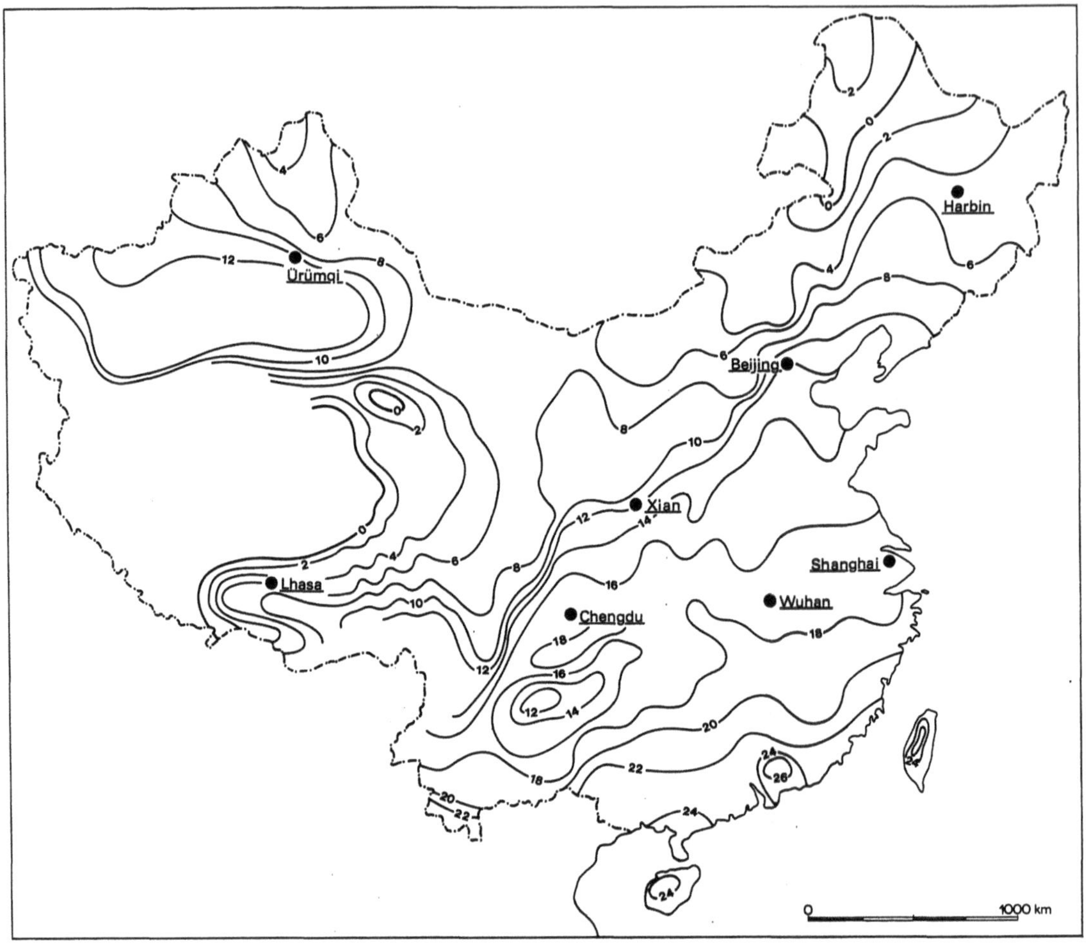

Fig. 4.5. Mean air temperature distribution in October

4.3 Annual Range and Annual Variation of Temperature

As can be deduced from the temperature distribution in January, April, July and October (Sect. 4.2), a striking seasonality in temperature and thus a marked annual temperature variation truly exists for China.

January and July represent the coldest and warmest months respectively, and clearly show the annual range of temperature which can be taken as an indicator for the severity of seasonality in China.

Considering the mean temperature range between January and July, mapped out by isolines (Fig. 4.6), remarkably large differences in seasonality over space are expressed. The smallest values of the mean annual range of air temperature occur in southern China, showing a minimum range of 10°C only in Hainan Island. With increasing latitude, the values of the mean annual temperature range

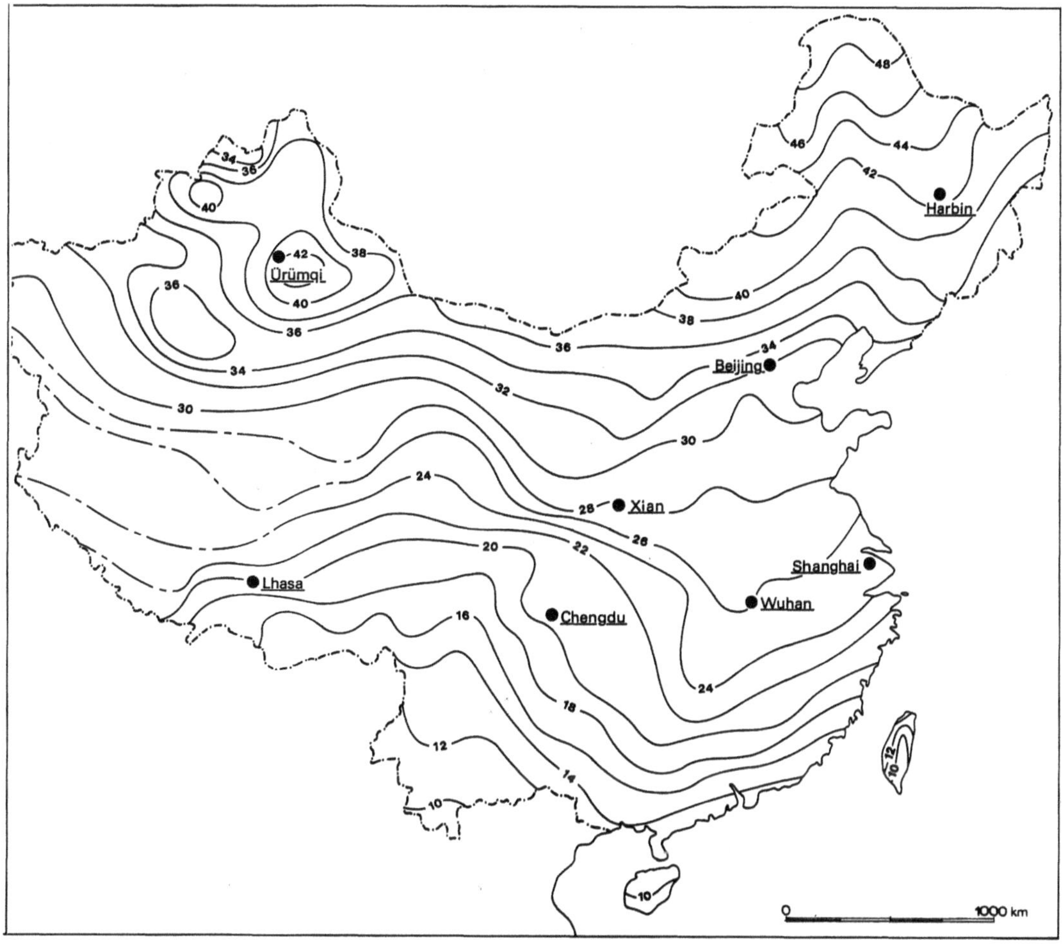

Fig. 4.6. Isolines of mean temperature range between January and July

progressively rise to a maximum value of 48° C in northeasternmost China. Indicated as a cellular type, in Xinjiang an extremely large annual range of temperature also occurs which, at the most, reaches 42° C.

In total, the mean isolines of temperature range between January and July are arranged in a prevailing latitudinal direction, thus giving evidence of the predominant effects of latitude and insolation on temperature.

As a result of the extremely different values of the mean temperature range between January and July, seasons are comparably weakly established in South China only, but strongly developed in all other parts. The most striking seasonality occurs in northeastern and northwestern China, expressed by a recorded temperature range >34° C between January and July. Thus, winter and summer are strikingly pronounced seasons.

The rapidly increasing seasonality from South to Northeast China is demonstrated by a south to north cross-section which gives the mean temperatures for

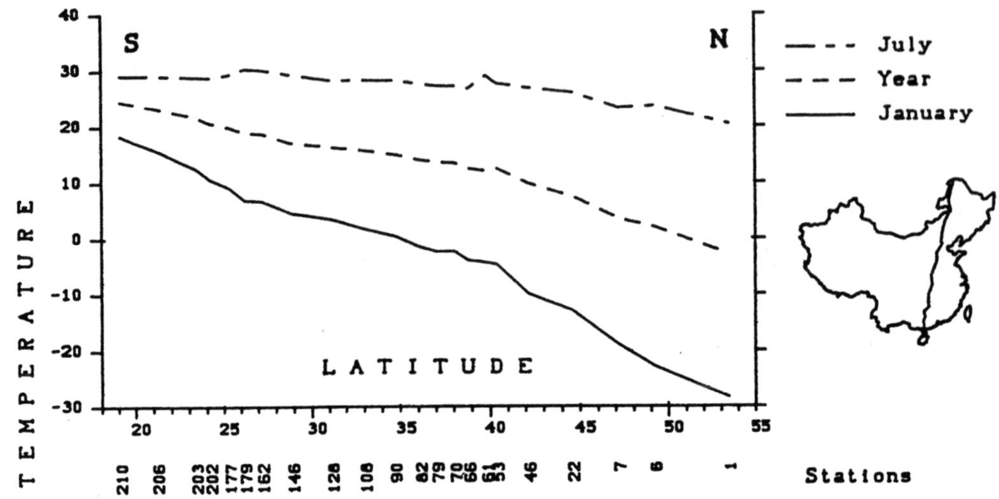

Fig. 4.7. South to North cross-section through eastern China, showing the sea-level mean temperatures (° C) in January and July, together with the annual mean

January and July, together with the mean annual temperature (Fig. 4.7). In all three cases, temperatures were reduced to sea level, referring to a lapse rate of 0.65° C 100 m^{-1}. The cross-section refers to eastern China only.

Among the three curves, January most clearly expresses a steep temperature gradient between North and South China, showing persistently increasing temperatures with decreasing latitude. The values of sea-level temperatures vary between −27.8° C (at Huma/2) and 18.3° C (at Dongfang/210), resulting in a latitudinal temperature range of as much as 46° C over a latitudinal distance of 32°. On average, January therefore experiences a temperature increase of 1.4° C for every degree latitude, which again clearly underlines the steep temperature gradient in eastern China in winter. In sharp contrast to the conditions in winter, in July a latitudinal variation of sea-level temperatures of less than 10° C can be observed. The curve for the mean annual temperature (Fig. 4.7) shows intermediate conditions between the curves for January and July, expressing a weak temperature gradient with slowly increasing values from North to South China.

A pronounced thermal seasonality also occurs in western China. Xinjiang shows the largest recorded annual temperature range which reaches its recorded maximum at Turpan/267: 42.2° C (July 32.7° C, January −9.5° C). In all of Xinjiang, the annual temperature variation amounts to between 30° and 40° C, for example at Ürümqi/269: 38.9° C. A remarkably lesser variation between 20° and 25° C occurs in Qinghai, for example at Madoi/245: 24.3° C (July 7.5° C, January −16.8° C). An even smaller variation is experienced in the Tibetan Plateau, namely between 15° and 20° C (Lhasa/239: 17.2° C; July 14.9° C, January −2.3° C).

The annual variation of temperature which so far has been shown to vary considerably over latitude, was further investigated according to longitudinal differences, by analyzing temperatures in January, July and for the annual mean

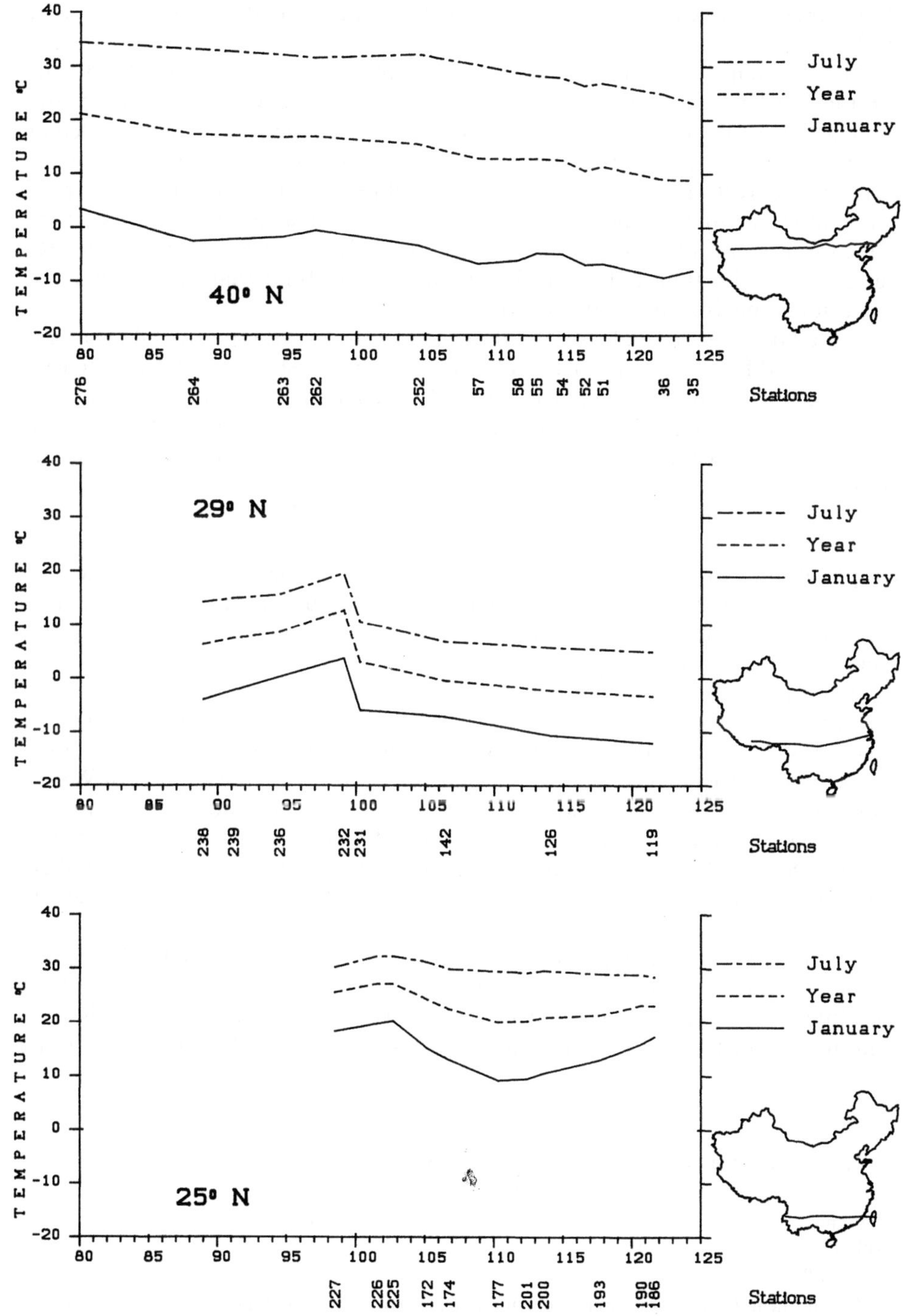

Fig. 4.8. West to East cross-section of mean temperatures in January and July and for the annual mean, given along 25°, 29° and 40° N

along three west to east cross-sections (Fig. 4.8) which were drawn at 25°, 29° and 40° N.

The two temperature cross-sections at 25° and 40° N represent sea-level temperatures, calculated with the common lapse rate of 0.65° C 100 m^{-1}. Since the vertical lapse rate of temperature is, however, not applicable above 2,000 m height, the temperature cross-section at 29° N contains recorded surface temperatures for the western part (Tibetan Plateau), while for the eastern part upper air temperature readings at the 600-mb level (about 4.2 km a.s.l.) were used; stations in the western part of the cross-section are Xigaze/238, Lhasa/239, Nyingchi/236, Batang/232 and Litang/231, while for the eastern part Shanghai/119, Wuhan/126 and Chongqing/142 were considered[14].

For all three latitudinal cross-sections, temperatures and the annual mean not only clearly differ between winter and summer, but also show distinct west to east differences. For 40° N, between Hotan/276 and Dandong/35, in a west to east distance of 45°, temperatures decreased by about 10° C. A similar temperature drop also took place for 29° N, between Xigaze/238 and Shanghai/119, if the whole west to east distance is considered; the temperature peak at Nyingchi/236 and Batang/232 is due to the lower elevation of both stations (in 3,000 and 2,589 m a.s.l., respectively) than Xigaze (3,836 m), which plausibly causes higher temperatures. For 25° N, there is a major change of temperature in a west to east direction in January only, which also affects the annual mean. As a whole, it is also worth noting the different annual temperature range which considerably increases with altitude.

Taking the seasonal thermal contrast as a major climate characteristic, a further characterization of the annual variation of temperature was attempted by classifying the annual temperature variation of all 279 stations under investigation according to a hierarchial cluster analysis, by applying the method of WARD. Considering a far-reaching spatial homogeneity, four types (Nos. 1 to 4, Fig. 4.9) have been determined, representing the characteristic annual temperature variation in China[15]. Of the 279 stations under consideration, the following numbers represent the different types: 43, 101, 72 and 63 stations for types 1 to 4 respectively.

Each type of annual temperature variation underlines a marked contrast between a warm summer and cold winter. Warm and cold are relative rather than absolute terms. All types also correspond in the warmest month, which is July, and in the coldest one, namely January. Less pronounced for all types are the

---

[14] The 20-year mean values of temperature at 600 mb were as follows:

	January	July	Year
Shanghai	−12.1	−4.9	−3.4
Wuhan	−10.7	−5.7	−2.3
Chongqing	− 7.2	6.8	−0.5

[15] Due to the nature of a hierarchial cluster analysis, two adjoining types of annual temperature variation may show a similar picture, in particular in the marginal region of each type. The boundary between two types of annual temperature variation can be regarded as a broader transitional belt.

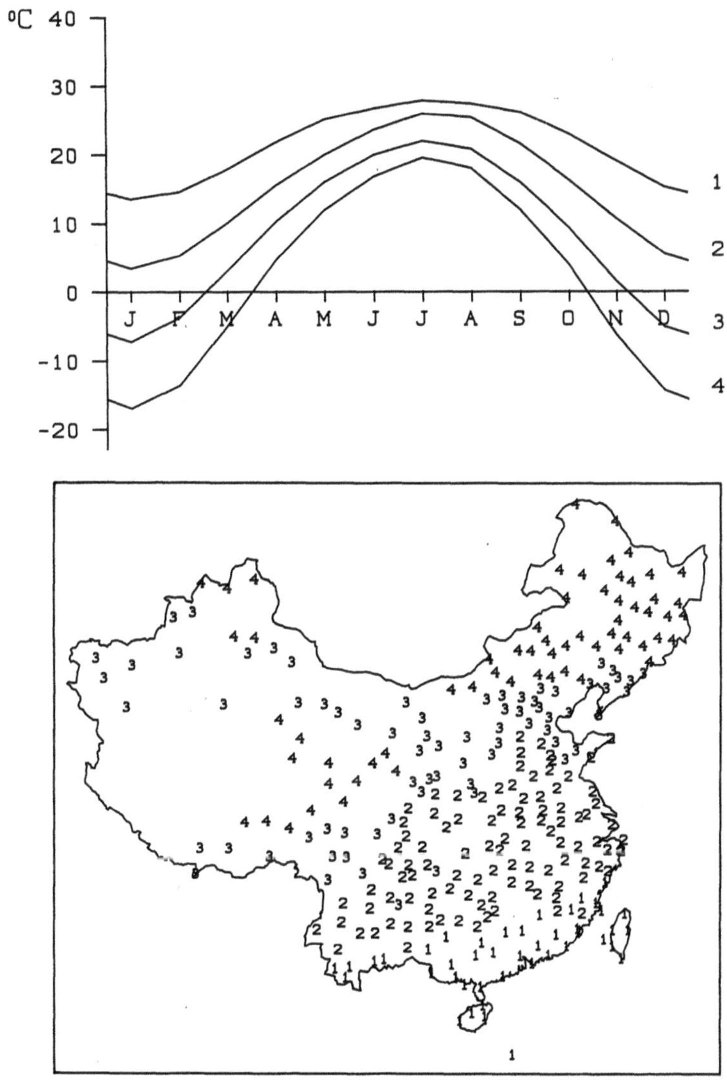

Fig. 4.9. Types of annual temperature variation and their distribution over space, computed by a hierarchial cluster analysis after WARD

intraseasonal differences in temperature during summer and winter, whereas all types correspondingly show a rapid temperature increase during spring and a decrease during autumn. Although valid, in principle, for all four types, all these characteristics are most significantly experienced for type No. 4 and then gradually weaken to types Nos. 3, 2 and 1. From the magnitude of temperature variation, valid for the four types, the temperature change gradually increases from type No. 4 to No. 1, a fact which is true to a larger extent in winter than in summer. This results in a wider range of temperature graphs of all four types (Fig. 4.9).

The spatial distribution of the four types of annual temperature variation shows a predominantly latitudinal direction, expressing the fact that insolation most strikingly effects temperature conditions, in general, and their seasonal variations, in particular. Evidently, the thermal contrast between summer and winter increases with latitude, so that from South to North China, gradually accelerating, the annual variation becomes more strongly established. The northern parts of Inner Mongolia as well as Northeast and Northwest China, which occupy Heilungkiang Province and Xinjiang respectively, show a maximum temperature range between summer and winter. The reasons for this are the extremely low winter temperatures, while the difference in summer temperatures is comparably small (Fig. 4.9). However, the temperature types Nos. 3, 2 and 1 do not record winter temperatures as extreme as the North China type (No. 4) which, therefore, leads to a lesser extent of thermal seasonality. Because of the mild winter temperatures, the South China type (No. 1) experiences the smallest annual temperature range, although summer temperatures are at a maximum for China as a whole[16].

It is particularly noteworthy that also in the central Tibetan Plateau, including the mountains of Qinghai and Gansu Provinces, the northern China type occurs which is characterized by a considerably large annual temperature variation (No. 4, see Fig. 4.9), whereas the southern Tibetan Plateau experiences the less pronounced type No. 2.

Each type of annual variation of temperature may, however, vary to a limited extent if all individual graphs are considered. This was shown in detail for the temperature types No. 1 and No. 3 (Figs. 4.10 and 4.11). Taking the individual graphs of the two temperature types under consideration, a temperature range of about 10° C can be observed for each type.

Besides the slight spatial modifications of each annual temperature type, each station also shows moderately differing curves of annual temperature variation if the graphs for each year of the 30-year observation period are considered (see Sect. 4.8.2).

For further use and individual interpretation the mean monthly and annual temperatures of all 279 reference stations are also given (Table 4.2, p. 94–101).

Contd. p. 101.

Fig. 4.10. Graphs of annual temperature variation for all 43 stations under temperature type No. 1 (*above*), respectively for all 101 stations under temperature type No. 2 (*below*); see Fig. 4.9

Fig. 4.11. Graphs of annual temperature variation for all 72 stations under temperature type No. 3 (*above*), respectively for all 63 stations under temperature No. 4 (*below*); see Fig. 4.9

[16] Because of the particular location of Turpan/267 in a depression (climate station at 34.5 m a.s.l.), the summer temperature reaches the absolute highest figure for China (mean temperature in July 32.7° C).

Fig. 4.10.

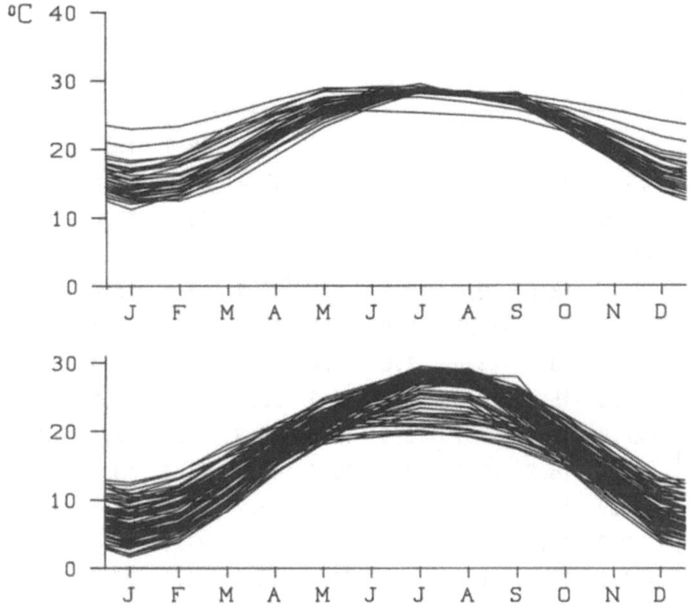

Fig. 4.11.

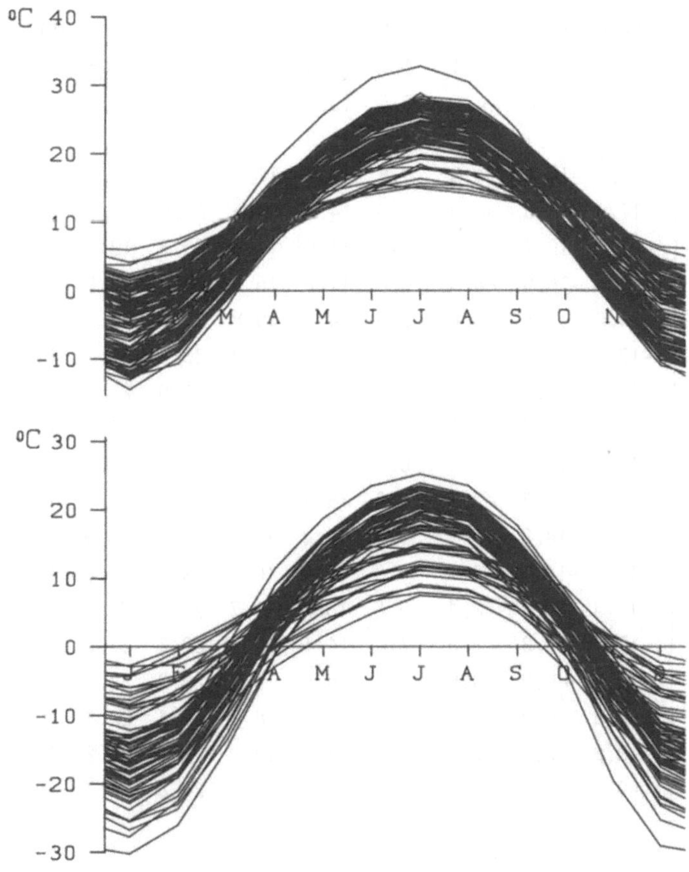

Table 4.2. Mean monthly and annual temperatures at all climate reference stations

No.	J	F	M	A	M	J	J	A	S	O	N	D	Mean
1	-30.3	-26.0	-14.8	-1.2	8.7	15.7	18.4	15.5	7.8	-3.3	-19.5	-29.1	-4.8
2	-27.8	-22.5	-11.5	1.2	10.6	17.4	20.2	17.6	10.2	0.1	-14.6	-25.3	-2.0
3	-25.5	-21.8	-11.2	1.7	10.3	16.4	19.3	17.0	10.1	0.6	-12.6	-22.9	-1.5
4	-25.5	-21.0	-9.5	2.8	11.7	18.1	20.6	18.4	11.3	1.7	-11.1	-22.1	-0.4
5	-21.3	-18.6	-10.0	0.6	9.5	15.4	17.8	15.6	9.0	0.3	-10.9	-18.3	-0.9
6	-26.8	-23.7	-12.7	0.9	10.1	17.1	19.6	17.1	9.8	0.1	-12.8	-23.2	-2.0
7	-25.6	-23.0	-13.6	-1.0	7.8	13.8	16.6	14.4	7.5	-1.1	-12.7	-21.8	-3.2
8	-19.5	-15.4	-5.3	5.4	14.1	20.2	22.8	20.9	14.0	4.7	-7.2	-16.5	3.2
9	-22.9	-18.6	-7.7	3.9	12.5	18.8	21.3	19.4	12.4	3.1	-9.5	-19.7	1.1
10	-22.6	-18.2	-7.3	4.0	12.4	18.8	21.4	19.5	12.7	3.5	-8.8	-19.3	1.3
11	-23.9	-19.0	-8.2	3.5	11.7	17.2	20.5	18.3	11.7	2.6	-9.7	-20.3	0.4
12	-20.2	-16.4	-6.5	4.4	12.7	18.5	21.5	20.4	14.2	5.0	-7.1	-17.2	2.4
13	-21.3	-17.0	-6.0	4.9	12.6	18.6	21.9	20.1	13.3	4.5	-7.2	-17.6	2.2
14	-19.9	-15.8	-5.5	5.5	14.1	20.2	22.9	21.0	14.3	5.0	-6.7	-16.9	3.2
15	-19.4	-15.4	-4.8	6.0	14.3	20.0	22.8	21.1	14.4	5.6	-5.6	-15.6	3.6
16	-20.9	-16.9	-6.1	4.7	12.6	18.2	21.6	19.8	12.8	4.2	-6.7	-16.9	2.2
17	-17.3	-13.8	-4.6	5.5	13.3	18.2	21.7	20.4	13.8	5.5	-5.4	-14.4	3.6
18	-17.1	-14.1	-5.8	4.1	11.3	15.5	19.2	18.4	12.1	4.3	-5.8	-14.1	2.3
19	-18.5	-14.5	-4.5	5.8	13.7	18.4	22.0	20.6	13.8	5.4	-5.5	-15.1	3.5
20	-17.6	-13.8	-4.0	6.5	14.8	20.5	23.4	21.7	15.1	6.2	-4.8	-14.2	4.5
21	-21.8	-19.1	-9.2	3.3	11.9	17.8	20.3	18.4	10.8	2.3	-10.2	-18.7	0.5
22	-19.4	-16.9	-8.3	3.0	11.2	16.5	19.5	17.5	10.7	2.3	-8.3	-16.3	1.0
23	-14.0	-11.5	-3.6	6.6	14.9	19.8	22.3	20.2	13.7	5.5	-4.8	-12.0	4.8
24	-13.3	-10.6	-2.7	7.5	15.8	20.8	23.4	21.7	15.3	6.5	-3.6	-11.1	5.8
25	-14.3	-10.9	-2.3	7.9	16.2	21.1	23.9	22.2	15.6	7.2	-3.2	-11.7	6.0
26	-16.4	-12.5	-3.5	6.8	15.0	20.4	23.2	21.6	15.1	6.5	-4.4	-13.4	4.9
27	-16.4	-12.7	-3.5	6.7	15.0	20.1	23.0	21.3	15.0	6.8	-3.8	-12.8	4.9
28	-14.8	-11.3	-2.2	7.7	15.7	20.5	23.4	22.0	15.6	7.6	-2.5	-11.2	5.9
29	-18.7	-14.9	-3.9	6.4	14.2	19.0	22.4	20.9	14.2	6.0	-4.2	-14.3	3.9
30	-17.4	-14.3	-3.4	4.5	11.9	16.3	19.8	18.8	12.2	4.4	-5.3	-13.9	2.8
31	-14.4	-10.8	-2.4	6.6	13.8	17.7	21.3	21.1	14.7	6.7	-3.2	-11.7	4.9
32	-17.1	-12.2	-2.3	7.0	14.1	16.6	22.2	21.2	14.4	6.8	-2.7	-13.3	4.6
33	-14.5	-9.9	-0.6	8.5	15.5	19.8	23.2	22.4	15.9	8.4	-0.5	-10.5	6.5
34	-12.8	-9.0	-0.8	7.6	14.5	18.9	22.4	22.1	15.9	8.6	0.0	-9.2	6.5
35	-8.2	-3.3	0.9	8.4	14.7	19.2	23.0	23.2	17.9	11.0	2.7	-5.2	8.7
36	-9.4	-6.6	0.8	9.4	16.7	21.4	24.8	24.1	18.7	11.0	2.0	-6.0	8.9

(contd.)

Table 4.2 continued

No.	J	F	M	A	M	J	J	A	S	O	N	D	Mean
37	-12.5	-8.9	-1.3	7.0	13.9	18.4	22.2	21.6	15.1	7.7	-1.0	-9.7	6.0
38	-12.0	-8.4	0.1	9.3	16.9	21.5	24.4	23.5	17.4	9.4	0.0	-8.5	7.8
39	-12.5	-8.9	-0.9	8.2	16.2	21.1	24.1	22.8	16.7	8.6	-1.3	-9.5	7.1
40	-14.2	-11.8	-4.3	5.8	13.9	18.5	21.0	19.2	13.2	5.3	-4.4	-11.8	4.2
41	-19.8	-17.0	-7.2	4.1	12.3	17.8	20.8	18.7	11.7	3.0	-7.6	-16.7	1.7
42	-22.0	-18.7	-8.3	3.4	11.6	17.4	20.3	18.9	11.3	2.2	-9.2	-19.0	0.7
43	-18.6	-15.0	-4.6	6.0	14.3	20.4	22.9	20.7	13.4	4.3	-6.9	-16.2	3.4
44	-15.3	-12.2	-3.6	6.4	14.2	15.7	22.0	20.1	13.5	5.5	-4.6	-12.9	4.1
45	-16.5	-13.4	-5.5	3.8	11.3	16.2	19.6	16.8	10.7	3.3	-6.3	-14.2	2.2
46	-18.1	-14.9	-6.2	3.9	11.3	15.9	18.7	16.8	10.6	3.1	-6.6	-15.3	1.6
47	-13.2	-10.2	-2.7	6.5	13.9	18.1	20.8	19.0	12.9	6.0	-3.2	-11.0	4.7
48	-11.7	-9.0	-1.3	8.5	16.5	20.8	23.3	21.6	15.6	8.0	-1.7	-9.3	6.8
49	-10.6	-7.2	0.7	10.5	18.3	22.2	24.6	23.0	17.3	9.7	0.0	-8.0	8.4
50	-8.8	-6.0	1.1	9.6	17.1	21.3	24.3	23.7	18.4	10.9	1.7	-5.9	9.0
51	-9.3	-5.8	2.0	11.3	18.4	22.2	24.4	22.8	17.2	10.0	0.8	-7.4	8.9
52	-11.3	-8.7	-0.7	8.9	15.9	19.9	22.1	20.2	14.2	7.2	-2.3	-10.1	6.3
53	-8.2	-5.3	1.8	10.9	18.1	22.2	24.1	22.4	16.9	9.6	0.7	-6.3	8.9
54	-9.7	-6.7	0.9	9.9	17.2	21.3	23.2	21.4	15.8	8.7	-0.5	-7.7	7.8
55	-14.0	-10.9	-3.4	5.3	12.7	17.2	19.1	17.3	11.7	4.6	-4.4	-12.0	3.6
56	-15.9	-12.4	-3.7	5.6	13.0	18.4	20.5	18.4	12.1	4.3	-5.5	-13.7	3.4
57	-15.2	-11.1	-2.5	6.6	14.3	19.9	21.9	19.9	13.4	4.8	-5.2	-13.4	4.4
58	-13.1	-9.0	-0.3	7.9	15.3	20.1	21.9	20.1	13.8	6.5	-2.7	-11.0	5.8
59	-11.3	-7.7	-0.1	8.3	15.4	19.9	21.8	20.1	14.3	7.5	-1.4	-8.9	6.5
60	-12.3	-8.4	-0.1	8.9	16.0	20.1	22.1	20.5	14.6	7.5	-1.6	-10.2	6.4
61	-4.6	-2.2	4.5	13.1	19.8	24.0	28.8	24.4	19.4	12.4	4.1	-2.7	11.8
62	-4.0	1.6	5.4	13.7	20.0	24.1	26.4	25.5	20.9	13.6	5.7	1.8	12.9
63	-6.5	-4.2	2.5	10.3	17.4	21.8	24.8	24.1	19.0	12.0	3.6	-3.7	10.1
64	-4.9	-3.4	2.1	9.1	15.5	19.4	23.0	23.9	20.0	13.6	5.8	-1.3	10.2
65	-3.9	-1.3	5.3	13.6	20.4	24.8	26.5	25.5	20.7	13.8	5.5	-1.5	12.5
66	-4.1	-1.4	5.5	13.9	20.5	25.2	26.6	25.1	20.2	13.2	4.7	-2.0	12.3
67	-8.5	-4.7	2.4	10.5	17.4	21.4	23.1	21.3	15.5	8.9	0.5	-6.8	8.4
68	-10.0	-5.5	2.5	10.2	16.7	21.4	23.4	21.6	15.5	8.7	0.1	-8.0	8.0
69	-6.6	-3.1	3.7	11.4	17.7	21.7	23.5	21.8	16.1	9.9	2.1	-4.9	9.4
70	-2.9	-0.4	6.6	14.6	20.9	25.6	26.6	25.0	20.3	13.7	5.7	-0.9	12.9
71	-3.4	-0.8	6.0	14.0	20.7	25.5	26.9	25.5	20.6	14.1	5.5	-1.0	12.8
72	-4.0	-1.6	5.1	13.0	20.1	24.7	26.3	25.2	20.0	13.6	5.5	-1.4	12.2

(contd.)

Table 4.2 continued

No.	J	F	M	A	M	J	J	A	S	O	N	D	Mean
73	-3.2	-1.0	5.1	12.5	19.1	23.6	25.9	25.2	20.2	14.2	6.5	-0.5	12.3
74	-1.1	-0.6	2.8	7.8	13.4	17.6	21.2	23.5	21.2	15.9	9.0	2.2	11.1
75	-1.2	0.1	4.5	10.2	15.7	20.0	23.9	25.1	21.4	15.9	8.8	2.0	12.2
76	-3.6	-1.2	5.3	12.7	19.1	23.4	25.3	24.4	19.2	13.2	9.6	-1.2	12.2
77	-1.7	0.9	7.3	15.1	21.8	26.3	27.6	26.3	21.7	15.8	7.8	0.8	14.1
78	-8.6	-6.7	-1.6	5.6	11.3	15.8	17.8	17.1	12.5	6.8	-0.2	-6.1	5.3
79	-2.9	0.1	6.9	14.7	21.1	25.8	26.7	25.2	20.3	13.8	5.9	-1.0	13.0
80	-5.1	-1.8	4.7	12.2	18.2	22.3	23.9	22.2	15.7	10.8	3.3	-3.2	10.3
81	-6.4	-2.5	4.5	11.4	16.9	21.1	22.9	21.5	15.7	9.7	2.5	-4.5	9.4
82	-2.0	0.8	7.1	14.6	21.0	25.9	27.0	25.5	20.7	14.5	6.6	0.0	13.5
83	-1.9	1.1	6.9	14.2	20.5	25.8	27.2	26.1	20.8	14.6	7.1	0.4	13.6
84	-1.9	0.9	7.1	14.3	20.3	25.4	26.9	26.0	20.8	14.7	7.2	0.5	13.5
85	-1.5	0.7	6.3	13.3	19.3	23.9	26.2	25.9	21.0	15.0	7.7	1.0	13.2
86	0.4	2.0	6.2	12.5	17.9	22.7	26.5	26.6	21.9	15.8	9.4	3.1	13.8
87	-0.7	1.6	7.3	14.2	20.0	25.0	26.8	26.3	21.3	15.4	8.3	1.9	14.0
88	-0.1	2.5	8.1	14.9	20.7	26.0	27.5	26.7	21.5	15.6	8.7	2.3	14.5
89	-0.1	2.5	7.9	14.5	20.3	25.7	27.2	26.1	21.0	15.1	8.2	2.1	14.2
90	-0.3	2.1	7.7	14.8	21.1	26.3	27.5	25.9	21.0	15.1	7.8	1.6	14.2
91	0.4	2.9	8.8	15.6	21.4	26.6	27.5	26.1	21.0	15.1	8.0	2.7	14.7
92	-1.5	1.4	7.2	13.5	18.5	23.6	25.6	24.3	18.7	13.2	6.9	0.9	12.7
93	-6.7	-4.5	0.6	6.4	11.0	15.5	17.5	16.6	11.7	6.6	0.5	-4.4	5.9
94	-2.1	1.5	8.3	14.8	20.5	26.0	27.3	26.3	20.3	14.2	6.3	-0.4	13.6
95	-1.9	2.1	8.1	14.1	19.1	25.2	26.6	25.5	19.4	13.7	6.6	0.7	13.3
96	-0.8	2.2	7.8	13.7	18.6	23.9	25.5	24.4	18.4	13.0	6.4	1.1	12.9
97	-2.8	0.4	6.4	12.3	16.7	20.7	22.6	21.6	18.2	10.8	4.4	-1.3	10.8
98	-5.6	-2.9	3.2	9.6	14.6	19.3	20.9	19.6	14.2	8.8	1.9	-4.0	8.3
99	-5.2	-2.2	3.8	10.1	15.0	19.2	21.0	19.7	14.4	8.8	2.1	-3.5	8.6
100	-8.8	-6.7	-1.4	4.3	9.0	13.0	14.9	13.9	9.2	4.0	-2.2	-6.7	3.5
101	2.8	5.7	10.9	16.0	19.6	22.7	24.8	24.2	19.4	14.9	9.3	4.2	14.5
102	-4.3	-1.7	2.7	6.6	9.8	12.1	14.5	14.0	10.8	6.3	0.9	-3.4	5.7
103	3.9	6.2	11.0	15.9	19.4	22.3	24.2	23.4	19.4	15.1	14.0	5.4	15.0
104	2.1	4.6	9.5	15.0	19.4	23.7	25.6	25.0	19.9	14.8	8.7	3.7	14.3
105	3.5	5.5	10.0	15.2	18.9	22.8	25.3	25.2	24.1	15.1	9.8	5.5	15.1
106	3.2	5.8	10.7	16.2	20.5	25.1	27.5	27.2	21.6	16.1	10.0	4.9	15.7
107	1.8	3.9	9.1	15.2	20.7	26.0	28.0	27.0	22.0	16.3	9.7	3.9	15.3
108	0.9	3.4	8.7	15.0	20.5	25.9	27.4	26.6	21.6	16.0	9.1	3.0	14.8

(contd.)

Table 4.2 continued

No.	J	F	M	A	M	J	J	A	S	O	N	D	Mean
109	1.4	3.6	8.8	15.1	20.6	25.4	27.9	26.9	21.6	15.9	9.4	3.8	15.0
110	1.0	3.2	8.5	14.9	20.4	25.8	27.5	26.5	21.4	15.9	9.2	3.2	14.8
111	0.7	3.1	8.4	14.0	20.5	25.7	27.9	27.1	21.9	16.2	9.5	3.3	14.9
112	1.6	3.7	8.9	15.2	20.5	25.4	27.9	27.2	22.1	16.6	13.1	4.0	15.5
113	2.0	4.0	9.1	15.4	20.1	24.6	27.6	26.8	21.5	15.7	9.8	4.2	15.1
114	1.9	4.2	9.2	15.3	20.7	25.1	28.5	28.2	23.0	17.0	10.6	4.6	15.7
115	0.7	3.1	8.3	15.0	20.6	25.6	28.3	27.7	22.5	16.3	9.9	3.5	15.1
116	2.0	3.8	8.4	14.8	19.9	24.5	28.0	27.8	22.7	16.9	10.5	4.4	15.3
117	1.4	3.0	7.1	13.3	18.5	23.4	27.2	27.1	22.3	16.4	10.1	4.0	14.5
118	0.1	1.9	6.9	13.6	19.0	23.9	26.3	26.7	21.6	15.7	9.1	2.7	14.0
119	3.3	4.6	8.3	13.8	18.8	23.2	27.3	27.8	23.8	17.9	12.5	6.2	15.6
120	5.4	5.9	9.1	14.1	18.4	22.6	26.8	27.2	24.3	19.2	14.2	8.5	16.3
121	5.3	5.7	9.0	13.9	18.2	22.5	26.3	27.0	24.1	19.2	13.5	8.4	16.1
122	4.2	5.6	9.8	15.9	20.2	24.2	28.6	27.8	23.6	17.9	12.5	6.9	16.4
123	3.6	5.0	9.2	15.1	20.3	24.3	28.7	28.2	23.5	17.4	12.1	6.1	16.1
124	-3.1	-1.7	2.7	7.9	11.7	14.9	17.7	17.3	13.7	8.9	3.5	-0.3	7.8
125	3.4	5.0	9.7	15.7	21.2	25.3	28.9	28.7	23.9	18.0	12.0	5.9	16.5
126	2.8	5.0	10.0	16.0	21.3	25.8	29.0	28.5	23.6	17.5	11.2	5.3	16.3
127	3.4	5.4	10.1	15.9	20.8	25.3	28.1	27.5	22.7	17.2	11.2	5.6	16.1
128	3.0	5.1	9.8	15.7	20.8	25.4	27.8	27.3	22.5	17.2	10.5	5.2	15.9
129	4.7	6.4	11.1	16.7	21.5	25.7	28.3	27.6	23.6	18.0	12.3	6.6	16.9
130	5.0	6.6	11.1	16.5	20.4	24.2	27.1	26.7	22.5	17.1	11.8	6.9	16.3
131	6.5	8.6	13.2	18.2	21.9	25.1	27.8	27.9	22.9	17.9	12.9	8.4	17.6
132	5.2	7.3	12.1	17.2	21.3	24.2	26.0	25.5	21.5	16.9	11.7	7.0	16.3
133	-0.8	2.4	6.6	10.3	12.9	14.4	18.4	15.9	13.1	8.7	3.6	-0.6	8.7
134	5.5	7.5	12.1	17.0	20.9	23.7	25.6	25.1	21.2	16.8	11.9	7.3	16.2
135	6.1	7.7	12.2	17.0	20.5	23.3	25.3	24.9	21.0	16.6	12.1	7.8	16.2
136	-6.0	-4.9	-0.7	3.3	6.1	9.0	11.8	11.2	7.8	3.4	-1.0	-3.8	3.0
137	0.7	3.0	6.6	9.7	12.9	14.0	15.2	14.6	13.0	9.5	4.7	1.2	8.8
138	9.5	11.7	16.4	19.8	21.5	21.0	22.7	22.2	20.0	16.7	13.1	10.2	17.1
139	7.9	9.7	14.4	19.1	22.4	24.5	27.0	26.6	22.9	18.1	13.9	9.7	18.0
140	7.1	9.0	13.9	18.7	22.1	24.5	27.1	27.0	22.5	17.8	13.3	8.9	17.7
141	7.7	9.5	14.3	18.9	21.9	24.4	27.3	27.2	22.5	18.1	13.7	9.5	17.9
142	7.5	9.5	14.1	18.8	22.3	25.2	28.6	28.5	23.8	18.6	13.9	9.5	18.4
143	-2.4	-0.9	4.0	8.8	12.0	15.1	17.8	17.3	13.9	9.1	4.2	0.1	8.3
144	3.7	5.0	9.5	14.9	19.0	22.8	25.4	24.8	21.1	15.8	10.6	8.0	15.0

(contd.)

Table 4.2 continued

No.	J	F	M	A	M	J	J	A	S	O	N	D	Mean
145	4.6	6.2	10.8	16.3	20.5	24.7	27.9	27.4	23.4	17.7	12.1	7.0	16.5
146	4.3	5.9	10.6	16.3	21.3	25.6	29.0	28.0	23.6	17.8	12.1	6.7	16.8
147	4.6	6.2	10.9	16.7	21.7	26.0	29.5	28.9	24.5	18.5	12.5	7.0	17.3
148	4.4	6.1	10.8	16.7	21.2	25.7	29.2	28.5	23.8	18.2	12.2	6.7	17.0
149	4.1	6.0	10.6	16.5	21.0	25.0	28.3	27.6	23.6	17.5	11.6	6.4	16.5
150	5.0	6.4	10.9	17.1	21.8	25.7	28.6	29.2	24.8	15.1	13.1	7.5	17.1
151	4.5	6.4	11.2	17.0	21.5	25.1	28.8	28.4	24.6	17.9	12.3	6.7	17.0
152	5.2	6.4	10.8	16.8	21.5	25.0	29.2	28.8	24.9	18.6	13.4	7.6	17.3
153	6.3	7.9	12.2	17.9	22.0	25.9	28.4	26.7	25.0	19.3	13.8	8.6	17.8
154	7.5	8.1	11.2	16.0	20.5	24.2	27.9	28.0	25.1	20.0	15.7	10.4	17.9
155	8.7	9.1	12.2	17.1	21.0	24.9	28.3	27.8	25.3	20.6	16.0	11.3	18.5
156	9.0	10.6	14.4	19.3	23.1	25.5	28.6	27.9	25.6	22.4	15.9	10.9	19.4
157	6.2	7.8	12.2	17.6	21.5	24.6	27.8	27.2	24.1	18.6	13.4	8.3	17.4
158	5.6	7.3	15.1	17.9	22.1	25.6	29.1	28.5	24.8	19.1	13.3	7.9	18.0
159	6.2	8.0	12.7	18.4	22.5	25.7	28.8	28.1	28.0	19.4	13.6	8.4	18.3
160	6.2	7.7	12.4	18.2	22.7	26.2	29.5	29.0	25.5	19.6	13.8	8.5	18.3
161	5.1	6.7	11.3	17.1	21.6	25.4	28.6	28.0	24.2	18.4	12.5	7.5	17.2
162	5.1	6.4	10.9	16.6	21.3	25.5	28.5	27.9	24.2	18.9	12.6	7.5	17.1
163	4.8	6.4	11.0	16.5	20.9	24.9	27.5	26.7	23.2	17.8	12.1	7.3	16.6
164	4.6	6.1	10.8	16.4	20.8	24.7	27.5	27.0	23.2	17.5	12.0	7.0	16.5
165	6.0	7.5	12.2	17.4	21.2	24.7	27.9	27.1	23.7	18.1	12.6	8.3	17.2
166	4.2	5.8	10.7	15.8	19.5	22.3	25.3	24.5	21.2	15.8	11.2	6.4	15.2
167	2.4	4.3	9.2	13.8	16.9	19.2	21.8	20.9	17.9	13.3	8.8	4.5	12.8
168	1.9	3.7	8.4	12.0	14.3	15.8	17.7	16.9	14.5	10.8	6.7	3.5	10.5
169	2.0	4.1	9.1	13.1	15.8	17.6	19.8	19.0	16.1	11.9	7.5	3.6	11.6
170	7.1	9.4	13.3	17.2	20.4	20.7	21.0	20.3	18.6	15.9	11.0	7.6	15.2
171	7.1	9.1	13.4	16.9	18.7	19.1	19.9	19.2	17.4	14.5	10.8	7.9	14.5
172	6.1	7.9	12.8	17.1	19.8	20.7	22.2	21.5	19.3	15.7	12.0	7.9	15.3
173	4.9	6.4	11.5	16.3	19.6	21.9	24.0	23.5	20.8	15.9	11.6	7.0	15.3
174	10.1	11.9	16.7	21.1	24.0	25.8	27.0	26.3	24.3	20.3	15.8	11.8	19.6
175	4.9	6.3	10.8	15.7	19.2	21.8	23.4	22.8	20.3	16.1	11.4	7.1	15.0
176	7.6	9.3	13.9	18.9	22.4	25.3	26.9	26.3	24.0	19.2	14.1	9.6	18.1
177	8.0	9.0	13.1	18.4	23.1	26.2	28.3	27.8	25.8	20.7	15.2	10.1	18.8
178	5.8	7.5	12.4	18.0	22.5	26.3	29.2	27.9	24.2	18.5	12.8	7.9	17.8
179	5.8	7.1	12.0	17.6	22.1	26.0	29.2	28.3	24.9	19.1	13.2	7.9	17.8
180	7.9	9.5	13.9	19.3	23.9	26.7	29.5	29.0	26.5	21.0	15.2	10.0	19.4
181	7.7	9.9	13.9	18.7	22.8	25.1	27.2	26.9	24.5	19.7	14.5	9.9	18.4

(contd.)

Table 4.2 continued

No.	J	F	M	A	M	J	J	A	S	O	N	D	Mean
182	9.8	10.8	15.1	19.7	23.2	25.6	28.0	27.2	25.0	19.9	14.9	10.6	19.2
183	10.4	10.6	13.4	18.1	22.2	25.3	28.7	28.3	26.0	21.6	17.8	13.1	19.6
184	10.9	10.5	12.8	17.3	21.4	25.2	27.9	27.7	26.2	22.4	18.2	13.8	19.5
185	14.9	15.5	17.6	21.4	24.6	26.8	28.6	28.4	26.9	23.5	20.4	17.2	22.2
186	17.1	17.8	19.6	22.3	24.9	26.8	28.3	27.8	26.6	24.1	21.5	18.5	22.9
187	20.3	21.1	22.9	25.3	27.4	27.9	28.3	27.9	27.4	26.0	23.9	21.7	25.0
188	16.7	17.8	20.7	24.3	27.2	28.1	28.6	28.3	27.7	25.4	22.0	18.7	23.8
189	16.1	16.5	19.1	22.7	25.5	27.2	28.3	28.1	27.3	24.8	21.8	18.3	23.0
190	15.4	16.3	18.9	22.7	25.7	27.1	28.3	27.8	26.9	24.2	20.9	17.4	22.6
191	4.1	5.5	8.9	12.5	15.3	17.3	19.2	18.6	16.9	12.4	8.6	5.9	12.1
192	12.6	12.5	14.9	19.0	23.2	26.0	28.3	28.3	27.0	23.2	19.7	15.2	20.8
193	12.7	13.2	15.8	20.1	23.8	26.4	28.7	28.3	26.7	22.8	19.0	15.0	21.0
194	13.2	13.8	16.4	20.3	24.3	26.6	28.2	28.0	26.6	23.1	19.3	15.3	21.3
195	11.2	13.5	17.5	21.8	25.2	26.9	28.6	28.3	26.7	22.9	18.1	13.7	21.2
196	14.4	14.9	17.7	21.6	25.3	27.0	28.2	28.0	28.2	24.2	20.7	16.7	22.2
197	12.0	13.5	17.4	21.7	25.1	28.8	28.2	27.8	26.5	22.3	18.2	13.9	21.3
198	14.1	14.9	18.4	22.1	25.7	27.3	28.2	27.7	26.6	23.7	19.7	15.9	22.0
199	13.4	14.2	17.7	21.8	25.8	27.2	28.3	28.2	27.0	23.8	19.7	15.2	21.9
200	10.0	11.3	15.3	20.1	24.8	26.9	29.1	28.6	26.6	22.1	16.9	11.9	20.3
201	8.8	10.5	14.7	19.6	24.1	26.5	28.5	28.0	25.7	21.0	15.6	10.9	19.5
202	9.7	11.1	15.1	19.9	24.1	26.3	27.8	27.3	25.7	21.5	16.3	11.5	19.7
203	12.3	13.5	17.2	21.4	25.6	27.3	28.6	28.1	26.9	23.5	18.6	14.5	21.5
204	12.0	12.8	16.6	21.1	25.3	26.9	28.3	27.9	26.6	23.1	18.4	13.8	21.1
205	14.6	15.4	18.8	22.4	26.1	27.3	28.1	27.6	26.7	23.8	19.9	16.3	22.3
206	15.6	16.2	19.3	23.2	27.1	28.2	28.9	28.2	27.3	24.8	21.0	17.5	23.1
207	17.2	18.2	21.6	24.9	27.4	28.1	28.4	27.7	26.8	24.8	21.8	18.7	23.8
208	18.0	19.1	22.3	25.1	27.5	28.1	28.3	27.5	26.6	24.5	21.7	19.2	24.0
209	22.9	23.3	25.1	27.1	28.9	28.9	28.7	28.3	27.9	26.8	25.5	24.0	26.5
210	18.3	18.9	22.0	25.7	28.6	29.1	29.1	28.2	27.0	25.2	22.4	19.7	24.5
211	16.9	18.3	21.6	24.9	26.9	27.5	27.5	26.7	25.6	23.7	20.6	16.1	23.0
212	14.3	15.2	18.9	23.0	26.5	28.1	28.7	28.1	27.1	24.3	20.1	16.5	22.6
213	13.6	14.4	17.9	22.3	26.2	27.6	28.3	27.9	26.9	23.8	19.6	15.6	22.0
214	13.9	15.5	19.1	23.1	26.6	27.8	28.1	27.5	26.2	23.1	18.5	15.4	22.1
215	12.9	13.9	17.3	21.9	26.0	27.4	28.2	27.9	26.7	23.3	18.9	14.8	21.6
216	10.7	12.1	16.3	20.9	24.5	26.6	27.9	27.3	25.8	21.7	16.8	12.8	20.3
217	13.3	15.3	19.5	23.8	26.8	27.9	28.6	27.8	26.3	22.9	19.6	14.5	22.2
218	8.2	10.6	15.3	19.2	21.5	22.0	22.5	21.6	19.9	18.8	12.8	9.7	16.8

(contd.)

Table 4.2 continued

No.	J	F	M	A	M	J	J	A	S	O	N	D	Mean
219	12.1	14.0	18.0	21.0	22.6	22.8	22.7	22.0	21.0	18.7	15.4	12.6	18.6
220	16.6	18.8	23.1	26.1	28.4	28.4	29.5	27.6	26.7	24.0	20.3	17.2	23.9
221	15.6	17.5	20.8	24.0	25.6	25.5	25.2	24.8	24.3	22.4	19.2	16.2	21.8
222	12.5	14.1	17.4	20.5	22.7	23.0	22.6	22.5	22.0	20.1	16.3	13.2	18.9
223	11.4	13.2	16.5	19.4	21.3	21.7	21.6	21.2	20.5	18.6	15.1	12.0	17.7
224	10.7	13.0	16.2	18.9	20.8	21.2	21.1	21.0	20.2	18.0	14.3	11.1	17.2
225	7.8	9.8	13.2	16.7	19.3	19.5	19.9	19.2	17.6	15.0	11.5	8.3	14.8
226	8.1	10.5	14.2	17.9	20.3	20.8	20.7	20.1	18.7	15.9	11.6	8.4	15.6
227	7.5	9.3	12.7	15.7	18.1	19.5	19.5	19.8	19.1	16.3	11.8	8.5	14.8
228	8.7	10.6	13.4	16.2	18.9	19.9	20.1	19.3	18.1	15.3	11.6	8.8	15.1
229	5.9	7.5	10.4	13.5	16.6	17.8	18.0	17.2	16.0	13.2	9.1	6.3	12.6
230	-2.9	-2.6	0.4	3.8	7.5	10.7	11.9	11.3	9.9	5.8	1.3	-1.2	4.7
231	-6.0	-4.1	-0.5	3.3	7.6	9.6	10.5	9.8	8.1	4.1	-1.3	-5.1	3.0
232	3.7	6.9	10.3	13.5	17.5	19.0	19.6	18.7	18.6	12.8	8.0	3.8	12.7
233	-4.4	-1.5	2.6	6.6	10.5	12.5	14.0	13.4	11.0	6.5	0.5	-4.1	5.6
234	-2.8	-0.1	3.5	7.3	10.8	12.9	14.5	13.9	11.4	8.7	1.6	-2.4	6.6
235	-2.5	0.3	4.3	8.5	12.4	14.9	16.3	15.2	13.1	8.2	2.1	-2.0	7.6
236	0.2	2.4	5.4	8.5	11.7	14.4	15.6	15.2	13.5	9.8	4.9	1.3	8.6
237	-9.1	-7.7	-4.5	-0.2	3.3	6.8	7.9	7.4	5.7	0.1	-4.5	-7.2	-0.2
238	-4.1	-0.5	2.9	7.7	11.9	14.7	14.2	13.0	11.8	6.5	0.4	-3.5	6.3
239	-2.3	0.8	4.3	8.3	12.6	15.5	14.9	14.1	12.8	8.1	1.9	-1.9	7.4
240	-14.4	-11.1	-6.5	-1.4	3.3	7.3	8.8	8.1	5.2	-1.3	-9.0	-13.7	-2.1
241	-9.9	-6.2	-2.2	2.3	6.0	9.1	11.2	10.7	7.6	2.0	-4.2	-8.9	1.5
242	-7.1	-4.5	-0.9	3.4	7.3	10.4	12.0	11.3	9.3	3.9	-2.2	-6.3	3.1
243	-7.8	-5.0	-0.5	4.0	7.7	10.6	12.5	11.6	8.7	3.3	-3.0	-7.2	2.9
244	-12.9	-10.3	-5.1	0.0	3.9	6.7	9.1	8.3	5.1	-0.4	-7.5	-12.4	-1.3
245	-16.8	-13.9	-8.3	-2.9	1.5	4.8	7.5	7.0	3.3	-3.2	-11.5	-16.4	-4.1
246	-13.4	-9.3	-3.3	2.5	6.7	8.4	11.7	11.0	6.9	0.5	-7.7	-12.7	0.1
247	-6.9	-2.3	5.2	11.8	16.6	20.3	22.2	21.0	15.8	9.4	1.7	-5.5	9.1
248	-7.4	-3.6	4.1	11.6	17.2	21.1	23.4	21.7	16.2	9.6	1.5	-5.6	9.1
249	-8.9	-5.4	2.1	9.8	15.8	20.4	22.3	20.5	14.7	8.0	-0.3	-7.1	7.7
250	-9.0	-4.8	2.8	10.6	16.9	21.4	23.4	21.6	16.0	9.1	0.9	-6.7	8.5
251	-10.6	-6.4	2.1	10.7	18.1	23.2	25.3	23.5	17.2	8.8	-0.9	-8.4	8.5
252	-12.1	-8.1	0.0	8.7	16.3	21.7	23.6	21.9	15.6	7.1	-2.8	-10.2	6.8
253	-10.2	-6.1	2.0	9.5	15.4	19.4	21.4	20.5	14.6	7.0	-1.4	-8.3	7.0
254	-9.6	-5.6	2.1	10.0	16.4	21.0	23.2	21.7	15.7	7.8	-0.9	-7.9	7.8
255	-12.2	-10.6	-5.4	0.2	5.1	9.1	11.3	10.4	5.7	0.5	-6.1	-10.1	-0.2

(contd.)

Table 4.2 continued

No.	J	F	M	A	M	J	J	A	S	O	N	D	Mean
256	-8.4	-4.9	1.9	7.9	12.0	15.2	17.2	16.5	12.1	6.4	-0.8	-6.7	5.7
257	-10.6	-7.3	-1.4	4.3	8.9	12.2	14.9	14.1	9.4	2.6	-4.9	-9.3	2.7
258	-10.9	-6.7	-0.2	6.5	11.5	15.3	17.6	16.7	11.5	3.8	-4.6	-9.9	4.2
259	-14.3	-10.6	-3.6	2.8	8.4	12.3	15.1	14.4	8.6	0.8	-7.4	-12.8	1.1
260	-13.1	-9.2	-2.5	4.2	10.1	14.5	17.0	16.4	10.4	2.0	-6.6	-11.9	2.6
261	-9.7	-5.9	1.8	9.4	15.6	20.1	21.8	20.7	15.0	7.4	-1.2	-7.9	7.3
262	-10.5	-6.4	1.5	9.1	15.5	20.0	21.6	20.6	14.8	7.0	-1.7	-8.4	6.9
263	-9.3	-4.1	4.5	12.4	18.3	22.7	24.7	23.5	17.0	8.7	0.2	-7.0	9.3
264	-8.5	-2.3	7.1	15.4	21.0	25.3	27.4	26.0	20.1	11.2	1.6	-6.3	11.5
265	-12.2	-5.8	4.5	13.2	20.2	25.2	27.2	25.9	19.1	9.9	-0.6	-9.0	9.8
266	-11.2	-5.8	2.7	11.5	18.6	24.3	26.3	25.0	18.4	9.1	-1.0	-8.7	9.1
267	-9.5	-2.1	9.3	18.9	25.7	31.0	32.7	30.4	23.3	12.6	1.8	-7.2	13.9
268	-18.3	-15.2	-3.8	9.1	16.2	21.4	23.5	21.9	15.4	6.0	-5.1	-14.8	4.7
269	-15.4	-12.1	-4.0	9.0	15.9	21.2	23.5	22.0	16.8	7.4	-4.2	-11.6	5.7
270	-17.0	-15.1	-6.1	7.0	14.9	20.4	22.1	20.5	14.6	5.8	-5.2	-14.1	4.0
271	-14.1	-11.7	-4.3	4.8	11.7	17.1	18.8	17.3	11.4	3.0	-5.8	-11.7	3.0
272	-12.6	-10.7	-2.3	8.7	15.3	20.1	22.2	20.7	15.1	6.8	-2.1	-9.2	6.0
273	-16.6	-12.3	0.2	11.5	18.8	23.5	25.2	23.5	17.7	8.2	-1.5	-11.0	7.3
274	-10.0	-7.0	2.6	12.1	16.9	20.5	22.6	21.6	16.9	9.3	0.9	-5.8	8.4
275	-8.4	-2.2	7.4	15.2	20.8	24.5	25.9	24.9	20.3	12.2	2.5	-6.1	11.4
276	-5.6	-0.3	9.0	16.5	20.4	23.9	25.5	24.1	19.7	12.4	3.8	-3.2	12.2
277	-7.4	-1.4	8.1	16.1	20.8	24.4	26.0	24.7	19.9	11.9	2.4	-5.5	11.7
278	-6.6	-1.2	7.8	15.5	19.7	23.8	25.4	23.7	18.9	11.5	2.7	-4.4	11.4
279	-6.6	-1.6	7.7	15.4	19.9	23.8	25.9	24.5	19.8	12.3	3.4	-4.2	11.7

4.4 Onset and End of Certain Limited Temperatures and Their Duration

Certain limited temperatures are not only important from a purely climatological viewpoint, because of their significance to a general thermal characterization, but also from the viewpoint of application for practical purposes. In this respect, certain limited temperatures are essential for agriculture since they control the suitability of cultivation of certain crops. A mean daily air temperature $\geq 10°C$ is regarded as the most crucial limited temperature since it controls the growing season of the most important crops in China, including rice, cotton and maize. Although a minor agroclimatic factor for China, a mean daily air temperature $\leq 0°C$, nevertheless, also represents an important limited temperature since it

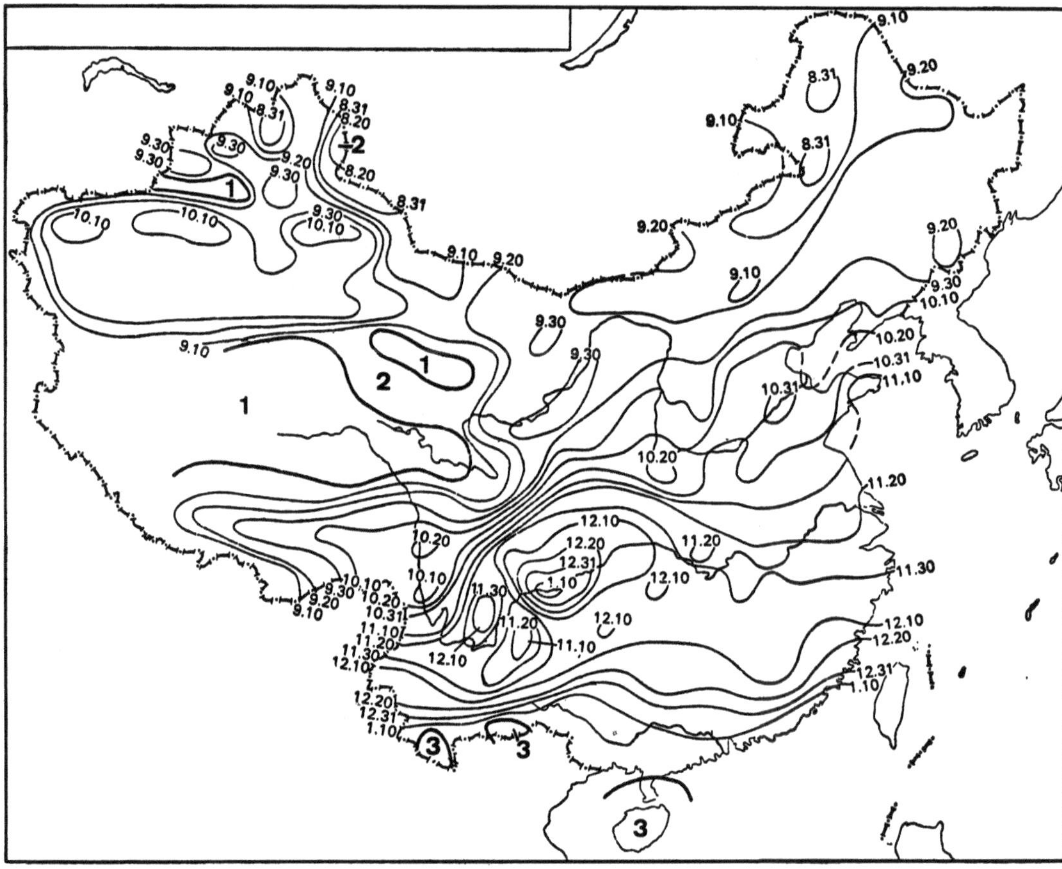

Fig. 4.12. Mean date of first frost, expressed by mean air temperature ≤0° C. (After Zhang Jiacheng and Lin Zhi-quang 1985) Isolines show month and date. 1 – permanent frost, 2 – much frost, 3 – frost-free

controls the growing season of winter wheat which is the leading crop in North and Northeast China. As for winter wheat, the plant enters the growing period through winter with a mean daily air temperature ≤0° C. The occurrence of frost is, however, the most injurious factor for all tropical crops which may even suffer from irreversible injuries. Also subtropical crops are sensitive to frost and strong cold outbreaks which affect the growing conditions, finally resulting in a harvest loss, both in terms of quantity and quality. The most common agroclimatic index is given by the accumulated temperature which represents the annual total of all positive daily temperature values for those days which record a mean ≥10° C.

The various limited temperature indices, mentioned above, are further discussed with particular reference to their variation over space. Additionally, a few other common indices of limited temperatures are also briefly described.

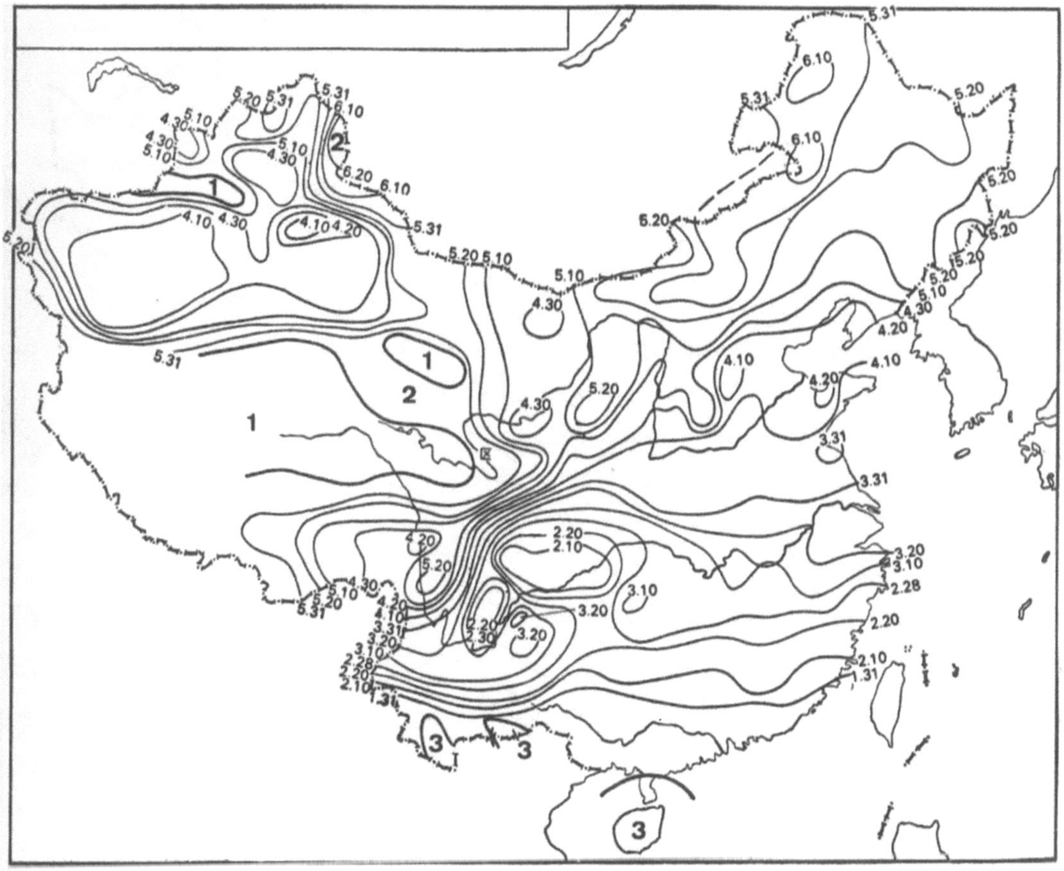

Fig. 4.13. Mean date of last frost, expressed by mean air temperature ≤0° C. (After ZHANG Jiacheng and LIN Zhi-quang 1985) Isolines show month and date. 1 – permanent frost, 2 – much frost, 3 – frost-free

4.4.1 Mean Daily Air Temperature ≤0° C

Since the major part of China records a pronounced thermal contrast between winter and summer, and the duration of winter is also very long, the occurrence of a mean daily air temperature ≤0° C, called frost, plays a main climatic role.

The mean date of first frost (Fig. 4.12) already indicates the occurrence of frost in practically all of China. Frost-free are only Hainan Island and Luichow Peninsula as well as two small, pocketlike areas in southernmost Yunnan Province. From North to South China, the mean date of first frost is gradually postponed from August 20 to January 10. Most likely, the isochrones of first frost prefer a latitudinal direction, with the exception of the central mountain "corridor" and some basins. To the first category belong the mountains of eastern Sichuan and Yunnan Provinces, to the second category the Sichuan Basin and the basins of Kunming and Tukow. These basins represent pocketlike, warm air cells,

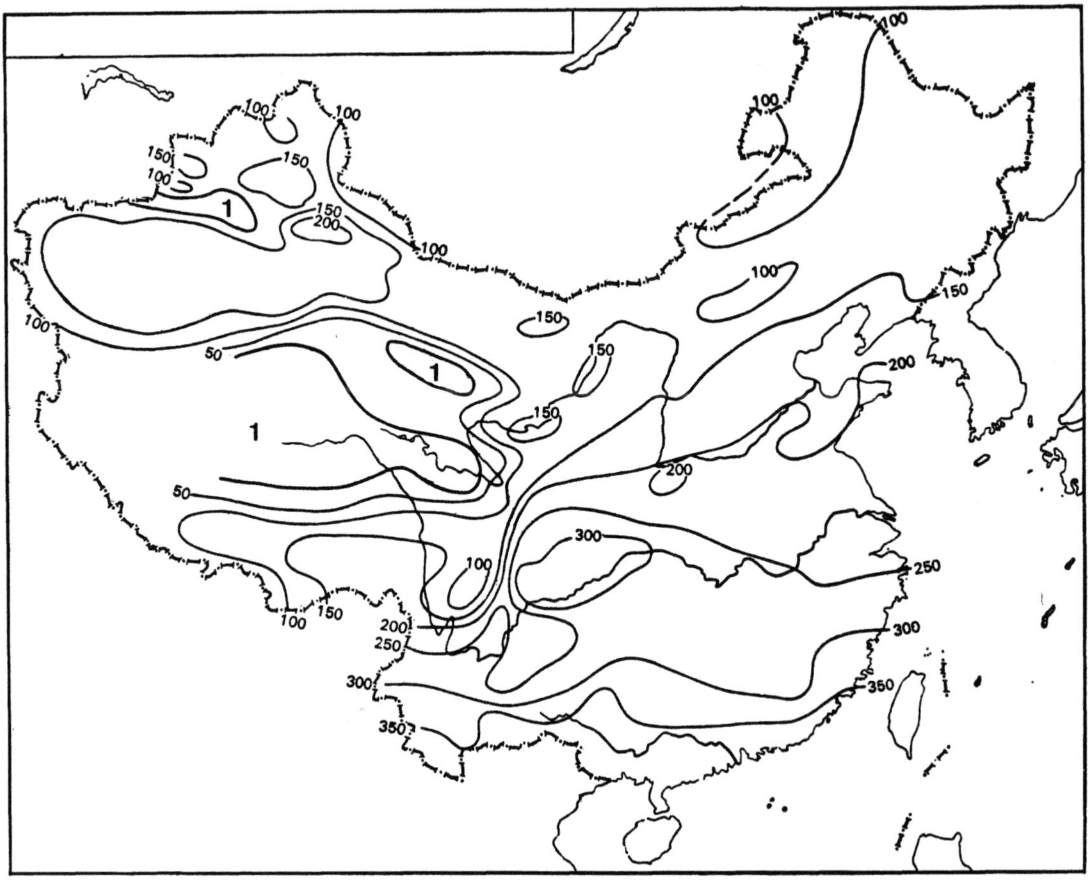

Fig. 4.14. Mean number of frost-free days. A frost day is expressed by a mean daily air temperature ≤0° C. *1* permanent frost. (After ZHANG Jiacheng and LIN Zhi-quang 1985)

which can be characterized as relative heat islands resulting in a 4–5-week later occurrence of the first frost (Fig. 4.12).

With regards to frost, the Qinghai-Xizang (Tibetan) Plateau represents a most striking exceptional case. As can be seen from temperature readings, the southern, southeastern and northeastern parts of the Plateau record an early date of the first frost which, on average, already falls on September 10. The reason for this is the high altitude of the Plateau at around 3,600–3,800 m a.s.l. A corresponding early date of the first frost is only recorded in northeastern and northwestern China which is, however, recorded at a latitude of 15° further to the north. In the Tibetan Plateau, the central, northern and western parts, which are located around 4,500 m a.s.l., record year-round permanent frost, although observation data are poor.

The mountains surrounding the Qinghai-Xizang (Tibetan) Plateau, e.g. the Himalayas and Kunlun Shan, are part of the permanent frost region of western

China. Additionally, also Nan Shan and Tian Shan show year-round permanent frost conditions; both are established as large-scale pockets under permanent frost conditions.

The mean date of last frost (Fig. 4.13) shows supplementary observations in comparison with the mean date of first frost. Subsequently, from South to North China, the mean date of last frost gradually develops from January 31 to June 20. Corresponding with the mean date of first frost (Fig. 4.12), the isochrones run mostly in a latitudinal direction, with the exception of the central mountains where the isochrones turn in a north to south direction, thus coinciding with the direction of the mountains. Also, for the mean date of the last frost, some "heat islands" are clearly established where frost dissipates about 4 weeks earlier than in other regions at the same latitude. Also shown by the mean date of the last frost, year-round frost-free conditions occur in a few parts of southernmost China only, while year-round permanent frost conditions occur in the main parts of the Qinghai-Xizang (Tibetan) Plateau, including the surrounding mountains as well as the Nan Shan and Tian Shan.

Resulting from the mean dates of first and last frost, the mean duration of a frost period and, vice versa, a frost-free period considerably varies over space. As far as the frost-free period is concerned (Fig. 4.14), its duration gradually increases from <100 days in northeastern and northwestern China to 350 days in southern China. Exceptional conditions are true for a few basins, such as the Sichuan Basin, which records >300 frost-free days, thus a 50- to 60-day longer duration than the adjoining regions at the same latitude. Of an even larger spatial extent, the Tarim Basin also represents a relatively large number of frost-free days (>150 days, for the Turpan depression even >200 days). In contrast to these favoured thermal conditions, the severest frost-stressed conditions, expressed by frost year-round, occur in the main parts of the Qinghai-Xizang (Tibetan) Plateau and in the high mountains.

The marked latitudinal gradient of the length, both of a frost period and a frost-free period, is demonstrated by a north to south cross-section which considers the mean dates of first and last frost (Fig. 4.15). Rather schematically, the length of the frost period decreases from north to south as the frost-free period increases. The frost period lasts from September 8 to July 7 at Mohe/1, while the corresponding dates were December 30 and January 26 at Guangzhou/199. Covering the remaining part of the year, the frost-free period decreases in its duration with increasing latitude. At Yulin, located on the southern coast of Hainan Island, year-round frost-free conditions were recorded, indicating that frost-free conditions only occur in the extreme south of China.

The significance of frost is not only apparent by considering mean figures, but also, and even more so, by referring to the recorded figures. Comparing mean and recorded figures for the first and last frost dates, together with the duration of the frost-free period, obvious differences can be observed (Table 4.3).

Comparing the data for the five stations under consideration, the north to south gradient of frost conditions is strongly underlined. Furthermore, a large variation between the mean and recorded extreme dates can be seen; obviously, the variation decreases with increasing latitude. It, therefore, can be assumed that the interannual variability of temperature may also decrease with increasing

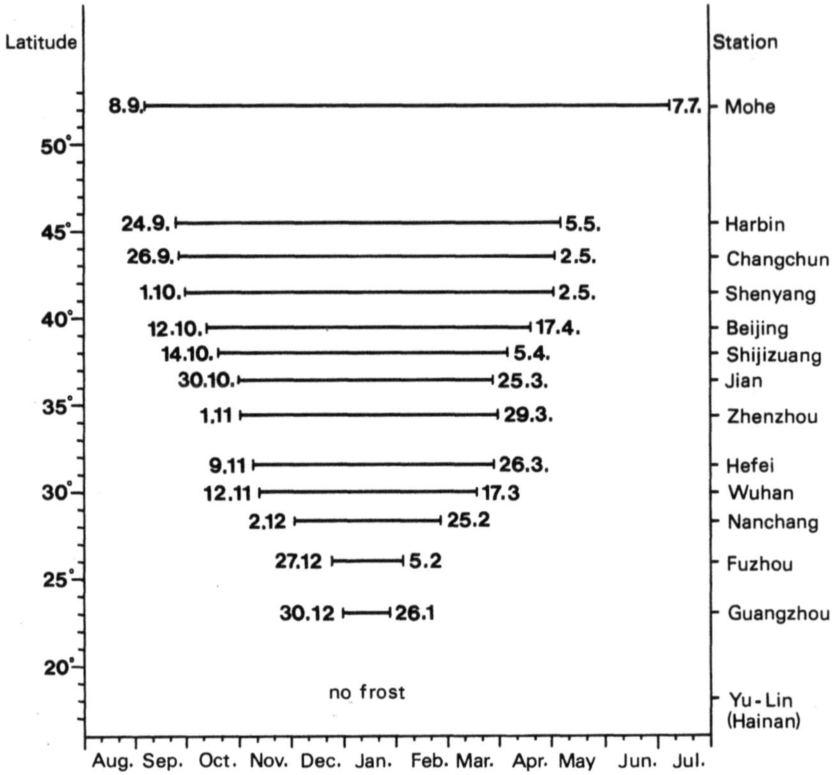

Fig. 4.15. Mean length of the frost period along a North to South cross section through eastern China

latitude (cf. Sect. 4.8). This is also shown by the different variation in the duration of the frost-free period, given both as a mean and recorded extreme figure.

The occurrence of frost must be regarded as a major climatic constraint. Reasons for the severity of frost are mainly the following: (1) the large extent of continentality, expressed by the distance from the sea; (2) the high altitude of a

Table 4.3. Frost observations at selected stations along a north to south cross-section[a]

	A	B	C	D	E	F
Mohe/1	Sept. 7	Aug. 11	June 5	June 28	93	63
Harbin/15	Sept. 29	Sept. 20	May 15	May 30	136	116
Beijing/61	Oct. 10	Oct. 4	Apr. 19	May 15	179	160
Shanghai/119	Nov. 21	Oct. 29	March 29	Apr. 16	236	213
Guangzhou/199	Jan. 9	Oct. 12	–	Feb. 12	353	304

[a] (A) Mean date of first frost; (B) date of recorded first frost; (C) mean date of last frost; (D) date of recorded latest frost; (E) mean duration of frost-free period (days); (F) duration of recorded shortest, frost-free period (days).

main part of China; (3) the unhindered passage of cold air masses, originating from Siberia, far south into China.

4.4.2 Mean Daily Air Temperature ≥10° C

Due to the strong effect of low winter temperatures, mean daily air temperatures ≥10° C occur delayed over space.

Mean dates of the onset of a mean daily air temperature ≥10° C, expressed in isochrones, show a progressive development from south to north, occurring between February 1 and June 1 (Fig. 4.16). In western China, isochrones show a rather disrupted course with a particular cellular development. This expresses the strong thermal impact, exerted by basins as "heat pockets", and mountains, as "cold pockets". This is clearly demonstrated, for example, by the Nan Shan which even shows the latest onset of a mean daily air temperature ≥10° C (July

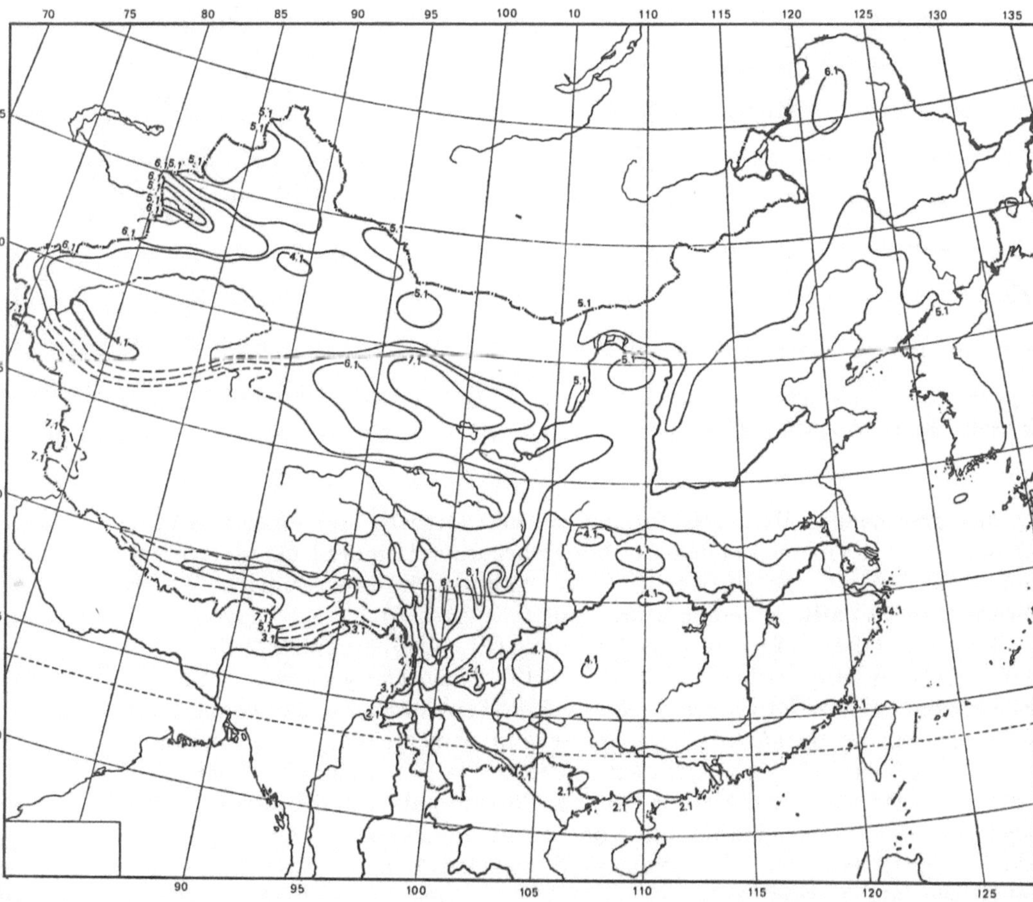

Fig. 4.16. Mean dates of the onset of a mean daily air temperature ≥10° C. (After ZHANG Jiacheng and LIN Zhi-quang 1985)

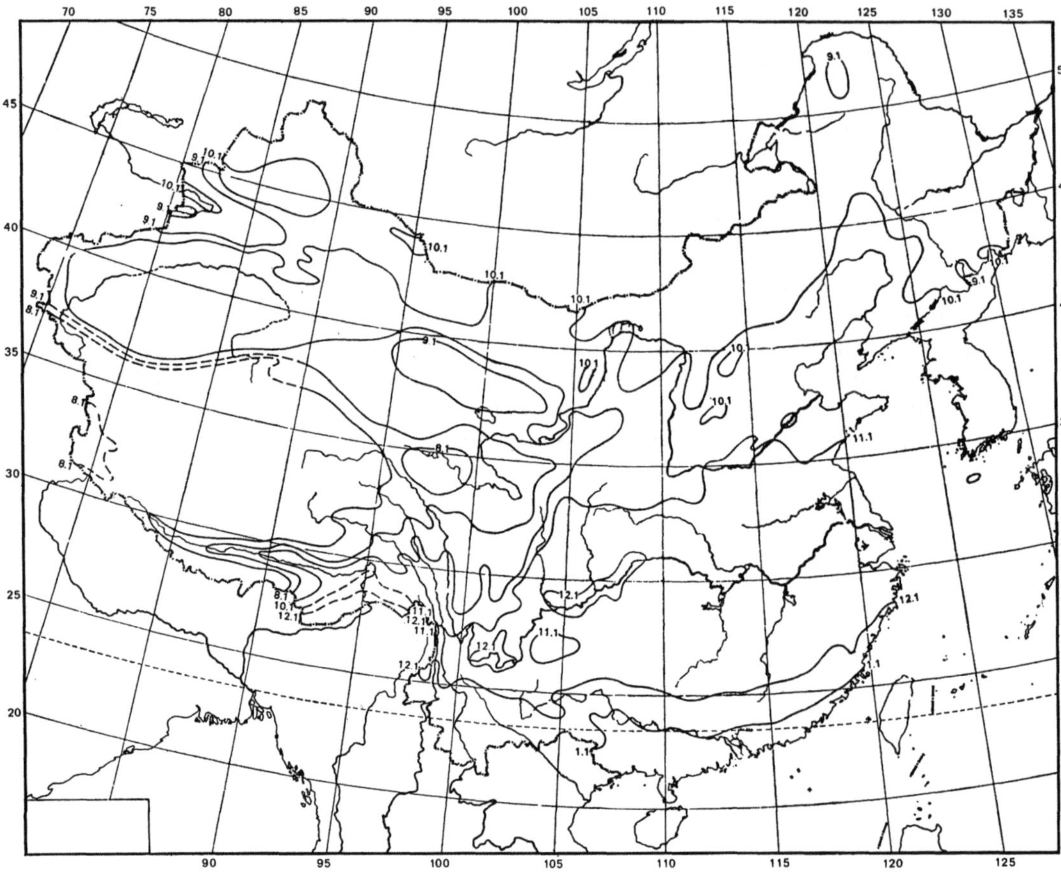

Fig. 4.17. Mean dates of the end of a mean daily air temperature ≥10° C. (After ZHANG Jiacheng and LIN Zhi-quang 1985)

1). The same date is also valid for the central, northern and western part of Qinghai-Xizang (Tibetan) Plateau, although accurate data are lacking. Expressing a "heat pocket", the Turpan Basin, for example, shows a comparably early onset of a mean daily air temperature ≥10° C (Fig. 4.16).

Opposite to the isochrones of the beginning (Fig. 4.16) are those of the end of a mean daily air temperature ≥10° C (Fig. 4.17). They clearly show a continuous development in a north to south direction, from September 1 to January 1. Even earlier than in northeastern China, the end of a mean daily air temperature ≥10° C takes place in most parts of the Tibetan Plateau where the isochrone of August 1 encloses nearly the whole Plateau. Corresponding to the distribution of the isochrones of the beginning, those of the end of a mean daily air temperature ≥10° C clearly show relative heat and cold pockets which are expressed by later and earlier dates respectively.

As a result of the considerably large spatial variation of the mean dates of both beginning and end of a mean daily air temperature ≥10° C, also the number

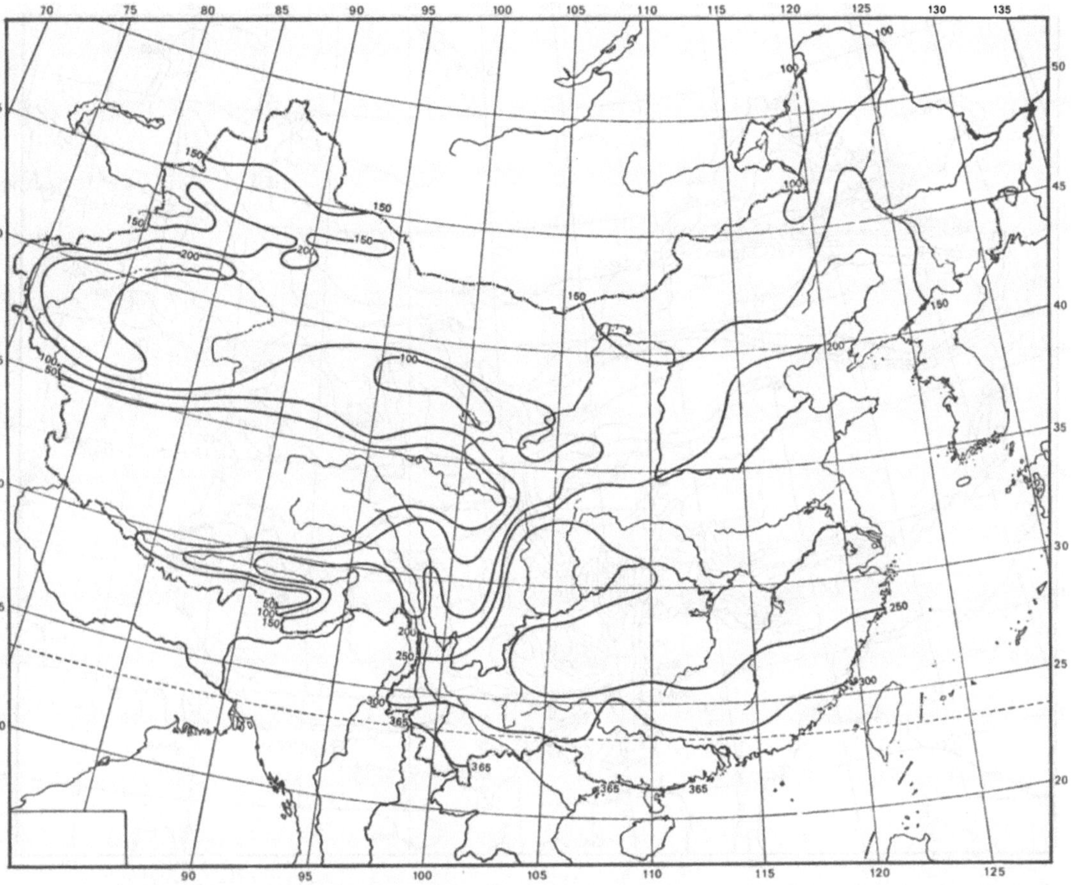

Fig. 4.18. Mean number of days with a mean daily air temperature ≥10° C. (After ZHANG Jiacheng and LIN Zhi-quang 1985)

of days with a mean daily air temperature ≥10° C shows extremely large spatial differences (Fig. 4.18). The corresponding value falls to <50 days, at a minimum, and rises to 365 days, at a maximum. The lowest figure is true for the main part of the Tibetan Plateau, while the maximum figure occurs in the southernmost parts of China. In eastern China, a continuous increase of the number of days with mean air temperature ≥10° C can be observed in a north to south direction, while in western China a rather complex distribution occurs. The reason for these observations are the effects of (1) latitude on radiation and temperature; (2) elevation upon temperature and (3) continentality.

The large spatial differences which have been shown for both a mean daily temperature ≤0° C and ≥10° C are also clearly expressed by the distribution of the mean daily *accumulated* air temperatures, referring to the annual total of all daily figures ≥10° C (Fig. 4.19). Underlining the tremendous spatial variation of temperature in China, the annual total of accumulated air temperatures varies between >9,000° and <500°. While the maximum figure occurs in the extreme

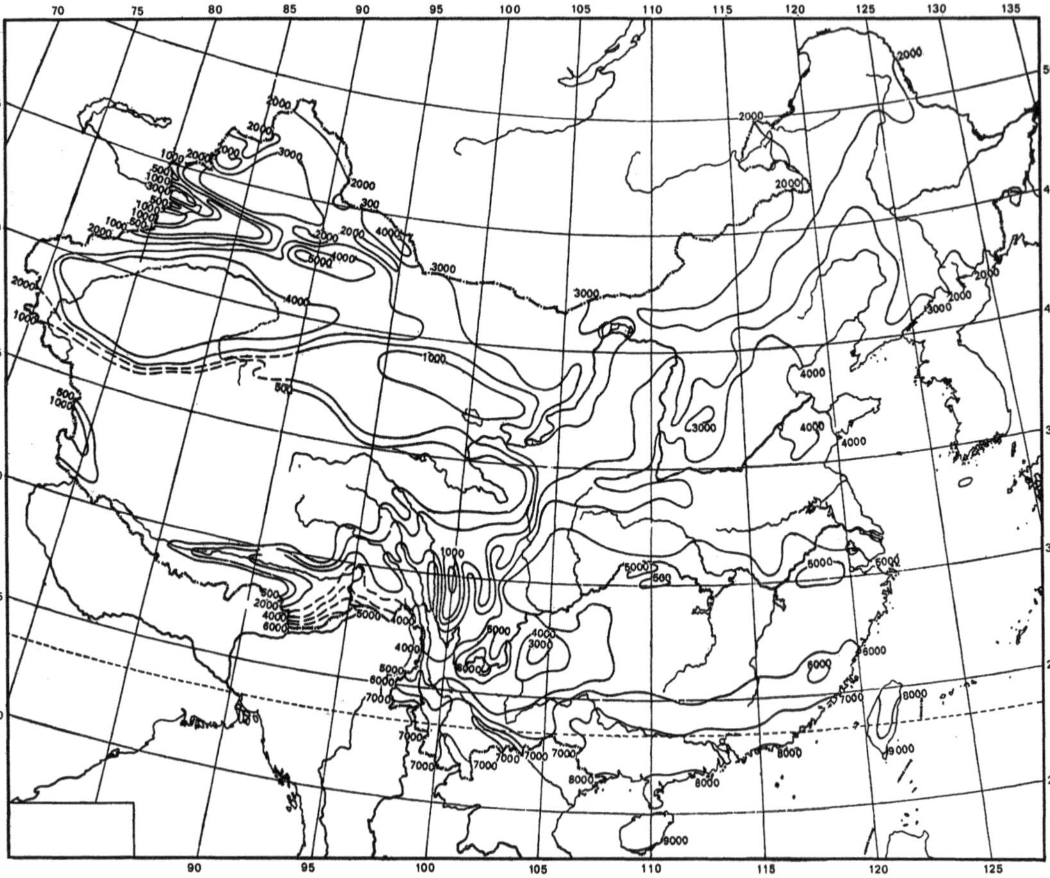

Fig. 4.19. Annual total of accumulated mean daily air temperatures with a mean temperature ≥10° C. (After ZHANG Jiacheng and LIN Zhi-quang 1985)

southernmost part of Hainan Island and in parts of Taiwan Island only, the minimum figure (<500°) occurs in the main parts of the Tibetan Plateau. However, it must be taken into consideration that the high mountains are excluded from this study due to missing data; logically, accumulated temperature in the glaciered and permafrost regions would fall to zero degrees. In total, the isolines of the annual total of accumulated air temperatures ≥10° C emphasize once more the complex and disrupted thermal conditions in China.

4.4.3 Maximum Daily Air Temperature ≥35° C

As a commonly used criterion, a maximum daily air temperature ≥35° C describes a "hot day".

As can be deduced from the long and cold winter regime in China, hot days logically cannot reach high values. The highest mean maximum figure amounted to 98.4 days $year^{-1}$ (at Turpan/267, mean value for 1951–1980), although this value represents an exceptional one, due to the location in a depression which causes a pronounced "heat island" effect. Although less marked, but valid for a larger area, the Tarim Basin as part of arid Xinjiang represents another heat island which is characterized by a larger number of hot days (> 30 $year^{-1}$). Other heat islands, of a smaller spatial area, are located in some inner parts of the South China Hills, in the Sichuan Basin and in northern Hainan Island as well as in parts of Guangxi. In contrast, the minimum number of hot days logically dropped to nil, as a result of the principally low temperatures valid for northeastern China, the Tibetan Plateau including the surrounding high mountains and other high mountains. In total, therefore, large parts in China show < 1 hot day, given as a long-term average value.

To further clarify the reasons for the spatial differences of hot days, their number per year was correlated with the geographical latitude (Table 4.4). Data for ten stations were used, based on the 30-year period 1951–1980. All reference stations were from eastern China.

Comparing the number of hot days with the latitude of the stations concerned, no definite relationship could be determined. Although at the stations with the highest latitudes (such as Mohe/1, Harbin/15, Changchun/27; Table 4.4), the lowest number of hot days was recorded, conversely the stations at lower latitudes did not record the highest values. Apparently, Guangzhou/199 even showed a rather low figure, but Nanning/215, located at the same latitude, a remarkably higher value. Most likely, factors other than the geographical latitude must also be taken into consideration, such as topo-climatological (-thermal) effects, caused by basins and depressions which represent heat cells, or by mountains which represent cells of coldness.

The number of hot days also seems to be controlled by air humidity; thus, a high humidity reduces temperature (as at Guangzhou/199 in comparison with Nanning/215, see Table 4.4), whereas low humidity leads to a rise in temperature and thus the number of hot days, for example at Xi'an/95: 24.3 hot days.

Table 4.4. Mean annual number of "hot days" at selected stations in eastern China; data for 1951–1980

	°N	No. of hot days
Mohe/1	53	0.5
Harbin/15	46	0.5
Changchun/27	44	0.5
Beijing/61	40	6.7
Shijiazhuang/70	38	16.2
Zhengzhou/90	35	20.8
Wuhan/126	31	21.0
Changsha/147	28	29.9
Guangzhou/199	23	5.2
Nanning/215	23	16.5

As a rule and resulting from the drop of temperature with increasing altitude, the high-altitude mountains and the Tibetan Plateau experience no hot days.

4.4.4 Other Extreme Limited Temperatures

To further characterize the distinct spatial and temporal variation of temperature in China, other extreme temperatures are commonly considered in China.

For a description of extreme *maximum* temperatures, daily maximum values of ≥40° C and ≥45° C were considered. Based on records for 1951–1980, some scattered locations in China recorded a few cases of a daily maximum temperature ≥40° C. The following regions are commonly regarded to be the hottest: (1) Henan Province; (2) the river valleys of the Loess Plateaux and the North China Plain; (3) the Tarim Basin and the Junggar lowlands. On a long-term average, the annual number of days with a maximum temperature ≥40° C drops to <1 case.

Of limited spatial occurrence only, Turpan/267 represents the largest number of days with a daily maximum temperature ≥40° C, thus establishing the hottest location in all of China. The reason for this is the strong topo-climatological heat island effect caused by a depression.

Expressed as monthly means from 1951–1980, Turpan recorded the following monthly mean number of days with a maximum temperature ≥40° C:

May	1.1	August	10.4	Annual total	37.5
June	9.7	September	0.5		
July	15.6	October-April	0.0		

The comparably northern location of the hottest regions in China can be explained by the low air humidity, associated with a well-established continentality, which more effectively influences the temperatures than a higher radiation supply in lower geographical latitudes. Evidence of this is also the rather northern position of the northernmost stations in China which so far have recorded a daily maximum temperature ≥40° C; they are Kelamayi (at 45° 36′ N) and Qiqihaer (at 47° 23′ N). For southern China, a daily maximum temperature ≥40° C has very rarely occurred so far.

Summer temperatures in China may occasionally rise to exceptionally high values which considerably exceed the mean temperature of the warmest month (Table 4.5). Observations from 14 stations clearly show a significant difference between the mean temperature of the warmest month, which is July in nearly all cases, and the recorded maximum temperature; observations refer to the period from 1930–1970.

The absolutely highest daily maximum temperatures, even ≥45° C, so far recorded in China were measured at the following stations:

Turpan/267:	49.6° C, on July 13, 1975
Xi'an/95:	45.2° C, on July 14, 1934
Yuncheng, Shanxi Province:	45.0° C, on July 21, 1942

Table 4.5. Mean temperature of the warmest month and recorded maximum temperature at selected stations, 1930–1970

Station	Warmest month		Recorded maximum temperatures	
	Time	Mean temperature (°)	(°)	(Date)
Nenjiang	July	20.4	38.1	20-07-1950
Dalian	Aug.	24.7	36.1	31-07-1939
Hohhot	July	21.6	38.0	06-1931 [a]
Beijing	July	26.0	42.6	15-06-1942
Zhengzhou	July	27.5	43.0	19-07-1966
Nanjing	July	28.2	43.0	13-07-1934
Haikou	July	28.4	40.5	02-05-1933
Yinchuan	July	23.5	39.3	08-07-1953
Anxi	July	25.3	43.0	13-07-1944
Turpan	July	33.0	47.6	24-07-1953 [b] 24-07-1956 [b]
Hotan	July	25.5	40.5	07-07-1953 12-07-1958
Changqing	July	28.6	42.2	19-08-1953
Guiyang	July	24.0	39.5	07-1930 [a]
Lhasa	June	15.5	30.5	25-06-1946

[a] Not precisely identified.
[b] Figures refer up to 1970 only.

A description of extreme *minimum* temperatures in China is commonly based on the following limited values for the daily minimum temperature: $\leq -10°$C, $\leq -20°$C, $\leq -30°$C and $\leq -40°$C. However, observations of extreme minimum temperatures must obviously be incomplete due to the missing data above 4,500 m (cf. Sect. 1.2). Based on the data available, observations of extreme minimum temperatures are most heavily concentrated at the highest possible geographical latitudes, and at high altitudes above sea level (Table 4.6).

As can be seen from the data, the extreme minimum temperatures may tremendously vary over space. The largest frequency and maximum severity were recorded at Mohe/1 which is the northernmost station. In contrast, the highest station with data available (Lhasa/239, 3,658 m a.s.l.) did not record the maximum frequency of extreme minimum temperatures, which may be due to the effect of an elevated heating surface in summer, exerted by the Tibetan Plateau. However, Lenghu/260 and Xining/256, both of which are high mountain stations at 2,733 and 2,261 m a.s.l. respectively, clearly expressed a greater frequency of extreme minimum temperatures, thus indicating the strong control of altitude on temperature.

The coldest areas in China are confined to the following areas: (1) northeasternmost China (Heilungkiang Province), (2) the northeastern part of Inner Mongolia, (3) northern Xinjiang, (4) the Qinghai-Xizang (Tibetan) Plateau including the surrounding high mountains.

Table 4.6. Mean number of days/year with extreme daily minimum temperatures ≤ 0°, ≤ −10°, ≤ −20°, ≤ −30° and ≤ −40° C. (Figures represent selected stations in eastern and western China, arranged in a north to south direction; for 1951−1980)

Station	°N Latitude	Altitude m a.s.l.	No. of days with a daily minimum temperature ≤ 0° C	≤ −10° C	≤ −20° C	≤ −30° C	≤ −40° C
Eastern China							
Mohe/1	53	296	237.7	175.2	141.5	93.5	25.1
Harbin/15	46	172	182.6	120.7	66.1	6.9	0.0
Changchun/27	44	237	174.8	111.1	45.3	1.6	0.0
Beijing/61	40	31	127.9	31.0	0.1	0.0	0.0
Shijiazhuang/70	38	82	113.0	17.9	0.2	0.0	0.0
Zhengzhou/90	35	110	92.8	3.3	0.0	0.0	0.0
Wuhan/126	31	23	43.8	0.7	0.0	0.0	0.0
Changsha/147	28	45	19.9	0.0	0.0	0.0	0.0
Guangzhou/199	23	6	0.0	0.0	0.0	0.0	0.0
Nanning/215	23	72	0.0	0.0	0.0	0.0	0.0
Western China							
Ürümqi/269	44	654	163.3	98.6	34.4	2.0	0.0
Turpan/267	43	35	124.6	59.0	4.6	0.0	0.0
Xining/256	37	2,261	169.1	79.8	3.0	0.0	0.0
Lenghu/260	39	2,733	232.2	148.7	53.4	1.2	0.0
Lhasa/239	30	3,658	169.8	36.7	0.0	0.0	0.0

The absolutely lowest daily minimum temperatures so far recorded, which even dropped to ≤ −50° C, were as follows:

Fuyun/Xinjiang, 47° N, 1,165 m a.s.l.: −51.5° C, on January 21, 1960;
Mohe/1, 53° N, 296 m a.s.l.: −51.2° C, on February 14, 1968;
Mianduhe, 49° N, 705 m a.s.l.: −50.1° C, on January 16, 1922.

4.5 Vertical Distribution of Temperature

Since large parts of China are occupied by mountains which even represent the highest elevations on earth, an extremely large vertical change of temperature must be logically expected. Its extent, however, cannot be given by precise records since climate stations were lacking >4,500 m a.s.l. Considering records from the mountain stations available, a significant drop of temperature could be shown (see Sect. 4.1), demonstrating negative annual means from about 4,000 m upwards. The lowest annual mean temperature recorded amounted to −4.1° C (Madoi/245, 4,272 m). With increasing altitude, temperatures would decrease further to a more serious extent.

As an approach towards a broad description of the vertical temperature distribution, the altitude and mean annual temperature were correlated for all climate stations >1,000 m a.s.l. (Fig. 4.20). The result is a relatively small gradient of temperature drop around 0.50° C 100 m^{-1} which is valid between 1,500 and 4,500 m a.s.l. Below 1,500 m a.s.l., no clear relationship between altitude and temperature can be seen; temperatures are widely scattered over a large range due

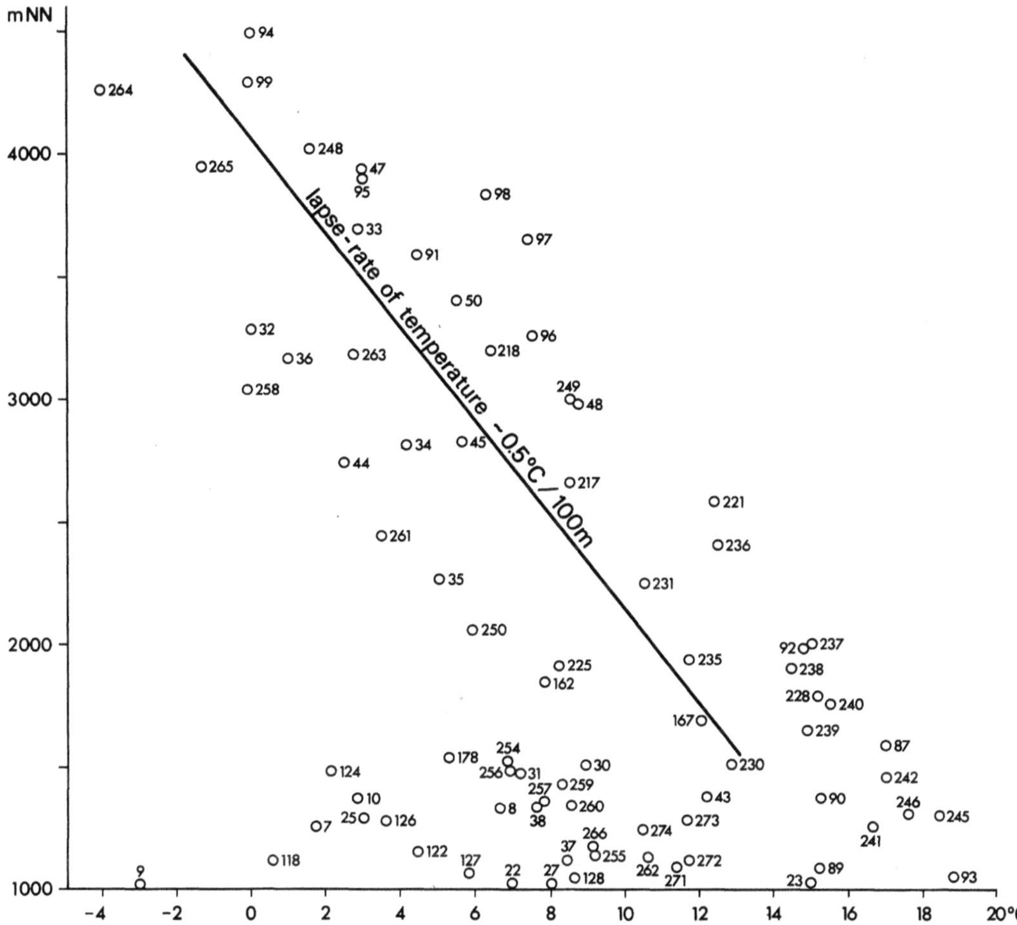

Fig. 4.20. Relationship between altitude and mean annual temperature considering all stations ≥1,000 m a.s.l.

to the different latitudes and locations of the stations not only in western (continental), but also eastern China.

The gradient of temperature drop is 0.15° C 100 m^{-1} smaller than the commonly accepted mean lapse rate of 0.65° C 100 m^{-1} which is valid for the free atmosphere and applicable on ground from sea level up to about 2,000 m a.s.l. The smaller vertical temperature gradient shown underlines the Tibetan Plateau as a large-scale, elevated heating surface and the effects of thermal continentality, which together lead to atypically high temperatures in the western parts of China, thus reducing the vertical temperature gradient.

Despite the rare observation data available, it was nevertheless attempted to arrive at a clearer picture of the vertical gradient of temperature decrease and its monthly variations, by correlating stations at the same latitude and correspondingly located in the central mountains between 90° and 105° E longitude. Accord-

Table 4.7. Monthly figures of vertical temperature gradients (°C 100 m^{-1}), computed for different vertical cross-sections

Cross-section[a]	J	F	M	A	M	J	J	A	S	O	N	D	Year
1 (a)	0.34	0.31	0.32	0.36	0.36	0.38	0.40	0.40	0.35	0.35	0.39	0.39	0.39
1 (b)	0.93	0.86	0.81	0.71	0.64	0.46	0.46	0.47	0.52	0.70	0.80	0.86	0.68
1 (c)	0.50	0.46	0.46	0.45	0.43	0.40	0.41	0.42	0.39	0.45	0.52	0.52	0.45
2	0.24	0.21	0.24	0.27	0.26	0.26	0.33	0.34	0.26	0.27	0.31	0.29	0.27
3	0.27	0.25	0.29	0.35	0.35	0.39	0.44	0.45	0.37	0.32	0.34	0.32	0.35

[a] 1. Chengdu–Nagqu, (a) Chengdu–Garze, (b) Garze–Nagqu, (c) Chengdu–Nagqu. 2. Chengdu–Lhasa. 3. Yibin–Juilong.

ingly, the mean monthly temperature gradients were computed along different vertical cross-sections:

1. Chengdu/134 (506 m a.s.l. – Garze/233 (3,394 m) – Nagqu/240 (4,507 m);
2. Chengdu (506 m) – Lhasa/239 (3,658 m);
3. Yibin/139 (341 m) – Juilong/137 (2,987 m), see Table 4.7.

As can be seen, the temperature gradients show in all cases remarkably low figures, the only exception being the distance from Garze to Nagqu in winter. Only for this mountain section under investigation could a marked seasonality of the vertical temperature gradient be shown. Thus, the mainly existing small temperature gradients valid throughout the year can be explained by relatively high temperatures in the mountains, due to the strong control of thermal continentality and effective heating effects in summer caused by the Tibetan Plateau. This is most clearly expressed by the extremely small gradients calculated for the profile from Chengdu to Lhasa (Table 4.7).

Taking the average annual gradient of temperature, calculated from the different figures of the five selected cross-sections, a value of 0.40° to 0.50°C 100 m^{-1} appears most realistic.

Applying this rate, the highest mountains of the Himalayas (>8,000 m a.s.l.) would record an annual mean temperature between −20° and −25° C, while the uppermost parts of the Tian Shan and Kunlun Shan would record around −20° C.

4.6 Comparison of Temperature at the Same Latitude

The thermal contrast between winter and summer in China as already pointed out (see Sect. 4.2) could also be demonstrated by a global comparison of temperature at the same latitude. Thus, global cross-sections of temperature were drawn at 40° N and 30° N for January, July and the annual mean (Figs. 4.21 and 4.22 respectively). For an objective comparison, sea-level temperatures (reduced by a gradient of 0.65° C 100 m^{-1}) were considered.

As can be seen for both latitudinal cross-sections, temperatures in China vary considerably between the western and eastern parts, showing mainly decreasing

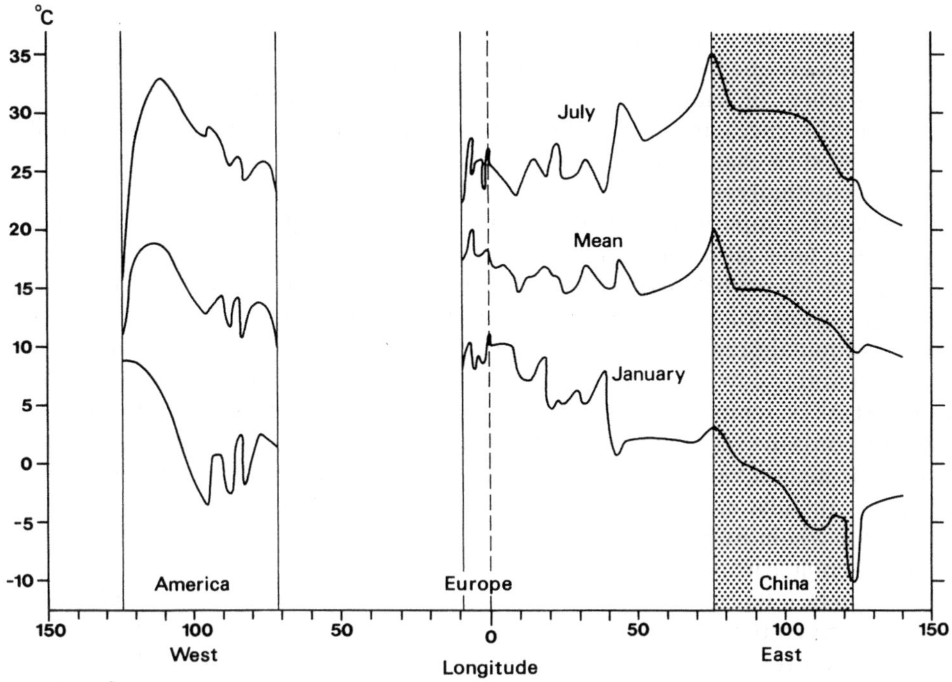

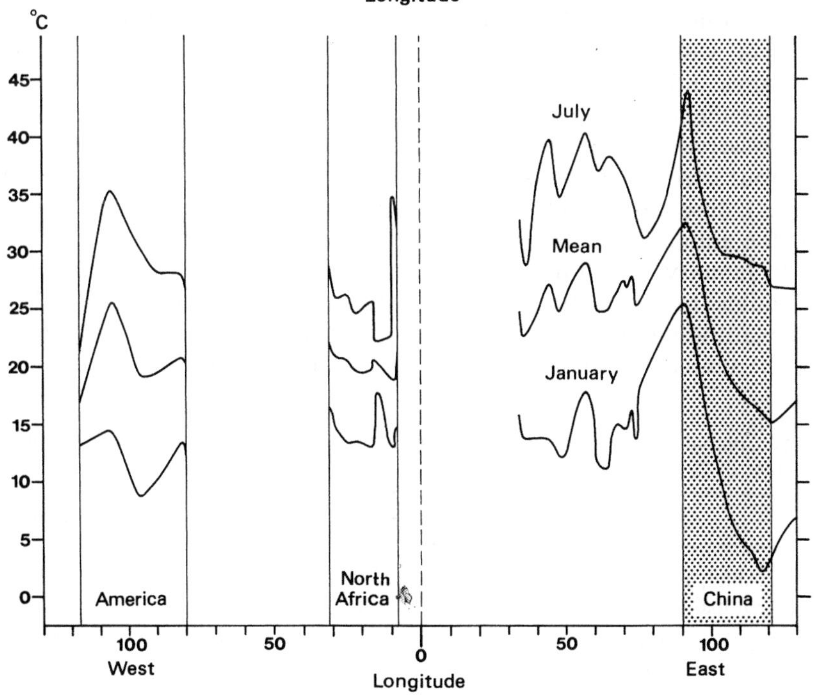

Fig. 4.21. Global cross-section of sea level temperatures at 40° N for January, July and the annual mean

Fig. 4.22. Global cross-section of sea-level temperatures at 30° N for January, July and the annual mean values from western to eastern China. This pattern is modified between January and July, on the one hand, and between 40° and 30° N, on the other. A strong west to east gradient of temperature occurs at 30° N, a weak one at 40° N. As far as seasonal differences are concerned, January shows at 40° N and 30° N a much stronger west to east gradient than July, while the annual mean logically shows intermediate figures.

A global comparison of temperatures at 40° and 30° N expresses a rather unique and unusual picture for China. At 40° N, January temperatures in China are the very lowest on earth, while July temperatures in most parts of China are the very highest for Eurasia as a whole and reach the same high level as for the Rocky Mountain region of North America. Mean annual temperatures in China are comparably low, as a consequence of the extremely low January temperatures. At 30° N, temperatures in China are extremely variable. In January, eastern China records the very lowest temperatures on a global scale, while western China shows the highest values, thus resulting in the most striking west to east gradient on earth. In July, eastern China records rather low temperatures which are the lowest for whole Eurasia, while western China experiences the highest values in a global comparison.

As can be inferred from the cross-sections of sea-level temperatures, the longitudinal temperature differences in China and in a global comparison appear to be strongly controlled by two factors: (1) the severe cooling effect and thus low temperatures in eastern China, and (2) the pronounced continentality and thus warming of western China. Both factors lead to either strikingly low or high temperatures and are thus responsible for the well-established west to east temperature gradients in China (Figs. 4.21 and 4.22).

This observation was also underlined by comparing the recorded temperatures at four selected stations in China with the global mean for the same latitude (Table 4.8); observations refer to the temperature means in January and July as well as the annual mean. Reference stations were selected at 20°, 30°, 40° and 50° N. In January, for all these latitudes, temperatures are extremely lower than the latitudinal mean. In July, however, temperatures in China are slightly higher than the latitudinal mean. Because of the extremely low winter temperatures in China, the annual mean temperature is also markedly below the latitudinal mean.

Table 4.8. Comparison of recorded temperatures in China with the global means of corresponding latitudes of 20°, 30°, 40° and 50° in January, July and the annual mean

		January	July	Year
50°:	Aihui, 166 m	−23.9	20.5	−0.5
	Global mean	−7.2	17.9	5.8
40°:	Beijing, 31 m	−4.7	26.0	11.6
	Global mean	5.5	24.0	14.1
30°:	Wuhan, 23 m	2.8	29.0	16.3
	Global mean	14.7	27.3	20.4
20°:	Haikou, 14 m	17.1	28.4	23.8
	Global mean	21.9	28.0	25.3

Table 4.9. Comparison of the annual range of temperature in China with the global figures at corresponding latitudes of 20°, 30°, 40° and 50° N

	50° N	40° N	30° N	20° N
China	Huma, 47.8	Beijing, 30.7	Wuhan, 26.2	Haikou, 11.3
Global latitudinal mean	26.0	20.0	13.0	6.5
Difference	21.8	10.7	13.2	4.8

As winter temperatures in China drop considerably below the latitudinal mean, and summer temperatures, however, rise slightly above it, the annual range of temperature in China significantly exceeds the latitudinal mean (Table 4.9). From the four reference stations in China, located at 20°, 30°, 40° and 50° N, it can also be deduced that the annual range of temperature in China differs most significantly from the latitudinal mean in North China (see stations Huma and Beijing), while it differs only slightly in South China (see station Haikou). This observation underlines a moderate annual temperature variation in southern China, which progressively increases to a very strong one in northeastern China.

4.7 Diurnal Range of Temperature

Expressing the range of temperature over 24 h of a day, the diurnal range of temperature can be given both as an absolute figure for a single day, and as a mean figure which expresses the difference between the daily mean maximum and mean minimum temperature, over a year or a month. The diurnal range of temperature is of basic climatic importance because it represents a climate characteristic for any climate region and it also expresses the common fluctuation of temperature through the day which helps to characterize the climate conditions of a location or region.

Table 4.10. Mean diurnal range of temperature at selected stations for different time intervals. (Data from M. J. Müller 1983)

	January	April	July	October	Year
Harbin	12.2	13.9	11.1	12.3	12.2
Beijing	10.7	13.3	9.3	11.9	11.2
Xian	9.4	11.7	8.3	8.4	9.5
Lanzhou	14.5	13.9	12.8	12.8	12.8
Shanghai	7.2	8.9	8.9	9.4	8.8
Chengdu	5.0	7.3	6.1	5.0	5.5
Guangzhou	7.8	6.1	7.2	8.9	7.8
Taibei	6.7	7.8	8.9	7.8	7.8
Hongkong	4.5	4.5	5.0	4.4	5.0
Frankfurt[a]	5.9	11.1	11.3	8.7	9.4

[a] Located in West Germany, 50° 07′ N, 08° 39′ E, 103 m a.s.l.

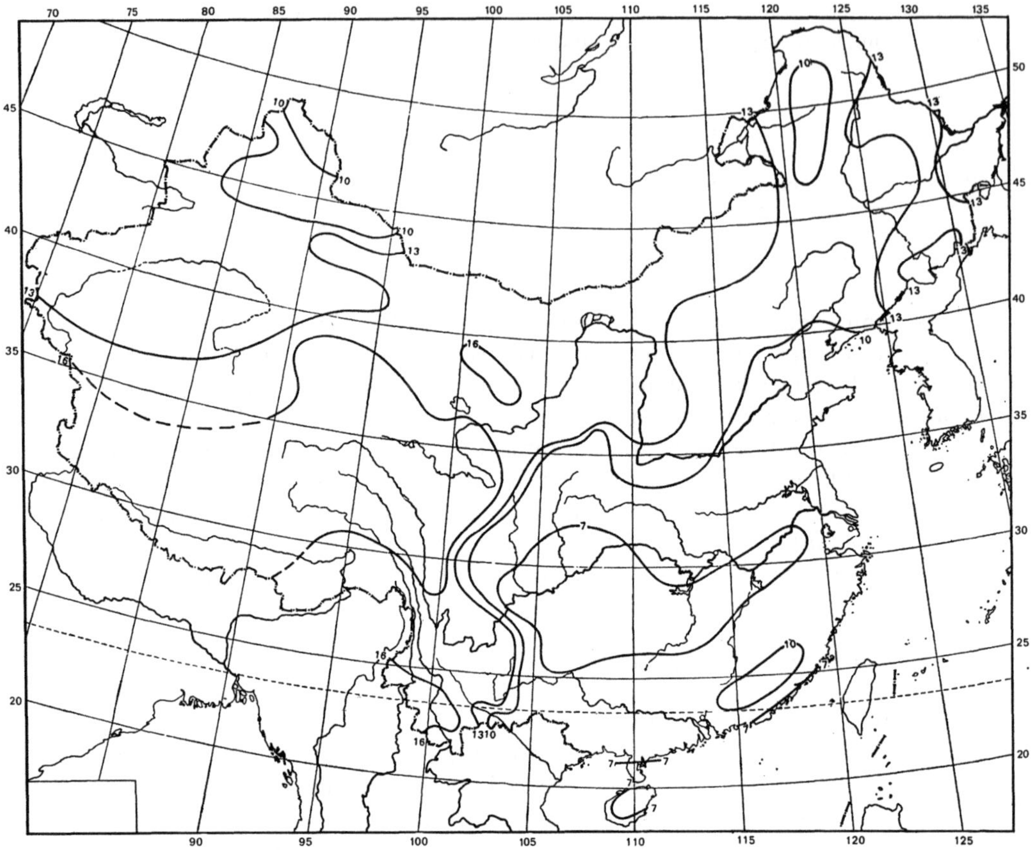

Fig. 4.23. Mean diurnal range of temperature in January. (After ZHANG Jiacheng and LIN Zhi-quang 1985)

Commonly used, the mean diurnal temperature range describes the usual range of temperature through a day. In order to show the diurnal temperature range in its spatial and temporal variations, figures of the mean 24-h range are given at nine selected stations for January, April, July and October, which, at the same time, can be regarded as the representative months of each season (Table 4.10). The corresponding figures at Frankfurt, FRG, were added for comparison with temperate mid-latitude conditions in western Europe.

As can be seen from the annual and monthly figures, remarkably large spatial differences of the diurnal temperature range are observed. The reason for this is the far-reaching effect of oceanity or continentality, in other words the proximity or distance to the China Seas. Oceanity places a moderating control on temperature, thus leading to a smaller diurnal range, for example, at Guangzhou, Taibei and Hongkong. These stations usually would expect much higher figures, according to the generally valid rule of a relatively large diurnal temperature range in the case of subtropical and tropical stations. Continentality, in contrast, causes a

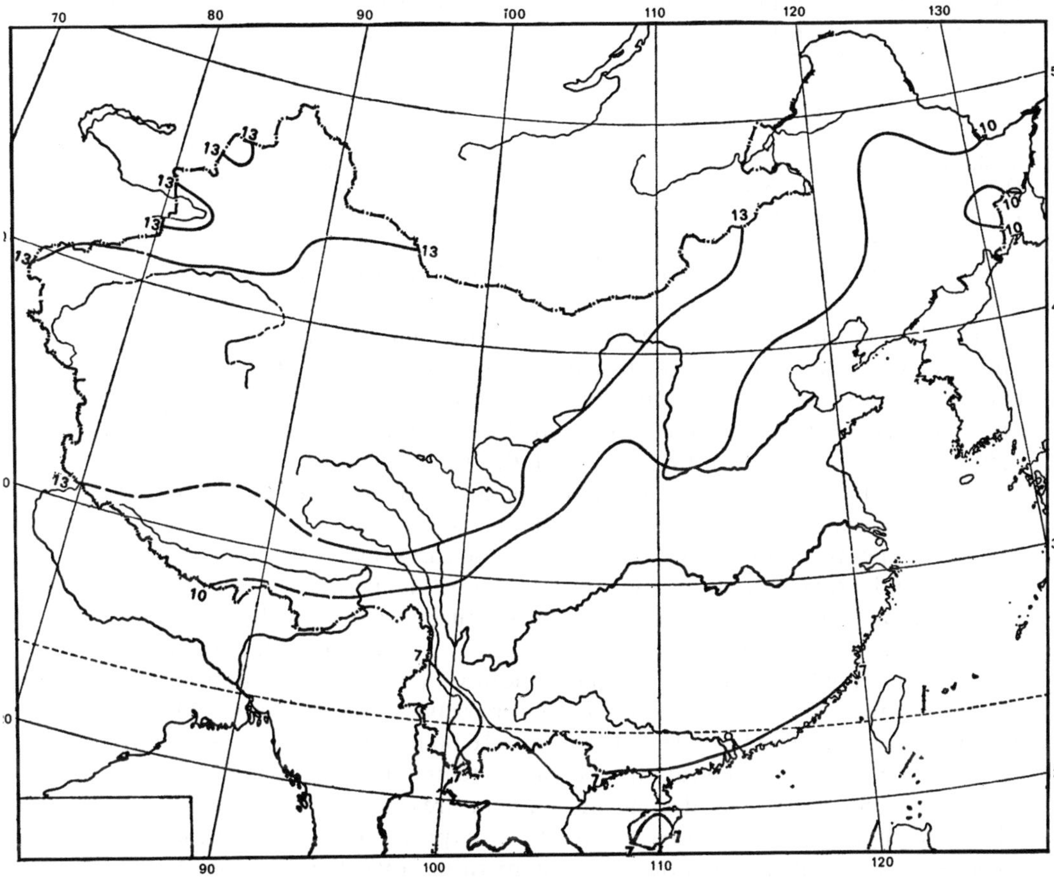

Fig. 4.24. Mean diurnal range of temperature in July. (After ZHANG Jiacheng and LIN Zhiquang 1985)

larger diurnal range, expressed e.g. at Harbin, Lanzhou, Xi'an and Beijing (Table 4.10). Intermediate figures are found at Shanghai, underlining the effective control of oceanity on the diurnal temperature range. An exceptionally small diurnal temperature range occurs at Chengdu which is mainly due to its location in the Sichuan Basin and the orographically induced stable and relatively warm air masses above it which also normally show great persistence. The figures for Frankfurt (Table 4.10) are of an intermediate order in comparison with China.

In each season, the well-established large spatial differences of the diurnal temperature range could also be shown by isolines for the mean diurnal temperature range. For this, January and July were selected as case studies (Figs. 4.23 and 4.24). In January, the isolines mostly run in a longitudinal direction, a fact which proves the far-reaching control exerted by oceanity and continentality on the diurnal range of temperature. In July, the distribution of isolines was less complex, but it still shows the effective control of oceanity and continentality.

Table 4.11. Mean diurnal temperature range at selected arid and high-altitude stations. (Data from M. J. MÜLLER 1983)

	January	April	July	October	Year
Lhasa	16.7	15.0	13.9	15.6	15.5
Kashgar	11.7	12.8	13.3	15.6	13.3
Ürümqi	11.1	13.4	13.4	10.6	11.1
Leh [a]	12.2	14.4	15.0	15.6	13.9

[a] Located in the Himalayas of India; at 34° 09′ N, 77° 34′ E; 3,506 m a.s.l.

Besides oceanity and continentality, two other factors, namely the altitude and aridity of a location, appear to control the diurnal range of temperature as can be shown by selected stations (Table 4.11). Kashgar and Ürümqi represent arid stations, while Lhasa and Leh[17] are high-altitude stations (in 3,658 and 3,506 m a.s.l.). Comparing the figures at the four reference stations with the figures given in Table 4.10, the diurnal temperature range increases with increasing aridity as well as with increasing altitude. In addition seasonal differences of the diurnal temperature range are negligible in these cases (cf. Tables 4.10 and 4.11).

As an individual case study, the diurnal range of temperature was investigated at Beijing by studying the mean 24-h temperature variation in January, April, July and October, using hourly figures (01.00, . . ., 24.00 h) for the observation period 1958–1964. For all four reference months, a distinct diurnal distribution of temperature is given, showing the minimum between 05.00 and 07.00 h, while the maximum occurred between 14.00 and 16.00 h. This diurnal temperature variation demonstrates the rapidly increasing temperatures in the morning and over noon, but the slowly decreasing temperatures in the late evening and at night. Beijing, therefore, can be regarded as a typical continental station which expresses an asymmetrical variation of temperature over the day (Fig. 4.25).

4.8 Interannual Variability of Temperature

Temperature variability expresses the variation of temperature over time for a given location or region. In statistical terms, variability is computed in relation to the annual or monthly means, by applying usually either the mean standard deviation or the coefficient of variability. Being more widely applied, the standard deviation is also taken in this chapter as an index for temperature variability, for both annual and monthly values. In both cases, observations were limited to five representative stations only, because of the meager climatic importance of temperature variability. The following stations selected refer to the four types of annual temperature variation in China (see Sect. 4.3): Haikou/207 (type 1 of the annual temperature variation); Hangzhou/123 and Anyang/82 (type 2); Nenjiang/4 and Ürümqi/269 (both type 4). Data of all five stations under investigation cover the 30-year period 1951–1980 (Table 4.12).

[17] Located in the Himalayas of India, but generally showing similar climate conditions as Lhasa.

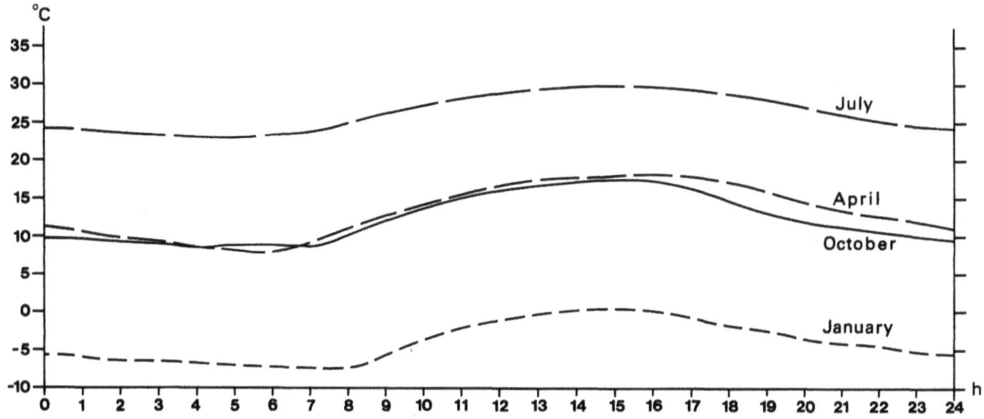

Fig. 4.25. Mean diurnal temperature variation per hour at Beijing, for January, April, July and October; 1958–1964

4.8.1 Variability of Annual Mean Temperature

The values of the annual standard deviation for the five stations under investigation (Table 4.12) are small and vary only slightly. They express a weak variability of the annual mean temperature, an observation which underlines rather equal annual temperatures from year to year.

The nearly constant annual mean temperatures are also shown by the range of temperature between the absolutely highest and lowest annual mean temperature for the 30-year observation period, given for each station under investigation (values in °C):

Station	Highest annual temperature	Lowest	30-year range of temperature
Haikou/207	24.6	23.2	1.4
Hangzhou/123	17.2	15.6	1.6
Anyang/82	14.3	12.5	1.8
Nenjiang/4	1.6	−2.5	4.1
Ürümqi/269	8.6	4.2	4.4

Table 4.12. Mean annual and monthly standard deviation at selected stations, in °C

	J	F	M	A	M	J	J	A	S	O	N	D	Year
Haikou/207	1.5	1.9	1.7	1.0	0.7	0.6	0.5	0.4	0.4	0.7	0.9	1.3	0.4
Hangzhou/123	1.4	1.8	1.3	1.2	0.9	1.0	1.2	1.1	1.2	1.0	1.4	1.6	0.4
Anyang/82	1.4	2.1	1.6	1.2	1.1	1.3	0.8	0.9	1.1	1.3	1.3	1.7	0.5
Nenjiang/4	2.6	3.0	2.8	1.7	1.4	1.0	0.9	0.9	0.9	1.1	2.6	3.1	0.9
Ürümqi/269	2.4	3.2	3.3	2.2	2.4	1.9	1.7	1.4	1.5	1.5	3.2	3.1	1.1

4.8.2 Variability of Monthly Mean Temperatures

The values of standard deviation expressing the monthly mean temperature variability are rather small, but nevertheless show for all stations moderate seasonal differences (Table 4.12), with the only exception of Hangzhou, which records a fairly even standard deviation year-round. In all other cases, the mean standard deviation is smallest during summer and largest during winter. Thus, year to year temperatures differ only slightly in summer, but more evidently during winter. In this sense, summer and winter overlap during the transitional seasons spring and autumn respectively.

To further demonstrate the seasonal differences in temperature variability, for Haikou/207 and Ürümqi/269 all the individual annual temperature graphs over the 30-year observation period 1951–1980 were plotted (Fig. 4.26). The graphs of each station underline the main valid type of annual temperature variation (types No. 1 and 4 respectively; see Sect. 4.3) each of which is characterized by a marked seasonal contrast of annual temperature variation. In addition, a rather uniform course of the 30 annual curves at both reference stations emphasizes a fair homogeneity of each type of annual temperature variation.

For reference purposes and individual interpretation, the monthly and annual temperature means at the five stations under investigation for the period 1951–1980 are given in Table 4.13. (continued p. 130)

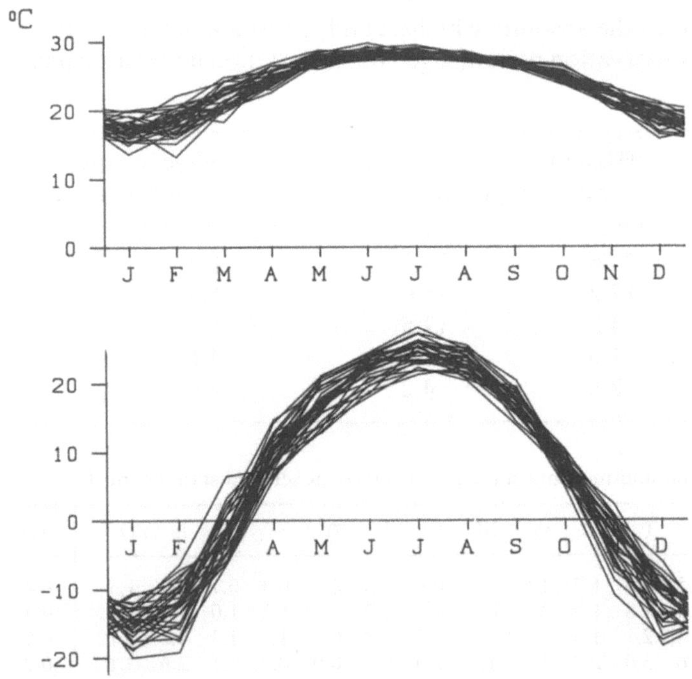

Fig. 4.26. Graphs of annual temperature variation at Haikou/207 and Ürümqi/269, 1951–1980

Table 4.13. Monthly and annual temperature means from 1951 to 1980 at (a) Haikou/207, (b) Hangzhou/123, (c) Anyang/82, (d) Nenjiang/4 and (e) Ürümqi/269

(a) Haikou/207

Year	J	F	M	A	M	J	J	A	S	O	N	D	Mean
1951	16.3	18.9	19.7	24.6	26.4	27.4	28.3	27.8	27.2	25.3	22.2	18.9	23.6
1952	18.4	20.3	24.8	25.7	28.6	28.8	28.9	27.6	26.8	25.3	22.7	16.8	24.6
1953	17.7	18.4	23.3	23.0	27.4	28.4	29.2	28.4	27.0	26.3	22.8	18.8	24.2
1954	20.0	19.3	18.3	25.7	28.1	29.1	29.5	28.2	27.0	24.5	22.4	17.5	24.1
1955	14.9	19.6	22.9	24.5	28.0	28.6	28.5	27.9	27.6	24.9	20.9	19.0	23.9
1956	16.0	19.4	22.3	25.9	27.7	28.5	28.6	27.7	27.3	25.0	20.6	16.7	23.9
1957	17.7	16.0	20.2	24.9	27.1	27.8	28.3	27.9	26.9	24.1	22.3	20.0	23.6
1958	17.3	16.6	23.1	25.8	28.2	27.5	28.1	27.4	26.9	24.1	20.9	18.7	23.7
1959	17.4	19.5	20.3	25.2	26.9	29.3	28.1	27.4	26.7	24.2	22.6	20.2	24.0
1960	17.4	18.6	24.3	24.1	26.8	27.9	27.9	27.7	26.6	24.4	22.6	18.2	23.9
1961	16.6	17.9	23.4	25.2	26.5	28.0	28.7	27.9	26.7	24.9	22.7	20.1	24.0
1962	15.3	17.5	20.3	23.7	27.7	27.8	28.5	27.8	26.7	24.6	21.3	17.8	23.3
1963	13.6	17.1	20.2	24.4	27.9	27.3	27.3	27.5	27.3	24.9	23.5	18.7	23.3
1964	18.4	16.4	21.0	26.5	27.0	27.8	27.8	27.4	26.8	25.6	20.7	17.7	23.6
1965	16.7	20.4	21.2	25.2	26.8	27.2	27.7	27.4	26.3	25.2	22.5	19.0	23.8
1966	19.8	20.3	22.9	26.5	26.0	27.9	27.5	27.5	25.9	25.3	22.3	20.7	24.4
1967	15.5	16.0	21.5	24.2	28.2	27.7	29.0	27.5	26.1	23.6	22.4	16.4	23.2
1968	17.6	13.2	20.2	22.6	26.9	27.4	28.9	27.9	26.4	24.2	23.2	21.0	23.2
1969	19.2	17.7	20.4	24.0	28.1	27.9	28.4	27.9	27.4	25.2	21.2	17.8	23.8
1970	16.4	20.0	20.3	23.8	27.5	28.4	28.6	28.0	27.0	24.2	21.2	19.4	23.7
1971	15.9	17.6	20.4	25.2	26.4	27.8	27.8	27.0	27.5	23.7	20.6	18.3	23.2
1972	16.5	18.3	20.8	24.0	27.1	28.1	28.1	27.0	27.0	26.6	22.3	19.8	23.8
1973	18.0	22.1	24.2	26.8	28.8	28.3	28.1	27.0	26.4	24.4	21.2	17.6	24.4
1974	17.0	16.6	19.7	23.5	27.3	27.9	28.6	27.5	27.1	24.5	20.7	18.3	23.2
1975	17.3	19.1	22.9	25.8	28.1	27.8	27.9	27.2	27.2	24.6	20.9	15.7	23.7
1976	16.3	19.5	20.5	24.8	27.0	27.7	27.5	27.0	26.8	25.2	20.1	18.6	23.4
1977	15.4	15.1	20.4	25.5	28.4	29.8	28.5	28.6	26.6	25.5	22.0	20.3	23.9
1978	18.0	18.1	22.2	25.0	26.8	28.1	28.6	27.6	26.6	24.2	21.9	19.8	23.9
1979	20.0	20.8	22.5	24.4	27.0	27.8	28.9	27.8	26.5	23.9	21.1	19.6	24.2
1980	19.1	16.7	24.2	25.8	27.1	28.3	28.3	28.4	26.8	25.8	22.6	19.9	24.4

Table 4.13 continued

(b) Hangzhou/123

Year	J	F	M	A	M	J	J	A	S	O	N	D	Mean
1951	3.5	5.3	8.2	13.2	21.4	23.8	27.2	28.8	22.4	19.0	11.5	8.2	16.0
1952	5.2	3.6	7.9	16.1	20.5	25.1	27.7	27.8	24.3	17.5	14.4	4.3	16.2
1953	4.0	6.4	9.9	14.4	21.7	25.8	31.1	30.3	22.9	19.2	12.8	7.8	17.2
1954	5.2	5.6	8.0	14.7	19.9	23.1	27.1	28.8	24.2	17.8	14.0	4.1	16.0
1955	1.2	6.7	8.7	15.2	20.4	24.6	27.9	26.9	25.0	15.7	9.9	7.8	15.8
1956	3.4	5.2	8.3	15.8	18.6	25.7	29.4	27.3	23.1	16.6	9.3	5.0	15.6
1957	4.5	2.1	8.4	15.0	19.3	23.8	29.4	27.2	20.7	16.5	12.9	7.4	15.6
1958	2.9	4.9	10.2	16.7	19.2	25.0	29.2	27.0	22.9	15.4	10.5	6.9	15.9
1959	2.4	5.6	11.1	15.2	19.3	24.3	29.1	28.6	23.0	17.6	11.6	6.8	16.2
1960	4.0	7.1	11.8	14.6	19.5	24.8	29.8	26.8	24.5	17.6	12.0	4.7	16.4
1961	3.7	5.3	9.8	16.1	20.2	25.4	30.2	29.0	24.4	18.2	13.6	6.6	16.9
1962	2.4	6.3	8.9	13.5	19.5	23.9	28.8	28.1	23.9	17.3	11.4	6.5	15.9
1963	2.1	4.2	10.6	12.9	22.0	23.7	29.2	28.3	23.9	16.4	13.0	6.8	16.1
1964	5.2	1.7	9.6	18.1	19.9	23.7	30.3	28.8	25.5	18.3	11.3	5.9	16.5
1965	5.2	6.4	8.3	13.2	21.0	23.1	29.1	27.0	22.1	17.8	11.7	5.1	15.8
1966	4.7	7.2	11.3	15.0	20.0	24.4	28.2	28.5	21.5	17.9	12.3	5.3	16.4
1967	2.5	4.6	9.8	14.7	20.9	25.0	28.7	30.1	23.5	17.8	12.2	1.7	16.0
1968	3.3	2.5	9.8	14.7	20.3	23.7	27.3	27.8	23.7	16.6	13.4	9.5	16.1
1969	3.7	2.0	8.2	16.2	21.8	23.2	27.6	28.0	24.9	17.3	10.1	5.0	15.7
1970	2.2	6.7	6.0	14.3	20.5	22.9	27.6	28.4	24.2	17.9	11.3	6.7	15.7
1971	3.7	4.7	8.3	15.3	20.4	26.1	30.7	29.1	22.9	16.0	12.5	6.7	16.4
1972	5.0	2.2	10.0	14.9	19.9	24.3	27.4	27.4	22.8	17.9	12.6	6.7	15.9
1973	4.7	6.9	10.9	17.3	19.2	23.2	28.2	28.5	21.5	17.8	12.3	5.9	16.4
1974	3.5	4.4	9.3	17.0	21.3	23.4	26.9	27.2	22.9	17.8	13.0	5.8	16.0
1975	5.7	6.3	9.5	15.4	19.1	23.7	28.0	28.3	26.0	19.0	12.1	5.0	16.5
1976	4.4	7.1	8.4	14.9	20.6	22.5	26.9	29.3	22.1	18.1	8.8	6.8	15.9
1977	0.0	4.0	10.8	16.9	19.2	23.6	28.9	26.7	23.8	19.6	12.2	8.9	16.2
1978	5.0	5.4	8.5	16.2	20.3	24.8	30.0	29.2	23.3	18.1	12.4	7.2	16.7
1979	5.8	8.0	9.7	15.0	19.3	25.6	29.5	27.0	23.0	18.7	11.0	8.2	16.7
1980	4.4	3.8	8.1	14.6	20.0	25.5	27.9	24.8	21.7	18.1	13.9	6.0	15.7

Table 4.13 continued

(c) Anyang/82

Year	J	F	M	A	M	J	J	A	S	O	N	D	Mean
1951	-5.5	-0.7	7.0	14.0	21.2	24.0	25.5	26.2	19.4	15.7	6.5	2.5	13.1
1952	-0.4	-2.0	6.2	15.0	20.8	26.7	26.9	24.1	21.3	14.6	5.7	-1.9	13.1
1953	-2.4	-0.1	8.9	15.0	20.9	24.3	27.5	24.0	21.5	16.2	6.5	0.3	13.7
1954	-0.7	1.2	6.4	14.9	20.7	24.4	25.6	24.5	20.7	14.0	8.4	-4.6	13.0
1955	-3.4	3.0	4.2	14.9	22.4	28.9	27.6	25.8	21.2	12.6	5.7	1.3	13.7
1956	-3.0	0.4	4.8	15.2	19.1	22.9	27.7	24.3	21.9	13.7	4.5	-1.9	12.5
1957	-3.1	-2.0	5.1	14.7	21.4	23.6	25.7	26.1	20.1	15.0	8.3	0.7	13.0
1958	-0.7	1.3	6.9	14.6	19.9	26.4	27.0	23.8	21.5	12.9	6.0	2.5	13.5
1959	-1.2	2.5	9.0	15.5	22.2	25.3	28.1	26.9	20.8	15.9	4.8	-0.1	14.0
1960	-1.2	4.6	7.8	15.1	19.6	27.0	26.3	25.0	21.2	13.5	6.2	-1.0	13.7
1961	-0.9	4.0	9.2	16.8	22.7	26.2	28.0	26.4	20.4	13.9	8.2	0.6	14.6
1962	-1.8	2.4	7.6	13.9	22.1	25.4	27.6	25.1	22.9	13.6	5.5	1.8	13.7
1963	-1.7	1.0	8.2	12.8	18.3	25.9	27.0	25.5	20.4	14.5	7.8	0.5	13.4
1964	-1.8	-4.2	6.6	12.9	19.9	26.3	26.2	25.6	19.7	14.2	7.4	1.3	12.8
1965	-0.7	2.1	8.2	12.4	21.8	26.5	27.8	26.5	22.0	16.3	7.9	1.0	14.3
1966	-0.7	3.9	7.3	15.0	21.1	26.1	27.1	26.9	19.7	15.0	7.5	-1.8	13.9
1967	-1.6	0.5	8.5	15.0	23.8	27.0	27.5	26.7	19.3	15.1	4.8	-3.0	13.6
1968	-1.4	-1.6	9.1	16.2	21.7	27.1	27.3	24.9	20.7	13.1	7.8	2.1	13.9
1969	-3.8	-2.7	5.5	13.9	21.2	26.0	26.9	25.9	20.1	14.9	5.9	-0.5	12.8
1970	-2.3	2.1	5.3	14.8	20.7	24.6	27.2	25.0	24.6	15.2	5.9	0.3	13.6
1971	-1.0	0.1	6.0	14.9	22.2	24.0	28.2	24.9	20.9	14.5	8.4	-1.5	13.5
1972	-3.2	-1.3	7.9	15.4	20.8	27.3	26.2	24.8	20.0	15.1	6.7	1.2	13.4
1973	-1.0	1.7	9.5	16.1	20.2	24.3	25.6	26.5	20.0	13.1	8.0	1.3	13.8
1974	-1.6	1.5	7.2	16.2	21.6	25.6	27.1	24.7	20.9	13.3	7.4	-0.3	13.6
1975	-0.7	2.2	9.7	15.4	21.2	26.5	26.1	26.4	21.8	14.5	7.6	-0.4	14.2
1976	-0.3	1.9	7.0	13.2	21.4	25.3	25.6	23.7	20.6	14.4	4.8	0.6	13.2
1977	-4.0	2.2	10.7	16.5	19.8	26.2	26.6	25.2	21.2	16.1	6.9	2.6	14.2
1978	0.3	1.1	7.7	16.9	21.6	27.0	26.9	26.0	21.0	13.9	7.0	1.1	14.2
1979	0.0	3.2	7.9	13.1	20.5	26.3	26.2	25.5	19.6	16.2	5.9	1.8	13.9
1980	-1.1	0.8	6.1	13.9	21.1	25.9	26.5	24.6	20.6	18.8	9.1	-0.1	13.6

Table 4.13 continued

(d) Nenjiang/4

Year	J	F	M	A	M	J	J	A	S	O	N	D	Mean
1951	-28.6	-21.1	-10.5	3.0	13.3	17.5	20.3	19.0	10.5	2.1	-7.6	-18.1	0.0
1952	-27.5	-26.5	-9.8	3.4	10.5	18.0	20.2	19.5	10.9	1.8	-13.6	-23.7	-1.4
1953	-26.3	-19.8	-7.8	3.4	10.5	19.6	20.7	17.9	10.9	2.8	-13.5	-24.3	-0.4
1954	-29.0	-20.1	-12.0	1.9	8.9	17.0	21.5	19.5	11.2	0.6	-11.9	-24.2	-1.4
1955	-22.0	-25.5	-15.0	0.2	11.7	17.4	20.9	20.1	11.7	1.4	-9.5	-15.9	-0.4
1956	-28.5	-25.7	-12.0	0.8	10.5	17.9	19.3	18.9	10.8	1.4	-15.6	-27.3	-2.5
1957	-24.8	-24.2	-16.4	2.9	11.9	17.4	19.2	16.9	9.5	1.4	-8.2	-19.8	-1.2
1958	-25.3	-20.3	-11.2	0.9	11.4	17.9	21.2	17.6	11.6	1.9	-9.0	-18.1	-0.1
1959	-25.0	-16.5	-3.7	3.8	11.7	18.0	19.0	19.3	14.0	3.2	-12.1	-20.1	1.0
1960	-27.6	-16.5	-9.6	1.3	9.4	17.3	20.9	19.0	11.8	1.8	-12.9	-25.9	-0.9
1961	-28.1	-22.0	-8.7	3.5	11.0	17.5	19.8	18.7	12.3	1.5	-7.7	-19.9	-0.2
1962	-23.7	-19.1	-10.4	3.3	11.6	17.6	20.5	18.0	11.7	1.0	-12.7	-19.5	-0.1
1963	-21.6	-17.2	-6.3	3.0	11.3	17.7	20.9	18.8	11.3	1.5	-11.4	-21.1	0.6
1964	-23.9	-23.9	-8.2	2.9	12.7	16.6	19.5	18.7	10.0	1.1	-12.5	-23.6	-0.9
1965	-27.4	-24.3	-12.0	1.7	10.8	18.7	19.6	17.3	11.0	3.0	-14.0	-26.7	-1.9
1966	-26.2	-23.6	-13.1	0.6	13.1	18.4	19.9	18.2	10.9	4.5	-14.5	-28.7	-1.7
1967	-23.7	-19.6	-8.8	5.3	13.4	17.9	21.4	18.2	10.9	3.8	-12.3	-20.0	0.5
1968	-28.8	-16.8	-11.4	6.8	13.0	18.6	22.2	18.3	10.3	0.5	-9.8	-21.9	0.1
1969	-28.1	-26.3	-10.1	2.4	9.8	17.6	20.8	18.0	11.7	2.1	-11.8	-25.0	-1.6
1970	-26.5	-20.4	-12.4	4.1	12.7	20.0	20.9	19.2	11.1	1.2	-10.0	-22.6	-0.2
1971	-25.3	-19.5	-10.9	2.6	12.0	19.6	20.0	17.9	11.9	3.6	-7.3	-18.2	0.5
1972	-22.9	-20.3	-5.3	4.4	10.9	16.2	19.7	16.4	10.7	1.0	-13.2	-22.2	-0.4
1973	-24.0	-18.6	-9.7	1.1	11.1	19.5	22.0	17.5	13.5	-0.1	-12.6	-22.0	-0.2
1974	-26.2	-23.2	-9.3	2.9	10.7	17.4	22.8	18.9	12.2	0.6	-12.3	-23.6	-0.8
1975	-21.1	-20.2	-5.2	6.7	13.4	19.1	19.9	18.6	11.7	2.5	-7.6	-18.5	1.6
1976	-20.0	-16.6	-7.8	2.9	11.5	16.6	21.3	16.4	11.7	0.7	-14.4	-24.6	-0.2
1977	-29.9	-22.5	-8.0	2.3	14.5	18.2	21.3	18.2	11.2	2.4	-14.3	-24.8	-1.0
1978	-27.4	-22.8	-9.9	4.1	11.6	19.5	19.4	18.5	11.7	0.2	-6.8	-20.4	-0.2
1979	-24.6	-19.2	-7.5	2.1	14.2	18.5	20.4	17.6	10.8	1.2	-14.8	-19.4	-0.1
1980	-26.8	-19.3	-9.6	0.4	11.2	19.3	21.1	19.9	10.3	0.1	-8.8	-22.1	-0.4

Table 4.13 continued

(e) Ürümqi/269

Year	J	F	M	A	M	J	J	A	S	O	N	D	Mean
1951	-17.6	-17.2	-6.2	10.2	19.8	22.0	24.6	22.1	17.4	8.7	-4.4	-8.5	5.9
1952	-14.2	-14.9	-4.8	8.4	15.7	19.7	23.3	24.0	16.7	4.7	-10.9	-17.6	4.2
1953	-15.8	-10.0	-2.0	9.1	18.0	22.6	24.7	20.8	19.5	6.4	-7.6	-12.5	6.1
1954	-14.8	-13.5	-6.0	8.9	12.9	18.8	21.3	22.4	15.8	7.4	-5.6	-18.4	4.1
1955	-17.7	-12.5	-6.2	6.1	16.6	23.5	23.1	23.7	15.9	6.0	-1.9	-8.2	5.7
1956	-18.9	-13.4	-4.5	10.3	15.9	18.5	25.5	21.3	15.4	7.2	-1.8	-12.9	5.2
1957	-16.2	-15.1	-1.0	6.5	13.6	23.5	23.4	21.6	16.7	7.6	-2.1	-8.2	5.9
1958	-15.0	-8.3	-0.9	8.9	14.0	18.1	22.3	20.4	13.9	6.8	-3.8	-9.5	5.6
1959	-15.4	-11.7	-5.7	10.9	13.8	20.9	21.9	21.3	19.0	8.0	-6.3	-15.0	5.1
1960	-14.8	-9.6	-6.0	7.3	13.1	19.8	21.9	21.7	16.0	5.8	-4.8	-11.1	4.9
1961	-16.9	-10.2	1.8	14.7	19.2	22.8	24.6	22.7	18.0	5.1	-2.1	-11.2	7.4
1962	-16.3	-11.5	3.3	10.7	21.2	24.3	27.3	25.5	19.1	9.1	-3.4	-11.0	8.2
1963	-12.9	-6.4	6.5	7.6	18.6	23.3	26.0	23.1	16.6	9.6	0.8	-9.7	8.6
1964	-14.8	-17.6	-1.3	10.5	17.1	22.9	25.3	24.7	17.4	6.1	-0.2	-11.2	6.6
1965	-12.6	-12.3	0.9	12.9	20.8	24.0	27.2	23.9	18.0	9.0	-1.4	-14.0	8.0
1966	-14.1	-9.4	0.9	8.3	16.0	24.3	25.3	25.2	20.4	8.1	-4.8	-15.8	7.1
1967	-18.7	-15.1	-0.8	12.5	20.4	24.0	25.2	22.3	16.6	9.8	-6.8	-13.5	6.3
1968	-13.0	-12.1	1.8	8.4	20.4	23.7	26.0	24.1	14.3	8.2	-2.4	-10.4	7.4
1969	-20.0	-19.2	-3.2	11.1	17.7	22.2	25.9	22.6	16.1	9.0	-1.2	-9.5	6.0
1970	-13.1	-7.9	-2.6	11.7	17.2	22.8	24.5	23.9	17.4	8.1	-4.5	-13.7	7.0
1971	-16.3	-12.3	-4.6	10.9	17.2	23.5	24.0	24.6	17.8	10.3	0.4	-6.1	7.5
1972	-14.2	-17.0	-1.8	11.6	19.7	23.5	23.1	23.7	16.5	8.1	-0.6	-10.2	6.9
1973	-11.9	-11.2	-0.9	12.2	16.8	23.2	26.3	24.5	17.6	7.1	-0.3	-11.0	7.7
1974	-14.4	-15.8	-1.1	14.3	21.1	24.5	28.2	24.8	17.2	6.8	-0.2	-17.0	7.4
1975	-10.5	-10.5	2.5	10.6	16.4	22.4	26.1	25.0	18.7	8.9	-6.2	-14.7	7.4
1976	-10.8	-8.8	-4.9	8.2	17.5	21.3	23.5	22.1	15.7	6.2	-8.8	-14.9	5.5
1977	-17.9	-11.9	-3.0	9.8	16.1	24.5	22.8	23.3	18.4	8.9	0.2	-9.8	6.8
1978	-14.0	-12.0	-1.1	12.9	16.5	21.8	24.7	20.6	17.7	6.9	-1.9	-7.5	7.1
1979	-12.2	-7.2	-3.7	6.8	14.5	20.7	23.0	22.7	15.6	10.2	-6.9	-10.3	6.1
1980	-15.1	-12.0	-6.2	8.8	17.5	20.2	24.5	22.4	16.8	8.2	2.3	-9.7	6.5

Table 4.14. Range of temperature between highest and lowest monthly means at selected stations, 1951–1980

	J	F	M	A	M	J	J	A	S	O	N	D
Haikou [a]												
(1)	20.0	22.1	24.8	26.8	28.9	29.8	29.5	28.6	27.6	26.6	23.5	21.0
(2)	13.6	13.2	18.3	22.6	26.0	27.2	27.3	27.0	25.9	23.6	20.1	15.7
(3)	6.4	8.9	6.5	4.2	2.8	2.6	2.2	1.6	1.7	3.0	3.4	5.3
Hangzhou [a]												
(1)	5.8	8.0	11.8	18.1	22.0	26.1	31.1	30.3	26.0	19.6	14.4	9.5
(2)	0.0	1.7	6.0	12.9	18.6	22.5	26.9	24.8	20.7	15.7	8.8	1.7
(3)	5.8	6.3	5.8	5.2	3.4	3.6	4.2	5.5	5.3	3.9	5.6	7.8
Nenjiang [a]												
(1)	−20.0	−16.5	−3.7	6.8	14.5	20.0	22.8	20.1	14.0	4.5	−6.8	−15.9
(2)	−29.9	−26.5	−16.4	0.2	8.9	16.2	19.2	16.4	9.5	−0.1	−15.6	−28.7
(3)	9.9	10.0	12.7	6.6	5.6	3.8	3.6	3.7	4.5	4.6	8.8	12.8
Ürümqi [a]												
(1)	−10.5	−6.4	6.5	14.3	21.2	24.5	27.3	25.5	20.4	10.3	2.3	−6.1
(2)	−20.0	−19.2	−6.2	6.1	12.9	18.1	21.3	20.4	13.9	4.7	−10.9	−18.4
(3)	9.5	12.8	12.7	8.2	8.3	6.4	6.0	5.1	6.5	5.6	13.2	12.3

[a] (1) Highest monthly mean, 1951–1980; (2) lowest monthly mean, 1951–1980; (3) range between highest and lowest monthly means.

For a broad description of the monthly temperature variability at a certain locality, the range for each month between the highest and lowest means of temperature over the 30-year observation period can also be considered. These figures are included in the climate tables (see Appendix). The temperature range may reach in every month very large values; however, figures in summer are generally smaller than in winter, thus showing a larger variability in winter than in summer. Over space, a remarkable observation can also be made which expresses an increasing variability of monthly temperature with latitude (Table 4.14).

4.9 Historical-Climatic Change of Temperature During the Last 5,000, 500 and 100 Years

Climate during historical times in China, covering the last 5,000 years, can be reliably reconstructed by the analysis of numerous Chinese historical documents which are rich in climatological and phenological events of the past. Considered so far as the most comprehensive approach towards reconstruction of the ancient climate, the famous Chinese climatologist Chu Ko-chen has successfully worked out a coherent and consistent picture of the main tendencies of temperature changes during historical times. The rather spectacular results are presented in Chu's widely appreciated contribution, *A preliminary study on the climatic fluc-*

tuations during the last 5,000 years in China (Scientia Sinica, vol. 16, 1973), CHU's major scientific findings are also reported in this section.[18]

According to CHU Ko-chen, the nature of available documents divides the past period of 5,000 years into four periods: (1) the archaeological period, from 3000 to 1100 B.C. when written records were lacking, with the exception of those carved on oracle bones; (2) the documental period, from 1100 B.C. to 1400 A.D. of which written documents, but not detailed regional reports were available; (3) the gazetteer period, from 1400 to 1900 A.D. when for most of the districts written reports on the regional history and geography were available; (4) the instrumental period, since 1900, which provided climatological observation data, precisely recorded.

Most available information for each of the four categories came, however, from the region east of 100° E and south of 40° N. In addition to this regional delineation of information, another restriction is concerned with the nature of the material which is mostly related to temperature, and that mainly in winter when the Mongolian anticyclone exerts its full control over the whole region.

The four categories of documents on the climatic, respectively temperature change during historical times delivered inestimable observations on historical climates in China. Their major contents shall be briefly dealt with for the four periods given.

1. Archaeological Period, from 3000 to 1100 B.C.

As can be deduced from the inscriptions of oracle bones, many prayers for rain were discovered during the Yin Dynasty at Anyang, located in Henan Province, north of the Yellow River. The records, most frequently concentrated on the first 5 months of the year, indicate that Anyang at that time had rather mild winters which may have exceeded the present winter by 5° C and the annual mean temperatures by 2° C. According to records, at that time rice could be planted about 1 month earlier than today. The certainly milder, subtropical temperatures are also proved by elephant subfossils, found in Yin-Hsu; this explains a more northern indigenous distribution of elephants.

Warmer temperatures for the period from 3000 to 1100 B.C. were also inferred from extensive excavations of two neolithic archaeological remains which took place during this century at Banpo (near Xi'an) and at Anyang. The rich discovery of remains of bones from subtropical animals such as water deer, bamboo rat, tapir, buffalo and oars has led to the recognition of a then existing warmer (and wetter) climate in the regions concerned than it nowadays occurs. Among floristic remains from the historical past, which are more difficult to preserve, only bam-

---

[18] Recently, ZHENG Sizhong anf FENG Liwen have presented a paper on *Historical evidence on climatic instability above normal in cool periods in China* (Scientia Sinica, Ser. B, vol. 29, 1986) in which the authors statistically correlated the change of temperatures in China since 1471 A.D. with the interannual variability of rainfall. The correlation coefficients show for the mid-latitudes of eastern China dominating negative values which provide evidence of climatic instability above normal in cool periods.

boo can be considered as a reliable plant. At present, the northern limit of bamboo groves coincides approximately with the mean January isotherm of 0° C which has retreated southward for about 3° latitude. The much more extensive distribution of bamboo groves during the neolithic age than today can be taken as proof of apparently warmer, subtropical temperatures which existed further northward than today.

2. Documental Period, from 1100 B.C. to 1400 A.D.

From the Chou Dynasty (1066–256 B.C.) onwards, written documents in bronze and later in bamboo books have led to a clearer description of climatic change than in the early part of history. At the beginning of this period, the extensive distribution of bamboo groves in the Yellow River valley can be regarded as an indicator of relatively mild temperatures, which therefore represent a continuation of the preceding warmer archaeological period. But during the early Chou Dynasty, temperatures were decreasing, proved, for example, by two cases of freezing in the Han River in 903 and 897 B.C. As can be inferred from classic poems, such as the Ode of Pin (which is a region close to Xi'an), a relatively sharp temperature drop in the early Chou Dynasty must have occurred which has probably led to similar temperatures as at present.

After a relative cold period of about a century, temperatures, however, again quickly increased, leading to a warm period from the ninth century B.C. to the Warring States period (480–222 B.C.). Indicators of this are classic inscriptions on the more northward distribution of plum (or "mei") trees and evidence of the wide use of fruits of the plum trees to season the food and to render it palatable since vinegar was then unknown. Also, written documents on the length of the growing seasons and the plants, which were cultivated in the Yellow River valley, gave evidence of a rather warm period until the third century B.C. Temperatures dropped from the second century B.C. onwards and continued for about 8 centuries up to the sixth century A.D. For this long period, a fair number of climatic indicators, particularly on the distribution of subtropical fruit trees, are found in the classic annals. Among the written documents, a report was given on the failure in the years 155–220 to grow oranges in the Royal Garden at Loyang, although earlier reports from the Royal Garden at Changan (near Xi'an) had shown the successful endeavour of orange cultivation.

Also for the period of "The Six Dynasties" (420–589 A.D.), much attention was given to phenological events in the then North and South Dynasty. For the latter, for example, reports on the dates of the full bloom of apricot, the first appearance of leaves on jujube or the withering of mulberry flowers showed a delay of 1 to 2 weeks, thus indicating lower temperatures than today. Also, other observations accurately described the colder conditions, for example, the first report on the freezing of the Huaie River (at Kuanling) in the year 225 A.D. or, for the same region, a larger number of frost events as late as in May, occurred between 280–289 A.D.

From the mid-seventh century onwards, lasting until the 11th century, temperatures again rose and led to a fairly warm period. For the winters of the years

650, 669 and 678 there were no reports of snow or ice at the capital of the Tang Dynasty (618–907), which was Changan, near Xi'an. As a proof of the then higher temperatures, many reports of the wide distribution of plum trees, for example, in the Royal Garden and in public parks, were given.

A drastic drop of temperature took place in the early 12th century. Occurrences of the freezing of the South Grand Canal and reports of snow on the Western Hills near Beijing in October seemed to be a common event during the 12th century. In addition, in 1111, for the first time on record, Lake Taihu was frozen solid enough to hold traffic. Low temperatures also killed the flourishing and famous mandarin trees of the region. The former capital Hangzhou (under the Sung Dynasty) often recorded snowfall in spring; on average, the latest date of snowfall was on April 9 which is a month later than at present. Also in South and Southwest China, low temperatures during the 12th century occurred, destructing in 1110 and 1178, on a large scale, the litchi trees which apparently are tropical fruit trees.

By the end of the 12th century, temperatures again increased, leading to a warm period until the second half of the 13th century. Evidence for this was given by the more northward distribution of bamboo culture. In addition, also records of ice- or snow-free winters at Hangzhou in the years 1200, 1213, 1216 and 1220 gave evidence of a warm, although short period of only one and a half century.

In the early 14th century, temperatures again dropped, because of vigorous winters. Proof was given by uncommon frost and ice events. It was reported, for example, that in 1209, the Grand Canal froze; the same happened with Lake Taihu in 1329 when it froze for the second time on record.

3. The Gazetteer Period, from 1400–1900 A.D.

Most luckily for a climate reconstruction of the past, China possesses numerous written reports and gazetteers, which also contain climatological events that are probably unrivalled by any country in the world for their comprehensiveness. Among the various climatological events, the severe colds (winters) were taken as a criterion for climate change during the past.

For a detailed investigation of temperature from 1400 to 1900, two criteria were particularly studied: (1) the freezing of lakes and rivers along the Yangtze, (2) snowfall in tropical South China. In detail, the three biggest fresh-water lakes in China (which normally remain unfrozen) were selected, i.e. Lake Poyang, Lake Tungting and Lake Taihu, all of them connected with the Yangtze. The Huai River at Yuyi and the Han River at Hsiangyang were chosen for the study. In total, from 665 district gazetteers the occurrences of freezing years and snowfall and frost years were compiled for the selected lakes and rivers (see Tables 4.15 and 4.16, both after Chu Ko-chen 1973).

As can be seen from the two indicators (Tables 4.15 and 4.16), cold years (winters) can be determined for the following periods: 1470–1520, 1620–1720 and 1840–1890, whereas warm years (winters) were found for the periods 1550–1600 and 1770–1830. Among the numerous observations, given on certain extreme temperature events, it is noteworthy that the cold period from

Table 4.15. Freezing years of lakes and rivers along the Yangtze between 1400–1900

Quarter of a century	Lake Taihu	Lake Poyang	Lake Tungting	River Han	River Huai
19th: 4th	1877, 1893		1877	1877, 1886, 1899	
3rd	1861	1861, 1865		1865, 1871	
2nd		1840		1830	1845
1st					
18th: 4th			1790		
3rd	1761				
2nd					
1st					1715, 1720
17th: 4th	1683, 1700		1690	1690, 1691	1690
3rd	1654, 1665	1670	1653, 1660	1653, 1660, 1670	1653, 1670, 1671
2nd					1640
1st			1621	1620, 1621	1619
16th: 4th	1578				
3rd	1568	1570			1564
2nd				1529	1550
1st	1503, 1513	1513	1510, 1513	1519	
15th: 4th	1476			1493	
3rd	1454				1454
2nd				1449	
1st				1416	
14th: 4th					
3rd	1353				
2nd					
1st					
13th:					1219
Before 13th:	1111			−879, −901	225, 515, 1186

1620–1720 was the absolutely coldest period during the last 500 years. Within this period, the period from 1650 to 1700 was the peak period of coldness; snow was often observed in the tropical regions of South China. It was also recorded that the orange and mandarin orchards in Jiangxi Province were practically annihilated in the cold waves of the years 1654 and 1676. Underlining the serious cold period from 1650 to 1700, it was reported that the Taibu, Han and Huai Rivers as well as the Poyang and Tungting Lakes were uncommonly frozen four times.

With regard to the two relative warm periods (1550–1600, 1770–1830) during the 500-year period from 1400 to 1900, at no time, however, did the high temperatures attain the warmth during the Tang and Han Dynasties, from the 8th to the 3rd century B.C. This is convincingly shown by many written documents in district gazetteers of the Yellow River valley, reporting many geographical names

Table 4.16. Years with snowfall and frost in tropical South China between 1400–1900

Quarter of the century		Years
19th:	4th	1878, 1882, 1893
	3rd	1854, 1856, 1862, 1864, 1871, 1872
	2nd	1831, 1832, 1835, 1840, 1846
	1st	1824
18th:	4th	1781
	3rd	1757, 1758, 1763, 1768
	2nd	1729, 1737, 1742
	1st	1711, 1723, 1731
17th:	4th	1681, 1682, 1683, 1684
	3rd	1654, 1655, 1656
	2nd	1635, 1636
	1st	1602, 1606, 1621
16th:	4th	1578
	3rd	
	2nd	1532, 1536, 1537, 1547, 1549
	1st	1506, 1512, 1522
Before 16th:		1245, 1415, 1449

of hills in memory of plum trees which formerly flourished there. For example, the gazetteer of Pingtuchow, Laichowfu, Shantung (36° 48′ N, 119° 54′ E) states that "seven li south of the city there is a hill called Ching-po, which was said to be covered with plum trees once. How could that ever happen, one often wonders" (cited after CHU Ko-chen 1973, p. 244).

4. *Instrumental Period, since 1900*

Although simple instrumental observations were already started much earlier, systematic measurements at a network of regular meteorological stations were set up by the Chinese government from 1911 onwards only. As far as instrumental records of temperature are concerned, the longest, continuous observations are available for Shanghai (since 1873), Hongkong (1874) and Tianjin (1891).

The temperature time series for Shanghai, expressing the annual mean from 1882 to 1985 (Fig. 4.27), shows a remarkably swinging distribution. By a broad interpolation of the curve, the minimum fell into the last quarter of the 19th century; from then onwards, temperatures progressively increased to a maximum in the 1940s and early 1950s, followed again by declining temperatures. On average, the fluctuation of temperature during the last 100 years showed an amplitude of 1° C.

This remarkably large temperature fluctuation in China is also shown by other indicators, particularly by the advance and retreat of snow lines and glaciers in the Tian Shan mountains, as it was reported by the Glacial and Snow Survey Expedition of Academia Sinica (1960–1963). Between 1910 and 1960, the snow

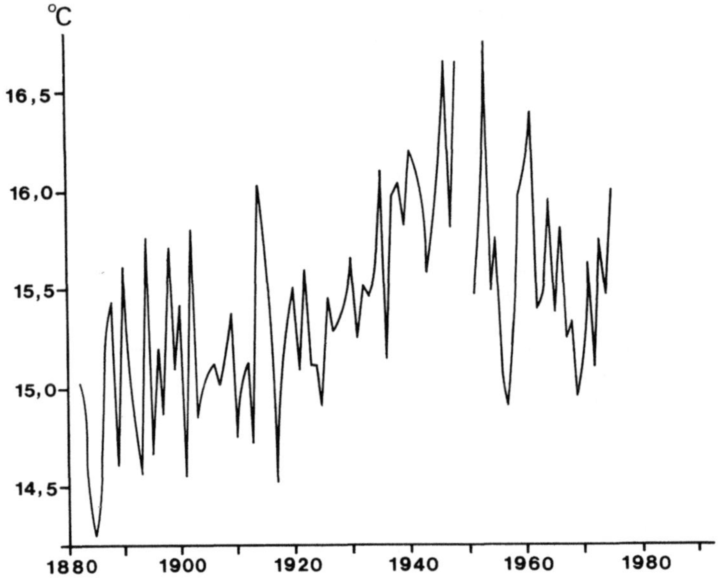

Fig. 4.27. Annual temperature time series at Shanghai; 1882–1985

line in Tian Shan ascended 40–50 m and glacier tongues in west Tian Shan retreated 500–1,000 m, in east Tian Shan 200–400 m. Synchronously, the ascent of the upper tree line in Tian Shan evidently proves a significant rise of temperature during the first half of this century.

As can be seen from the annual temperature time series at Shanghai (Fig. 4.27), temperatures may show an extremely abrupt interannual variation. In fact, extremely cold winters and very warm summers may alternate; many individual measurements and indicators of those events are available. Being one of the coldest winters during the recent past, exact temperature records for January 1955 have led to the lowest values ever recorded at many places in southern and central China, for example: in central China: Chengyangkuan, Anhwei −24.1° C; Hsuchow, Kiangsu −16.7° C; Hankow −14.6° C; Nanking −14.0° C; in South China: Swatow 0.4° C; Nanning −2.1° C; Tingan, Hainan Island −0.3° C; Yangkiang, Hainan Island −1.4° C.

As the only event in this century so far, the Huai River was frozen from January 1–25, 1955; the Han River from January 1–20, the Tungting Lake from January 3–6 of the same year.

Other spectacular events were also reported in order to express the occurrence of a severe interannual temperature drop in winter. It was reported for Tianjin, for example, that the important port as well as the outlet of the River Hai were ice-bound from the beginning of February until early March, 1936. This had never happened before. Also, temperatures dropped to an extremely low mean of −6.7° C and −4.4° C (for February and March 1936 respectively).

For the instrumental period, a temperature time series was worked out for China which shows trends of fair similarities with a global temperature time series (see Fig. 4.28). In both cases, 5-year mean values were considered. Two observa-

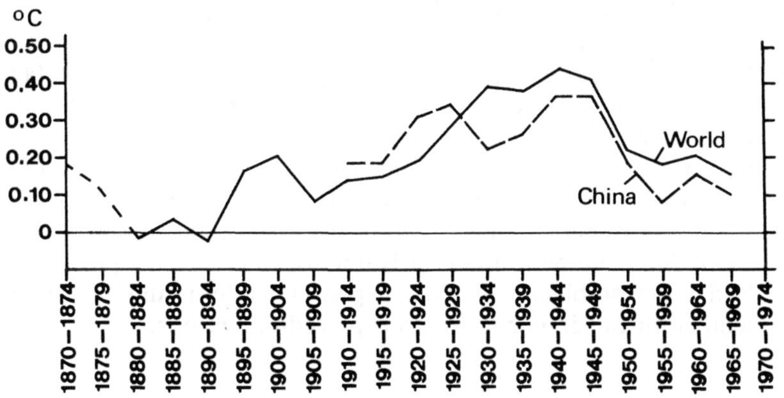

Fig. 4.28. Temperature deviation (°C) from normal time series for China and as a global mean, expressed as 5-year mean values from 1910/1914 resp. 1870/1874 onwards

tions are noteworthy: (1) temperatures in China are normally lower than the global means, but they are higher for the 5-year mean values from 1910/14 to 1925/29; (2) in contrast to the global temperature curve, temperatures from 1925–1929 dropped in China, while the global mean increased; also from 1930–1934, both temperature curves showed an opposing development.

Temperature change during the last 5000 years in China can be summarized such that four periods of alternating high and low temperatures can be distinguished (Fig. 4.29); the temperature fluctuations given are tentative values above or below the present-day mean. In a broad approach, the four alternating warm and cold periods are as follows:

1. From 3000–800 B.C. The first 2000 years most likely showed the highest temperatures until today. The annual mean temperature was about 2°C higher than the present one; January temperatures were probably even 3° to 5°C higher. A sudden and sharp temperature drop occurred towards the end of the period, from 1000–800 B.C.

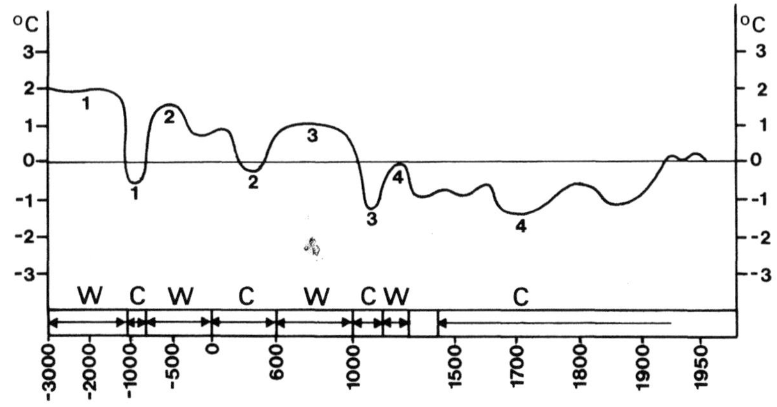

Fig. 4.29. Temperature change during the last 5,000 years in China, expressed as deviation from the present-day mean

2. From 800 B.C.–600 A.D. during which a short, warm period (until 200 B.C.) is followed by a longer period with declining temperatures.
3. From 600–1200 A.D., showing a longer warm period (until 1000 A.D.) which alternated with a shorter, but severe cold period (1000–1200 A.D.).
4. From 1200 A.D. onwards. This long period is characterized by a short, significant warm period (around 1200–1300), followed by a series of swinging temperatures with cold periods from 1470–1520, 1620–1720 and 1840–1890, while warm periods occurred from 1550–1600 and 1770–1830. Between 1900–1940, temperatures clearly developed to another peak period which may lead to the establishment of a fifth period of alternating spells with high and low temperatures, similar to the four preceding periods.

5 Precipitation

The spatial and temporal variations of precipitation in China have to be considered under two major aspects: (1) the effects of seasonally alternating air masses of either a dry continental or moist oceanic origin (cf. Chap. 3); (2) the impact upon precipitation from the large extent over space and the extremely contrasting landforms. Both aspects have to be verified with particular reference to exact observation data; however, there is a lack of available data from remote areas.

5.1 Mean Annual Precipitation Distribution

The annual total (in mm) clearly shows an extremely variable precipitation distribution in China (Fig. 5.1). Mean annual totals vary from <25 mm to >2,000 mm, thus representing a large range of precipitation and providing evidence of a remarkably uneven distribution over space. With the exception of a cellular type of isohyets in some parts of China, most isohyets roughly run in a southwest to northeast direction, clearly announcing an increasing precipitation pattern from northwest to southeast over China.

Despite the country's vast area, China can be divided into the following two major precipitation regions:

1. A dry western part which mainly occupies Xinjiang and the Qinghai-Xizang Plateau, but also includes Inner Mongolia.
2. A semi-wet to wet eastern part of China.

These two major parts can be demarcated by the 500-mm isohyet which, in a generalized way, crosses China in a southwest-northeast direction from the easternmost parts of the Tibetan Plateau roughly via Lanzhou and Taiyuan to Harbin (see Fig. 5.1). This dividing isohyet approximately coincides with the prevailing direction and location of some major mountains.

The broad division into two major precipitation regions corresponds to the regime of two contrasting systems of air masses which are, on the one hand, moist air masses originating from the Pacific Ocean and establishing the semi-wet and wet eastern part of the country, while on the other hand, continental, dry air masses are responsible for the dry western part of China. The high mountains of inner China act as orographical barriers for the advancing air masses and thus represent major limits of precipitation distribution over space. From the annual precipitation totals, China as a whole is commonly regarded as a comparably dry country, expressed by a country-wide mean annual total of 630 mm. However, in practice an extremely large spatial variation of precipitation exists which can best be seen by considering either the driest or wettest regions in China, as far as they are represented by precise observation data.

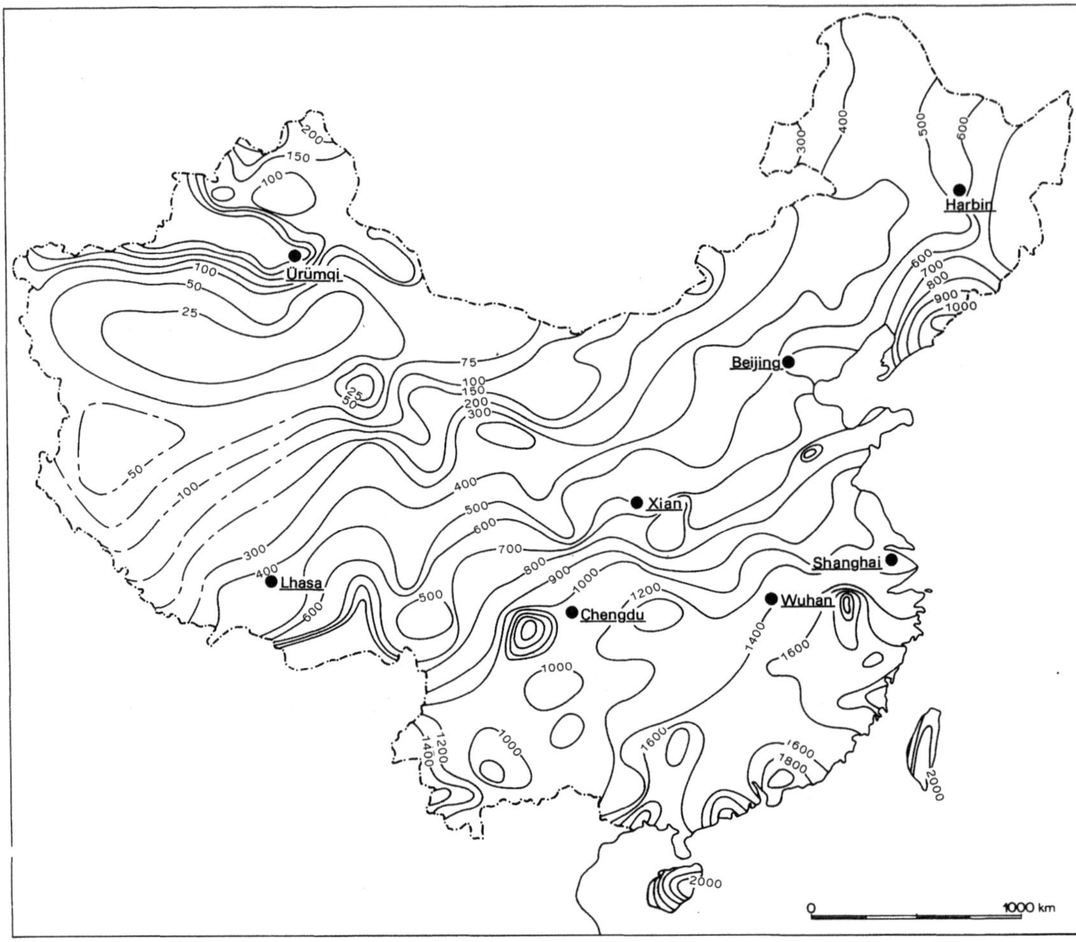

Fig. 5.1. Distribution of mean annual precipitation

The driest recorded parts which receive a mean annual total <25 mm occupy the southern parts of Xinjiang and Gashun desert which together represent a huge intramontane region, sheltered by high mountain ranges, in particular Tian Shan and Kunlun Shan. Smaller regions with extremely dry conditions are the Dsungaria, Qaidam and western Gobi. Despite the lack of accurate precipitation data, also the western parts of the Tibetan Plateau must be included with the driest regions in China. As a mean figure, both Ruoqiang/264 and Lenghu/260 recorded an annual total of only 17 mm each and, at the same time, they experienced the lowest precipitation totals in whole China.

In contrast, the wettest parts which received an average annual total >2,000 mm are several small, pocketlike areas in southern China (Fig. 5.1). The large precipitation totals are mostly caused by orographical lifting of the air masses which thus yields large amounts of orographical precipitation on the wind-exposed slopes. Due to the considerably varying landforms in southern

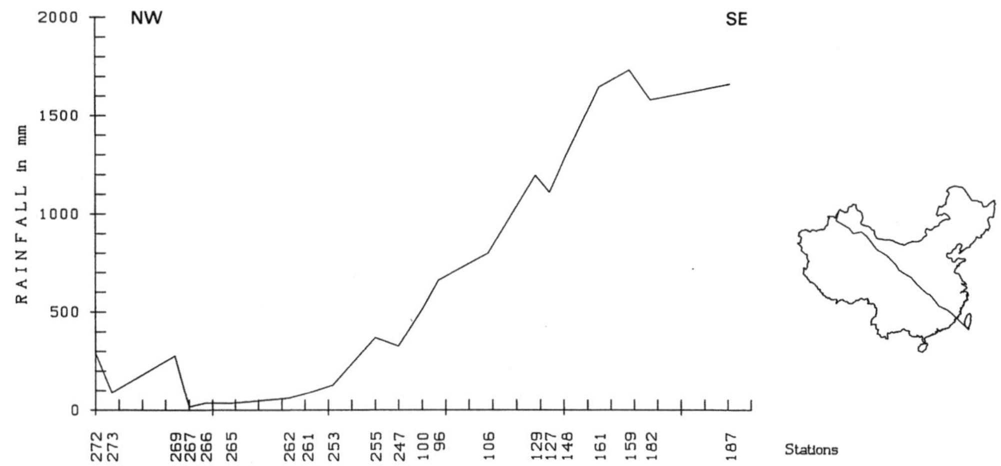

Fig. 5.2. Northwest to Southeast cross-section showing annual precipitation totals (in mm) at selected stations

China, precipitation zonation is widely characterized by a cellular type which may be established more clearly on a meso-scale basis. As a mean figure, the largest annual total in China was recorded at Huangshan/124 (1,840 m a.s.l.) which experienced 2,399 mm [19]. The station represents a mountain station which experiences ample precipitation due to orographical lifting of the air masses arriving from the Pacific Ocean. A few other stations where the average annual total amounted to >2,000 mm are scattered in South China (Fig. 5.1).

The spatial variation of annual precipitation in China, as determined in a southeast-northwest gradient of decreasing totals, can be shown by a cross-section from southern Taiwan Island and the southeastern China coastline to Xinjiang (see Fig. 5.2). It demonstrates the division of the country into a wet eastern and southern part, on the one hand, and a dry western and northwestern part, on the other. The 500-mm isohyet which runs from the Da Hinggan Mountains to the Sino-Butanese border (see Fig. 5.1) may be considered as the boundary.

5.2 Mean Seasonal Precipitation Distribution

Since winter, spring, summer and autumn are commonly considered as the principal seasons in China (see Sect. 2.5), a representative month for each season may reliably describe the seasonal precipitation pattern in China. Thus, January (winter), April (spring), July (summer) and October (autumn) were selected.

[19] Large annual totals, as shown by even >4,000 mm in Taiwan Island (by ZHANG Jiacheng and LIN Zhi-quang 1985), could not be confirmed by precise data. Also, in the case of the seasonal precipitation distribution (see Sect. 5.2), the corresponding larger totals in Taiwan Island could not be shown by observation data.

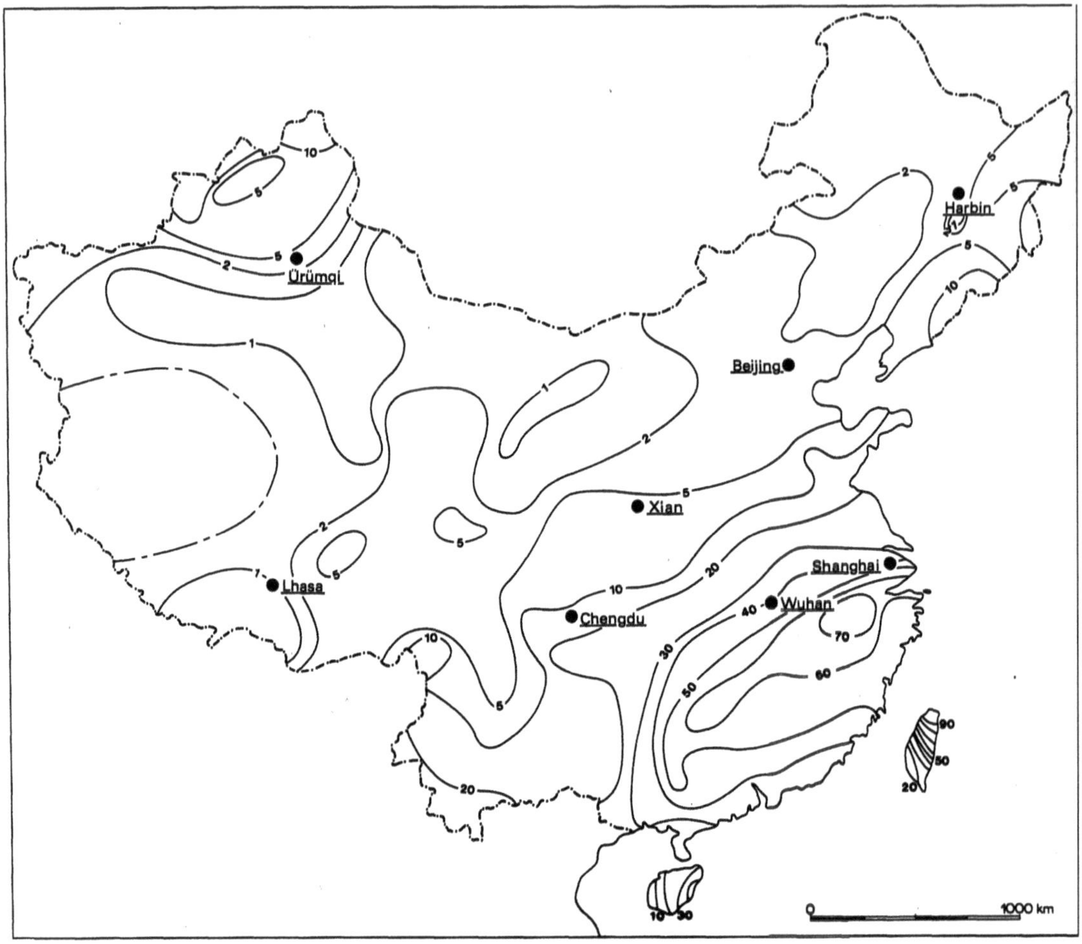

Fig. 5.3. Distribution of mean precipitation in January

January (Fig. 5.3)

For China as a whole (Table 5.1, p. 143–151), January represents a dry month, both in absolute and relative terms when compared with the annual total (see Fig. 5.1). Nevertheless, a considerably large variation of precipitation over space is still experienced in January, ranging from <1 mm only to >70 mm in mainland China, and at a maximum even to >150 mm on Taiwan Island. In most parts of China, dry conditions occur which are expressed by a monthly total of <5 mm in all of western China and also including northern and northeastern China. The driest conditions occur in southern Xinjiang and the western and southern Tibetan Plateau where precipitation in January sometimes even dropped to <1 mm only. In contrast, the comparably wet parts are limited only to the South China Hills where the Wuyi Shan, due to orographical rains, recorded a maximum of >70 mm in January.

(continued p. 151)

Table 5.1. Mean monthly and annual totals of precipitation at all climate reference stations

No.	J	F	M	A	M	J	J	A	S	O	N	D	Total
1	3	4	8	21	33	60	105	93	49	17	11	6	410
2	4	4	8	23	36	70	110	100	63	23	12	8	461
3	8	5	8	21	49	87	136	114	79	24	9	7	547
4	4	4	6	18	36	72	130	112	68	18	5	6	479
5	3	3	4	17	34	76	138	112	60	16	5	4	472
6	3	3	5	13	28	55	95	82	39	12	4	4	343
7	7	5	10	21	40	80	116	85	47	20	9	8	448
8	2	2	6	13	31	64	127	99	49	16	4	3	416
9	3	3	6	17	38	74	157	105	66	19	5	4	497
10	3	4	7	22	46	97	144	114	79	23	6	5	550
11	4	5	10	25	53	95	166	147	78	26	10	7	626
12	5	6	11	27	47	77	109	110	76	36	12	7	523
13	6	6	11	29	61	89	153	127	74	35	12	8	611
14	2	2	7	15	32	65	137	94	58	14	4	4	434
15	4	5	11	24	38	78	161	97	66	28	7	6	525
16	6	8	15	34	54	104	166	143	77	37	16	8	668
17	5	7	8	23	52	80	114	118	66	38	14	8	533
18	6	9	11	26	57	87	102	128	72	43	19	10	570
19	4	6	10	26	50	79	108	128	65	35	14	8	533
20	2	3	7	13	34	84	139	94	49	22	5	3	455
21	2	2	4	10	19	50	79	65	26	7	3	2	269
22	3	3	16	11	27	60	96	80	32	17	5	2	352
23	1	2	4	7	23	70	139	85	30	14	5	1	381
24	.9	2	4	7	24	72	130	96	30	18	3	1	388
25	1	4	6	11	31	69	124	80	37	25	4	2	394
26	2	3	7	15	39	72	158	94	48	23	7	4	472
27	4	5	9	22	42	91	184	128	61	34	12	4	596
28	6	6	10	33	56	101	183	142	67	37	14	7	662
29	9	10	16	41	65	109	196	155	69	44	23	12	749
30	5	8	11	29	63	95	150	142	65	32	16	7	623
31	4	5	8	25	50	85	96	129	55	26	15	6	504
32	10	13	19	47	77	112	213	199	80	39	31	17	857
33	11	14	22	49	68	109	237	246	94	45	36	18	949
34	12	14	21	50	68	114	311	322	106	62	39	18	1137

(contd.)

Table 5.1 continued

No.	J	F	M	A	M	J	J	A	S	O	N	D	Total
35	10	12	20	57	57	110	296	258	104	97	31	14	1066
36	7	6	11	36	46	68	171	164	81	45	21	9	665
37	9	11	19	44	61	115	296	230	87	57	25	14	968
38	7	8	13	40	56	89	196	169	82	45	20	11	736
39	2	3	5	24	43	74	162	110	56	30	8	3	520
40	2	2	7	10	25	68	132	93	28	13	3	1	384
41	3	3	6	9	26	48	88	69	26	13	4	2	297
42	2	2	4	8	21	43	66	58	22	12	3	2	243
43	2	2	3	5	10	17	39	39	16	7	2	.9	143
44	2	4	6	7	16	31	65	58	22	12	3	2	228
45	3	4	9	12	26	51	89	83	35	18	5	2	337
46	3	4	11	14	30	12	109	88	42	17	5	2	337
47	2	4	8	15	37	75	138	103	48	21	5	2	458
48	2	3	5	11	32	68	99	84	35	16	4	1	360
49	2	3	4	19	35	80	161	109	43	22	7	3	488
50	4	5	3	28	41	85	171	128	62	33	10	5	575
51	2	6	8	18	49	94	154	143	55	23	6	1	559
52	2	3	7	14	31	84	147	133	50	20	6	1	498
53	3	4	8	15	24	64	115	116	43	21	6	1	420
54	3	4	10	16	31	55	120	115	46	20	7	1	428
55	3	4	8	18	24	56	97	104	40	17	5	2	378
56	3	3	5	10	15	29	71	72	32	12	5	2	259
57	2	2	4	6	12	25	61	61	21	10	3	.6	208
58	3	6	10	18	27	66	102	126	46	24	7	1	436
59	3	4	10	19	29	45	95	94	51	24	8	2	384
60	3	6	8	26	35	78	219	164	37	23	11	4	614
61	3	7	8	19	33	78	193	212	57	24	7	3	644
62	3	6	6	21	31	69	190	162	43	25	9	4	569
63	3	4	8	21	33	63	106	96	48	26	8	2	418
64	8	8	13	36	44	86	176	153	69	35	22	11	661
65	3	7	7	20	29	64	165	190	42	27	9	4	567
66	4	6	7	24	32	79	218	170	48	27	11	5	631
67	2	5	7	21	26	52	113	129	65	24	9	2	455
68	3	5	10	24	26	34	97	116	59	27	11	3	415

(contd.)

Table 5.1 continued

No.	J	F	M	A	M	J	J	A	S	O	N	D	Total
69	3	6	10	24	30	53	118	104	64	31	13	3	459
70	4	8	11	27	29	62	152	156	50	35	18	5	557
71	4	7	9	29	31	75	188	145	48	32	17	5	590
72	4	7	8	32	29	74	191	150	48	36	18	6	603
73	8	13	18	38	36	91	200	135	63	36	24	11	673
74	14	11	19	49	43	84	197	172	88	43	30	17	767
75	10	13	22	43	48	102	179	162	104	48	31	11	773
76	8	12	18	63	37	88	237	145	80	32	22	11	753
77	6	10	16	36	37	74	214	148	61	33	30	8	673
78	16	17	30	55	58	143	338	238	119	68	35	17	1134
79	3	2	11	24	33	49	139	169	59	32	17	5	543
80	4	6	15	27	33	56	122	104	73	36	14	4	494
81	3	6	15	32	45	62	123	118	84	40	16	4	548
82	4	8	15	31	42	60	177	166	62	36	24	5	630
83	8	9	21	45	44	83	208	129	81	34	27	8	697
84	9	11	19	43	39	86	234	149	64	36	21	12	723
85	.9	15	26	55	51	109	273	194	94	34	27	14	893
86	25	29	45	66	78	112	252	187	113	43	41	21	1012
87	13	19	33	60	62	107	248	156	87	36	29	18	868
88	18	22	39	61	64	88	216	146	75	42	26	16	813
89	17	18	35	63	57	84	184	120	78	48	29	15	748
90	9	13	29	51	46	68	135	135	67	41	34	8	636
91	7	13	25	45	47	66	142	94	75	46	31	9	600
92	6	9	25	50	62	71	141	102	83	54	27	7	637
93	15	18	46	77	89	92	170	132	124	88	41	12	904
94	5	7	22	45	51	111	122	93	80	49	25	4	614
95	8	11	25	52	63	52	99	72	98	62	32	7	581
96	6	9	25	59	61	57	121	101	123	70	26	4	662
97	4	6	16	41	54	67	99	85	93	48	15	3	531
98	4	7	16	40	60	55	119	104	93	44	19	4	565
99	3	4	13	33	48	56	115	104	80	41	14	2	513
100	5	7	15	34	54	64	97	106	77	40	12	4	515
101	2	3	15	38	57	61	93	81	75	39	9	.9	474
102	7	10	31	62	111	110	108	87	113	75	13	4	731
103	3	6	17	50	90	105	213	180	132	47	15	2	860

(contd.)

Table 5.1 continued

No.	J	F	M	A	M	J	J	A	S	O	N	D	Total
104	8	10	33	68	88	103	159	126	150	83	37	8	873
105	5	10	40	94	130	132	223	178	193	103	38	10	1156
106	5	10	34	73	90	90	134	116	132	78	33	7	802
107	20	29	51	76	94	61	122	145	101	63	48	19	829
108	14	17	34	69	65	95	189	145	87	50	30	11	806
109	33	43	68	100	115	134	231	171	77	53	57	28	1110
110	23	28	52	78	80	125	202	168	98	54	42	14	964
111	23	31	52	74	86	136	203	113	77	44	33	19	891
112	31	41	71	101	112	147	220	118	89	53	48	30	1061
113	47	65	102	134	169	176	206	194	118	75	65	40	1391
114	34	50	75	106	106	96	182	114	80	43	53	32	971
115	27	33	61	33	75	106	206	154	97	38	40	22	892
116	31	50	73	94	100	167	184	113	96	46	48	29	1031
117	29	39	56	80	78	139	218	183	131	47	41	28	1069
118	22	30	45	59	71	115	246	170	106	40	35	21	960
119	44	63	81	111	129	157	142	116	146	47	53	39	1128
120	61	83	90	116	155	166	109	136	174	83	68	56	1297
121	52	82	109	128	192	199	83	119	188	97	67	63	1379
122	49	78	104	125	179	203	117	121	144	64	47	44	1275
123	64	84	117	130	186	192	132	136	183	67	61	49	1401
124	70	127	185	258	329	379	291	297	206	109	84	64	2399
125	42	68	117	166	216	247	188	108	76	43	57	36	1364
126	36	61	105	144	161	218	179	133	81	53	57	34	1262
127	26	41	80	122	163	163	151	127	83	80	49	28	1113
128	24	31	57	98	127	134	171	105	86	60	42	21	956
129	22	28	65	106	140	158	236	195	101	74	49	26	1200
130	25	38	73	123	153	211	222	168	173	115	70	33	1404
131	16	15	31	80	118	136	171	138	163	96	38	19	1021
132	7	11	20	52	78	136	239	206	143	42	18	5	957
133	3	6	18	44	107	147	124	102	134	65	9	2	761
134	6	11	21	51	90	111	236	234	118	46	18	6	948
135	19	29	53	93	142	168	399	448	226	115	62	22	1776
136	17	27	55	115	174	239	400	470	235	122	49	21	1924
137	2	6	12	37	89	204	174	139	155	70	10	2	900

(contd.)

Table 5.1 continued

No.	J	F	M	A	M	J	J	A	S	O	N	D	Total
138	5	8	9	25	80	215	223	191	174	88	21	4	1043
139	18	30	41	74	112	157	251	257	130	88	43	21	1222
140	15	17	23	57	94	156	215	191	139	76	30	14	1027
141	28	30	35	65	137	170	186	179	139	93	50	30	1142
142	20	20	39	90	158	166	142	138	136	97	48	25	1079
143	29	33	62	123	202	215	189	180	156	129	74	36	1428
144	24	30	71	146	210	225	176	150	138	130	65	31	1396
145	43	96	104	190	254	242	183	126	74	101	62	39	1514
146	44	67	115	171	223	175	156	148	72	99	58	48	1376
147	53	87	152	199	245	185	123	106	69	85	70	48	1422
148	47	71	129	175	202	228	95	119	54	73	63	47	1303
149	59	91	157	204	261	293	149	121	71	65	61	46	1578
150	58	95	164	226	302	291	126	103	76	53	53	47	1594
151	76	106	175	250	321	315	160	134	71	57	65	50	1780
152	74	117	180	211	312	280	128	92	91	59	70	52	1666
153	52	87	118	165	230	248	114	122	140	59	48	44	1427
154	50	87	135	152	211	247	146	224	255	75	65	50	1697
155	47	79	116	150	218	225	126	261	254	88	46	46	1656
156	59	99	166	213	295	306	133	151	106	56	45	49	1678
157	63	126	183	231	326	334	140	121	98	61	47	49	1779
158	64	108	179	230	302	277	152	116	77	63	54	56	1678
159	60	108	177	244	332	321	113	115	81	74	58	52	1735
160	55	91	150	210	266	212	112	116	73	66	57	48	1456
161	70	109	176	220	282	220	133	128	69	76	70	96	1649
162	52	74	118	191	239	179	107	115	55	84	72	44	1330
163	58	69	114	185	225	194	110	132	75	94	82	50	1388
164	37	49	84	168	235	183	150	124	59	103	69	36	1297
165	21	26	50	116	191	203	156	115	92	108	61	26	1165
166	22	24	39	89	151	199	149	130	91	114	60	27	1095
167	17	17	21	60	122	153	169	167	109	64	34	16	949
168	11	11	13	42	107	183	183	161	130	74	26	11	952
169	6	6	11	31	74	149	154	126	100	58	18	6	739
170	7	6	9	16	66	232	268	213	186	96	23	8	1130
171	13	15	14	31	105	207	184	182	115	88	40	14	1008
172	21	27	29	65	188	248	216	200	116	115	50	26	1301

(contd.)

Table 5.1 continued

No.	J	F	M	A	M	J	J	A	S	O	N	D	Total
173	20	22	33	108	192	213	179	142	83	89	56	26	1163
174	14	17	34	85	281	206	199	168	85	81	41	18	1229
175	26	33	58	123	225	226	211	156	98	103	60	28	1347
176	27	37	56	125	200	211	168	150	66	87	57	29	1213
177	56	76	134	280	319	316	224	167	66	97	83	56	1874
178	70	93	138	190	210	218	105	146	84	91	70	61	1476
179	64	97	147	213	235	174	92	121	67	69	81	60	1420
180	54	88	150	201	226	221	105	130	85	73	51	53	1437
181	59	96	175	239	308	307	132	148	86	62	45	52	1709
182	59	92	168	199	274	285	109	139	103	63	45	49	1585
183	53	80	121	136	210	224	119	142	155	31	29	29	1329
184	47	72	97	118	160	242	84	130	115	33	27	28	1153
185	94	119	143	139	204	285	227	253	253	106	83	73	1979
186	69	76	80	106	196	176	193	297	303	213	166	78	1953
187	18	24	27	66	169	406	359	380	149	28	24	13	1663
188	29	39	39	67	100	179	151	186	107	36	26	19	978
189	22	17	18	91	172	363	377	502	341	112	97	24	2136
190	36	50	60	103	231	386	243	318	136	17	18	24	1622
191	57	82	137	164	261	271	173	247	200	94	55	50	1791
192	37	82	74	109	162	206	150	116	98	18	23	19	1094
193	45	70	89	137	200	290	200	205	151	63	41	31	1522
194	33	51	69	123	231	339	234	205	147	60	38	26	1556
195	50	78	117	161	233	229	116	188	145	67	33	31	1448
196	23	58	61	134	268	388	284	299	253	59	48	18	1893
197	48	83	110	215	347	399	203	196	155	67	32	37	1892
198	28	45	52	132	215	336	312	353	269	98	33	26	1899
199	39	62	92	159	267	299	220	225	204	52	42	20	1681
200	52	83	150	226	240	264	128	139	91	57	49	44	1523
201	57	78	123	232	292	244	125	162	81	78	49	53	1574
202	47	63	91	216	303	323	196	224	101	80	50	45	1739
203	56	67	102	213	277	294	187	232	108	78	56	48	1718
204	39	54	102	197	227	269	177	215	115	56	40	34	1525
205	38	55	79	234	394	401	267	381	258	73	41	31	2252
206	26	33	56	125	186	238	197	283	281	96	39	21	1581

(contd.)

Table 5.1 continued

No.	J	F	M	A	M	J	J	A	S	O	N	D	Total
207	24	30	52	93	188	241	207	240	303	172	98	38	1686
208	42	41	71	117	176	253	170	309	384	291	150	73	2077
209	35	14	17	26	70	173	234	239	245	261	146	47	1507
210	9	16	19	29	58	130	131	279	197	82	21	5	976
211	21	25	36	88	210	213	231	300	350	196	81	34	1785
212	29	36	60	99	142	256	306	401	191	63	32	21	1636
213	38	46	74	111	175	393	415	431	207	92	49	28	2059
214	26	30	43	96	196	237	201	243	140	74	28	30	1344
215	40	42	63	84	183	241	180	204	110	67	44	25	1283
216	27	39	62	120	225	273	249	243	91	80	49	34	1492
217	19	17	29	65	168	197	185	196	109	81	34	16	1116
218	14	16	26	62	136	184	188	204	116	66	30	17	1059
219	13	17	26	44	91	127	157	160	86	52	28	16	817
220	14	14	18	40	93	133	126	140	74	73	40	20	785
221	23	12	21	47	137	190	221	244	142	94	43	24	1198
222	30	10	14	32	166	255	368	341	170	170	63	34	1653
223	29	14	22	42	155	236	328	318	154	138	35	28	1499
224	15	12	18	29	97	187	258	227	137	120	44	15	1159
225	10	10	14	20	78	182	216	195	123	95	34	16	993
226	10	9	12	14	63	131	172	187	103	76	32	10	819
227	20	27	42	68	131	251	289	261	155	157	44	21	1466
228	19	26	32	20	67	188	179	229	157	117	35	12	1081
229	2	6	12	18	56	168	249	217	140	69	13	3	953
230	12	20	30	55	69	70	145	141	56	59	11	8	676
231	1	5	9	18	57	150	173	158	108	36	4	2	721
232	.1	.7	5	12	33	91	125	106	77	22	2	.7	475
233	4	9	15	32	72	128	110	95	113	47	7	4	636
234	2	4	8	22	57	124	130	114	111	36	4	2	614
235	2	3	7	13	45	98	113	107	73	28	3	2	494
236	2	5	18	46	77	135	119	112	101	36	5	1	657
237	5	5	18	23	23	44	96	109	49	23	2	.4	397
238	.6	0	.9	2	9	66	144	154	58	4	.2	0	439
239	.2	.1	2	4	21	73	142	149	57	5	.8	.3	454
240	2	1	4	6	24	81	100	101	63	18	3	3	406

(contd.)

Table 5.1 continued

No.	J	F	M	A	M	J	J	A	S	O	N	D	Total
241	5	4	7	15	49	131	130	109	89	28	3	2	572
242	3	7	12	15	53	143	148	148	98	42	9	3	681
243	4	5	7	11	57	99	105	90	76	26	3	1	484
244	10	5	11	21	59	102	112	58	87	32	5	3	505
245	3	4	6	9	24	55	72	65	45	18	3	2	306
246	2	3	6	17	60	73	97	84	56	25	4	1	428
247	1	2	8	17	36	33	64	85	49	25	5	1	326
248	1	1	4	12	18	21	43	86	36	15	5	.9	243
249	1	3	7	17	24	28	58	83	45	21	8	2	297
250	1	2	6	12	15	19	44	56	27	14	5	.7	202
251	.9	1	2	6	6	13	29	22	13	6	2	.1	101
252	.7	.5	2	4	7	13	28	28	10	4	2	.6	100
253	2	1	3	5	13	19	28	31	18	5	2	2	129
254	.7	.5	3	5	8	12	20	37	19	8	2	.4	116
255	2	4	11	21	40	61	91	54	62	18	5	2	371
256	1	2	5	20	45	49	81	82	55	25	3	.9	369
257	6	7	7	8	19	37	36	30	16	8	4	3	181
258	.6	.5	.9	1	4	7	9	8	5	1	.9	.3	38
259	2	2	2	1	11	20	20	16	7	1	.8	.6	83
260	.5	.1	.2	.2	1	5	5	5	.2	0	.1	.1	17
261	2	2	3	5	9	12	21	17	11	2	2	1	87
262	1	2	5	5	7	8	15	9	5	2	2	.9	62
263	.8	2	1	3	2	7	12	5	2	1	1	.7	38
264	1	.6	.6	1	1	3	6	2	.5	.2	.4	.7	17
265	2	1	1	3	3	6	6	5	3	2	2	2	36
266	.6	.1	.4	.7	4	7	11	8	4	.7	.4	.6	38
267	2	.3	1	.4	.5	3	2	3	1	1	.4	1	16
268	17	18	25	27	28	27	24	14	13	20	24	21	258
269	9	11	21	34	35	39	22	24	26	24	19	15	279
270	12	14	9	14	15	18	19	16	15	14	18	17	181
271	3	3	5	9	12	30	39	19	13	5	4	4	146
272	16	17	19	36	28	28	31	18	14	27	33	24	291
273	3	3	9	11	13	12	12	10	6	5	3	4	91
274	4	6	11	20	17	24	22	19	21	14	11	8	177
275	2	2	2	4	6	12	13	11	5	3	3	1	64

(contd.)

Table 5.1 continued

No.	J	F	M	A	M	J	J	A	S	O	N	D	Total
276	2	3	.8	3	7	7	4	3	3	.6	.4	.7	35
277	.8	2	.2	2	6	10	11	8	3	1	.4	.4	45
278	1	3	1	3	10	6	8	6	3	1	.9	.7	44
279	3	7	5	5	13	6	7	7	6	2	1	2	64

From precipitation totals in January, China can be divided into two major parts. The by far larger one, which occupies western as well as northern and northeastern China as a whole and even includes parts of the Great Plains, represents a dry part where monthly precipitation does not exceed 10 mm. In this part, precipitation distribution over space does not vary to a great extent. In quite a remarkable contrast, the small South China Hills experience, however, comparably wet conditions, clearly showing a gradually increasing amount of precipitation from the marginal and lower Hills to the central parts which, at a maximum, record around 60 to 70 mm for January.

The 10-mm isohyet can be considered as the boundary between the two major precipitation zones. It roughly divides China from northeast to southwest, starting from Shantung Peninsula, crossing the southern foothills of Tapa Shan, following the middle section of the Yangtze through the Sichuan Basin and finally reaching Yunnan Province, which borders Burma.

April (Fig. 5.4)

Precipitation distribution underlines the contrasting moisture conditions which are generally valid in China. Although the division of China into a dry western and northern part, on the one hand, and a comparably wet eastern and southern part, on the other, widely corresponds to the division in January (see Fig. 5.3). The range of precipitation in April is, however, much larger, amounting to <1 mm to >250 mm in mainland China. The boundary between the two major precipitation areas can be drawn for April by the 50-mm isohyet which approximately coincides with the 10-mm isohyet in January (cf. Fig. 5.3).

July (Fig. 5.5)

Precipitation totals throughout China in July are not only considerably higher than in both January and April, but also more variable over space, as seen particularly on a medium scale. The reasons for this are orographical effects of the uplifting or descending of air masses and thus the foehn effect which can be clearly observed in the South China Hills. Proof of this is the obviously occurring cellular type of isohyets in several cases which produce comparably small, pocket-like regions of large precipitation activity. These were mostly represented by one station only, for example, Huangshan/124 (291 mm in July), Tai Shan/78 (338 mm) and Lancang/222 (368 mm). In other cases, high precipitation pockets

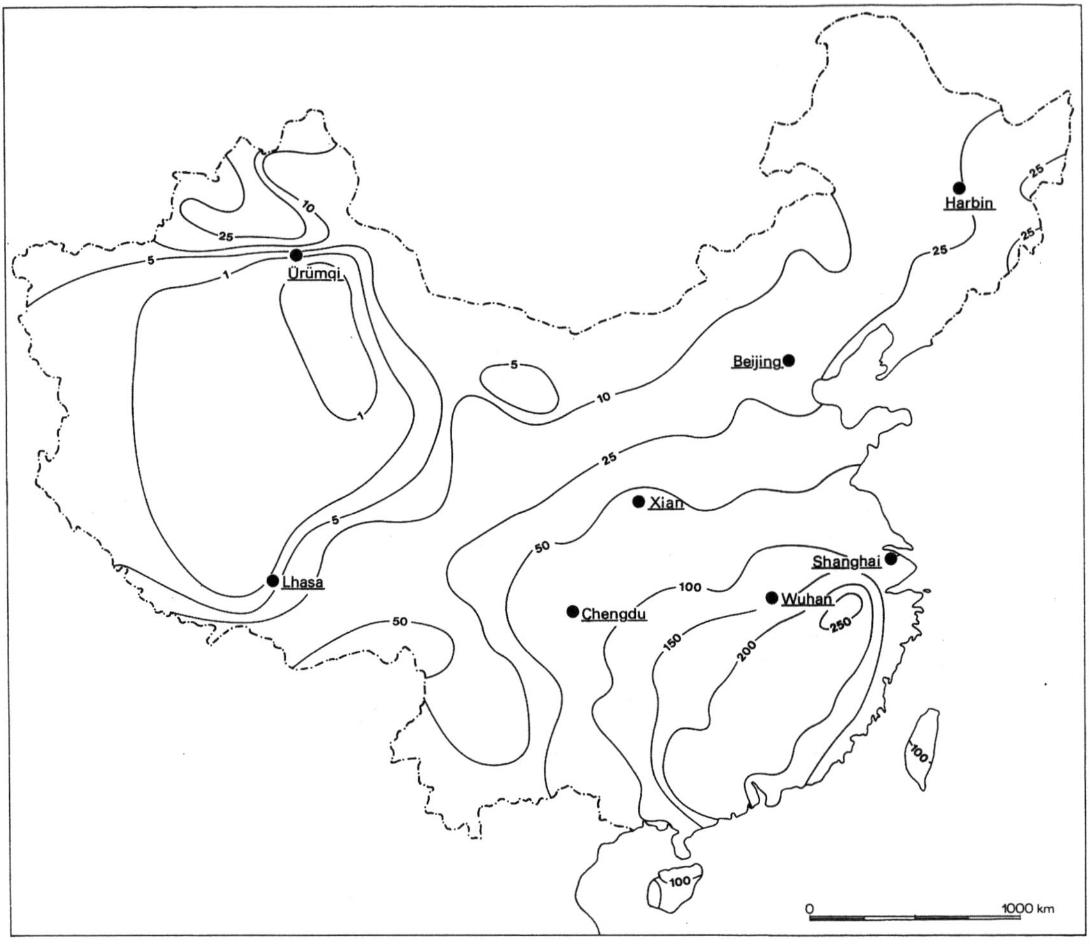

Fig. 5.4. Distribution of mean precipitation in April

were also represented by two stations each, for example, Ya'an/135 and Emei Shan/136 (399 and 400 mm respectively), Qinzhou/213 and Beihai/212 (415 and 306 mm respectively), Shenzen/198 and Haifeng/196 (312 and 284 respectively). These pocketlike regions of larger precipitation totals are scattered in the South China Hills.

The amount of precipitation in July and its distribution in China again justifies a division into two parts, similar to the division worked out for January and April (see Figs. 5.3 and 5.4). However, it is not as easy as in the preceding cases to demarcate the dry western and northern part from the wet eastern and southern part, although the 150-mm isohyet may best represent the boundary. With some simplifications, it runs according to the previously described characteristic boundary isohyets in a northeast to southwest direction from the Northeast China Plain to Yunnan Province.

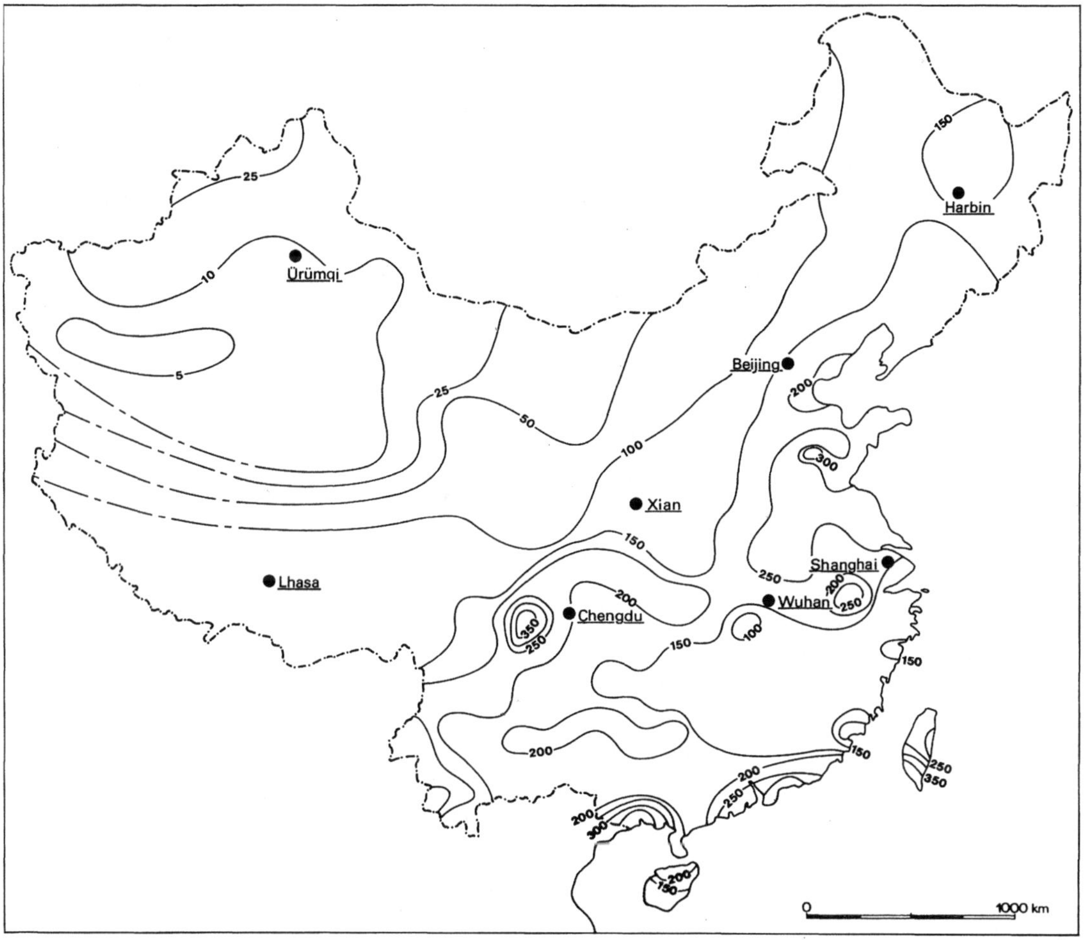

Fig. 5.5. Distribution of mean precipitation in July

October (Fig. 5.6)

Analogous to the three preceding months, the amount and distribution of precipitation for October also vary widely over space, confirming likewise for autumn a division of China into a dry western and northern part and a wet southeastern part. Taking the 50-mm isohyet as the boundary, it runs from the Yangtze mouth through the Great Plains to the northern foothills of the Chin Ling Shan, then turning south to the Sichuan Basin and from there across the mountains to the Chinese-Burmese borderline.

The remarkably large spatial variation of precipitation which was already clearly experienced in January, April and July (see Figs. 5.3, 5.4 and 5.5), also occurs in October (Fig. 5.6). The driest part of China, which does not even receive a precipitation total of 1 mm in October, occupies the whole Tarim Basin, including the Taklamakan Desert, and the western part of the Tibetan Plateau. In

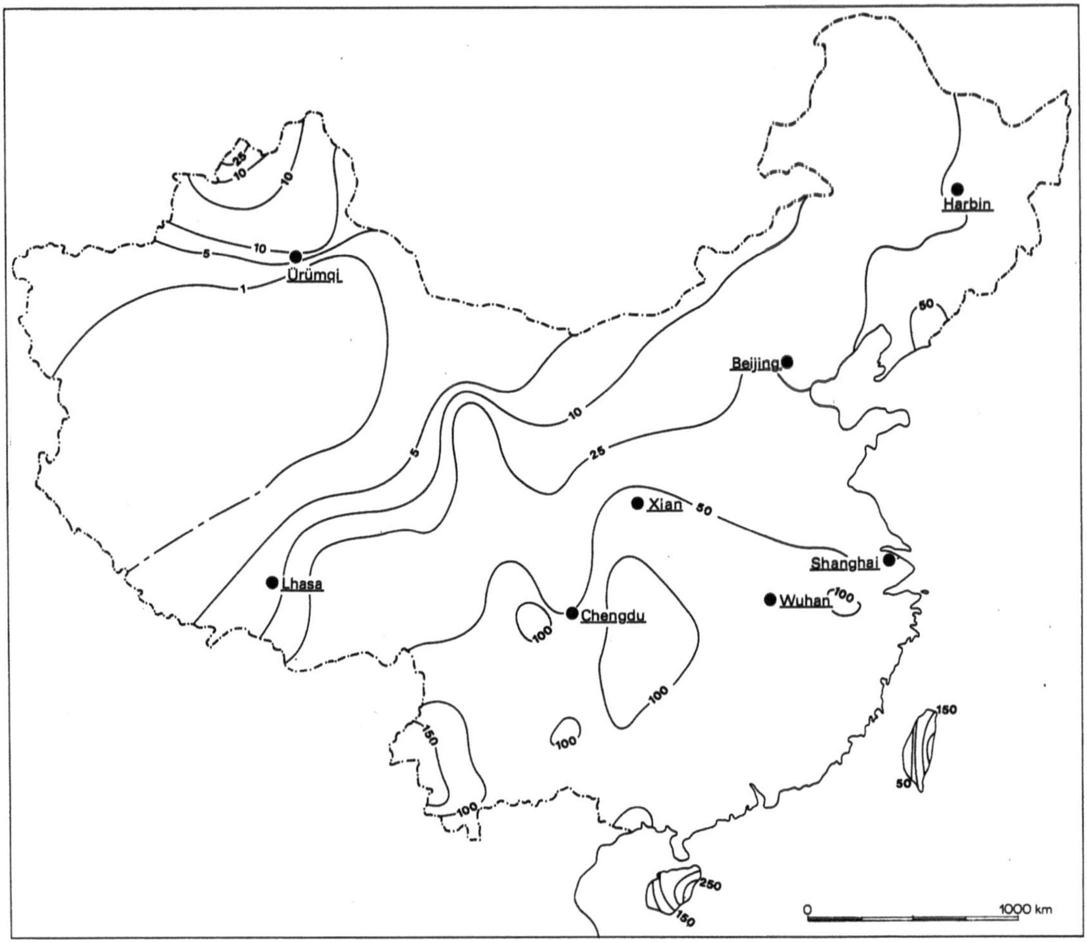

Fig. 5.6. Distribution of mean precipitation in October

contrast, the maximum amount of precipitation in October was recorded in pocketlike, marginal regions of China which are represented by the southwestern mountains of Yunnan Province as well as Hainan and Taiwan Islands which, due to the well-established foehn effect, experienced the highest precipitation on the eastern rather than the western side. The precipitation maximum for October amounted to even 291 mm at Qionghai/208 in Hainan Island.

Comparing the amount and distribution of precipitation for the four months (Figs. 5.3–5.6), three major findings can be summarized:

1. Throughout the year, the division of China into a dry western as well as northern and northeastern part, on the one hand, and a semi-wet to wet eastern and southern part, on the other, is well established. The wet part mostly covers the South China Hills and Yunnan Province, including Hainan and Taiwan Islands, partly extending over the North China Plain.

2. In general, most isohyets run in a predominantly southwest to northeast direction, markedly indicating a steep precipitation gradient from Northwest and West China to the southeastern parts of the country.
3. In most parts of China, July (see Fig. 5.5) experiences the maximum amount of precipitation in comparison with January, April and October. Thus, a characteristic annual precipitation variation can be expected which shows a summer maximum and a winter minimum, while spring and autumn experience transitional conditions. For the South China Hills, however, April rains exceed precipitation in July; both are rather wet months and indicate both a wet spring and summer, while January and October experience, less, but still a reasonable amount of precipitation.

Already at this stage, a different type of annual precipitation variation for the South China Hills including Yunnan Province than for all other parts of China can be determined. It can also be seen that the South China Hills record not only the highest totals in China, but less pronounced seasonality of precipitation also seems to exist, whereas all other parts of China obviously confirm a well-established wet season in summer as opposed to a long and pronounced dry season in winter, which may even start in autumn and extend to spring (cf. Sect. 5.3).

5.3 Annual Variation of Precipitation

5.3.1 Specific Precipitation Types and Their Distribution

From the observations given on the mean seasonal precipitation distribution (see Sect. 5.2), a distinct annual variation of precipitation could be determined, showing a pronounced seasonality between a wet summer and a dry winter, with transitional conditions in spring and autumn. Hitherto, particular attention was given to characteristic types of annual precipitation variation with a far-reaching spatial homogeneity. Therefore, the annual precipitation variations of all 279 stations under investigation were classified according to a hierarchial cluster analysis, by applying the method of WARD.

Three types of annual precipitation variation were determined, each of which represents far-reaching homogeneity in terms of the precipitation variation over the year (Fig. 5.7). From all 279 stations, the following numbers represent each type: 120, 84 and 75 respectively. The three types may be characterized and named after their maximum precipitation conditions in summer:

Type 1: Weak Mid-Summer Precipitation Type

This type shows a summer peak in July and August, but June and September are still part of a fairly wet summer. The summer peak, however, is not as clearly established as for the two other types, so that this type may be characterized as a weak mid-summer precipitation type only. In relation to the whole year, the wet summer covers only a minor part of the year, while during the major part dry and sometimes even extremely dry conditions occur.

The weak mid-summer precipitation type is widely distributed over space and is exhibited in western, northern and northeastern China, it extends to a bound-

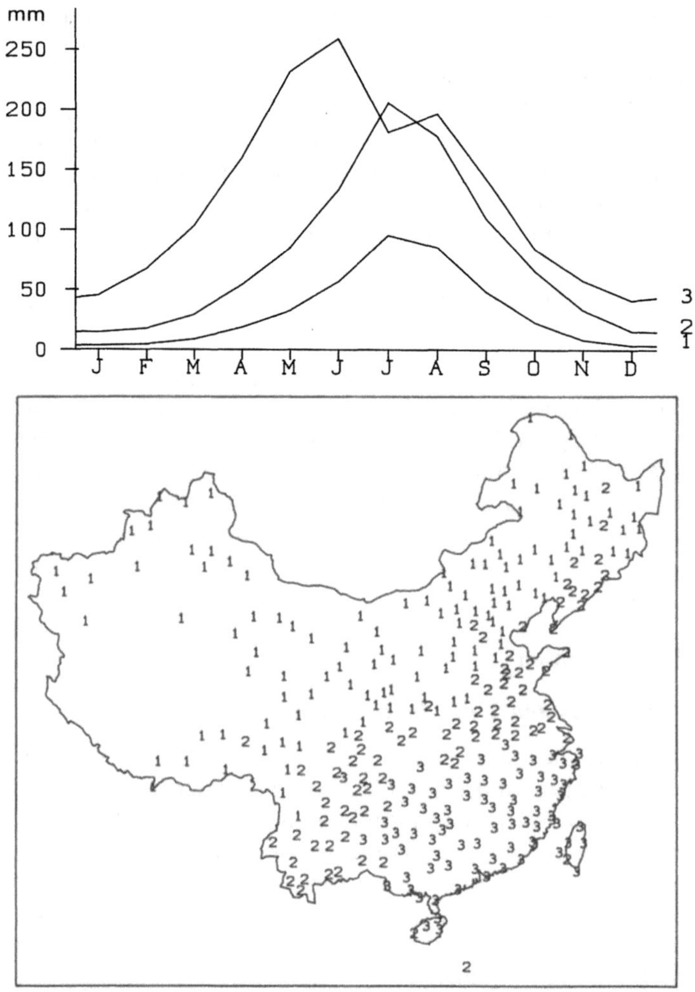

Fig. 5.7. Types of annual precipitation variation and their distribution over space; computed by a hierarchial cluster analysis after WARD

ary which roughly runs from the mouth of the Huang He in a northeast to southwest direction, finally terminating in the upper mountains of Yunnan Province.

Type 2: Strong Mid-Summer Precipitation Type

This type is similar to the weak mid-summer type, showing maximum precipitation in July and August and also producing substantial precipitation in the preceding months of May and June as well as in the succeeding month of September and sometimes even in October. But in sharp contrast, the strong mid-summer precipitation type is characterized by a much stronger precipitation regime in summer which, on the one hand, shows a much higher summer peak and also a longer duration of the wet summer season, on the other.

In regional terms, the strong mid-summer precipitation type represents a transitional type between the two other types, namely it extends in a roughly northeast to southwest direction, from the Great Plains along the northern foothills of the South China Hills up to the Sichuan Basin, thus widening southwards and covering Yunnan, Guizhou and Guangxi Provinces.

Type 3: Strong Early-Summer Precipitation Type

Maximum precipitation occurs in May and June, although a long and pronounced wet season continues from March or April through September and sometimes even until October. Significantly established is a secondary precipitation peak in August, thus the long wet season is characterized by a double-peak precipitation distribution. As can be deduced from the long period of summer precipitation, the winter season is of a short duration only. As far as the amount of precipitation in winter is concerned, the monthly totals substantially exceed the low winter totals, which are valid for the two other precipitation types. This type of annual precipitation variation is restricted to southeastern China, including Hainan and Taiwan Islands.

It is worth noting that the strong early-summer precipitation type is largely related to the "mei-yu" depressions and heavy rains associated with them in May and June. These rains, which are also known as "plum" rains, are associated with depressions passing seawards along a trough lying east-west somewhere between the Yangtze and the southern coast.

The three types of annual precipitation variation, which were statistically computed by applying a hierarchial cluster analysis (after Ward) and which aimed for the widest possible regional homogeneity, may show modifications to a certain extent if the annual variation of precipitation is considered for all stations of each type individually. As a case study, the graphs of precipitation variation over the year for each of the 75 stations of type No. 3 are shown in Fig. 5.8. It can be seen that despite limited deviations of the annual precipitation variations the principally valid precipitation type still is true.

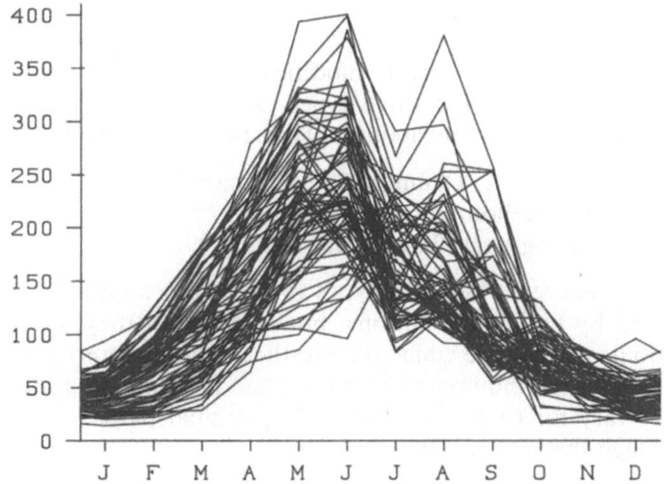

Fig. 5.8. Graphs of annual precipitation variation for all 75 stations under the "strong early-summer precipitation type"

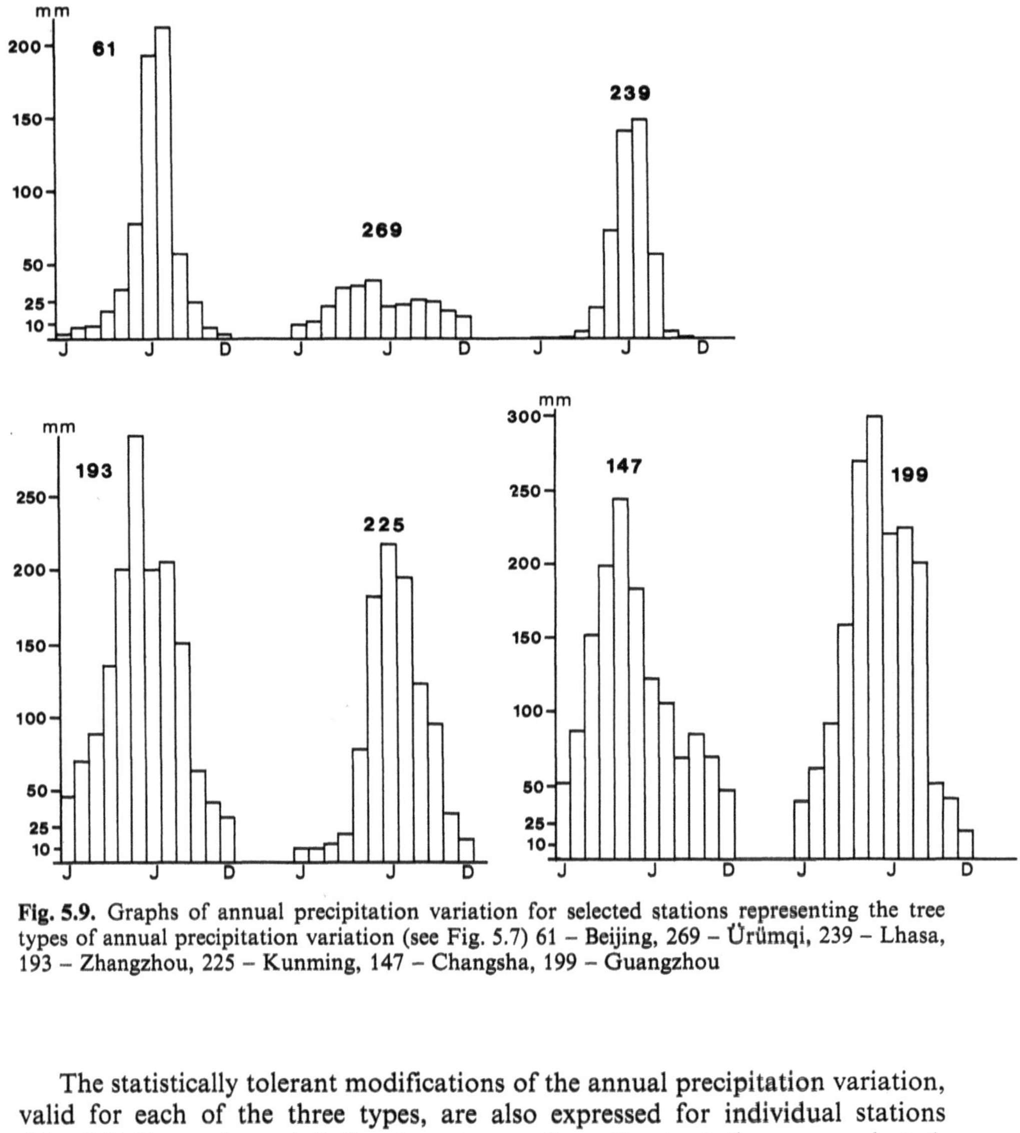

Fig. 5.9. Graphs of annual precipitation variation for selected stations representing the tree types of annual precipitation variation (see Fig. 5.7) 61 – Beijing, 269 – Ürümqi, 239 – Lhasa, 193 – Zhangzhou, 225 – Kunming, 147 – Changsha, 199 – Guangzhou

The statistically tolerant modifications of the annual precipitation variation, valid for each of the three types, are also expressed for individual stations (Fig. 5.9). For each precipitation type, two or three climate stations were selected:

Type 1: Beijing/61, Ürümqi/269 and Lhasa/239.
Type 2: Zhangzhou/193 and Kunming/225.
Type 3: Changsha/147 and Guangzhou/199.

The existence of three types of annual precipitation variation is thus confirmed by individual stations, but in addition they also prove certain modifications of each type which refer to the maximum precipitation during summer as well as to the minimum during winter. Thus, the annual range of precipitation may vary for each type, while the principal pattern of precipitation variation through the year remains the same.

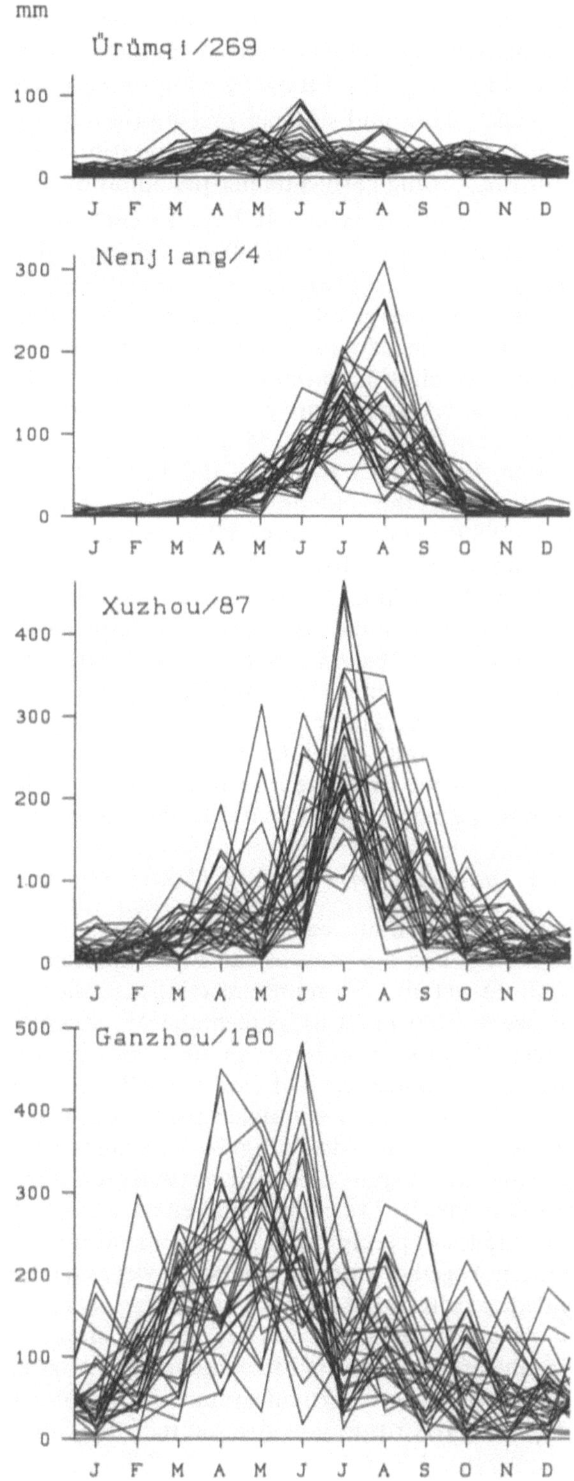

Fig. 5.10. 30-year graphs of annual precipitation variation at selected stations, representing the three types of annual precipitation variation in China

The three types of annual precipitation variation which are valid for all of China were further considered by studying for one representative station of each type all the 30 individual graphs of annual precipitation variation over the period given (Fig. 5.10). The following stations were selected:

- for the "weak mid-summer precipitation type": Nenjiang/4;
- for the "strong mid-summer precipitation type": Xuzhou/87;
- for the "strong early-summer precipitation type": Ganzhou/180.

Because of the extremely large extent over space of the weak mid-summer precipitation type, Ürümqi/269 was considered as an additional station for this type.

The individual graphs of annual precipitation variation for each station under investigation justify a collectively valid and statistically classified type of the annual precipitation variation, although remarkably large deviations may also occur (Fig. 5.10) which are largest during the wet summer, but smallest during the dry winter.

For the reader's own use and interpretation, for the four stations under investigation, the individual monthly and annual precipitation totals for 1951–1980 are given in Table 5.2 (a: Nenjiang/4; b: Xuzhou/87; c: Ganzhou/180; d: Ürümqi/269); p. 161–164.

For all of China, the three principal precipitation types and the individual graphs of the annual variation of precipitation underline a marked moisture regime which is characterized by an alternating wet summer and dry winter. The pronounced summer, however, is comparably short compared to the well-established, long winter. Both seasons vary regionally in duration. From south to north, the length of winter increases from about 3 to even 8 months, while summer, in the opposite direction, decreases from 9 to 4 months. Spring and autumn are normally comparably short and less marked.

5.3.2 Variation of Wet and Dry Months over Space and Time

Despite the pronounced seasonality of precipitation in China, the annual variation of wet and dry months can be considered as a major climate characteristic.

Since wet and dry represent relative rather than absolute terms, they can be defined by the monthly percentage of precipitation against the annual total. Under the assumption of an even precipitation distribution from January through December, each of the 12 months would receive an 8.33% portion. Thus, a month which receives a higher percentage can be called "wet", a month with a smaller percentage "dry". Figures determined accordingly may express a relative precipitation surplus or deficit respectively, if an even precipitation distribution from January to December is taken as a basis. Applying this criterion, monthly percentages of precipitation against the annual total were calculated for selected stations, together with the number of wet and dry months which were subsequently deduced from the monthly percentages (Table 5.3).

As can be seen, wet and dry months show a distinct seasonal variation. Wet months are concentrated in summer, dry months in winter. The number of wet months varies between a maximum of seven and a minimum of three; the majority of stations records four or five wet months. Considering the regional distribution

(continued p. 166)

Table 5.2. Monthly and annual precipitation totals at (a) Nenjiang/4, (b) Xuzhou/87, (c) Ganzhou/180 and (d) Ürümqi/269 from 1951 to 1980

(a) Nenjiang/4

Year	J	F	M	A	M	J	J	A	S	O	N	D	Total
1951	2	2	2	33	1	73	198	310	100	18	1	4	744
1952	3	.2	2	30	43	62	194	164	91	8	4	.5	602
1953	7	4	6	2	75	35	206	118	29	7	4	13	506
1954	4	.5	12	13	12	101	31	22	81	64	11	2	354
1955	9	11	18	21	33	81	57	62	138	16	4	22	472
1956	8	5	5	13	30	42	187	49	99	48	8	2	496
1957	10	8	10	11	60	51	127	77	52	7	5	10	428
1958	3	3	14	38	16	78	90	143	87	13	.5	5	491
1959	4	.3	.7	11	35	62	123	221	93	34	9	3	596
1960	4	5	9	15	50	92	129	64	100	.7	12	3	484
1961	3	1	14	2	32	22	150	264	89	21	1	10	609
1962	.6	.6	3	.4	36	67	153	124	63	.9	.1	5	454
1963	2	.9	12	29	46	94	114	43	109	23	8	6	487
1964	4	0	7	6	18	90	31	121	33	13	12	4	339
1965	2	5	4	6	49	22	142	96	19	49	4	7	405
1966	9	16	3	47	29	116	110	151	39	32	9	2	563
1967	.6	2	3	27	73	29	154	94	68	3	3	6	463
1968	3	6	5	14	42	65	90	100	38	4	7	8	382
1969	4	1	9	37	34	35	197	260	18	13	3	5	616
1970	.5	5	2	4	39	40	103	69	85	.5	5	7	360
1971	3	3	8	15	32	88	84	147	75	15	1	4	475
1972	2	2	5	7	39	81	164	87	73	43	7	4	514
1973	1	5	9	47	41	25	140	36	16	10	20	2	352
1974	3	2	.2	14	20	46	87	171	105	6	3	4	461
1975	3	2	3	17	59	101	145	39	44	30	6	.8	450
1976	2	3	.1	13	68	74	172	46	16	3	3	7	407
1977	.9	2	11	10	19	83	146	56	34	30	4	6	402
1978	2	1	6	33	39	156	132	18	77	18	1	8	491
1979	3	10	5	13	2	100	82	99	55	13	4	9	395
1980	4	1	1	17	7	108	162	115	108	8	3	1	535
Mean	4	4	6	18	36	71	130	112	68	18	5	6	478

Table 5.2 continued

(b) Xuzhou/87

Year	J	F	M	A	M	J	J	A	S	O	N	D	Total
1951	12	27	7	20	107	25	211	111	18	17	35	31	621
1952	.8	29	68	22	28	103	106	153	98	38	97	18	761
1953	.4	24	42	37	24	85	150	149	20	33	16	16	596
1954	26	45	4	33	236	36	245	65	16	68	71	35	880
1955	1	11	22	23	16	37	354	261	112	16	4	13	870
1956	4	.6	71	62	60	303	178	160	30	17	3	4	893
1957	56	8	8	50	40	86	336	112	0	19	29	28	772
1958	22	3	29	77	33	184	357	347	99	59	71	16	1297
1959	12	30	70	25	64	181	210	67	87	12	64	50	872
1960	4	0	51	13	45	156	287	326	173	21	51	11	1138
1961	11	1	21	32	60	116	87	189	149	11	34	18	729
1962	0	9	2	32	4	82	206	240	248	68	101	23	1015
1963	0	.3	58	79	314	21	453	217	51	2	12	8	1215
1964	44	33	20	137	73	55	277	101	140	79	6	6	971
1965	11	18	6	71	12	90	464	183	26	62	39	8	990
1966	6	22	102	66	6	100	219	10	22	24	14	22	613
1967	10	37	35	42	7	202	164	117	80	17	47	0	758
1968	4	0	25	51	20	19	301	62	62	57	3	34	638
1969	47	49	8	56	110	49	297	126	218	.7	5	4	970
1970	1	45	6	76	44	136	212	52	139	71	4	12	798
1971	16	18	39	31	13	263	178	265	31	9	10	17	890
1972	23	11	50	13	68	38	446	41	72	57	54	4	877
1973	26	28	21	131	57	77	296	91	36	45	1	0	809
1974	5	25	45	105	169	25	276	214	39	50	6	62	1021
1975	2	23	25	192	3	67	157	55	151	60	17	29	781
1976	0	57	3	84	48	53	197	161	74	3	23	9	712
1977	.7	0	38	98	42	40	217	39	47	112	30	33	697
1978	.1	22	20	6	8	102	230	210	62	28	22	6	716
1979	43	19	61	61	32	254	222	62	158	1	8	36	957
1980	17	2	41	41	90	127	102	204	20	128	5	0	777
Mean	14	20	33	59	61	104	248	146	83	39	29	18	854

Table 5.2 continued

(c) Ganzhou/180

Year	J	F	M	A	M	J	J	A	S	O	N	D	Total
1951	46	66	193	428	128	160	81	80	79	53	179	43	1536
1952	23	139	203	250	316	138	301	124	171	40	.6	75	1781
1953	131	105	192	242	304	182	195	116	72	127	60	183	1909
1954	24	37	58	344	388	268	28	147	54	2	2	87	1439
1955	16	45	175	52	176	136	201	209	66	2	68	46	1192
1956	65	143	178	149	184	221	47	53	19	10	57	31	1157
1957	21	187	177	186	275	214	39	88	117	184	25	40	1553
1958	59	83	74	131	270	164	231	32	58	12	.7	17	1132
1959	23	298	163	150	212	341	42	246	107	9	47	68	1706
1960	30	.2	229	134	290	158	55	227	140	63	73	35	1434
1961	23	117	200	290	290	366	127	285	255	17	138	76	2184
1962	3	17	52	201	295	482	36	118	27	79	76	8	1394
1963	4	73	111	101	33	301	82	86	80	22	131	49	1073
1964	194	98	138	149	84	363	12	112	83	59	0	2	1294
1965	16	43	129	189	197	252	151	85	25	122	54	46	1309
1966	35	34	172	188	88	247	33	47	34	78	25	60	1041
1967	36	110	158	142	217	18	108	152	59	7	29	29	1065
1968	25	88	107	122	184	474	214	112	19	71	39	66	1521
1969	176	94	211	74	185	168	60	80	64	159	40	17	1328
1970	108	54	216	259	203	206	129	131	132	138	29	137	1742
1971	40	57	40	56	148	165	111	226	87	31	2	63	1026
1972	11	128	48	310	186	146	64	78	33	137	122	121	1384
1973	98	40	111	449	373	199	99	221	92	105	70	.5	1858
1974	44	97	67	122	180	224	106	100	5	126	48	70	1189
1975	90	121	260	229	343	109	96	169	94	216	119	56	1902
1976	6	125	124	228	218	397	125	135	48	158	15	10	1589
1977	92	42	22	259	357	198	120	41	81	18	8	80	1318
1978	64	32	239	139	311	156	77	115	6	60	43	25	1267
1979	66	52	259	144	188	100	31	105	265	.1	20	15	1245
1980	66	113	207	317	174	67	133	185	100	77	10	19	1468
Mean	55	88	150	201	227	221	104	130	82	73	51	52	1435

Table 5.2 continued

(d) Ürümqi/269

Year	J	F	M	A	M	J	J	A	S	O	N	D	Total
1951	5	12	8	13	20	45	4	.3	32	43	23	19	224
1952	5	3	11	56	25	88	19	9	9	37	26	10	298
1953	14	25	62	17	14	8	17	57	7	44	37	9	311
1954	2	3	10	16	35	61	22	61	27	.1	17	21	275
1955	.5	7	23	33	57	9	40	21	15	26	22	12	266
1956	5	13	12	43	49	71	6	8	21	21	3	15	267
1957	8	11	8	41	15	13	34	24	22	44	10	10	240
1958	27	17	36	49	37	38	58	62	43	2	9	23	401
1959	4	11	43	51	39	33	44	30	27	39	20	21	362
1960	17	7	22	26	52	40	44	7	16	16	10	10	267
1961	13	2	14	9	37	14	21	19	8	26	16	3	182
1962	9	7	4	23	1	20	3	1	29	24	18	7	146
1963	.8	4	24	15	15	76	13	62	6	17	16	5	254
1964	3	4	45	48	36	17	8	13	20	7	4	20	225
1965	5	1	5	9	.2	33	15	21	4	28	20	12	153
1966	12	3	27	37	28	43	13	9	17	12	25	14	240
1967	5	8	8	13	32	11	31	16	8	13	13	.3	158
1968	.8	.6	26	35	4	12	19	9	20	10	10	3	149
1969	4	6	19	20	39	53	14	10	24	21	5	4	219
1970	2	5	15	18	58	12	27	30	7	15	26	7	222
1971	7	9	18	58	46	2	30	3	19	11	10	16	229
1972	10	7	20	29	3	35	38	12	37	21	20	11	243
1973	6	23	3	33	16	20	10	3	10	20	13	1	158
1974	6	10	28	10	7	12	5	21	12	12	5	2	130
1975	.2	12	10	22	7	91	20	11	17	17	10	10	227
1976	1	7	7	41	60	21	5	13	28	31	21	12	247
1977	10	16	9	16	27	7	11	5	9	14	11	27	162
1978	13	5	21	5	42	95	17	12	19	16	18	10	273
1979	19	9	19	53	35	46	25	6	66	3	37	8	326
1980	9	9	30	44	37	17	0	23	34	23	6	8	240
Mean	7	9	20	29	29	35	20	19	20	20	16	11	236

Table 5.3. Monthly percentages of precipitation against the annual total, together with the number of wet and dry months respectively, at selected stations

No. of months															
Wet	Dry	Station		J	F	M	A	M	J	J	A	S	O	N	D
5	7	2	Huma	0.8	0.8	1.8	4.9	8.3	15.2	23.9	21.8	13.7	4.9	2.7	1.8
4	8	15	Harbin	0.7	0.9	2.2	4.5	7.2	14.9	30.7	18.6	12.7	5.3	1.3	1.1
4	8	21	Don Ujmqin Qi	0.9	0.9	1.6	3.6	7.2	18.5	29.3	24.0	9.8	2.5	1.0	0.7
4	8	61	Beijing	0.5	1.1	1.3	3.0	5.1	12.1	29.9	33.0	8.8	3.7	1.0	0.4
4	8	252	Bayan Mod	0.7	0.5	1.6	4.3	7.0	13.3	28.5	28.5	9.6	4.2	1.7	0.6
4	8	69	Taiyuan	0.7	1.3	2.2	5.2	6.6	11.5	25.7	22.5	14.0	6.7	3.7	0.7
4	8	90	Zhengzhou	1.4	2.0	4.6	8.0	7.3	10.7	21.2	21.3	10.6	6.4	5.4	1.2
7	5	123	Hangzhou	4.6	6.0	8.3	9.3	12.3	13.7	9.4	9.7	13.1	4.8	4.4	3.5
6	6	126	Wuhan	2.8	4.8	8.3	11.5	12.8	17.3	14.2	10.6	6.4	4.2	4.5	2.7
5	7	147	Changsha	3.7	6.1	10.7	14.0	17.2	13.0	8.7	7.5	4.9	6.0	5.0	3.4
5	7	134	Chengdu	0.6	1.2	2.6	5.4	9.5	11.6	24.9	24.8	12.5	4.9	2.0	0.6
5	7	215	Nanning	3.1	3.2	4.9	6.6	14.3	18.8	14.0	15.9	8.9	5.2	3.4	1.9
5	7	225	Kunming	1.0	1.0	1.4	2.0	7.9	18.3	21.8	17.7	12.4	9.6	3.4	1.6
6	6	199	Guangzhou	2.3	3.7	5.4	9.4	15.9	17.8	13.0	13.4	12.1	3.0	2.5	1.2
5	7	210	Dongfang	0.9	1.6	2.0	3.0	6.0	13.3	13.4	28.6	20.2	8.4	2.2	0.5
7	5	269	Ürümqi	3.1	3.8	7.7	12.4	12.6	14.1	7.7	8.5	9.3	8.8	6.7	5.3
7	5	276	Hotan	4.5	8.4	2.4	8.4	20.4	20.6	11.4	10.2	8.7	1.8	1.2	2.1
4	8	239	Lhasa	0.0	0.0	0.3	1.0	4.5	16.1	31.2	32.8	12.6	1.1	0.2	0.1

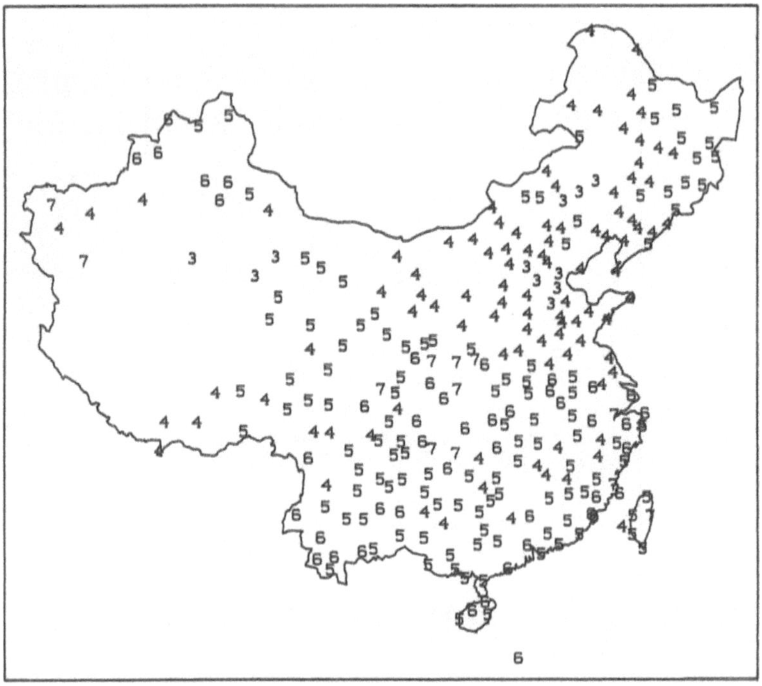

Fig. 5.11. Distribution of the mean number of wet months/year. A wet month is defined by > 8.33% of precipitation against the annual total

of wetness and dryness, no clear pattern of either a prevailing latitudinal or longitudinal distribution can be observed (Fig. 5.11). Instead, the number of wet months shows regional differences on a medium scale, indicating between four and six wet months over most parts of China.

The division of the year into a short, wet summer and a long, dry winter appears as the leading characteristic of the annual variation of precipitation. This observation was further confirmed by considering the total of all 279 stations under investigation and determining for each month the number of wet and dry cases respectively; the corresponding figures are as following:

	J	F	M	A	M	Jn	Jl	A	S	O	N	D
Wet:	1	2	27	81	106	277	256	265	218	39	5	2
Dry:	278	277	251	198	173	2	23	14	60	239	274	277

As can be deduced from the monthly totals of wet and dry cases respectively, only the four months from June to September show an overwhelming majority of wet cases while, in contrast, for the 8-month period from October to May an extremely large majority of dry cases is noted. Thus, a short, wet summer and a long, dry winter are strikingly alternating and governing the pattern of the annual precipitation variation.

As can be inferred from the pronounced annual precipitation variation, the marked division of the year into a 4-month wet and an 8-month dry period can be considered as the principal feature of the annual variation of precipitation in China. Only for April and May (which show a majority of 198 and 173 dry cases respectively of all 279 stations) a fair number of wet cases has to be noted; this observation gives evidence of a longer duration of the wet summer over larger parts of China, lasting 5 or 6 months. Such conditions particularly occur under the strong early-summer precipitation type which mainly applies to southern China.

Considering the common division of the year into a wet summer and dry winter, the extent of a relative precipitation surplus and deficit respectively, however, varies over space. As can be inferred from the monthly precipitation percentages for selected stations (Table 5.3), the maximum values for a wet month may even rise to 33% of the annual total, but may fall for a dry month to 0% (see, for example, Beijing/61 in August, and Lhasa/239 in January and February; Table 5.3). Also worth noting are moderate percentages for both the wet summer and dry winter at the same station, which express a fairly even precipitation distribution throughout the year, for example at Hangzhou/123, Changsha/147 and Ürümqi/269 (Table 5.3).

The main contrasting conditions between a wet summer and a dry winter are also expressed by diagrams showing the monthly percentages of precipitation against the annual total for selected stations (see Fig. 5.12). Values >8.33% can

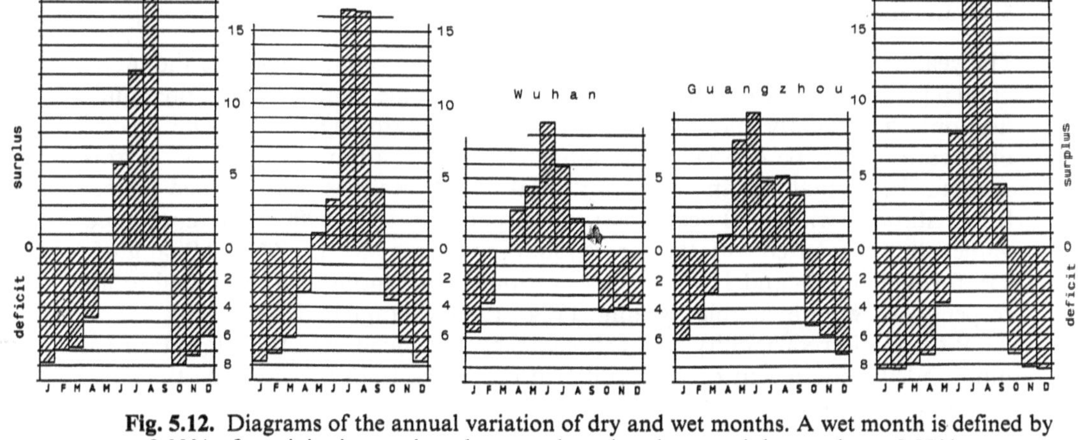

Fig. 5.12. Diagrams of the annual variation of dry and wet months. A wet month is defined by >8.33% of precipitation against the annual total, a dry month by a value <8.33%

be regarded as a relative precipitation surplus, while a precipitation deficit is given in case of figures <8.33%. As can be seen from the diagrams, the division of the year into an alternating wet and dry season shows minor modifications in two directions: (1) the length of the wet season may be extended (see the diagrams for Wuhan and Guangzhou); (2) the percentages of precipitation, valid for a wet summer, may differ considerably, thus underlining a highly fluctuating annual precipitation variation.

Irrespective of the type of annual precipitation variation, in most parts of China the wet summer covers 4 months (June to September) and the comparably long winter 8 months (October to May). Undoubtedly the annual variation of precipitation shows a very unbalanced picture in that a well-established, long dry season governs the extremely uneven precipitation distribution throughout the year.

5.3.3 Summer Precipitation

Underlining the distinctly established annual variation of precipitation and the prevailing summer rains in China, the percentage of summer precipitation against the annual total can also be considered. "Summer", in this respect, was confined to the 3-month period from June to August. The percentage figures of precipitation for summer are presented in Fig. 5.13.

The strong prevalence of summer rains can be seen, since in most parts of China between 50 and 70% of all precipitation occurs during the 3-month period from June to August; only the South China Hills and the middle and lower Yangtze valley experienced a 30–50% portion of precipitation from June to August. At the most, a 83% portion of summer rain was recorded at Xigaze/238, while a minimum value of 27% only was experienced at Lingling/179. The only medium to smaller percentages of summer precipitation over the South China Hills, including the Yangtze valley, underline a less pronounced summer peak and thus less annual precipitation variation. In the major parts of China, however, a strongly established summer peak of precipitation is once again confirmed (see Sect. 5.3.1).

5.4 Interannual Precipitation Variability

The year to year variation of precipitation, which is commonly called the interannual precipitation variability, is usually related to the annual total of precipitation only whereas the monthly and seasonal totals are neglected. More precise observations can be expected, the longer the observation period lasts. In this section, the annual and monthly precipitation variabilities were studied for selected stations with a 30-year observation period (1951–1980). The computed values were compared with the corresponding values at Beijing/61 and Shanghai/119 for which 257- and 108-year observations of the monthly and annual totals respectively, were available (Beijing, 1724–1980; Shanghai, 1873–1980).

Precipitation variability was expressed by the mean standard deviation (in mm) and the coefficient of variability (in per cent), both representing the commonly applied indices (Tables 5.4 and 5.5).

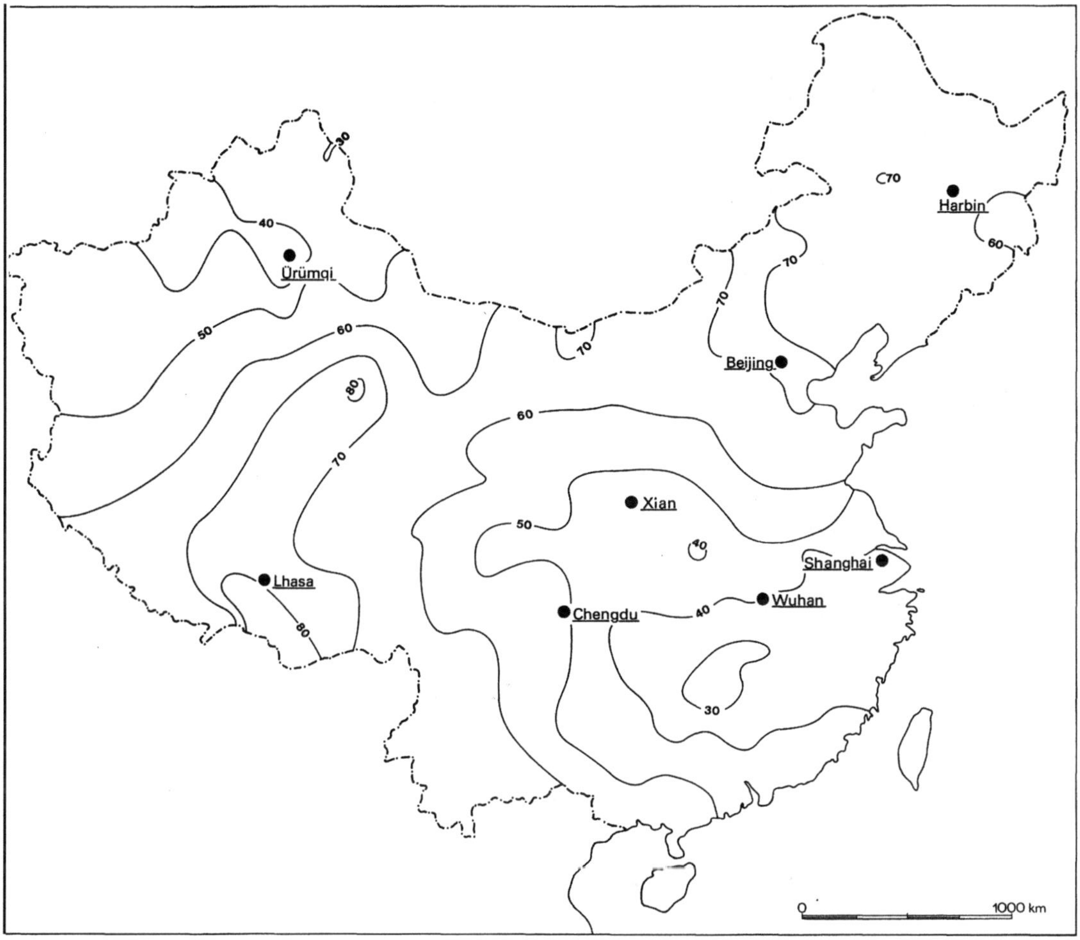

Fig. 5.13. Percentage of summer rain (June to August) against the annual total

5.4.1 Variability of Annual Precipitation

As the mean annual precipitation totals vary considerably over space (see Sect. 5.1), both the mean standard deviation and the coefficients of variability also show remarkably large regional differences (Tables 5.4 and 5.5). It can be seen from the figures given for selected stations that the mean annual standard deviation (also called variance) reached its highest values in the case of a high annual precipitation total, while the lowest values were found by the lowest annual totals. The close relationship between the annual precipitation total and the mean standard deviation was proved by correlating both parameters for the stations under consideration (Fig. 5.14). It can be seen that, as a rule, the mean standard deviation increased with increasing annual precipitation totals and vice versa. Taking the mean standard deviation as a measure for the interannual precipitation variability, precipitation varies to a greater extent with higher totals,

(continued p. 173)

Table 5.4. Annual and monthly mean standard deviation of precipitation (mm) at selected stations; 1951–1980

No.	Station	J	F	M	A	M	J	J	A	S	O	N	D	Year
4	Nenjiang	3	4	5	13	19	32	47	74	33	16	4	4	93
9	Keshan	2	4	4	11	25	42	78	57	42	16	4	5	122
36	Yingkou	7	10	9	27	27	39	109	91	43	33	17	8	153
61	Beijing	5	9	6	28	33	60	94	144	47	26	5	4	242
82	Anyang	5	10	15	22	35	60	91	119	53	29	19	8	191
87	Xuzhou	16	16	25	42	70	76	99	88	64	33	28	15	173
119	Shanghai	28	33	34	45	51	61	88	96	83	43	32	31	197
123	Hangzhou	45	55	56	39	91	88	81	73	99	59	37	37	292
127	Jiangling	18	24	39	58	65	99	105	85	86	53	35	19	261
140	Neijiang	8	10	14	31	37	84	107	116	81	34	15	8	227
180	Ganzhou	48	59	69	100	88	112	69	65	62	61	47	41	296
217	Bose	20	12	21	40	88	92	100	81	65	69	38	19	198
207	Haikou	19	22	38	66	71	119	127	108	176	143	124	43	347
269	Ürümqi	6	6	14	16	18	27	14	18	13	12	9	7	65
263	Dunhuang	1	2	2	5	3	9	15	6	3	2	2	1	21
255	Tianzhu	3	4	10	11	21	23	32	41	32	9	5	2	95
279	Kashi	3	12	8	10	18	6	13	10	8	4	3	3	29
61	Beijing (1724–1980)	5	5	8	17	25	46	121	94	37	18	10	4	193
119	Shanghai (1873–1980)	34	33	41	43	46	86	88	84	84	63	41	34	201

Table 5.5. Annual and monthly coefficients of precipitation variation (%) at selected stations; 1951–1980

No.	Station	J	F	M	A	M	J	J	A	S	O	N	D	Year
4	Nenjiang	73	102	74	72	54	45	36	66	49	89	80	78	20
9	Keshan	76	133	71	65	66	57	50	55	63	81	72	108	24
36	Yingkou	101	121	79	75	59	58	64	55	53	72	81	94	23
61	Beijing	161	93	72	144	94	76	47	64	81	103	87	154	36
82	Anyang	121	108	103	74	93	104	49	81	93	78	88	126	32
87	Xuzhou	117	82	76	42	70	76	99	88	64	33	28	15	173
119	Shanghai	19	16	13	13	12	11	20	23	17	26	20	23	5
123	Hangzhou	72	62	49	30	50	45	64	53	56	76	68	69	21
127	Jiangling	68	57	49	48	40	60	70	67	103	66	71	69	23
140	Neijiang	55	61	60	54	39	54	50	61	59	45	48	58	22
180	Ganzhou	89	67	46	50	39	51	66	50	76	84	91	78	21
217	Bose	107	74	72	61	52	47	55	41	59	86	111	120	18
207	Haikou	81	73	73	71	38	49	61	45	58	83	127	112	21
269	Ürümqi	83	69	69	54	61	78	70	94	65	60	56	63	28
255	Tianzhu	124	92	88	51	54	37	35	44	51	49	98	126	23
279	Kashi	138	180	168	189	136	110	185	145	132	226	193	174	46
61	Beijing (1724–1980)	151	89	94	98	81	64	57	54	65	92	119	142	32
119	Shanghai (1873–1980)	72	54	49	45	44	49	61	62	62	91	77	87	18

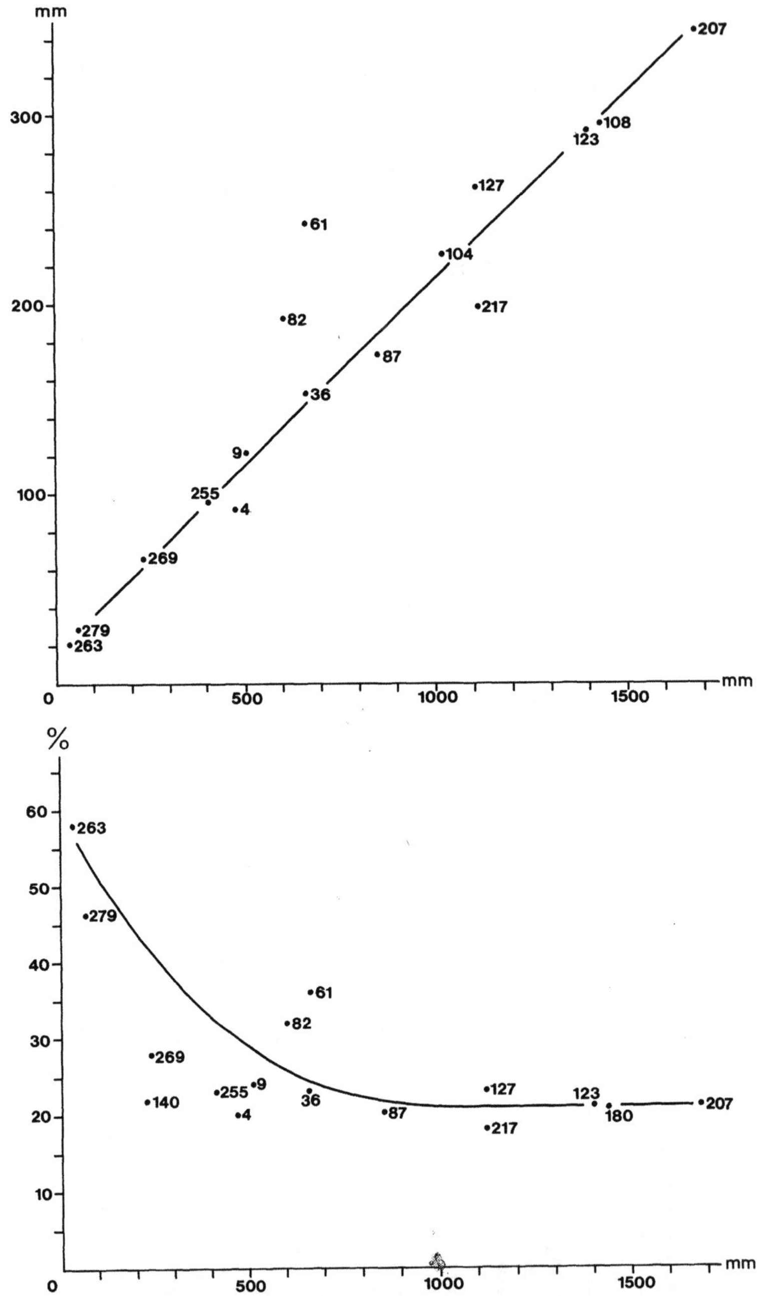

Fig. 5.14 a, b. The relationship between annual precipitation total and (**a**) mean standard deviation (cf. Table 5.4); (**b**) coefficients of variability (cf. Table 5.5)

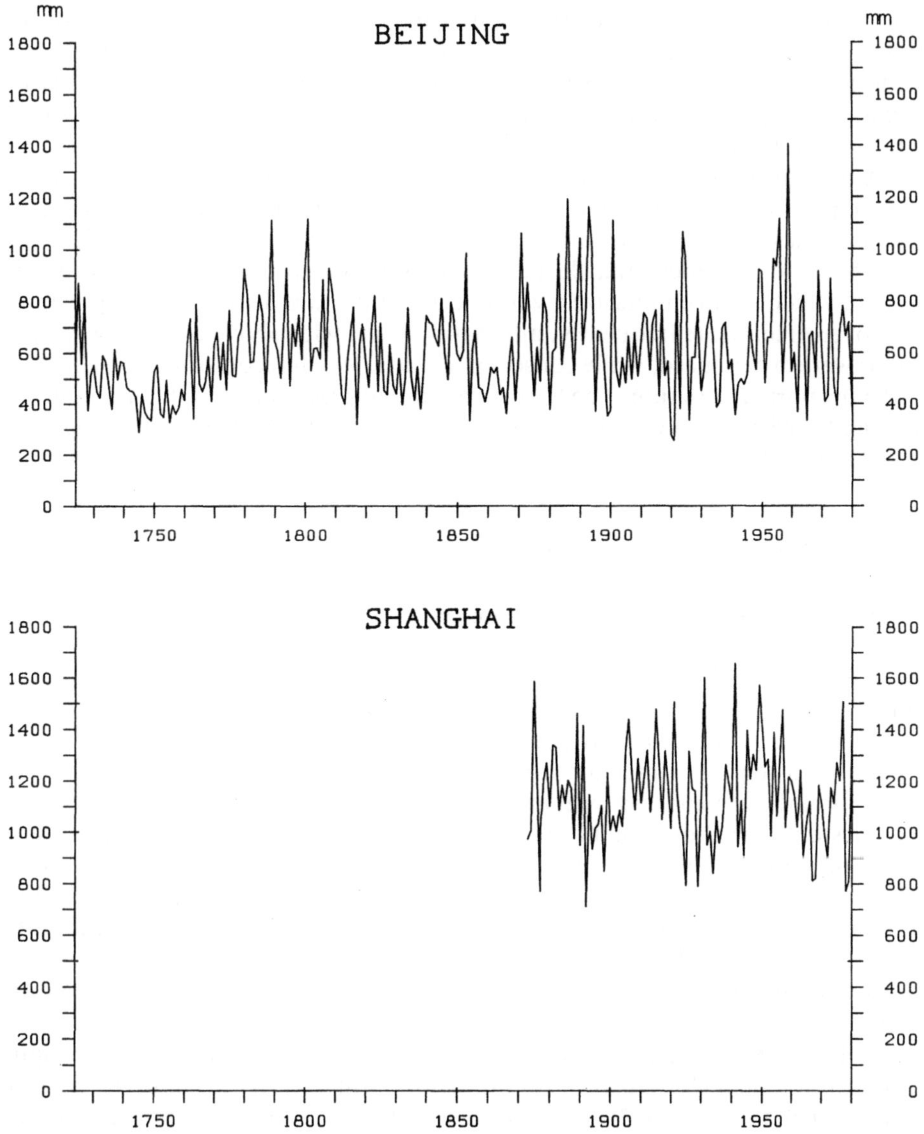

Fig. 5.15. Annual precipitation series at Beijing from 1724–1980 and Shanghai from 1873–1980

but to a lesser extent with smaller totals. Therefore, large precipitation totals are highly variable and less stable, while small precipitation totals are less variable and more reliable.

If precipitation variability were, however, expressed in relative terms and thus measured by the coefficient of variability, the opposite observation would be true (Table 5.5). The coefficient of variability represents the quotient between the mean standard deviation and the mean precipitation total (expressed in percent).

The close relationship between the coefficient of variability and the mean precipitation total is shown in Fig. 5.14 b and clearly expresses the increasing coefficients with decreasing precipitation totals and vice versa; this underlines, in relative terms, the largest interannual variability in cases of small precipitation totals, while a small variability is observed in cases of high precipitation totals.

Although without using any statistical index, the disrupt variation of the annual precipitation totals over time can also be inferred from the annual precipitation series at Beijing from 1724–1980 and at Shanghai from 1873–1980 (Fig. 5.15). It is evident that large year to year variations of precipitation occurred, which again emphasize a variable and uncertain distribution of precipitation over time. More frequently, alternating years with above and below normal totals were experienced, while consecutive years, which record totals either above or below normal, are rare. The long-term annual precipitation total amounted at Beijing to 609 mm (in comparison with 644 mm for 1951–1980); at Shanghai to 1,144 mm (1,124 mm for 1951–1980).

5.4.2 Variability of Monthly Precipitation

Similar to the annual values of the mean standard deviation and coefficient of variability (Sect. 5.4.1), the monthly values also showed remarkable differences, both for each station throughout the year and from station to station (Tables 5.4 and 5.5).

In all cases, the standard deviation for the summer months showed higher values than for the winter months (see Table 5.4). This is particularly true for stations of the wet and semi-wet eastern part of China, whereas the stations of the dry western part show small differences of the standard deviation between summer and winter. Thus, it can be deduced that the mean standard deviation is closely related to the precipitation total of any month under consideration. Furthermore, as in the case of the annual standard deviation, increasing monthly values occurred with increasing monthly precipitation totals, and vice versa. Therefore, the interannual variability of precipitation is greater for wet months, but smaller for dry months.

Taking the coefficient of variability as a measure for the interannual precipitation variability, a close relationship to the monthly precipitation totals could be proved such that the coefficients increase with decreasing totals of precipitation, and vice versa (Table 5.5). Thus, in relative terms, small precipitation totals are most variable, large totals, however, least variable. According to this rule, the rather small winter precipitation is also extremely uncertain and variable from year to year while, in contrast, the heavy summer rains are less variable and more stable.

5.4.3 Variability of Annual and Monthly Precipitation at Beijing

As a case study, the variability of precipitation was considered at Beijing for a 257-year period from 1724–1980, due to the fortunate opportunity that continu-

Table 5.6. Annual and monthly variability of precipitation at Beijing/61 for the 257-year observation period from 1724–1980, expressed by various calculations (all values are rounded, given in mm)[a]

	J	F	M	A	M	J	J	A	S	O	N	D	Year
1	4	7	11	24	44	89	271	224	70	27	11	4	717
2	0	3	3	6	13	43	124	106	35	7	3	1	467
3	8	11	17	36	61	125	364	303	101	39	18	8	877
4	0	0	1	1	5	36	84	76	23	3	1	0	380
5	49	31	69	142	146	282	872	575	292	133	113	29	1,406
6	0	0	0	0	0	1	28	23	0	0	0	0	256
7	4	5	8	17	31	72	210	174	57	20	8	3	609

[a] 1. Upper quartile value (mm); 2. lower quartile value (mm); 3. upper decile value (mm); 4. lower decile value (mm); 5. recorded maximum value (mm); 6. recorded minimum value (mm); 7. arithmetic mean value (mm).

ous data were available for such an exceptionally long observation period. It must be taken into consideration, however, that the data series may be inhomogeneous to a slight extent, due to the fact that the location of the meteorological station at Beijing was shifted a few times.

For the observation period 1724–1980, the mean standard deviation for the annual precipitation total amounted to 193 mm which is considerably lower than the value calculated for the 30-year observation period (242 mm, see Table 5.4). The annual coefficient of variability amounted to 32% which also differs from the corresponding value of the 30-year observation period (36% see Table 5.5). Furthermore, the monthly coefficients of variability of the 257-year period are in most cases smaller than for the 30-year observation period; the same is true for the values of the monthly mean standard deviation (see Tables 5.4 and 5.5).

For a further clarification of the annual and monthly variability of precipitation, other calculations were considered, such as the upper and lower quartile values and upper and lower decile values, again referring to the 257-year observation period from 1724–1980. The thus resulting interquartile range and interdecile range of precipitation respectively, express a 50 and 80% probability of precipitation respectively. Moreover, the recorded maximum and minimum values as well as the means for the given period were considered (Table 5.6).

For the reader's own use and interpretation, the recorded annual and monthly totals of precipitation at Beijing for a 100-year observation period from 1881–1980 are given in Table 5.7a without applying any calculations. For comparison, the corresponding figures at Shanghai[20] were added (Table 5.7b). All values are rounded, with the exception of a precipitation total <1 mm.

(continued p. 182)

[20] Among all climatological observatories in China, Shanghai has had the longest observation period with a completely homogeneous data series since 1881, which is due to the permanent site of the meteorological station.

Table 5.7. Monthly and annual precipitation totals at (a) Beijing and (b) Shanghai from 1881 to 1980

(a) Beijing/61

Year	J	F	M	A	M	J	J	A	S	O	N	D	Total
1881	.1	3	.2	57	13	135	74	228	74	5	12	.5	602
1882	1	5	1	9	34	69	169	168	117	28	18	.5	620
1883	3	15	11	4	46	6	631	188	42	29	9	0	984
1884	1	0	14	3	16	1	232	188	75	17	7	.7	555
1885	3	3	9	34	15	92	254	208	31	18	5	3	675
1886	5	3	2	33	89	110	522	279	44	86	13	7	1193
1887	4	4	9	9	60	59	197	322	45	30	4	.7	744
1888	6	10	4	21	0	43	148	215	45	10	9	.7	512
1889	6	5	9	14	3	53	271	243	119	31	16	0	770
1890	0	2	.5	9	21	15	872	90	2	27	7	0	1046
1891	5	4	6	24	6	157	273	97	33	21	5	.7	632
1892	14	.8	3	19	20	64	447	108	73	1	19	0	769
1893	8	0	7	24	61	237	583	139	90	7	8	0	1164
1894	7	.5	10	35	15	282	323	244	54	23	14	2	1010
1895	1	7	8	56	5	11	134	66	17	62	0	4	371
1896	.7	.3	18	33	32	86	295	163	46	7	1	3	685
1897	16	0	27	58	12	59	372	70	42	14	4	0	674
1898	1	9	5	8	1	56	257	129	72	.5	19	0	558
1899	4	16	30	22	23	87	129	23	0	0	13	5	352
1900	2	3	5	11	19	44	130	106	35	12	5	2	374
1901	4	5	.8	6	46	120	408	310	154	39	19	.7	1113
1902	4	3	10	14	14	54	246	105	57	12	5	9	533
1903	.8	3	13	26	4	37	89	166	110	12	6	0	467
1904	3	5	8	17	29	68	203	165	54	18	8	3	581
1905	5	.9	35	13	48	97	111	92	36	44	0	.1	482
1906	4	1	25	.7	47	46	246	183	58	54	.1	0	665
1907	.2	.8	.1	35	3	13	278	80	36	42	9	1	498
1908	49	8	.5	2	4	58	231	137	172	15	1	0	678
1909	3	4	7	15	25	59	178	145	48	16	7	2	509
1910	5	0	5	16	13	118	261	139	39	9	23	0	628
1911	21	0	69	24	10	40	217	302	44	5	21	.9	754
1912	4	6	10	21	37	86	255	208	68	23	10	3	731
1913	3	5	7	15	27	62	186	151	50	17	8	2	533

(contd.)

Table 5.7 continued

Year	J	F	M	A	M	J	J	A	S	O	N	D	Total
1914	1	10	40	0	9	117	256	86	34	96	71	.8	721
1915	3	4	3	.5	58	90	451	92	33	19	5	4	763
1916	2	6	3	15	28	46	81	179	33	8	21	7	429
1917	.2	.2	3	0	20	38	438	120	122	39	0	2	782
1918	.9	.8	15	21	117	68	113	142	5	.6	27	1	511
1919	4	0	8	4	24	91	264	88	26	52	4	2	567
1920	3	6	5	1	13	42	144	27	25	.9	4	7	278
1921	.6	0	4	.4	35	45	99	62	9	1	.1	0	256
1922	1	4	3	11	8	135	298	337	36	.5	3	.3	837
1923	.3	.3	16	12	4	9	138	156	36	4	1	4	381
1924	.4	6	3	0	20	55	648	254	37	21	15	7	1066
1925	0	0	3	43	61	134	448	197	64	.9	6	0	957
1926	0	2	2	14	38	72	111	55	28	.4	10	2	334
1927	20	.5	11	7	15	36	330	142	15	.3	0	5	582
1928	16	0	7	22	20	64	145	193	86	12	11	7	583
1929	17	.8	2	9	10	30	346	296	12	15	.5	29	767
1930	0	27	0	7	56	51	106	113	29	60	.1	2	451
1931	2	2	6	15	51	44	154	199	48	.5	2	12	536
1932	.6	.9	0	7	41	77	400	42	118	.2	1	0	688
1933	.4	.6	16	32	60	282	183	116	51	4	12	5	762
1934	1	13	3	10	46	162	116	202	99	6	.1	3	661
1935	1	0	3	0	20	66	125	139	2	11	19	0	386
1936	2	21	0	9	30	38	116	53	127	0	7	6	409
1937	.6	4	17	22	18	52	201	273	94	6	5	.3	693
1938	.2	.8	10	7	110	59	304	124	27	62	.5	11	716
1939	.2	4	17	3	30	36	176	172	19	58	20	.7	536
1940	1	.3	.1	13	24	74	177	118	46	.3	113	5	572
1941	3	6	1	.8	7	115	56	89	57	6	9	5	355
1942	.1	3	2	26	9	30	248	123	17	11	8	0	477
1943	4	.1	1	16	62	11	48	182	127	2	41	6	500
1944	.2	1	4	27	31	59	231	70	42	1	9	0	475
1945	.4	3	6	22	34	25	63	260	82	1	15	2	513
1946	3	0	38	41	14	24	367	117	61	34	19	.6	719
1947	10	12	9	0	12	41	195	169	141	.8	0	12	602
1948	.2	7	4	33	85	49	219	101	17	5	10	3	533

(contd.)

Table 5.7 continued

Year	J	F	M	A	M	J	J	A	S	O	N	D	Total
1949	0	6	3	7	58	113	417	208	69	17	7	18	923
1950	0	7	.6	132	53	59	289	334	20	7	9	.6	911
1951	8	10	1	.9	146	53	73	123	26	36	5	.8	483
1952	.1	4	4	9	88	90	141	259	26	28	6	2	657
1953	.1	4	4	9	88	90	141	259	26	28	6	2	657
1954	.8	7	5	21	30	232	223	393	14	16	16	4	962
1955	0	.2	9	10	103	53	173	395	144	33	12	.9	933
1956	0	12	11	9	26	252	131	476	129	54	16	0	1116
1957	6	8	7	29	3	54	103	225	39	0	13	0	487
1958	4	0	7	14	38	37	243	172	131	35	5	6	692
1959	.6	31	20	11	4	84	511	575	149	16	3	2	1407
1960	11	3	6	3	10	38	281	91	52	24	6	1	526
1961	4	4	20	7	13	28	255	128	119	10	11	2	601
1962	2	18	2	13	27	50	201	34	18	0	2	.1	367
1963	.9	.8	12	33	52	6	162	492	8	5	4	.2	776
1964	8	6	11	142	13	37	125	358	81	39	.1	0	820
1965	0	6	.1	24	9	42	110	109	20	6	8	0	334
1966	.5	12	18	9	42	58	186	308	18	8	0	2	662
1967	4	7	9	30	24	90	283	176	30	25	2	0	680
1968	.7	0	5	11	5	17	199	77	107	60	13	9	504
1969	.8	4	17	30	41	44	311	320	123	6	18	.1	915
1970	.3	7	13	28	48	67	258	152	26	30	0	0	629
1971	0	10	21	3	15	102	114	97	42	5	.5	0	410
1972	12	7	.2	3	9	15	225	27	100	22	11	0	431
1973	20	6	12	6	20	81	278	327	99	30	7	0	886
1974	.4	2	2	4	28	26	248	98	41	10	4	13	476
1975	.1	.3	1	.4	11	56	192	109	14	2	7	0	393
1976	0	11	4	.7	8	144	160	274	40	38	4	2	686
1977	.5	2	5	15	68	179	260	122	7	133	3	16	811
1978	0	6	7	3	37	106	115	290	81	14	2	4	665
1979	.6	22	11	74	40	135	228	191	4	5	.5	7	718
1980	.6	9	15	24	27	114	31	98	27	34	0	3	383
Mean	4	5	9	19	32	74	237	178	56	20	10	3	648

Table 5.7 continued

(b) Shanghai/119

Year	J	F	M	A	M	J	J	A	S	O	N	D	Total
1881	.7	54	140	120	89	170	140	257	155	141	47	28	1342
1882	108	47	15	86	111	231	275	215	91	22	105	26	1332
1883	15	91	57	96	174	123	125	184	65	40	100	16	1086
1884	36	62	75	64	102	126	120	152	147	149	147	5	1185
1885	50	43	98	136	107	290	91	53	143	32	19	52	1114
1886	31	44	75	65	94	310	3	343	91	134	9	4	1203
1887	197	38	35	37	91	280	168	61	236	16	10	4	1173
1888	70	94	117	59	56	86	95	56	161	96	64	24	978
1889	43	58	72	75	64	152	276	243	140	304	30	7	1464
1890	29	91	128	88	60	197	115	92	49	8	13	76	946
1891	28	78	49	89	35	63	240	333	253	162	25	62	1417
1892	12	71	110	99	160	66	7	28	73	15	63	5	709
1893	72	30	60	65	103	144	91	332	158	80	7	6	1148
1894	47	17	145	95	136	114	92	99	51	76	46	16	934
1895	19	53	85	99	57	225	129	177	81	19	50	23	1017
1896	29	50	110	41	149	246	95	51	22	160	56	23	1032
1897	71	20	152	86	108	19	234	172	114	72	36	22	1106
1898	29	79	101	135	159	54	27	151	37	41	31	7	851
1899	21	82	55	65	77	134	172	290	111	80	55	92	1234
1900	81	29	48	123	38	159	138	90	167	29	88	19	1009
1901	166	0	42	85	70	190	296	12	72	106	14	11	1064
1902	19	9	66	149	97	67	231	181	41	55	18	73	1006
1903	23	41	138	125	102	231	306	28	40	25	25	1	1085
1904	11	25	126	213	111	42	110	74	139	138	8	26	1023
1905	97	30	154	116	125	77	231	278	69	68	3	84	1332
1906	107	181	65	91	118	197	197	150	202	83	29	22	1442
1907	59	61	99	54	71	137	204	198	59	175	109	10	1236
1908	46	34	49	156	62	130	195	130	84	136	24	43	1089
1909	54	46	145	45	23	325	88	109	186	169	50	49	1289
1910	133	29	160	70	104	285	82	51	40	37	103	22	1116
1911	34	32	133	95	105	182	177	143	123	63	31	100	1218
1912	42	39	117	94	71	298	218	258	54	47	53	31	1322
1913	54	71	49	144	119	183	228	28	109	1	63	28	1077

(contd.)

Table 5.7 continued

Year	J	F	M	A	M	J	J	A	S	O	N	D	Total
1914	3	80	85	97	69	240	85	105	211	83	121	26	1205
1915	27	103	52	125	64	254	271	123	96	220	144	0	1479
1916	19	69	80	128	116	225	235	173	88	68	43	27	1271
1917	13	19	51	34	68	302	218	133	52	74	68	19	1051
1918	0	19	109	74	58	238	175	177	69	51	196	153	1319
1919	81	48	120	47	70	308	288	92	56	32	24	25	1191
1920	28	108	66	94	84	139	148	86	93	19	25	128	1018
1921	17	38	80	150	130	257	152	285	304	52	26	16	1507
1922	73	91	52	52	102	231	70	134	283	57	12	4	1161
1923	7	91	86	128	126	166	210	75	48	10	65	10	1022
1924	31	103	63	32	142	235	9	81	198	66	21	7	988
1925	63	32	67	22	144	42	170	94	95	5	50	12	796
1926	21	28	58	56	64	251	123	261	243	61	73	80	1319
1927	16	113	138	125	41	206	127	212	116	31	36	10	1171
1928	98	33	78	82	12	215	125	115	321	1	43	37	1160
1929	57	18	9	28	106	70	128	153	48	9	26	140	792
1930	48	32	100	116	99	179	139	123	110	98	43	57	1144
1931	33	148	39	90	185	140	370	112	342	24	70	51	1604
1932	8	35	23	123	213	182	57	155	78	18	21	40	953
1933	83	82	67	129	83	110	35	140	188	57	16	13	1003
1934	41	28	93	106	91	42	35	20	233	35	81	37	842
1935	26	73	70	74	25	217	95	173	86	72	85	66	1062
1936	54	57	45	105	94	112	210	117	43	.1	52	69	958
1937	83	90	72	97	78	113	17	87	110	103	122	52	1024
1938	73	42	104	56	155	469	66	114	40	71	16	60	1266
1939	43	66	103	77	58	104	287	123	86	151	98	.8	1197
1940	56	108	79	34	93	94	37	125	141	311	8	36	1122
1941	77	114	60	68	133	292	287	311	101	36	153	27	1659
1942	33	83	66	134	141	198	30	78	91	20	67	2	943
1943	37	64	96	115	117	152	130	180	114	48	68	3	1124
1944	52	64	40	51	127	140	41	144	113	72	50	15	909
1945	60	35	72	98	60	328	130	75	381	51	82	27	1399
1946	22	60	290	59	72	89	230	96	178	24	27	63	1210
1947	106	31	45	17	171	234	256	133	115	101	9	87	1305

(contd.)

Table 5.7 continued

Year	J	F	M	A	M	J	J	A	S	O	N	D	Total
1948	21	50	109	55	176	142	188	140	147	74	15	126	1243
1949	.3	52	130	132	162	153	278	44	59	296	175	91	1572
1950	64	94	106	111	77	240	127	103	353	42	67	32	1416
1951	27	79	85	181	77	135	176	149	146	86	50	66	1257
1952	57	114	98	66	138	176	177	153	195	67	29	17	1287
1953	54	67	90	56	83	143	116	45	93	84	99	58	988
1954	122	84	44	93	217	288	224	189	17	16	31	66	1391
1955	40	81	132	106	66	168	324	48	11	8	35	42	1061
1956	27	19	111	74	181	245	202	150	217	39	10	12	1287
1957	71	80	77	98	125	212	334	184	168	55	39	36	1479
1958	26	50	60	147	128	92	18	250	147	79	11	13	1021
1959	25	120	60	195	199	105	121	36	172	4	79	101	1217
1960	60	18	124	126	127	210	38	127	229	7	113	19	1198
1961	72	59	70	54	133	228	55	38	156	158	105	16	1144
1962	15	56	13	103	83	101	142	164	225	45	56	16	1019
1963	.1	26	47	99	195	179	43	208	299	30	95	22	1243
1964	81	62	37	171	115	178	33	37	138	34	6	18	910
1965	6	109	40	132	36	176	190	82	12	169	24	54	1030
1966	23	38	123	107	70	178	222	51	171	5	55	79	1122
1967	17	45	111	177	159	42	86	.2	42	10	117	5	811
1968	44	13	61	99	140	56	68	83	91	19	48	100	822
1969	100	81	90	101	146	55	173	275	125	5	24	10	1185
1970	19	59	140	38	168	167	105	52	262	18	45	35	1108
1971	27	68	31	42	119	236	41	107	206	70	4	40	991
1972	25	105	55	75	52	163	24	118	78	79	67	62	903
1973	52	119	84	127	142	205	42	79	282	40	4	.3	1176
1974	58	51	56	63	102	145	173	130	43	100	71	123	1115
1975	43	108	57	98	74	229	244	39	208	96	28	48	1272
1976	14	64	89	80	63	139	143	232	253	62	36	28	1203
1977	67	31	93	167	232	101	257	184	233	21	61	63	1510
1978	55	30	59	97	149	105	53	16	95	40	43	32	774
1979	27	8	61	177	77	115	101	100	54	1	49	42	812
1980	66	36	147	54	91	197	101	455	145	62	33	6	1393
Mean	47	60	84	95	106	174	148	137	135	68	53	39	1147

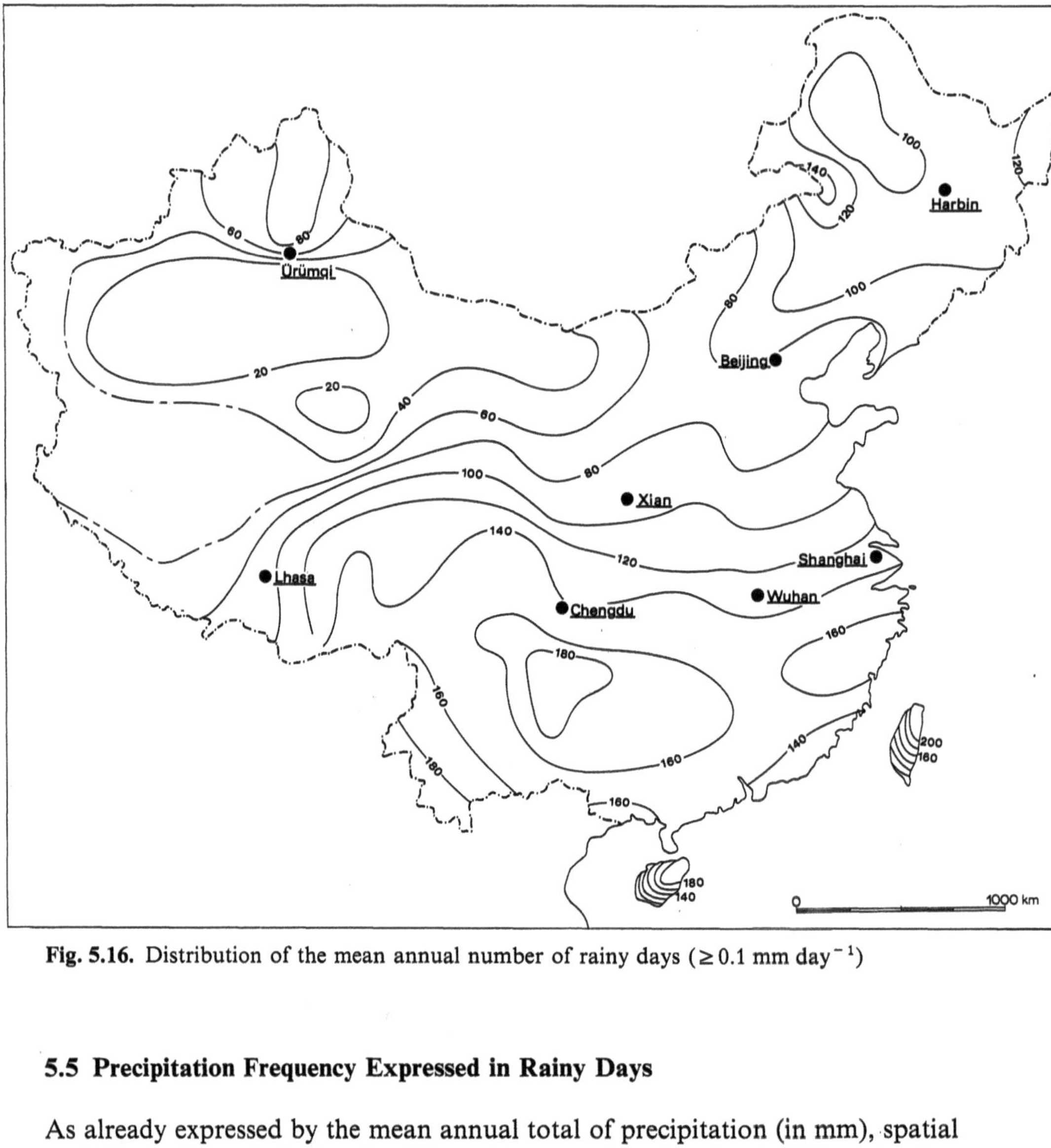

Fig. 5.16. Distribution of the mean annual number of rainy days (≥ 0.1 mm day^{-1})

5.5 Precipitation Frequency Expressed in Rainy Days

As already expressed by the mean annual total of precipitation (in mm), spatial variations of precipitation in China show considerably large differences (see Sect. 5.1). Evidence is also given by the mean annual *number of rainy days*. A rainy day is defined by a daily total of ≥ 0.1 mm precipitation.

The map of the mean annual number of rainy days (Fig. 5.16) shows isolines for quite variable totals. Although not as clear as in the case of annual isohyets (Fig. 5.1), China can nevertheless be broadly divided into two major parts according to the number of rainy days. The dry western part records a comparably small number of rainy days, mostly < 60, with the exception of the southeastern parts of the Tibetan Plateau, where even a maximum figure of 158 rainy days per year was recorded (at Dengqen/242). The wet and semi-wet eastern part experiences a remarkably larger number of rainy days which mostly varied between 80 and 160. In all of China, the average annual number of rainy days ranges from a minimum of 12.6 only (for Ruoqiang/264), located in the driest parts of western China, to a maximum of 180 to 190 in certain parts of southern China. The absolutely highest number of rainy days per year amounted even to 191.8 (at Lancang/222)[21].

5.6 Precipitation Intensity

As an indicator of the nature and effectiveness of precipitation, its intensity expresses the amount of precipitation over a certain wet spell, normally a day or less.

Roughly, the mean precipitation intensity can be given by the quotient between the precipitation total and the number of rainy days, either on an annual or monthly basis. Accordingly, mean precipitation intensity per day was calculated on a monthly and annual basis for selected stations (Table 5.8).

Observations on the mean precipitation intensity per day show considerably large differences both over time and space; the values vary between 1.1 and 11.2 mm day^{-1}. It can be clearly seen that low intensities are found for stations with low annual precipitation totals, while high intensities are valid for stations recording high totals of precipitation (Table 5.8). It can be deduced that the amount of precipitation and its intensity are directly related to each other in that precipitation intensity per rainy day increases with increasing precipitation totals while, conversely, it decreases with decreasing totals.

This relationship is given by the mean precipitation intensity for selected dry and wet stations in China:

	Annual total (mm)	Mean precipitation intensity (mm day^{-1})
Ruoqiang/264	17	1.2
Tinghe/273	91	1.3
Hotan/276	35	1.9
Jingdezhen/151	1,780	11.8
Qinzhou/213	2,059	12.1
Haifeng/196	1,893	14.9

Considering the monthly values of the mean daily precipitation intensity (Table 5.8), it can be seen that precipitation intensity increases with increasing precipitation totals. This is shown by the high intensities during summer, particularly for stations in Southeast and South China. The maximum daily intensities were recorded at Dongfang/210 (August, 22.0 mm rainy day^{-1}). In contrast,

[21] Proof of larger annual totals of rainy days, as shown by even >200 days in Taiwan Island (by ZHANG Jiacheng and LIN Zhi-quang 1985), could not be given by records.

Table 5.8. Mean precipitation intensity per rainy day on a monthly and annual basis for selected stations in mm/day

No.	J	F	M	A	M	J	J	A	S	O	N	D	Year
2	0.5	0.8	1.4	3.1	3.4	5.0	7.6	7.2	4.9	3.2	1.6	0.9	4.0
4	0.6	0.6	1.3	2.8	4.0	5.4	6.8	8.2	5.6	3.0	1.1	0.8	4.6
6	0.5	0.5	0.9	1.9	3.6	4.7	6.2	6.0	3.7	2.1	0.8	0.5	3.3
7	0.4	0.4	0.9	2.2	3.8	5.3	6.5	5.6	3.8	2.2	0.8	0.5	2.8
8	0.6	0.6	1.9	2.6	4.0	5.7	8.8	8.9	4.8	3.4	0.9	1.0	5.2
15	0.6	0.9	2.0	3.4	3.6	6.0	10.3	7.7	5.9	3.6	1.2	0.9	4.9
17	0.8	1.2	1.3	2.6	4.0	6.3	7.7	8.4	5.5	4.3	2.3	1.2	4.5
19	0.8	1.2	1.8	3.1	4.1	5.1	6.8	9.1	5.1	3.9	2.2	1.3	4.6
21	0.5	0.7	1.0	2.3	3.1	4.0	5.6	5.1	3.3	2.2	0.6	0.5	3.3
27	0.6	1.0	1.7	3.0	4.3	6.4	11.3	9.9	6.5	4.4	2.1	0.9	5.8
31	1.3	1.1	1.7	3.2	4.1	5.5	6.1	9.5	4.9	3.4	2.5	1.6	4.8
38	1.7	2.0	2.5	5.8	6.0	7.4	13.4	14.4	9.4	6.3	3.8	2.9	7.9
51	1.5	2.1	1.9	3.5	6.0	7.4	10.1	10.7	6.9	4.4	2.6	1.0	7.0
56	0.7	0.9	1.1	2.3	2.5	2.7	5.3	6.1	4.5	2.9	1.3	0.5	3.4
59	1.2	1.3	2.3	3.6	4.1	4.2	6.3	7.7	5.5	4.2	2.6	0.8	4.8
61	1.0	2.4	2.1	4.6	5.6	8.0	13.7	16.1	8.4	4.4	1.8	1.6	8.7
62	3.9	2.3	1.9	4.5	5.6	7.6	14.4	14.0	6.8	5.9	1.2	1.7	8.4
68	1.2	1.4	2.4	4.5	4.4	4.4	8.5	9.8	6.2	4.8	3.3	1.3	5.7
69	1.3	1.9	2.4	4.6	4.9	5.2	8.2	8.1	8.7	5.0	2.9	1.8	5.8
72	1.6	2.0	2.2	5.6	5.2	8.9	13.5	13.6	6.8	7.3	3.7	2.0	8.1
77	2.3	2.6	3.3	5.9	6.2	9.6	14.7	12.7	8.7	6.5	5.4	2.4	8.6
81	1.3	1.6	2.9	4.5	5.5	6.4	8.9	8.9	7.6	5.2	3.4	1.7	6.2
87	3.3	3.3	5.0	7.0	7.9	12.3	16.7	14.3	10.4	5.9	5.2	3.7	9.4
90	2.8	2.9	4.6	6.9	6.2	9.4	11.0	12.5	7.6	5.9	5.6	2.5	7.6
95	1.8	1.9	3.6	5.8	6.6	6.2	9.1	8.1	8.2	6.3	4.4	1.6	6.0
101	1.0	1.4	2.0	3.4	4.4	4.9	6.7	6.9	5.7	3.3	1.9	0.7	4.5
114	4.2	5.7	6.3	8.6	8.9	9.4	14.9	11.3	7.8	5.7	5.9	4.3	8.1
116	4.0	5.2	6.7	7.8	8.9	15.2	14.8	10.1	9.3	6.0	5.9	4.3	8.7
119	4.9	6.2	6.1	8.2	8.6	12.0	12.5	11.6	12.6	5.6	5.9	4.6	8.5
123	6.1	6.6	7.6	8.4	11.2	13.3	11.2	10.8	13.8	6.8	5.8	4.9	9.1
126	4.4	6.0	7.7	9.8	11.4	18.6	18.6	15.7	10.0	6.0	5.5	4.1	10.0
129	3.3	3.3	5.0	8.0	9.2	12.0	15.8	14.1	8.7	6.7	4.6	3.1	8.6
134	1.0	1.4	2.0	3.7	5.6	7.0	13.5	15.1	7.3	3.1	2.2	1.0	6.4
138	2.5	2.0	2.3	3.6	6.1	10.4	11.8	10.6	10.3	6.1	4.1	1.8	8.3
142	2.1	2.1	3.5	6.7	8.9	10.5	12.8	12.8	9.8	5.9	3.5	1.8	7.1
147	4.6	6.1	8.8	11.4	14.1	15.4	13.8	10.3	8.9	7.3	6.4	4.5	9.5
154	4.5	5.9	7.3	8.5	10.4	13.4	12.1	14.8	18.0	8.1	6.5	4.6	9.9
173	1.4	1.6	2.3	7.2	9.9	12.6	11.2	9.3	7.3	5.2	4.0	2.1	6.5
177	4.4	5.2	6.9	13.9	15.4	17.9	14.2	11.5	7.6	10.5	7.8	4.9	10.7
183	4.6	5.7	7.3	8.5	11.5	13.1	12.0	11.4	12.2	4.7	3.8	3.4	8.8

(contd.)

Table 5.8 continued

No.	J	F	M	A	M	J	J	A	S	O	N	D	Year
199	5.0	5.5	6.2	10.5	15.0	14.7	13.2	13.9	15.6	8.7	7.5	3.2	11.2
210	2.4	2.7	3.5	6.3	9.2	14.6	16.1	22.0	13.2	9.0	3.7	1.5	11.1
215	4.2	3.8	4.2	6.0	11.4	13.6	11.1	12.3	9.3	8.1	5.0	3.0	8.4
222	3.1	1.9	2.9	3.8	9.3	10.0	13.2	12.8	8.0	8.5	4.5	3.0	8.6
225	2.4	2.5	2.7	3.6	6.8	9.6	10.2	8.3	8.1	5.9	4.9	3.5	7.3
230	1.7	1.5	2.3	4.3	4.6	3.9	6.8	6.3	3.5	5.6	2.0	2.0	4.3
233	1.2	1.6	2.0	2.8	4.3	5.7	4.9	4.8	5.8	3.7	2.0	1.6	4.3
238	1.2	0.0	1.0	1.1	2.2	5.5	7.2	6.6	4.5	2.0	0.4	0.0	5.6
239	0.4	0.2	0.8	1.2	2.7	5.1	6.9	7.0	4.2	1.7	1.3	1.5	5.2
240	1.0	0.7	0.7	1.4	2.3	4.1	4.8	5.0	3.7	2.4	1.0	1.0	3.6
243	1.3	1.2	1.2	1.3	3.4	4.5	4.9	4.9	4.0	2.3	1.1	0.8	3.6
247	0.7	1.0	2.1	2.8	4.3	3.5	5.4	7.4	4.5	3.8	2.1	0.9	4.3
250	0.9	1.4	3.2	4.3	3.4	3.5	5.9	6.3	4.3	3.6	2.9	0.9	4.4
252	0.5	0.5	0.9	1.7	2.3	2.7	3.3	4.3	2.4	2.4	1.3	0.5	2.7
256	0.4	0.7	1.2	3.2	3.9	3.9	5.3	5.7	4.1	3.5	1.2	0.4	3.9
258	0.5	0.5	0.9	0.9	1.4	1.5	1.7	2.0	2.2	1.3	0.9	0.5	1.5
259	0.8	0.8	1.1	0.9	3.2	3.9	3.0	2.8	2.6	1.2	0.7	0.5	2.4
260	0.7	0.3	1.0	0.7	1.8	2.1	1.6	2.2	0.3	0.0	0.3	0.3	1.5
261	0.8	0.9	1.3	2.4	3.1	2.0	2.5	2.7	3.1	1.5	1.0	0.6	2.1
264	0.7	0.9	1.5	1.8	1.2	1.6	2.0	1.5	1.3	1.0	1.0	0.7	1.4
265	0.9	1.0	1.1	2.0	1.8	2.1	1.6	1.7	1.9	2.1	2.0	0.7	1.6
267	0.7	1.0	3.5	0.8	0.7	1.5	0.9	1.2	0.8	3.7	1.3	0.8	1.1
269	1.1	1.2	2.9	4.5	4.9	4.6	2.8	3.8	4.9	5.1	2.5	1.5	3.1
270	1.8	2.0	1.7	2.3	2.4	2.5	2.5	2.2	2.6	2.3	2.2	1.9	2.2
273	0.5	0.6	1.5	2.1	2.0	1.6	1.6	1.6	1.5	2.0	0.8	0.5	1.3
276	0.8	1.2	1.6	2.8	3.0	3.2	1.7	1.8	2.6	2.0	0.7	0.5	1.9

minimum intensities were experienced during winter when in many cases daily intensities <1 mm were recorded. Due to the different precipitation intensities which normally distinctly differ in summer and winter, a clearly established annual variation of the precipitation intensities could be observed in most cases. Hence, the annual variation is less established for stations with small precipitation totals; in these cases, the maximum intensities in summer were remarkably reduced so that the annual variation was only moderately established.

Precipitation intensity can be recorded more accurately through instrumental readings, which show the amount of precipitation for a certain time period. It is known that such individual figures may reach extremely large totals. For example, at Hongkong precipitation totals of 101 and 256 mm have been recorded in spells of 1 and 3 h respectively, while at Paishih (24° 33′ N, 121° 13′ E) precipitation of 437, 771 and even 1,248 mm have been recorded in 6, 12 and 24 h

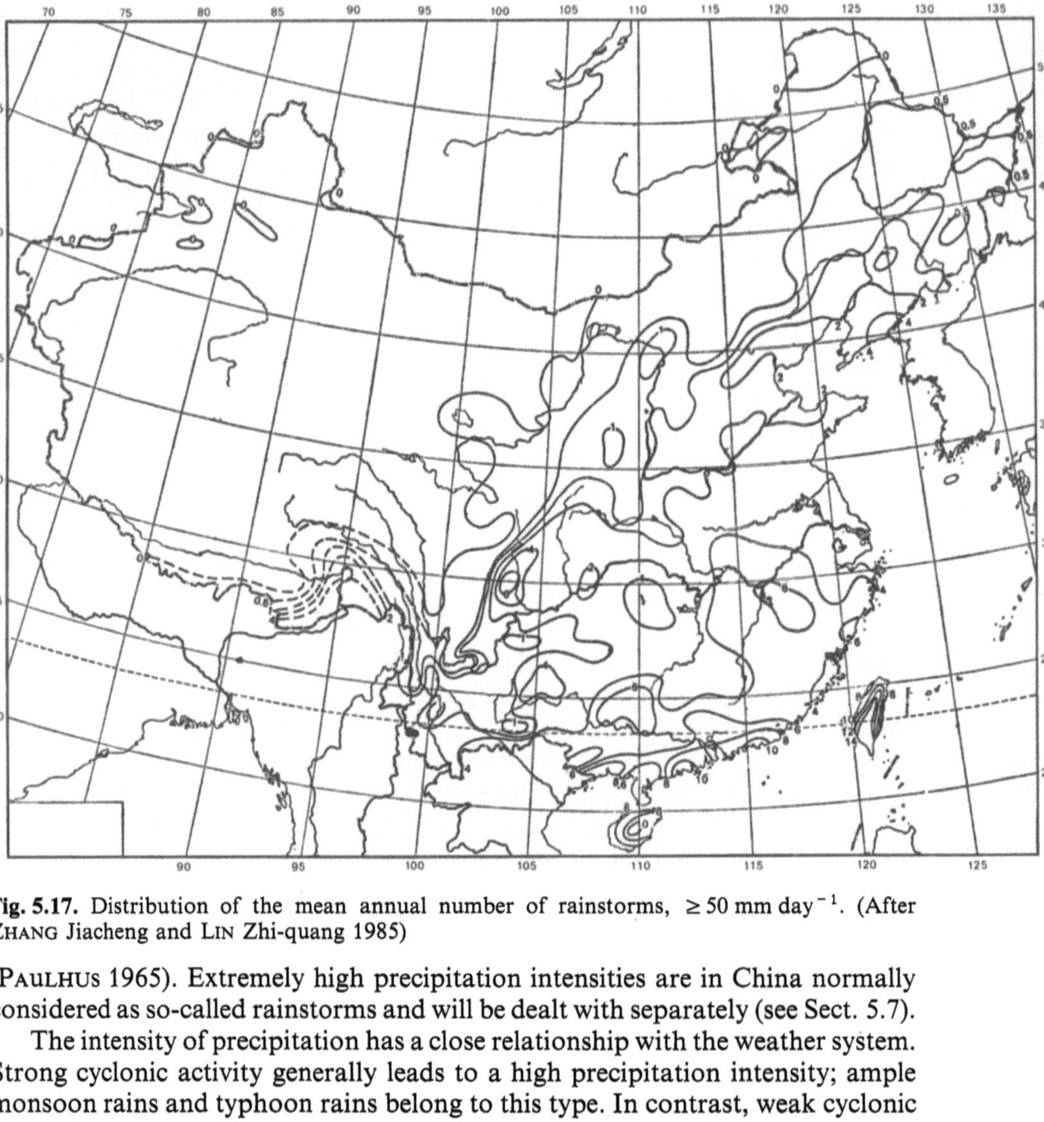

Fig. 5.17. Distribution of the mean annual number of rainstorms, ≥ 50 mm day^{-1}. (After ZHANG Jiacheng and LIN Zhi-quang 1985)

(PAULHUS 1965). Extremely high precipitation intensities are in China normally considered as so-called rainstorms and will be dealt with separately (see Sect. 5.7).

The intensity of precipitation has a close relationship with the weather system. Strong cyclonic activity generally leads to a high precipitation intensity; ample monsoon rains and typhoon rains belong to this type. In contrast, weak cyclonic activity and prevailing anticyclonic regimes prevent efficient precipitation and thus lead to a low precipitation intensity. Local modifications of air masses, for example by orographical lifting, can lead to higher values of precipitation intensity. Convectional activity, although occurring strongly in the vast deserts of China, however, does not necessarily produce a higher intensity, because of the lack of atmospheric vapour.

5.7 Rainstorms and Certain Events of Heavy Rainfall

Considering the climate of China, Chinese climatologists and meteorologists gave special attention to rainstorms and certain events of heavy rainfall. Representing

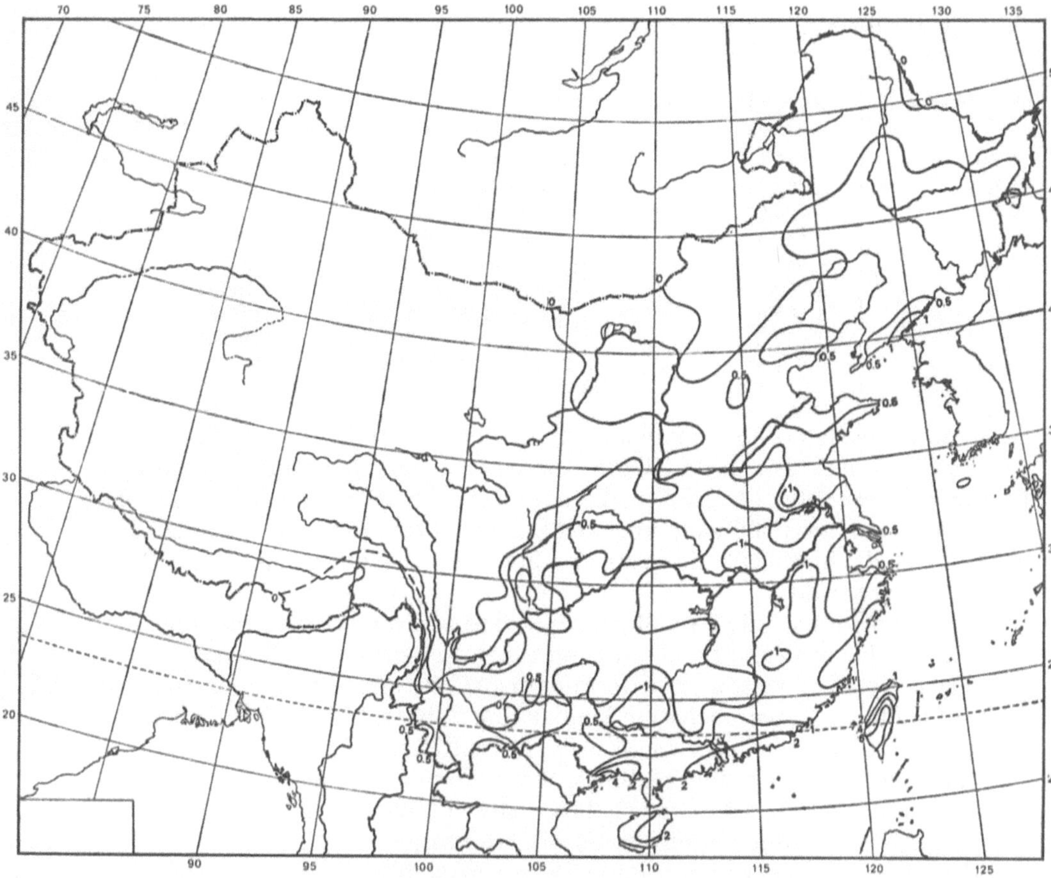

Fig. 5.18. Distribution of the mean annual number of strong rainstorms, ≥ 100 mm day^{-1}. (After ZHANG Jiacheng and LIN Zhi-quang 1985)

the leading investigation, TAO Shiyan et al. have presented a comprehensive handbook on *Severe rainstorms in China* (1980, 225 pp, in Chinese). The observations given in the following section are based mainly on TAO Shiyan[22].

In China, a daily rainfall total >50 mm is commonly defined as a rainstorm. Rainstorms are regarded as strong rainstorms when a daily rainfall total >100 mm is recorded. Rainstorms are normally considered according to the annual or monthly number of cases (days) which record a rainfall total >50 and >100 mm; however, certain events of heavy rainfall can also be considered according to the amount of rainfall over a certain time interval which may be represented by 5 min, 1 h, several hours or even up to 1 day.

The distribution of the mean annual number of rainstorms and strong rainstorms is given in Figs. 5.17 and 5.18 respectively. It can be clearly seen that rainstorms are restricted to the eastern, wet and semi-wet parts of China while the

[22] Further comments were given by TAO Shiyan in his lecture on "Rainstorms in China", presented at the German-Chinese Workshop on "The Climate of China" at the Institute of Geography, Mainz University, FRG, August 25–September 5, 1986.

Table 5.9. Maximum recorded rainfall for certain time intervals; observation period 1953–1983; Taiwan is excluded. (After TAO Shiyan et al. 1980)

Time interval	Rainfall total (mm)	Date	Province
5 min	53.1	July 1, 1971	Shanxi
1 h	267.0	June 20, 1981	Shanxi
3 h	494.6	August 7, 1975	Henan
6 h	830.0	August 7, 1975	Henan
12 h	954,4	August 7, 1975	Henan
24 h	1,060.0	August 7, 1975	Henan
3 days	1,605.0	August 5–7, 1975	Henan
5 days	1,631.0	August 3–7, 1975	Henan
7 days	2,051.0	August 2–8, 1963	Hebei

western, dry parts do not experience rainstorms. The 0-day line of rainstorms divides China into two parts. Within the eastern part, the largest number of rainstorms (generally >6 cases, sometimes up to 14 cases per year) occurs in the coastal lowland areas of Guangdong and Guangxi Provinces and on Hainan and Taiwan Islands. In central and southern Taiwan the maximum numbers of rainstorms were recorded.

Of all rainstorms, only a small part appeared as strong rainstorms (Fig. 5.18 after ZHANG Jiacheng and LIN Zhi-quang 1985). In most parts of eastern China, <1 case per year of a strong rainstorm occurred; the main areas are the coastal areas of southern China, including Hainan and Taiwan Islands. The maximum number of six strong rainstorms occurred in central and southern Taiwan.

According to instrumental readings, the maximum recorded rainfall for certain time intervals reached extremely high values (Table 5.9), figures refer to the period 1953–1983 (observations in Taiwan are missing).

The severe intensity of precipitation which may affect China can be deduced from the highest recorded daily maximum amount of rainfall which was 1,672 mm, recorded in Taiwan as a 24-h total on October 17, 1967. This is considered to be the second highest rainfall maximum in the world. The 3-day absolute maximum amount of rainfall in China amounted to 2,749 mm (October 17–19, 1967, on Taiwan Island). Also the second highest, so far recorded 24-h and 3-day maximum amount of rainfall in China was experienced in Taiwan and amounted to 1,248 and 1,684 mm respectively, on September 11, 1963 and September 11–13, 1963.

The province given for the distribution of rainstorms (Table 5.9) does not mean a provincewide occurrence; rainstorms are apparently restricted to small areas. The 7-day maximum recorded rainstorm, from August 2–8, 1963 in Hebei Province, was concentrated in an area of 5,560 km^2.

Besides the maximum recorded rainfall totals (Table 5.9), other cases of extremely heavy rainfall amounts were reported. Among them, it is worth noting a severe rainstorm which occurred in Maowusu Desert, located between Shanxi Province and Inner Mongolia Autonomous Region, on August 1–2, 1977 when a 24-h maximum total of 1,050.0 mm was recorded. This amount is regarded as the highest amount of rainfall which was ever recorded in a desert on earth. It is also noteworthy that a severe rainstorm occurred in Shanghai on August 21, 1977

Table 5.10. Mean monthly number of rainstorms for eight representative stations in eastern China. (After ZHANG Jiacheng and LIN Zhi-quang 1985)

	J	F	M	A	M	J	J	A	S	O	N	D	Year
Harbin							0.5	0.2					0.7
Beijing				0.1	0.1	0.2	0.6	1.4	0.3				2.7
Zhengzhou				0.2	0.1	0.4	0.5	0.8	0.2	0.1	0.1		2.4
Wuhan		0.1	0.2	0.4	0.5	1.2	1.0	0.7	0.4	0.1	0.1		4.7
Changsha			0.1	0.4	1.0	0.9	0.8	0.3	0.3	0.2			4.0
Guangzhou	0.1		0.2	0.6	1.2	1.6	0.7	0.9	1.1	0.1	0.2		6.7
Yangjiang	0.1	0.1	0.3	1.5	1.4	2.6	1.4	2.0	1.3	0.2	0.2	0.1	11.2
Kunming					0.2	0.4	0.5	0.4	0.3	0.1			1.9

with a 24-h rainfall total of 581.3 mm. For several locations and regions in eastern China, strong summer rainstorms were reported which lasted for some weeks and thus led to unusually high rainfall totals between 1,500 and 2,000 mm for the summer period. Observations of this occurrence were given particularly for the middle and lower sections of the Yangtze River basin, especially for 1954 and 1980.

Due to their frequency and severity, rainstorms and other events of heavy rainfall are a characteristic phenomenon and, at the same time, the most dangerous occurrences to the climate of China. Among the countries of the world, which suffer from severe rainstorms, China is commonly regarded as one of the most seriously stricken countries. Severe rainstorms in China are normally associated with serious, damaging floods which particularly hit the fertile lowlands of eastern China and which thus may devastate agricultural lands especially during the growing season. In the coastal regions, rainstorms are often associated with extremely high tides.

Rainstorms and certain events of heavy rainfall predominate in summer, as can be seen from the annual variation of rainstorms for eight representative stations in eastern China (Table 5.10; stations are arranged from Northeast to South). Expressed as the mean monthly number of rainstorms, the maximum frequency occurs in June, July and August, while winter shows a negligible number. As evident (Table 5.10), the annual frequency of rainstorms remarkably increases from Northeast to South China. The period, possibly affected by rainstorms, increases from 2 months only (July and August) up to even 8 months (March–October).

The distinct prevalence of events of heavy rainfall in southern China, in regional terms, and in summer, in temporal terms, is underlined by considering the total number of cases of "heavy rainfall" for 10-day intervals from April to October (April 1–10 – October 21–31), based on observations in northern and southern China from 1953–1983 (Table 5.11). Heavy rainfall, in this respect, is defined by a 24-h rainfall > 500 mm which has occurred during a 10-day interval. As can be seen from the annual total, southern China is affected more than three-fold by events of heavy rainfall than northern China. It can be inferred that events of heavy rainfall are practically limited to July and August in northern China, whereas they occur from May to September in southern China; even April and October may be affected (Table 5.11).

Table 5.11. Total number of cases of "heavy rainfall" (≥ 500 mm day^{-1}) for 10-day intervals from April to October in northern and southern China from 1953–1983

Month	April	May	June	July	August	September	October
10-Day period	2 3	1 2 3	1 2 3	1 2 3	1 2 3	1 2 3	1 2
Northern China	0 0	0 0 0	0 0 2	2 4 7	7 4 0	1 0 0	0 0
Southern China	1 1	3 6 6	2 4 5	5 9 4	8 9 4	8 4 5	0 2

The duration of a rainstorm or an event of heavy rainfall is rather variable and may last from a 5-min interval only to several days or even several weeks if consecutive rainstorms are combined. According to the different time intervals, also different scales of weather and circulation systems influence the formation of rainstorms. Whereas small and mid-scale disturbances may cause rainstorms up to a 3-h duration, rainstorms of a 3-h to 1-day duration originate from distinct synoptical systems, such as typhoons, cyclones and upper low pressure systems. Extremely long-lasting rainstorms (>3 days) are related to anomalies of the atmospheric circulation. Since rainstorms are restricted to the summer months, also the prevailing pressure systems during summer are the major reason for the formation of rainstorms; these include the summer monsoon system and its interaction with westerlies (which leads to fronts and cyclonic vortices), on the one hand, and the typhoons, on the other (see Chap. 3). As a result of substantial research work carried out by scientists of the Institute of Atmospheric Physics, Academia Sinica, Beijing, several types of heavy rainfall-producing systems have been found, which must particularly consider the interaction between surface, low and upper-level pressure and wind conditions in the atmosphere.

5.8 Diurnal Variation of Precipitation

The distribution of precipitation over a day, i.e. the time of when precipitation is experienced, varies over space and time. In most parts of China and for the major part of the year, precipitation is most frequently concentrated in the afternoon and evening, while morning and night experience little precipitation.

However. some regional modifications have to be taken into consideration. It is noteworthy to observe over the islands and the coastal regions a distinct precipitation maximum during the night, while the minimum occurs during the daytime. During May and June, precipitation most frequently occurs in the morning along the coast as well as in the lowlands and hinterland up to a width of about 100 km (Ramage 1952). In July, the strong morning maximum is weakened and shifted to the afternoon where it is well established from August onwards, thus representing the characteristic diurnal variation of precipitation in China. During May and June, a modification of the diurnal precipitation distribution was also observed in the western parts of China, particularly in Xinjiang and the Qinghai-Xizang (Tibetan) Plateau, showing a night or early morning maximum (Hanson-Lowe 1941; Lui 1957; Watts 1962).

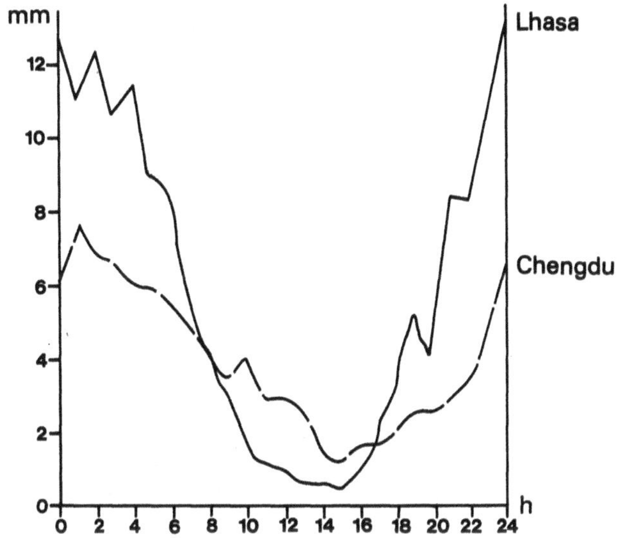

Fig. 5.19. Diurnal variation of mean precipitation at Chengdu and Lhasa

The diurnal variation of precipitation also appears to be influenced by the orography, according to LIN Zhi-quang (1981). On high mountains, the diurnal variation was characterized by a precipitation peak during daytime and a depression at night while, in contrast, in large valleys, precipitation was normally less during daytime and greater at night. This rule can be modified in two cases: for large valleys in Yunnan and Xizang, night rains followed a fine day, while in Sichuan Basin, precipitation at night occurred after an overcast day (Fig. 5.19). This case is considered as the only one in China where over a large area night precipitation significantly dominates.

5.9 Influence of Topography and Elevation on Precipitation

For two reasons, the influence of topography and elevation on precipitation has to be considered with particular attention: (1) the landforms in major parts of China are of a mountainous nature, (2) strong advectional activity prevails year-round in China, due to the persisting air masses. Depending upon the location of mountains, the influence of topography and elevation on precipitation can predominantly be expected on a regional and local scale only; in contrast, effects on a large scale seem to be of minor importance.

The influence of topography and elevation on precipitation is chiefly confined to two aspects: (1) the influence of the exposition of slopes, (2) the effects of elevation. For both aspects, the total amount of precipitation as well as the number of rainy days, together with various expressions of them, can be considered.

5.9.1 Influence of the Exposition of Slopes on Precipitation

This aspect particularly refers to the well-known foehn effect which is commonly defined as a dry and warm wind descending the lee side of a mountain range, when stable air is forced to flow over the mountain range by the regional pressure

Table 5.12. Observations on the foehn effect for Qinshan Mountain of Changbeishan system, Northeast China

	Windward slope Dandong	Windward slope Kuandian	Lee side slope Shenyang	Lee side slope Yingkou
Annual amount of precipitation	1,085	1,202	755	689
Annual number of rainy days	105	119	92	85
Precipitation total in July, August	624	697	398	361
No. of rainy days in July, August	33	35	27	25
Maximum precipitation per day	414	306	179	193
Annual No. of rainstorms	5.1	4.7	2.5	2.3

gradient. Related to this process is a typical pattern of atmospheric moisture and precipitation. On the windward slope of a mountain range, the ascending and thus cooling air masses lead to a significant loss of moisture by precipitation, while on the lee side both the relative and absolute air humidity are lowered so that condensation cannot take place. The impact of the foehn effect on precipitation, therefore, refers to the strong contrast in the amount of precipitation between a moist windward side of a mountain range and a comparably dry lee side. The amount of precipitation depends on the vapour pressure of the air masses and the elevation of the mountains above ground.

The contrasting precipitation conditions on the windward and lee side of a mountain range were shown by observations in different parts of China. For example, observations for Qinshan Mountain of Changbeishan system, Northeast China, expressed the well-established foehn effect (Table 5.12).

The marked foehn effect was also shown by observations given for the windward and lee side of Taihang Shan Mt. (Table 5.13).

As another example for the well-established foehn effect, observations from the mountains in Northwest China were reported (Table 5.14).

In the examples given so far, the windward slope and lee side act predominantly as such year-round, although opposite regimes of air masses were also proved to a fair extent by the precipitation conditions on the lee side. However, also seasonally alternating windward slopes and lee sides were observed in China, depending upon the seasonal variation of air masses governing the climatic condi-

Table 5.13. Observations on the foehn effect for Taihang Shan Mountain

	Annual total	Annual No. of rainy days	Precipitation July, Aug.	Rainy days July, Aug.	Max. precipitation per day	No. of days ≥ 25 mm
A[a] Shijiazhuang	599	78	342	26	251	6.0
B[a] Taiyuan	467	81	224	27	184	2.7

[a] (A) Windward side; (B) lee side.

Table 5.14. Observations on the foehn effect for mountains in Northwest China

	Tian Shan		Zhungeer Basin	West mountains
	Xiaaquzi [a]	Bayinbuluke [b]	Tacheng [a]	Kelamayi [b]
Annual precipitation total	536	275	280	97
Annual No. of rainy days	118	111	97	65
Maximum precipitation per day	55	46	57	21
No. of days ≥ 5 mm	32.9	16.0	15.7	4.7
No. of days ≥ 10 mm	16.7	5.7	5.5	1.3
No. of days ≥ 25 mm	2.3	0.8	0.6	0.1

[a] Station on the windward mountain side.
[b] Station on the lee side.

tions. This was shown for Yuyishan mountain which recorded a winter maximum of precipitation on the north side which then represents the windward side while, in contrast, the south side gets maximum precipitation during summer.

5.9.2 Effect of Elevation on Precipitation

As a principal rule, generally being observed for non-tropical regions, precipitation increases with altitude, due to prevailing advectional processes. Evidence of this was also given in China, as simply seen from a comparison of precipitation totals between a top mountain station and a valley station and the resulting vertical precipitation gradient (mm 100 m^{-1}). Observations were made for selected mountains in eastern China (Table 5.15); precipitation totals refer to the annual total as well as to the totals for January, April, July and October. As can be seen for each mountain under consideration, the positive vertical precipitation gradient varies over the year, thus indicating an increasing amount of precipitation with altitude which, however, varies seasonally. In all cases, summer represented the wet season, and, at the same time, the precipitation gradient showed the largest values in comparison with all other seasons; for the dry winter, the gradients were smallest. It can be deduced that the increasing amount of precipitation with altitude was positively influenced by principally moist conditions, but weakened under less moist conditions (Table 5.15).

This observation was proved for the less moist mountainous regions in Northwest China where smaller precipitation gradients were observed (Table 5.16). The precipitation gradients are much lower than in the wet eastern parts of China, although again the gradient for the wet summer is larger than for the other drier seasons. As can be seen from the observations for Northwest China, also drier mountains experienced increasing precipitation totals with elevation.

Table 5.15. Annual and monthly precipitation totals for top mountain stations and valley stations in selected mountain ranges in eastern China, together with the resulting vertical precipitation gradients (mm 100 m^{-1})

Region	Station/ gradient	January	April	July	October	Year
	Huangshan	72	247	294	96	2,339
Huangshan	Denxi	72	208	154	57	1,643
	Gradient	0	2.3	8.3	2.3	41.3
	Lushan	58	211	188	78	1,834
Lushan	Jinjiang	58	169	133	59	1,397
	Gradient	0	4.0	5.1	1.7	40.7
	Emeishan	17	115	413	118	1,960
Emeishan	Emei	13	93	364	82	1,594
	Gradient	0.1	0.8	1.9	1.4	14.1
	Taishan	15	60	354	61	1,164
Taishan	Tainan	6	37	229	32	726
	Gradient	0.7	1.6	8.9	2.0	31.2
	Huanshan	16	72	178	89	925
Huanshan	Xian	8	53	106	62	604
	Gradient	0.5	1.1	1.6	3.7	19.2
	Wutaishan	14	61	231	51	966
Wutaishan	Yuanping	2	22	121	26	469
	Gradient	0.6	1.9	5.4	1.3	24.5

Table 5.16. Annual and monthly precipitation totals for top mountain stations and valley stations in selected mountain ranges of Northwest China, together with the resulting vertical precipitation gradients (mm 100 m^{-1})

Region	Station/ gradient	January	April	July	October	Year
Mid-Tianshan	Xiaoquzi	6	50	100	27	536
	Ürümqi	6	23	16	17	195
	Gradient	0	1.8	5.5	0.6	22.7
West slope of the basin	Hebukesaier	3	5	46	6	141
	Kelamayi	2	7	25	6	97
	Gradient	0.1	−0.2	2.5	0.1	5.1
East Tianshan	Yiwa	1	3	34	2	86
	Naomaohu	1	0	3	0	13
	Gradient	0	0.2	2.5	0.1	5.9

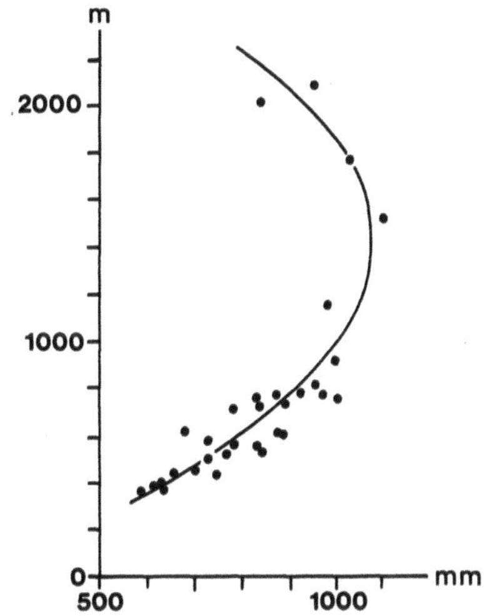

Fig. 5.20. Distribution of annual precipitation on the windward (northern) slope of Qinling mountain. (After ZHANG Jiacheng and LIN Zhi-quang 1985)

The close relationship between precipitation and elevation cannot only be expressed by the amount of precipitation alone, but also by the number of rainy days (≥ 0.1 mm 24 h^{-1}). In eastern China, the annual gradient (of rainy days) was given as 1.5–3.5 days 100 m^{-1}; the gradients show a clear variation between higher values during the moist summer and smaller values during the less moist remaining seasons. In Northwest China, the gradients were smaller (0.88–2.57 days 100 m^{-1}), but still emphasized the seasonal variation.

Keeping in mind the relationship between increasing precipitation and elevation, Chinese climatologists also placed strong emphasis on the elevation of the maximum precipitation. On the basis of annual precipitation totals, it was found for the windward, resp. north slope of Qinling mountain that maximum precipitation occurred at about 1,400 m above sea level; a steady increase of precipitation was observed below this elevation, but a decrease above it (Fig. 5.20).

For other parts of China, different observations were made, however, a principally valid rule about the vertical precipitation distribution cannot be formulated. It is worth mentioning that, for example, for the north side of Tian Shan the maximum amount of precipitation was found at an elevation of 2,000 m above sea level (which, however, occurred in July at almost 3,000 m), but in contrast the elevation of maximum precipitation dropped to 350 m only for Taihang Shan and to even 170 m for Yan Shan; both mountains are located in Hebei Province.

5.10 Historical Change of Precipitation

The enormous Chinese historical writings contain abundant descriptions of the climate so that the past climate of China and its fluctuations can be reconstructed with fair reliability (cf. also Sect. 4.9 on the historical change of temperature). Chinese climatologists and meteorologists have spent much effort in processing more than 2,200 local (county) annals and many other historical writings in order to elucidate the climate of the past. As far as precipitation is concerned, a compre-

Table 5.17. The driest and wettest years of the 510-year observation period 1470–1979, together with their dryness/wetness grades

Wet years/"d" value		Dry years/"d" value	
1528	4.15	1569	1.92
1640	4.15	1613	2.02
1641	4.11	1849	2.13
1721	3.97	1762	2.14
1589	3.90	1889	2.20
1785	3.84	1668	2.21
1778	3.84	1823	2.25
1972	3.81	1931	2.26
1484	3.81	1848	2.29
1965	3.80	1853	2.29

hensive atlas has been compiled with annual maps of dryness/wetness in China for the 510-year period 1470–1979 [23]. Maps were presented in the scale 1:36 million. For map compilation on the basis of the climatic observations from the literature, the territory of China was divided into 120 regions, each with an area approximately corresponding to one or two prefectures of the present administration. Dryness/wetness was estimated for each region on an annual basis, by distinguishing five grades: 1 – very wet, 2 – wet, 3 – normal, 4 – dry, 5 – very dry. For the grades of dryness/wetness, the following indices "d" were defined: very dry: $d<3.45$, dry: $d<3$, normal: $d=3$, wet: $d>3$, very wet: $d>3.45$.

For the 510-year period under consideration, certain periods experienced an extremely large frequency of serious dryness/wetness conditions, for example the period from 1521–1650 and 1720–1779, both with prevailing dryness, and from 1841–1890 with prevailing wetness. The ten driest and wettest years from the 510-year observation period 1470–1979, together with their dryness/wetness grades are given in Table 5.17.

Dry and wet years do not necessarily affect all of China, but may be concentrated on certain parts. In order to eliminate regional differences of dryness/wetness for the observation period 1470–1979, China was divided into three major zones: (1) north zone, north of the Huanghe River, north of 35° N; (2) middle zone, around Yangtze and Huaihe Rivers, between 27°–35° N; (3) south

Table 5.18. Number and percentage of dry and wet years for the north, middle and south zones of eastern China from 1470–1979

	Eastern China			Entire region
	North zone	Middle zone	South zone	
Dry years	98 (19%)	43 (9%)	40 (8%)	52 (10%)
Wet years	72 (14%)	84 (17%)	111 (21%)	43 (8%)

[23] Academy of Meteorological Science, State Meteorological Administration of China: Yearly charts of dryness/wetness in China for the last 500-year period. Beijing, Map Press; 1981 (in Chinese).

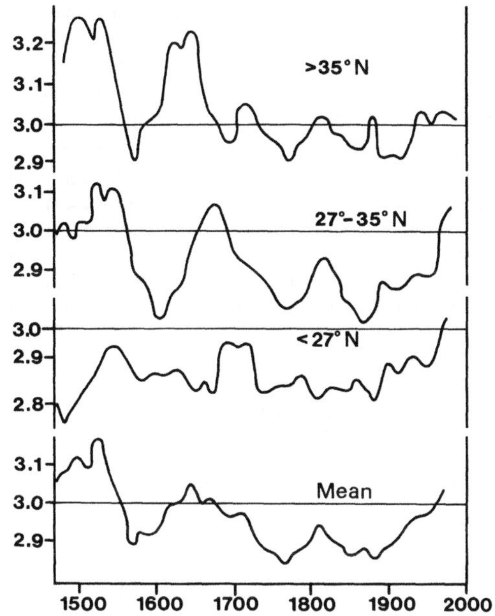

Fig. 5.21. 50-year running means of wetness/dryness grades for North, mid- and South China from 1470–1979. (After ZHANG Jiacheng and LIN Zhi-quang 1985)

zone, south of Nanling mountain, south of 27° N. The frequency of dryness/wetness for these three zones from 1470–1979 (Table 5.18) showed remarkably large differences. For the north zone the frequency of dry years was much higher than for wet years, while in the middle and south zone wet years clearly exceeded the frequency of dry years.

On the basis of the wetness/dryness index, the fluctuation of precipitation over time since 1470 has also been investigated for the three different zones in China, which have been introduced as North, mid- and South China (in latitudinal terms north of 35° N, between 27°–35° N, and south of 27° N respectively). Fluctuations over time were studied by 50-year running means of wetness/dryness grades (Fig. 5.21). As can be seen, a steep north to south gradient of decreasing wetness, respectively increasing dryness, is marked. The wetter conditions in North China are confirmed against the drier conditions in South China. Besides the variations over space, remarkably large differences also occurred over time. As can be seen from the curve of the mean precipitation fluctuation since 1470 (Fig. 5.21), a marked periodicity was experienced, which showed peaks in the early 16th and mid- to second half of the 17th century, while depressions occurred in the second half of the 16th and 18th century. During the last century, comparably dry conditions were experienced which continued to the 20th century, but which then slightly changed to wetter conditions.

In the three different parts of China the occurrence of extreme dry and wet years was also studied (Table 5.19, "d" values are given in parentheses). It can be seen that the north zone recorded a large frequency of extremely dry years which was much greater than in the two other zones; the south zone recorded just one case of an extremely dry year. However, in the middle and south zone, the number of wet years was greatest.

Table 5.19. Variation over space of the occurrence of extremely dry and wet years in China; the "d" values are given in parentheses

North zone		Middle zone		South zone	
Dry	Wet	Dry	Wet	Dry	Wet
1640 (4.83)	1569 (1.39)	1589 (4.55)	1931 (1.29)	1544 (4.36)	1485 (1.27)
1484 (4.56)	1604 (1.67)	1966 (4.46)	1849 (1.39)		1478 (1.50)
1639 (4.48)	1652 (1.68)	1641 (4.38)	1954 (1.44)		1582 (1.57)
1528 (4.46)	1653 (1.69)	1978 (4.31)	1831 (1.61)		1713 (1.71)
1877 (4.41)	1822 (1.74)	1835 (4.29)	1848 (1.75)		1647 (1.85)
1722 (4.37)	1848 (1.75)	1544 (4.23)	1915 (1.89)		1839 (1.86)
1616 (4.33)	1864 (1.76)	1785 (4.19)			
1721 (4.33)	1613 (1.78)				
1965 (4.32)	1553 (1.80)				
1615 (4.26)	1760 (1.80)				
1638 (4.25)	1819 (1.90)				
1689 (4.20)					
1900 (4.20)					

Dryness/wetness, as studied for the three major parts of China, are in specific cases not only confined to 1 year only, but may extend to 2 and more consecutive years. For the observation period 1470–1979, the frequency of consecutive dry or wet years showed large differences for the three different zones of China (Table 5.20). In the north zone, a larger frequency of consecutive dry years than wet years was experienced while, in contrast, in the mid- and south zones a larger frequency of consecutive wet years could be observed. The north zone, for example, recorded five consecutive dry years from 1637–1641, while the south zone experienced nine consecutive wet years from 1477–1485.

From the observations of dryness/wetness, ZHANG Jiacheng et al. (1983) made a comprehensive study on droughts and floods in China since the 15th century. With the exception of the 20th century, serious floods and droughts were proved for the preceding centuries. The authors described 20 serious floods and droughts since 1470, of which only three occurred during the 20th century. A severe drought, extending over 5 years, occurred from 1637–1641, while a flood period was recorded for 17 consecutive years, from 1839–1855. In regional terms, droughts more frequently affected North China, while floods affected South China. With regard to the 1970s, when considerable unusual climatic calamities in

Table 5.20. Frequency of consecutive dry and wet years in different zones of eastern China from 1470–1979

Duration (years)	North zone		Middle zone		South zone		Entire region	
	Dry	Wet	Dry	Wet	Dry	Wet	Dry	Wet
2	15	8	5	13	8	19	7	5
3	5	6	1	2	1	9	2	2
4	4	2	0	2	0	1	0	0
5	1	1	0	0	0	1	0	0

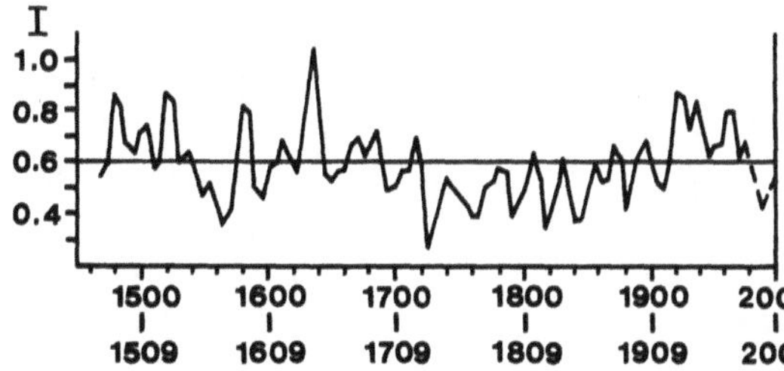

Fig. 5.22. Mean index of dryness (I), given as 10-year running means from 1470–1977 in eastern China. (After ZHANG Xiangong 1981)

many parts of the world occurred, the climate of China showed neither an unusual extent of variation nor extreme conditions (TAO Shiyan 1984). Thus, a recent climatic change is not evident.

From the observations on dry and wet years, given for more than a 500-year period, no definite results on particular precipitation fluctuations or periodic cycles in historical times can be obtained. However, there are several noteworthy contributions on the question of specific precipitation cycles in China. WANG Shaowu and ZHAO Zongci (1982) found that there is an oscillation of wetness within a period of 36 years which, in its mechanism, is closely related to the oscillation of the general circulation in the central Pacific Ocean. ZHANG Xiangong (1981), who defined an index (I) of dryness and wetness, determined the curve of the mean index of dryness/wetness for 1470–1979 in eastern China, by using 10-year running means (Fig. 5.22). It was found that a predominantly dry period occurred from 1479–1691, followed by a wet period from 1692–1890 which changed to a dry period from 1891 onwards. In each period short cyclic fluctuations occurred at a 10- to 20-year periodicity.

As a further attempt to investigate the question of a specific periodicity of precipitation fluctuations in China, 10-year running means of annual precipitation were presented for North China, the middle and lower sections of the Yangtze and for South China. Observations refer to the period from 1850–1974 (Fig. 5.23). As evident from the three curves, precipitation distribution over time

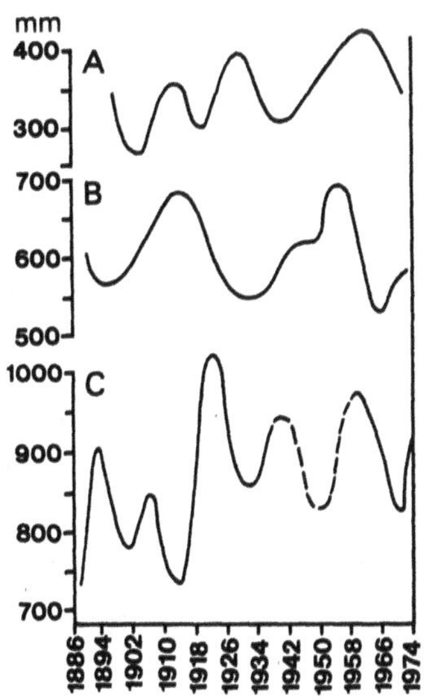

Fig. 5.23. 10-year running means of annual precipitation in North China (*A*), the middle and lower sections of Yangtze (*D*) and South China (*C*)

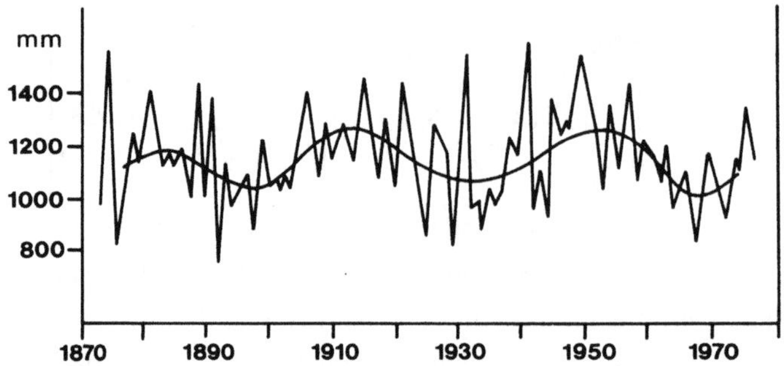

Fig. 5.24. 10-year running means of annual precipitation at Shanghai (from the 1870s to 1970s)

fluctuated in each case to a considerably large extent, however, the time intervals of fluctuation varied between a 35-year period in the Yangtze region, only a 7- to 8-year periodicity in North China and a 14- to 18-year periodicity in South China.

Proof of a 35-year periodicity in the Yangtze region is given by the 10-year running means of annual precipitation at Shanghai; observations are based on the period 1873–1976 (Fig. 5.24) which also contain the absolute annual precipitation totals.

For Shanghai and Beijing, ZHANG Jijia (1984) has determined the occurrence of the wettest and driest periods which were experienced in recent times (Table 5.21). Additionally, the mean annual precipitation totals for the periods concerned were calculated. It can be seen that for both stations the annual totals of precipitation between the wet and dry periods differ considerably from each other. Furthermore, Beijing experienced much lower annual totals during the wet as well as dry periods. In total, however, the difference between the annual precipitation totals during the wet and dry periods, which is valid for each station, confirms the great extent of precipitation variation. A periodic fluctuation of precipitation every 30 to 40 years can be deduced from the observations given.

Using the 257-year observation period (1724–1980) at Beijing, the 30-year running means of precipitation were taken in order to study the precipitation

Table 5.21. The occurrence of wet and dry periods, together with the corresponding rainfall totals for Shanghai and Beijing

	Wet period	Mean amount of precipitation	Dry period	Mean amount of precipitation
Shanghai	1875–1882	1202	1892–1898	971
	1912–1921	1243	1924–1934	1071
	1945–1957	1226	1964–1972	989
Beijing	1841–1845	706	1874–1878	606
	1892–1896	799	1917–1921	478
	1922–1926	715	1941–1945	464
	1955–1959	927	1968–1972	569

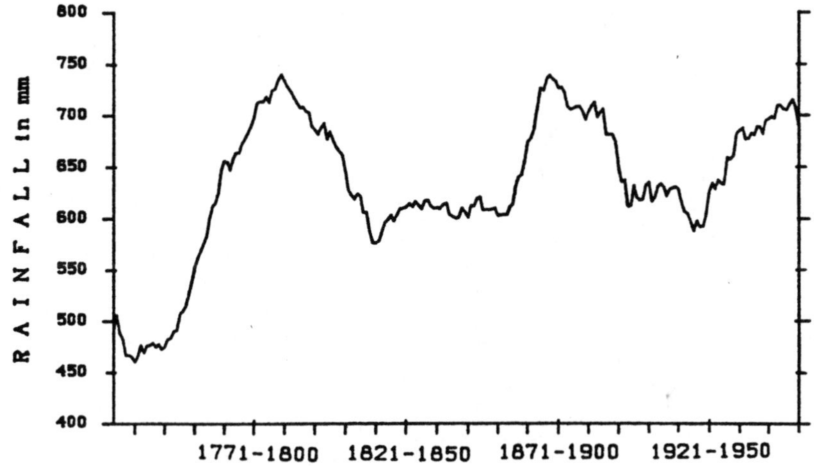

Fig. 5.25. 30-year running means of precipitation at Beijing (from 1724–1980)

fluctuation in recent times (Fig. 5.25). The graphs give evidence of a clearly established precipitation periodicity which expressed three peak and depression periods each. Precipitation peaks were recorded from about 1770–1800, 1860–1890 and 1930–1945. Precipitation depressions were experienced from 1730–1745, 1810–1850 and 1890–1920. It is worth noting the rather large differences in the annual precipitation totals between peak and depression periods and the normally long and significant periods of either ample or scarce precipitation.

5.11 Snow

Representing a type of precipitation during frost temperatures, snow can be expected to depend directly on temperatures below zero which may result from both a higher latitude and a greater elevation above sea level. Both criteria are obviously fulfilled in major parts of China, due to the far-reaching northern extent and the extremely high altitudes. Furthermore, if the vapour content of the air is sufficient for sublimation, conditions for snowfall would be ideally satisfied, thus leading to a greater climatic significance of snow for China.

This assumption is supported by observations of various snow indices, which will be dealt with in detail. Long-term monthly and annual records of snow conditions are included in the climate tables (see Appendix): (1) the number of days of the snow cover period; (2) the number of snowfall days; (3) the maximum depth of snow.

5.11.1 Mean Length of Snow Cover Period

Considering the mean length of the snow cover period, which means the number of days between the (mean) first and last dates of snow cover, considerably large differences over space occur (Fig. 5.26). It can be seen that for all of China the

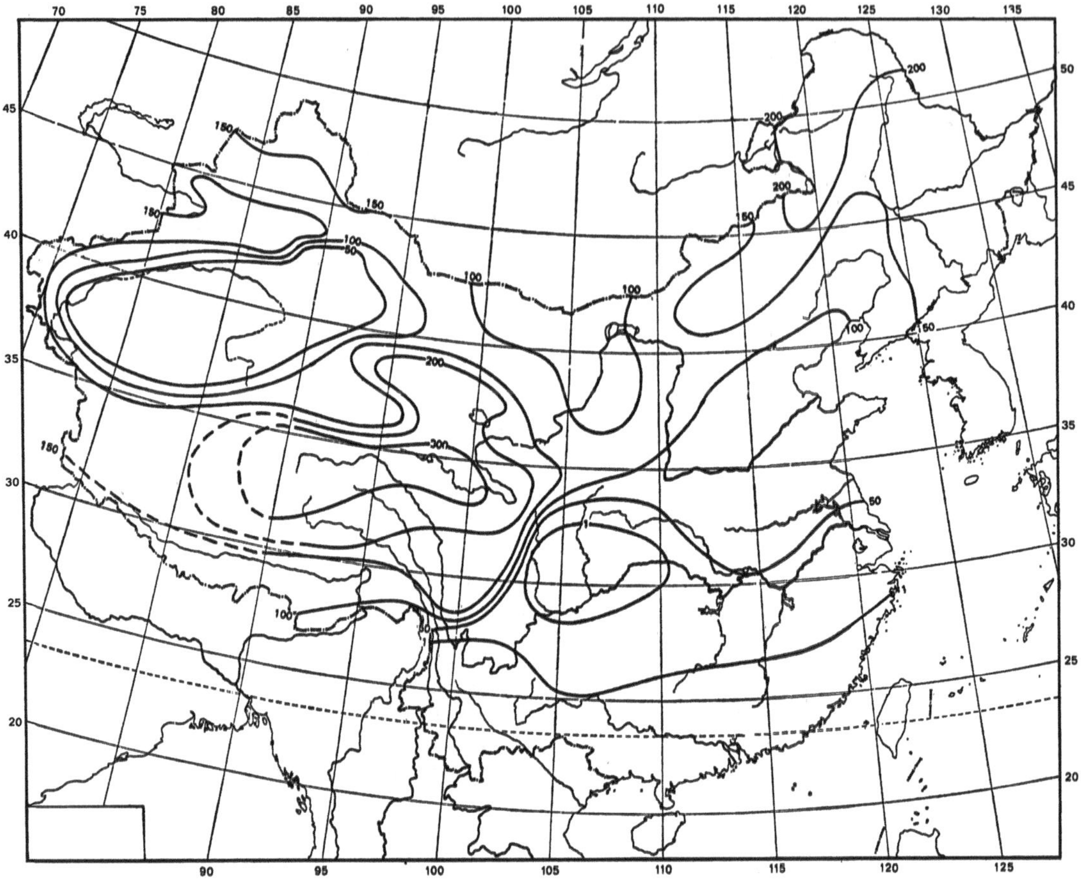

Fig. 5.26. Mean length of the snow cover period, expressed by the number of days between the mean first and last date of snow cover

mean length of the snow cover period varies between less than 1 day and more than 300 days; the recorded maximum figure even amounts to 365 days, valid for the year-round glaciered high-altitude mountains. It thus can be expected that the largest possible range of the length of the snow cover period is valid for China, i.e. completely snow-free conditions throughout the year, on the one hand, to fully snow-covered conditions year-round, on the other. The mean length of the snow cover period at any location plausibly depends on two aspects: (1) its latitudinal position and (2) its elevation above sea level. As can be inferred from the map of the mean length of the snow cover period (Fig. 5.26), a strong latitudinal effect is established over the *eastern* parts of China, showing an increasing length of the snow cover period from south to northeast. While completely snow-free conditions occur in South China, a more than 200-day period of snow cover is valid in Northeast China. A considerable variation of the length of the snow cover period is also observed in the *western* part of China, mostly dependent upon altitudinal differences. The maximum length of the snow cover period, exceeding 200 days (sometimes even 300 days), occurs in Nan Shan, Kunlun Shan and the Tibetan Plateau, while the minimum number of less than 50 snow cover days is observed in the Tarim Basin. The contrasting elevation and vapour content between the two regions lead to completely different thermal and hygric conditions wich are responsible for the formation and type of precipitation.

Excluding the highest mountain regions (which experience a year-round snow cover period), the longest recorded duration of the snow cover period occurs in the mountains of Qinghai and (West) Sichuan as the following stations demonstrate:

Station	Elevation (m a.s.l.)	Mean date of first/ last snow	Days
Da Qaidam/259	3,173	Sept. 30, June 13	257
Tongde/246	3,289	Sept. 16, June 22	280
Yushu/243	3,703	Sept. 15, June 11	270
Litang/231	3,949	Sept. 03, June 17	288

Due to the high elevation of the stations under consideration, the altitudinal impact on snow conditions may be responsible for the great length of the snowfall period, although the mean annual number of snowfall days does not clearly underline this assumption (cf. Sect. 5.11.2). However, the considerably large difference between the length of the snow cover period and the number of snowfall days can be taken as proof of the great variability of snowfall as far as the first and last dates of snowfall are concerned.

While the observations given so far refer to the *possible* maximum lengths of the snow cover period only, the recorded number of days with snow cover shows much smaller values and represents a more realistic picture of snow conditions. The distribution of the mean number of days with snow cover (Fig. 5.27) shows an extremely variable pattern over space. In eastern China, the recorded number of snow cover days decreases from >150 in northeasternmost China to <0.1 in southern China. In addition, also in western China the number of days with snow cover vary between >100 and <10. As in the case of the mean length of the snow cover period (cf. Fig. 5.26), also the distribution of the number of snow cover days shows a strong impact with altitude. Locations at higher latitude and high altitude experience a larger number of snow cover days, while locations at lower latitude and low altitude record a small number; in this case the number of snow cover days even falls <0.1.

Corresponding to the considerably large differences over space with regard to the length of the snow cover period and the recorded number of days with snow cover, the dates of the first and last snow also vary considerably, both in terms of latitude and altitude above sea level. Generally, snow is experienced earlier and lasts longer with increasing latitude. The same observations are true for increasing elevation above sea level.

The latitudinal differences in the first and last dates of snow can be clearly seen along a north to south cross-section through the eastern parts of China (Table 5.22).

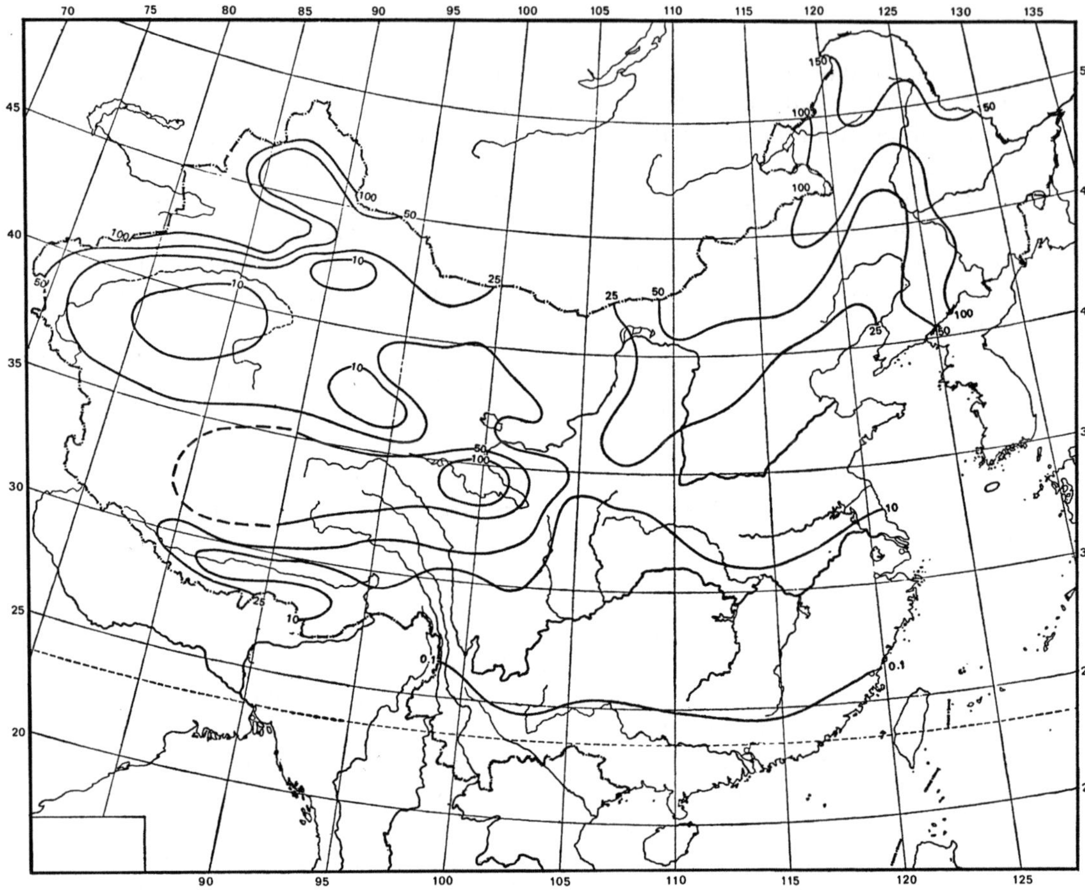

Fig. 5.27. Mean number of days with snow cover

Table 5.22. First and last dates of snow, mean length of the snow cover period and number of snow cover days at selected stations along a north to south cross-section through the eastern parts of China

Station/No.	Lat. °N	First snow	Last snow	Snow cover days	Snowfall days
Mohe/1	53	27.09.	14.05.	175.9	47.2
Harbin/15	46	15.10.	19.04.	105.1	33.1
Changchun/27	44	14.10.	21.04.	88.4	27.1
Shenyang/38	42	25.10.	13.04.	61.5	20.5
Beijing/61	40	26.11.	19.03.	15.6	9.5
Shijiazhuang/70	38	27.11	14.03.	18.4	10.6
Jinan/77	37	30.11	22.03.	14.6	9.3
Zhengzhou/90	35	1.12	15.03.	14.8	10.9
Nanjing/116	32	14.12.	10.03.	8.9	8.4
Shanghai/119	31	5.01.	11.03.	3.2	5.5
Hangzhou/123	30	20.12.	11.03.	7.8	9.8
Wuhan/126	30	6.12.	4.03.	8.9	9.2
Changsha/147	28	20.12.	28.02.	6.1	8.8
Fuzhou/183	26	–	–	0	0.8
Guangzhou/199	23	–	–	0	0

5.11.2 Number of Snowfall Days

As a result of the disrupted pattern of temperature and precipitation in China, the number of snowfall days varies to a large extent over space (see the corresponding figures in the climate tables, Appendix). A snowfall day is characterized by any precipitation of a snow type.

The distribution of the (mean) number of snowfall days corresponds to the same two principles as for the length of the snow cover period, i.e. the clear relationship with both latitude and altitude (see Sect. 5.11.1). Southern China records the smallest number of snowfall days (<1), whereas in Northeast China at the most <60 snowfall days are experienced. The absolutely highest figure (>100 snowfall days) is valid for the eastern part of the Kunlun Shan. It is, however, worth noting that the number of snowfall days over western China varies considerably over space. The absolutely smallest number is true for the Turpan depression (<1 snowfall day), but also the whole Tarim Desert as well as the Gobi Desert and even parts of the southeastern Tibetan Plateau record <10 snowfall days.

The increasing number of snowfall days with increasing latitude in eastern China is clearly shown for selected stations at different latitudes (see Table 5.23). At the same time, the different annual variation of snowfall days is underlined. It is also noteworthy that the peak months of snowfall shift from November and December (at Mohe/1) to December and January (at Harbin/15), and to February and March (at Beijing/61 and Shanghai/119). Guangzhou/199 experiences no snowfall at all.

Table 5.23. Mean monthly number of snowfall days and length of the snow cover period at selected stations at different latitudes

	S	O	N	D	J	F	M	A	M	Annual No. of snowfall days	Length of snow cover period (days)
Mohe	0.9	6.1	9.0	8.8	6.0	6.5	5.6	5.1	1.0	49.0	230.4
Harbin	0.1	1.6	5.1	6.0	7.0	4.9	4.8	1.6		31.1	183.1
Beijing			0.8	0.8	1.6	3.3	3.3	0.1		9.9	109.0
Shanghai				0.5	1.7	3.1	0.8	0.2		6.3	78.0
Guangzhou										0	0

Looking individually at the number of snowfall days for the stations of the mountainous western parts of China (see climate tables, Appendix), snowfall conditions vary considerably and thus underline a strong influence of the altitude and landforms on the snow conditions as can be seen at the following stations:

Station	Elevation (m a.s.l.)	Mean annual No. of snowfall days
Songpan/102	2,828	47.3
Tongde/246	3,289	32.4
Garze/233	3,394	33.4
Deqen/230	3,589	56.4
Yushu/243	3,703	37.3
Litang/231	3,949	46.4

In contrast, stations at a high elevation may also experience a small number of snowfall days only as also stations at a comparably low elevation may record a large number of snow days. Both observations can be seen for stations in central and western China:

Station	Elevation (m a.s.l.)	Mean annual No. of snowfall days
Lenghu/260	2,733	2.4
Xigaze/238	3,836	6.2
Golmud/258	2,808	7.2
Lhasa/239	3,658	8.3
Da Qaidam/259	3,173	12.8
Jiulong/137	2,987	13.8
Altay/270	735	37.8
Ürümqi/269	654	46.5

Moreover, it is worth noting the extremely low number of snowfall days even at very high altitudes (see Xigaze/238, 3,836 m; Lhasa/239, 3,658 m). In the Tibetan Plateau, the surrounding high mountains act as barriers against the cold northerlies, thus leading to poor snow conditions. Evidence of the severe influence of cold waves on the snow conditions is given at Ürümqi/269 and Altay/270, which lie at a rather low altitude (654 and 735 m above sea level respectively), but due to their windward location they still record a large number of snow days (46.5 and 37.8 respectively).

The strong heating effect, exerted by orographical depressions and hence leading to a heat surplus, contributes to a small number of snowfall days, proof of which is given at Turpan/267 (at 34.5 m above sea level, 4.5 snow days) and Ruoqiang/264 (888 m, 4.1 snow days). Additionally, precipitation in the desert is very low and subsequently the snowfall activity is depressed.

5.11.3 Maximum Depth of Snow

The data for maximum depth of snow are also included in the climate tables (see Appendix). It can be deduced that the maximum depth of snow varies over space, yet the variation does not clearly correspond to the variations already shown in the case of the length of the snow cover period and the number of snowfall days (cf. Sect. 5.11.1 and 5.11.2).

When considering the distribution of the maximum depth of snow over space, the largest values occur in the Tian Shan and Altay (> 50 cm). It is noteworthy that the Tibetan Plateau, including the surrounding high-altitude mountains, records a comparably small value for the maximum depth of snow, mostly around 10 cm. With the exception of northernmost China, there are no major differences in the maximum depth of snow, which may result from the different altitudes and landforms in western China. Over eastern China, the distribution of the maximum depth of snow shows major variations over space, but a close relationship to latitude cannot be found. Principally, however, remarkably large differences over space occur for the maximum depth of snow showing larger values for Northeast China (up to >40 cm) and the smallest values for South China (<1 cm), yet a true correlation with latitude cannot be proved.

5.11.4 Altitude of the Snow Line

For such an extremely mountainous country as China, the snow line represents a most important hygro-climatic index. It also relies on the permanent snow and ice coverage, including glaciers. The drastic extent of snow and ice is emphasized by the fact that the Qinghai-Xizang Plateau with its surrounding high mountains, including the Himalayas, Karakorum, Pamir, Kunlun, Qilian and Hengduan Mountains, is the most extensive area of mountain glaciers on earth. In the last three decades, a number of Chinese expeditions of the glaciers in Qilian, Central Himalayas, East Pamir and Karakorum have been carried out and have enabled a comprehensive report on the basic features of the glaciers on the Qinghai-Xizang Plateau and the surrounding mountains (see in particular Shi Ya-feng and Li Ji-jun, their results are particularly quoted in this section).

The total area above the snow line, representing the glaciered area, amounts to about 56,500 km^2, of which 46,640 km^2 (83%) belong to the glaciers of the Qinghai-Xizang Plateau in the Chinese territory (Table 5.24, according to Shi Ya-feng and Li Ji-jun 1981). In comparison with the total glaciered area of High Asia, which was estimated at about 100,000 km^2 (von Wissmann 1959), about one-half belongs to the Chinese part of the Qinghai-Xizang Plateau.

The development and distribution of glaciers in the Qinghai-Xizang Plateau and its surrounding mountains are influenced greatly by the spectacular elevation, topography and particular climate of the Plateau. The height of the snow line varies from 4,400 in East Qilian Mountain and southeastern Xizang to 6,000–6,200 m in southern and western Xizang, and the isolines of the snow line show some irregular concentric circles with their centre in southwestern Xizang (Fig. 5.28). These peculiar variations of the snow line are mainly due to the rapidly decreasing precipitation as well as intensification of the thermal plateau effect between some outer mountains and the inner part of the Qinghai-Xizang Plateau. The annual total of precipitation in the glaciered area of East Qilian Mountain amounts to 800 mm, but 2,000 mm in southeastern Xizang, decreasing

Table 5.24. Glaciered area on Qinghai-Xizang Plateau in China. (After SHI Ya-feng and LI Ji-jun 1981)

Mountains	Height of snow line	Area of glacier (km^2)
Qilian	4,300 – 5,200	1,973
Kunlun	4,700 – 5,800	11,639
Pamirs	5,500 – 5,700	2,258
Karakorum	5,100 – 5,400	3,265
Qiangtang Plateau	5,600 – 6,000	3,188
Tanggula	5,400 – 5,700	2,082
Gandise	5,800 – 6,000	2,188
Nyainqen tanglha	4,200 – 5,700	7,536
Hengduan Mountains	4,600 – 5,600	1,456
Himalayas	4,300 – 6,200	11,055
	Total:	46,640

to 200–300 mm in central and western parts of the Plateau (SHI Ya-feng and LI Ji-jun 1981). As can be seen, the snow line on the southern slope of the Himalayas is lower than on the northern which is due to the differences in precipitation (Fig. 5.28). The peculiar conditions of solar radiation in the Qinghai-Xizang Plateau, in comparison with the surrounding lowlands, apparently influences the height of snow lines. Due to the vast extent of the Plateau, extremely high totals of radiation are absorbed by the Plateau which in its central and western parts record the highest values in all of China, so that the height of the snow lines is elevated.

The distribution of the glaciered area in the Qinghai-Xizang Plateau is uneven. The largest glaciered areas, which occupy the Karakorum, Pamir and West Kunlun, are located in the western borderlands of the Plateau where its considerable elevation compels the prevailing westerlies with high vapour content which leads to much precipitation (snow) in the glaciered areas. Although the deep valleys are characterized as very dry areas, the annual total of precipitation at 5,500 m elevation of the Fedschenko Glacier of West Pamir and the Batura Glacier of West Karakorum reaches about 1,500 mm or more. The second glaciered area, which includes Nyainqentanglha and the eastern end of the Himalayas, presents favourable conditions for the passage of moisture-laden air masses of the Indian Southwest monsoon. Therefore, from May through September clouds cover this area densely and heavy snowfalls frequently occur in the high mountains. Through intense glaciological research in China, it was also found for the southern slopes of the Himalayas as well as the northern slopes of Tian Shan that two maximum precipitation belts exist in the glaciered mountain areas. The lower belt of maximum precipitation is at 1,500–2,000 m above sea level. Above this elevation, the amount of precipitation decreases, but increases again and reaches its second belt of maximum precipitation in the glaciered height of more than 4,200 m above sea level. This interesting distribution of precipitation results from the conditions of local convection by strong solar radiation in the day-time (according to investigations by the Lanzhou Institute of Glaciology 1975).

The intensification of glaciological research in China has led also to interesting results on the variation of glaciers. Mainly based on the comparison of aerial

Fig. 5.28. The heights of the snow line in China. (After SHI Ya-feng and LI Ji-jun 1981)

maps of the 1960s with Landsat images of the 1970s, it was found that among the 116 glaciers on the Qinghai-Xizang Plateau, 35 are advancing, 62 retreating and 9 are stable or without a major variation (SHI Ya-feng and LI Ji-jun 1981). According to studies on the distribution of glaciers and the height of the snow line in the Qilian Shan for the last 1,000 years (LIU Guangyuan et al. 1984), glaciers advanced between 1500 and 1900, retreated between 1920 and 1960 and again increased since the 1970s. Accordingly, the snow line varied in its elevation.

6 Cloudiness and Sunshine

6.1 Mean Annual Cloudiness and January and July Amount

Records on mean monthly and annual cloudiness are contained in the climate tables (see Appendix). Despite regional gaps of data sources, the attempt was nevertheless made to work out some principal observations on cloudiness.

Mean *annual* cloudiness in China (Fig. 6.1) mostly varies between 4/10 and 8/10. (A value of 4/10 represents a cloud amount of 40% while 8/10 express 80%.)

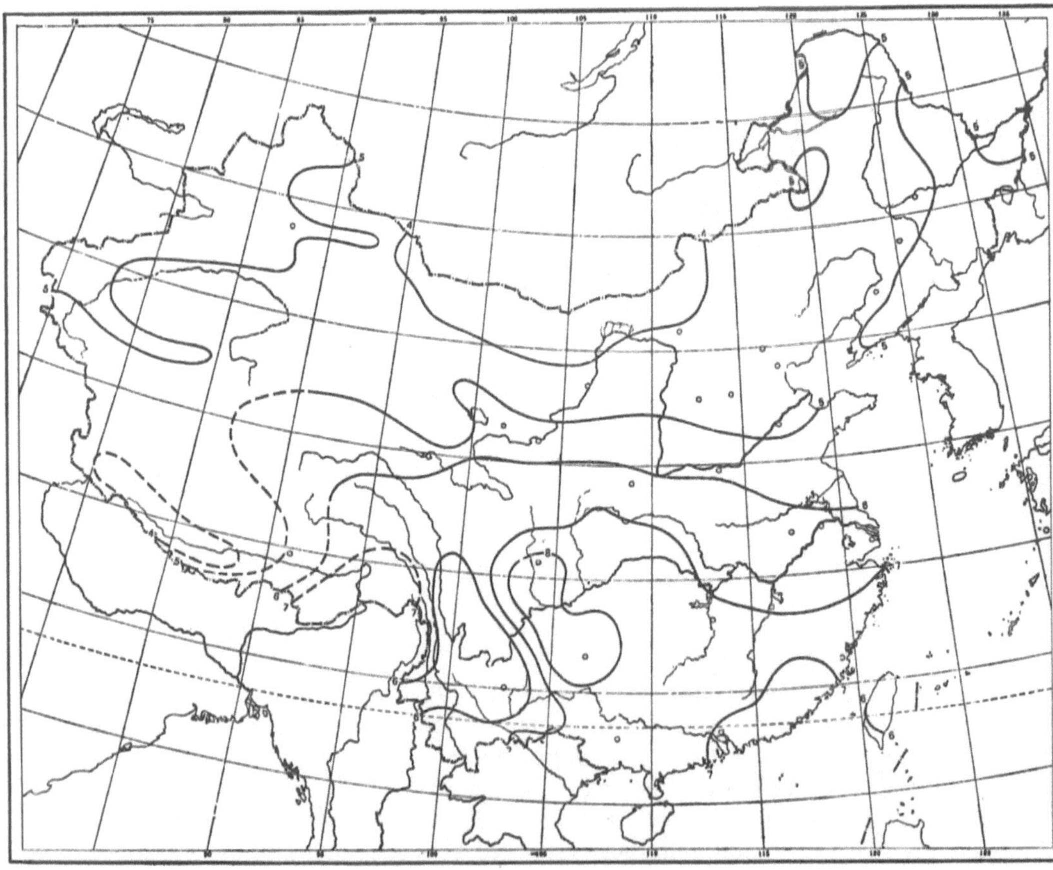

Fig. 6.1. Distribution of mean annual cloudiness, expressed as percentage values, 1 . . . 9 = 1/10 . . . 9/10 cloud amount (After ZHANG Jiacheng and LIN Zhi-quang 1985)

These figures only partly meet the common mid-latitude conditions of cloudiness which are characterized by fairly low mean values between 4/10 and 6/10 that are due to the intensive effect of continentality. The partly recorded higher values of cloudiness in China (7/10 and 8/10) express a strong oceanic influence which results from moist air masses of a Pacific Ocean origin. It is worth noting that the arid regions, such as Xinjiang and Xizang, record comparably high amounts of cloudiness which vary between 4.5/10 and 5.5/10, whereas deserts and arid regions commonly record cloudiness <3/10 only.

From the annual mean, cloudiness in China varies distinctly on a macro-scale. Lower and medium amounts of cloudiness, namely between 4/10 and 6/10 occur throughout the western and northern parts, while higher values, from 6/10–8/10, are experienced over the southeastern sector of China (see Fig. 6.1). The 6/10 isoline of cloudiness may be considered as a dividing line between a more

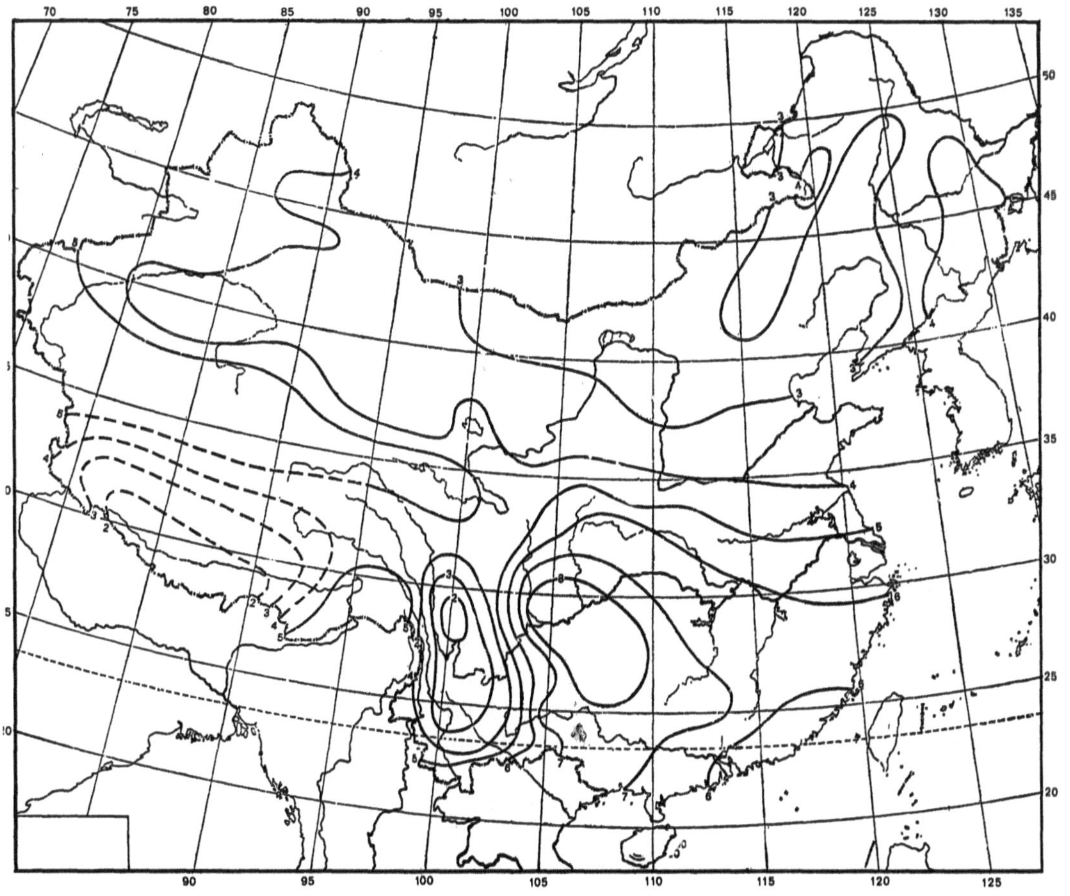

Fig. 6.2. Distribution of mean amount of cloudiness in January, expressed as percentage values, 1 . . . 9 = 1/10 . . . 9/10 cloud amount. (After ZHANG Jiacheng and LIN Zhi-quang 1985)

continentally influenced part, on the one hand, and a strongly affected oceanic part, on the other.

The *seasonal* amounts of cloudiness show some distinct differences which are significantly expressed by comparing cloudiness conditions in the two major seasons, i.e. winter and summer; thus, January and July respectively, can be taken as representative months.

January (Fig. 6.2)

Cloudiness shows a rather similar distribution in China as the annual mean. Thus, China can be roughly divided into two parts, according to the different amounts of cloudiness. The southern region, located south of the Yangtze and extending as far as the Yunnan Plateau in the west, records high means of cloudiness which, in January, even exceed 8/10. In contrast, in all other parts of China, considerably lower values of cloudiness occur which are mostly in a medium order of 3/10 to 4/10.

The comparably low means of cloudiness in most parts of China in January express the prevailing dry and continental air masses which govern the climate of China.

July (Fig. 6.3)

Regional differences of cloudiness in China are less pronounced than in January. Monthly means vary mostly between 6/10 and 8/10, without giving evidence of a distinct division of China according to cloudiness. In comparison with January, southern China records about the same amounts of cloudiness, while in all other parts cloudiness reaches higher values, which at the most increase in northeastern China. Harbin/15 can be taken as representative; it records 3.1/10 in January, but 6.7/10 in July.

In macro-scale terms, the strong influence of oceanic tropical and subtropical air masses in China in summer is evident according to the large values of cloudiness in July.

6.2 Sunshine

Records of mean monthly and annual sunshine duration are contained in the climate tables (see Appendix). The figures show two variables of sunshine: (1) the mean daily duration on both a monthly and annual basis, (2) the percentages of sunshine duration recorded against the maximum possible duration, resulting from the radiation budget.

Sunshine conditions in China were studied by determining the variations over space, on both an annual as well as a seasonal basis, referring to the major seasons winter and summer for which January and July respectively, were taken as representative.

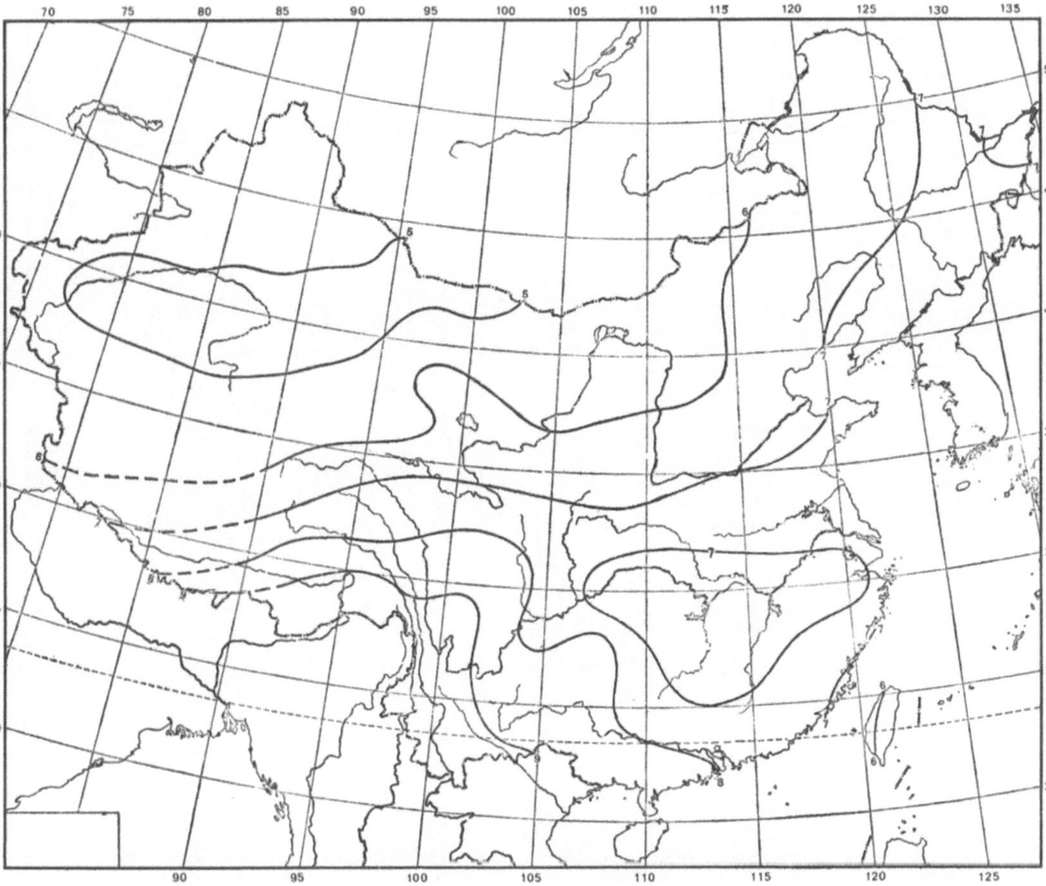

Fig. 6.3. Distribution of mean amount of cloudiness in July, expressed as percentage values, 1 . . . 9 = 1/10 . . . 9/10 cloud amount. (After ZHANG Jiacheng and LIN Zhi-quang 1985)

6.2.1 Annual Sunshine Duration

As evidently expressed by the annual total of sunshine hours in China (Fig. 6.4), sunshine duration varies considerably over space. This can be simply shown by the range between the stations experiencing the largest and smallest annual totals; the corresponding values amount to 3,551 and 1,143 sunshine hours per year (for Lenghu/260 and Yibin/139 respectively). Expressed as the mean daily sunshine duration, the corresponding values would be 9.7 and 3.0 h respectively. Therefore, the large amplitude of sunshine over space is clearly underlined by a 3.2 times higher sunshine duration at Lenghu than at Yibin.

From the map of annual sunshine duration (Fig. 6.4), also some remarkable observations on the spatial differences of sunshine can be given. As a whole, the western parts of China, which cover the Tibetan Plateau and Xinjiang, are extremely favoured by high sunshine totals. However, a small belt of large sunshine

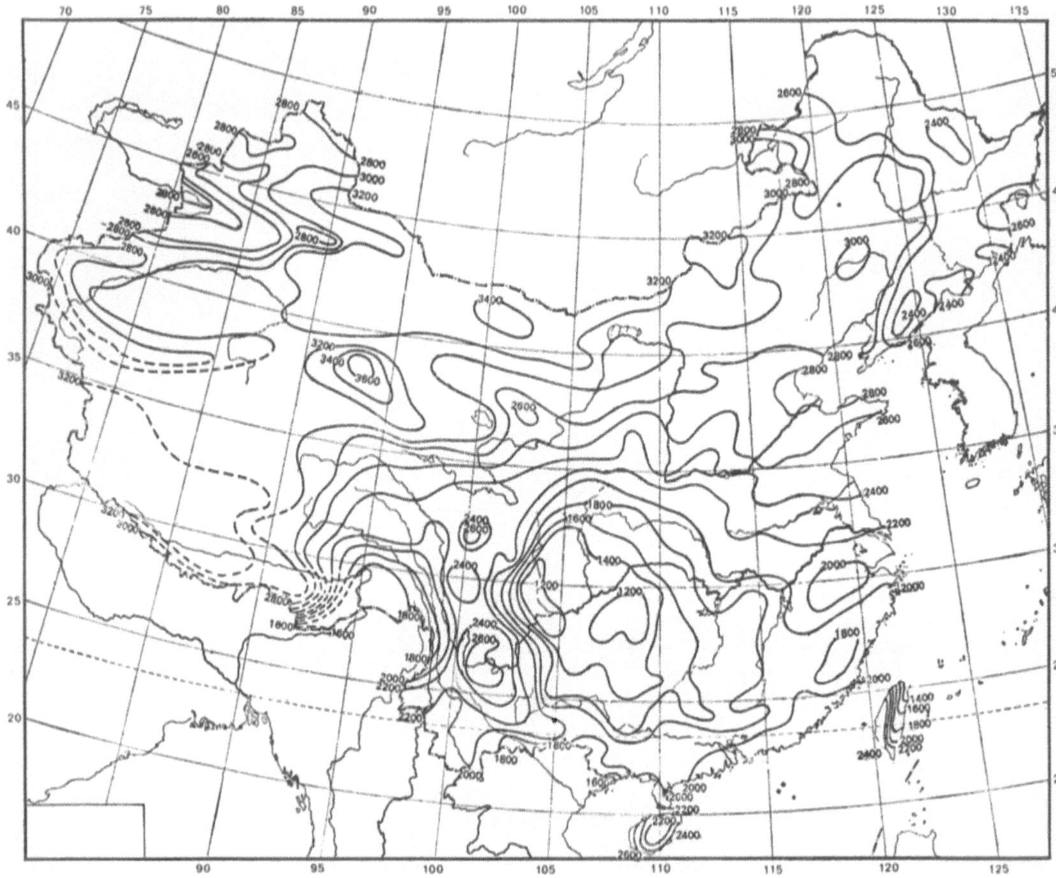

Fig. 6.4. Distribution of mean annual sunshine duration, expressed as the annual total of sunshine hours. (After ZHANG Jiacheng and LIN Zhi-quang 1985)

totals extends from Xinjiang via Inner Mongolia to northern Mandzuria. Annual sunshine totals in western and northern China are mostly between 2,500 and 3,200 h. Remarkably lower totals are surprisingly recorded in the southeastern Tibetan Plateau (for example, at Nyingchi/236, 1,978 h). From here, a small belt of also comparably low sunshine figures extends over the south Yunnan Plateau (2,100–2,300 h). The absolutely lowest sunshine totals occur in the Sichuan Basin and Guizhou where, on average, the annual total even drops to 1,150–1,300 h; Yibin/139: 1,143 h, Zunyi/166: 1,218 h, Chengdu/134: 1,228 h and Chongqing/142: 1,244 h are the stations with the lowest values. But also throughout the South China Hills, sunshine duration is relatively low and the annual total rarely exceeds 2,000 h. On the South China coasts and markedly pronounced in Taiwan and Hainan Islands, the annual sunshine duration reaches higher values, climbing to a maximum of 2,775 h (at Dongfang/210, in Hainan Island).

The large variations of sunshine in China, expressed by the mean annual totals, are also proved by two north to south cross-sections through the eastern

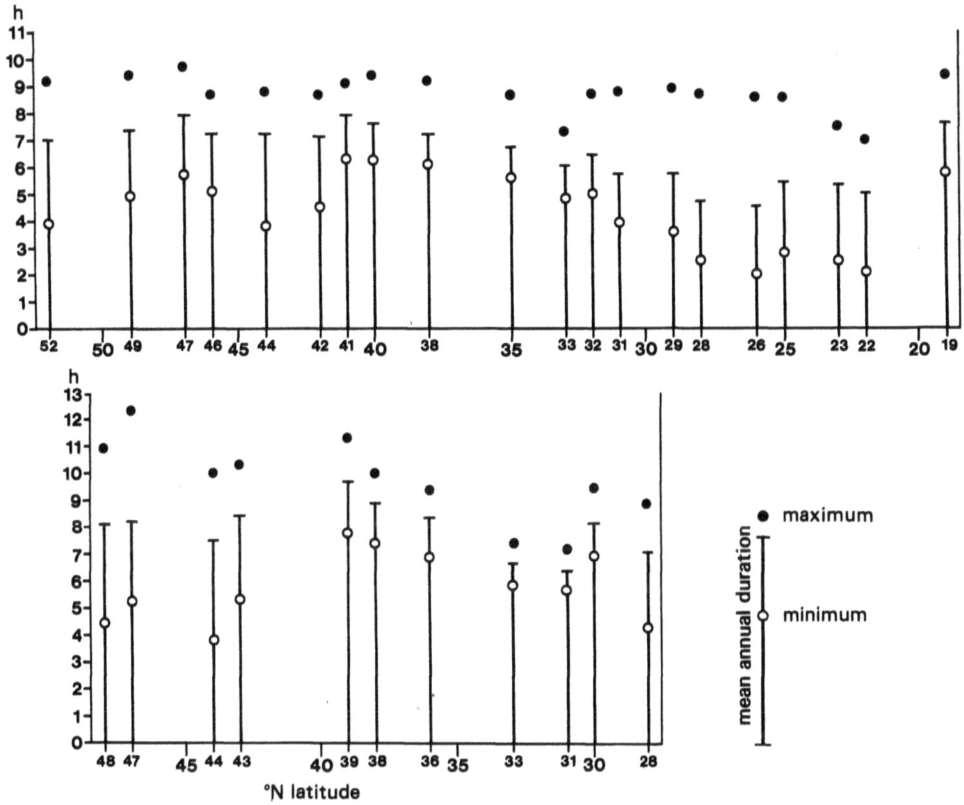

Fig. 6.5. Two north to south cross-sections through eastern China (top) and western China (bottom), showing the mean annual sunshine duration in comparison to the mean maximum and minimum months of sunshine duration; figures are given in h · day^{-1}. The following stations are representative for the different geographical latitudes:
for eastern China: *52° N* Huma/2, *49* Nenjiang/4, *47* Qiqihar/8, *46* Harbin/15, *44* Changchun/27, *42* Shenyang/38, *41* Chengde/51, *40* Beijing/61, *38* Taiyuan/69, *35* Zhengzhou/90, *33* Benghu/115, *32* Hefei/114, *31* Wuhan/126, *29* Jingdezhen/151, *28* Changsha/147, *26* Lingling/179, *25* Shaoguan/200, *23* Guangzhou/199, *22* Qinzhou/213, *19* Dongfang/210;
for western China: *48° N* Altay/270, *47* Hoboksar/271, *44* Ürümqi/269, *43* Turpan/267, *39* Lenghu/260, *38* Da Qaidam/259, *36* Golmud/253, *33* Yushu/243, *31* Qamdo/235, 30 Lhasa/239, *28* Yadong/237

and western part of China (Fig. 6.5). The cross-sections not only show the mean annual sunshine totals, but also the mean maximum and minimum months of sunshine duration (figures are given in h day^{-1}). As can be deduced from the two cross-sections, the spatially varying figures of both the annual totals and the minimum and maximum months of sunshine duration result from two major factors: (1) the different amounts of incoming solar radiation according to latitude; (2) the amount of cloudiness which results from the type and origin of air masses and the circulation conditions.

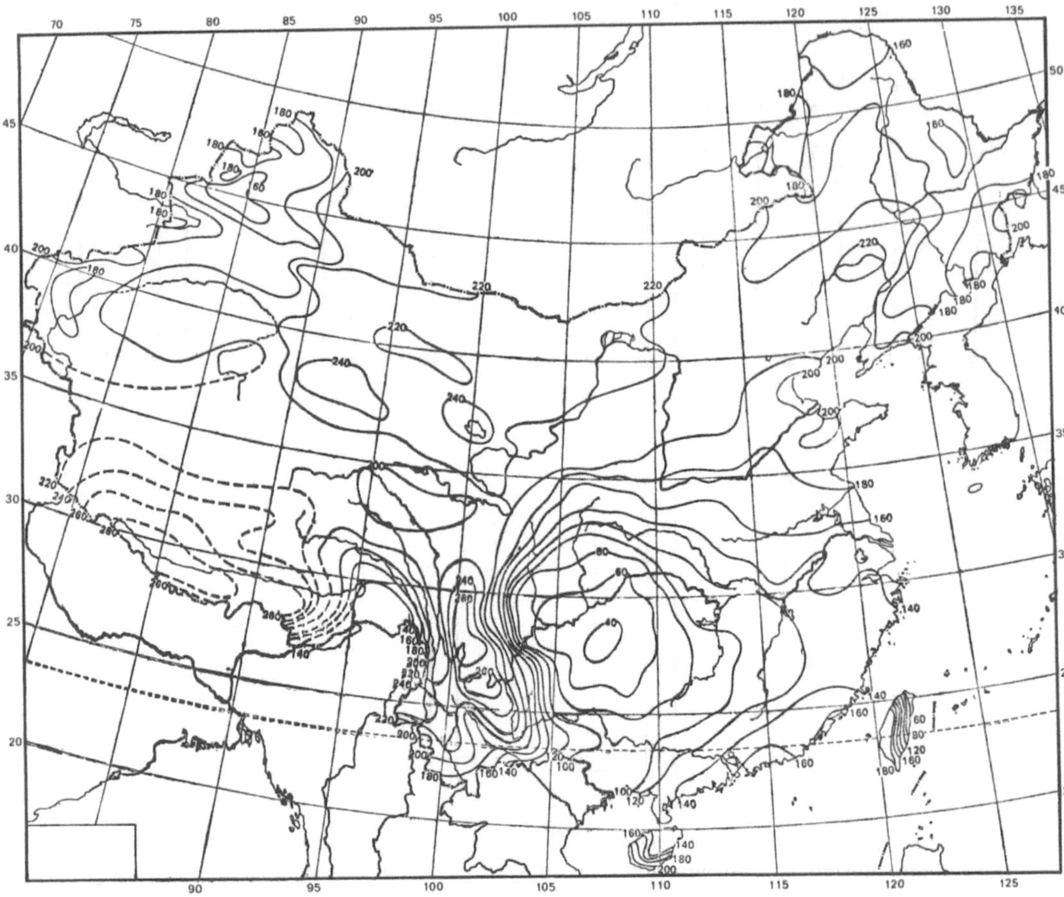

Fig. 6.6. Distribution of mean sunshine duration in January, expressed as the monthly total of sunshine hours. (After ZHANG Jiacheng and LIN Zhi-quang 1985)

6.2.2 Sunshine Duration in January and July and Annual Variation

In *January*, the duration of sunshine, expressed by the mean monthly total of sunshine hours, varies to an even greater extent (Fig. 6.6) than for the annual totals (Fig. 6.4), when considering the relative variation. For January, mean maximum and minimum totals in China are 266 (Xigaze/238) and 36 h (Zunyi/166) respectively; this means a 7.4 times higher maximum than minimum value. Variation of sunshine over space is in January similar to the annual variation. Again, western and northern China records remarkably larger totals of sunshine duration, while the Sichuan Basin and Guizhou recorded the minimum; however, also the South China Hills experience rather low values. Expressed as mean daily hours, sunshine duration varies between about 6.5–8 h in western and northern China and only 1.5–4 h in southeastern China.

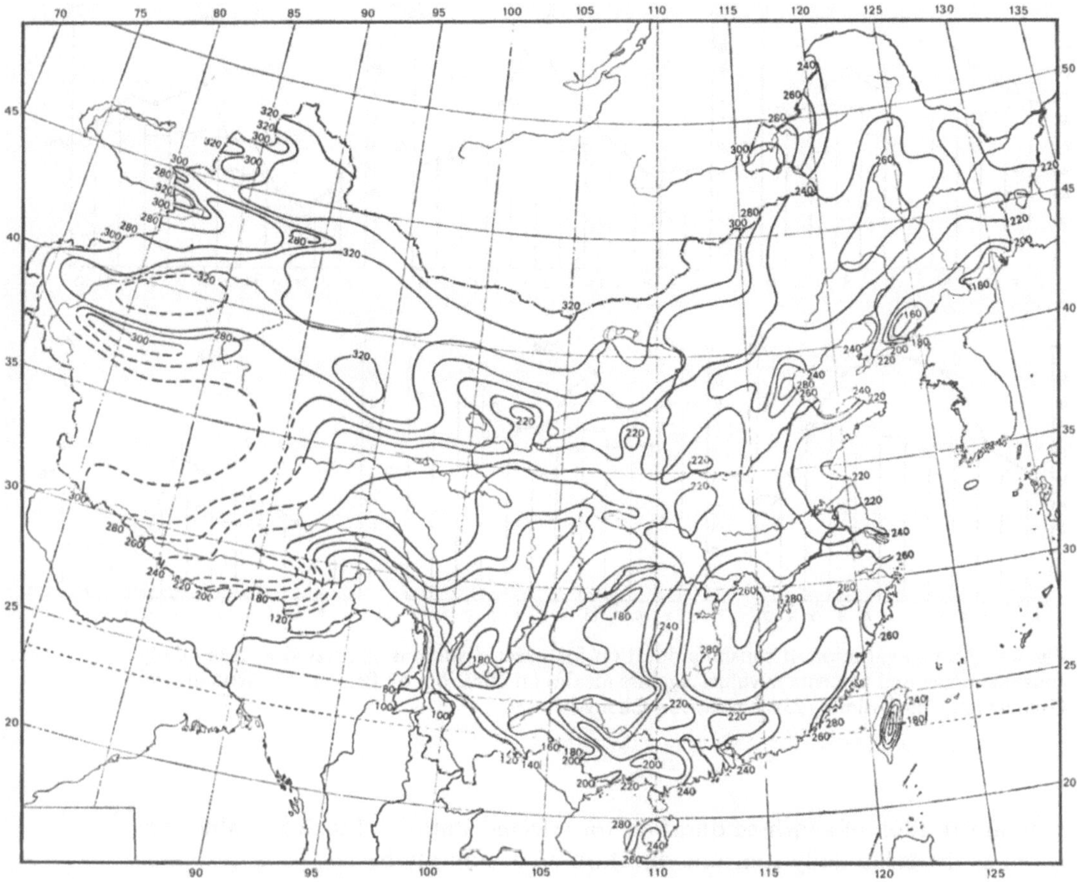

Fig. 6.7. Distribution of mean sunshine duration in July, expressed as the monthly total of sunshine hours. (After ZHANG Jiacheng and LIN Zhi-quang 1985)

In *July*, sunshine duration is for most of China higher than in January; the only exceptions are the southeastern parts of the Tibetan Plateau and the south Yunnan Plateau (Fig. 6.7). As a daily average of sunshine duration, the values for July vary between 3.5 and 11 h, although most parts of China record high values of 7 to 9 sunshine hours per day. Taking into consideration the corresponding percentage values of sunshine duration as compared to the maximum duration, it is surprising to note that in most parts of China only medium and fair values between 50 and 65% occur. It can be inferred that a fairly high amount of cloudiness remarkably reduces the duration of sunshine which is basically controlled by solar radiation conditions.

Expressed by the mean annual total and the monthly totals of both January and July, sunshine duration strikingly varies over space. However, comparing the monthly sunshine totals for January and July, also considerably large seasonal variations of sunshine are indicated which can be clearly determined by the

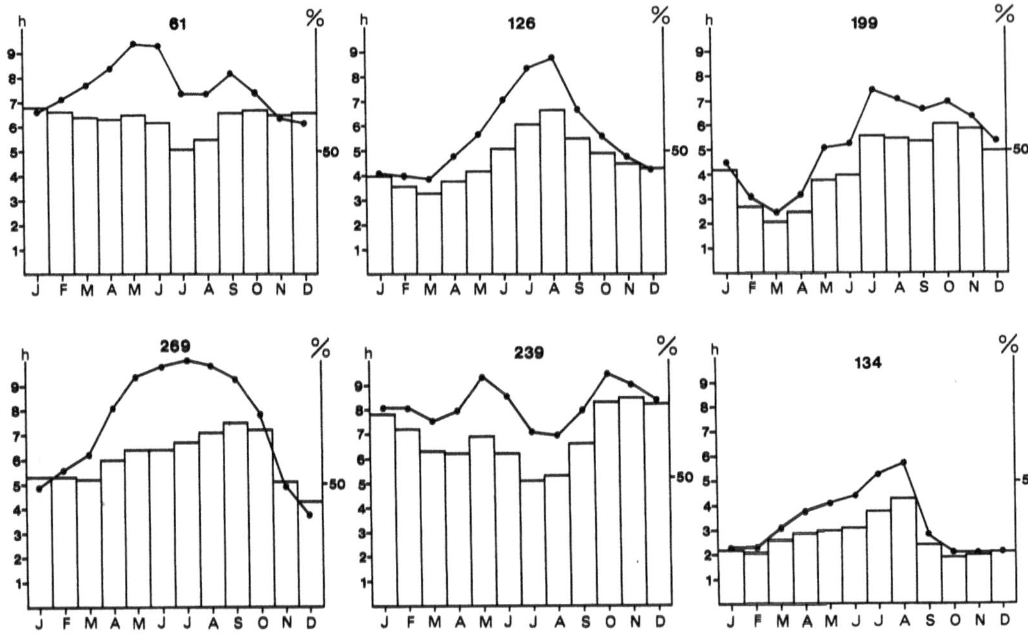

Fig. 6.8. Annual variation of sunshine duration for selected stations, expressed as mean daily sunshine hours and percentage values against maximum duration. *61* Beijing, *126* Wuhan, *199* Guangzhou, *269* Ürümqi, *239* Lhasa, *134* Chengdu

annual variation of sunshine duration for selected stations (Fig. 6.8); values are given as the mean daily sunshine duration on a monthly basis.

6.3 Global Radiation

Solar radiation respectively insolation, expressed in kcal cm^{-2}, was briefly considered as mean annual and monthly totals in order to express the variation in China over space and time.

The annual totals of solar radiation obviously show a large variation over space. The range of solar radiation extends from a minimum of <80 kcal cm^{-2} to a maximum of >220 kcal cm^{-2}. While the maximum supply of solar radiation occurred in the western and southwestern Tibetan Plateau, the minimum values were recorded for Sichuan and Guizhou Provinces, showing the absolute minimum total in the Sichuan Basin. In total, however, over the whole South China Hills and in the North China Plains, solar radiation drops to rather low values (mostly <120 kcal cm^{-2} $year^{-1}$). From the annual supply of solar radiation, China can be divided into two major parts; (1) the western part comprising Qinghai-Xizang, Xinjiang and Inner Mongolia which record large totals of radiation; (2) the remaining eastern part with lower totals. As the threshold value between the two parts an annual total of 160 kcal cm^{-2} can be taken.

Table 6.1. Mean solar radiation for selected months and stations (kcal cm^{-2})

	Huma	Harbin	Beijing	Shanghai	Guangzhou	Xisha
January	3.2	4.5	6.3	6.2	7.2	10.0
April	11.7	12.2	13.2	10.0	8.1	15.2
July	14.6	14.3	14.3	14.4	13.7	14.9
October	6.6	7.8	9.3	8.9	11.2	12.2
Year	108.2	118.1	130.1	113.8	116.1	154.8

The annual variation of solar radiation emphasizes a well-established seasonality in that during summer not only the maximum supply of solar radiation, but also the least variation over space was recorded. Values of solar radiation vary mostly in July between 14 and 24 kcal cm^{-2}. In all other seasons, the variation over space is greater, in January from 3 to 15, in April from 9 to 24, in October from 6 to 20 kcal cm^{-2}. Corresponding to the annual total, also the monthly and seasonal totals of solar radiation express the division of China into a western part with a high supply of solar radiation, and an eastern part with a low supply. In seasonal terms, the variation of solar radiation is more pronouncedly established in the eastern than western part. These observations were also confirmed by comparing the mean totals of solar radiation for selected months and the annual total at six selected stations (Table 6.1).

The distinct pattern of the annual variation of solar radiation was shown by the annual graphs for selected stations. Two major types of annual radiation distribution could be determined: (1) a well-established summer peak and winter depression type (Fig. 6.9a), as can be seen from the graphs at Lenghu, Beijing, Harbin and Jinhua; (2) a double-peak type (Fig. 6.9b), which shows a pre-summer peak, on the one hand, and a late or post-summer peak, on the other; for this type, Kunming, Guangzhou, Emli and Pali were taken as reference stations.

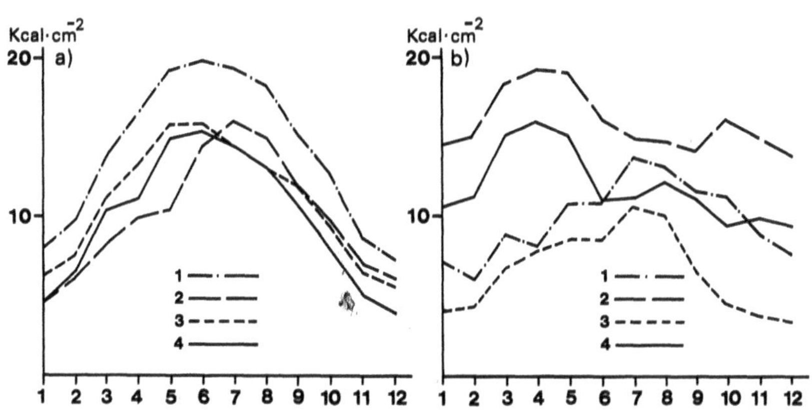

Fig. 6.9a, b. Annual variation of solar radiation for selected stations, showing two types: **a** Summer peak and winter depression type: *1* Lenghu, *2* Beijing, *3* Harbin, *4* Jinhua. **b** Double-peak type: *1* Guangzhou, *2* Pali, *3* Emei, *4* Kunming. Values in kcal cm^{-2}, given on a monthly basis

Comparing the two major types of annual radiation variation, the first type shows a larger range of radiation, while the second one underlines a less pronounced seasonality. As can be deduced from the location of the reference stations, in the main parts of China summer and winter are well established as alternating seasons with either a large or a small supply of solar radiation, thus expressing the seasonal differences in the supply of solar radiation, due to the annual variation of the insolation budget. In southern China (Fig. 6.9), winter records the minimum totals of solar radiation, while from spring through autumn a spatially variable variation of solar radiation occurs which may be due to larger totals of cloudiness (see Sect. 6.1) which results from the moist summer monsoon.

6.4 Fog

Fog, which is defined by a visibility <1 km, occurs in China as both the advection and radiation type. While the occurrence of advection fog is particularly associat-

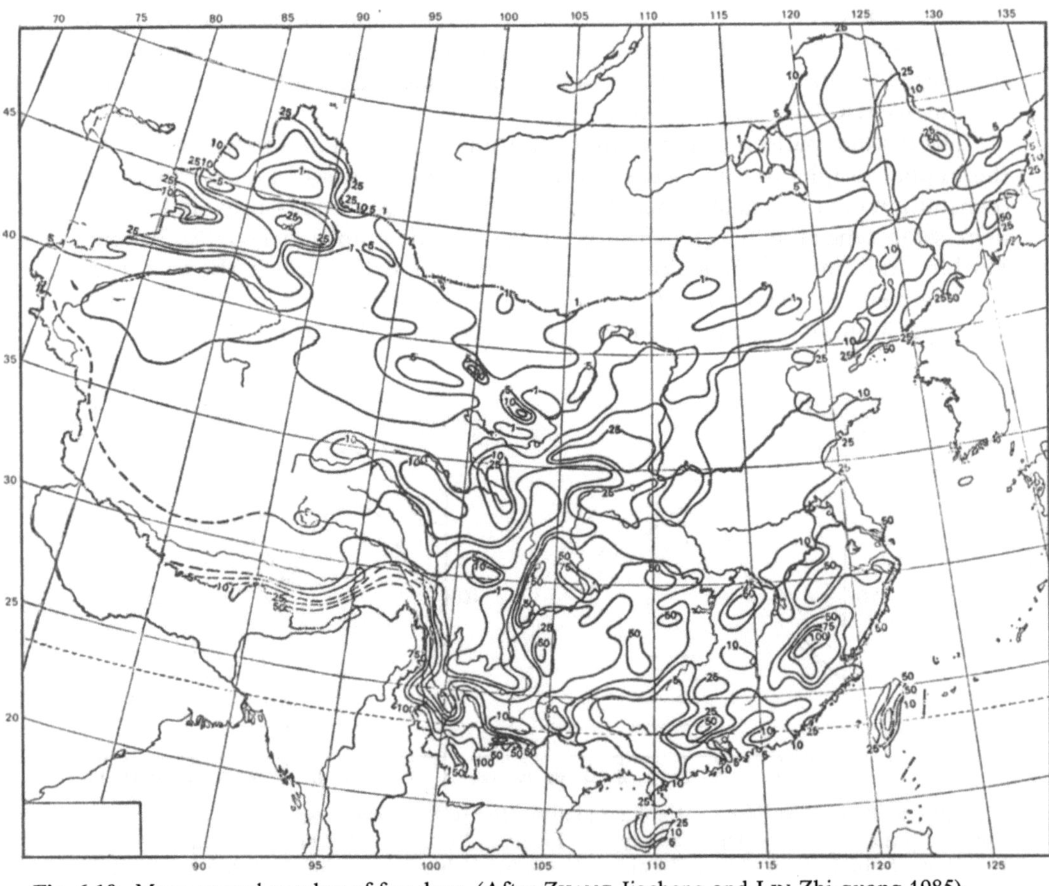

Fig. 6.10. Mean annual number of fog days. (After ZHANG Jiacheng and LIN Zhi-quang 1985)

ed with the moist and warm summer monsoon, radiation fog represents a typical phenomenon during winter and is thus associated with continental, cold air masses. Both types are small-scale rather than large-scale weather phenomena. For most parts of China, fog has to be considered as a hazardous weather phenomenon, because of the comparably large number of cases.

The annual number of fog days shows a great variation over space (Fig. 6.10), without a significant relationship to either the landforms or to the proximity respectively distance to the China Seas. Despite a rather irregular pattern of the distribution of the annual number of fog days, the maximum figures were recorded in the South China Hills, Sichuan Basin and southern Yunnan. The absolutely highest figures amounted to 186 fog days year^{-1} and were recorded in southern Yunnan. The minimum number of fog days year^{-1} (<5) was experienced over major parts of Qinghai-Xizang, Xinjiang and Inner Mongolia, including Qaidam and Talimu Basins as well as Gansu Province.

It is worth noting that the coastal lowlands and the islands of the China Seas, including Hainan and Taiwan Islands, did not experience a maximum frequency of fog days, but a medium frequency mostly between 5 and 25 fog days year^{-1}. On the other hand, notable medium-scale differences in the number of fog days were observed betweeen top mountain stations, on the one hand, and valley stations, on the other, expressing an increasing number of fog days with increasing altitude. Also, the exposure of a mountain was proven to effect the formation of fog such that the wind-exposed slope received a greater frequency of fog, while, in contrast, on the lee side of a mountain less frequency was recorded.

Obviously, medium-scale variations of both humidity and air temperature, including their interactions, seem to be responsible for the formation of fog. As far as the type of fog is concerned, advection fog mostly affects the coastal lowland areas, while radiation fog predominantly occurs in the inland areas.

The frequency of fog days also shows a clear annual variation which differs between the coastal lowlands and the inland and mountainous areas. For the coastal lowlands, summer represents the major fog season, but the frequency of fog days is stepwise shifted from pre- and early summer in southern China to late summer in northern and northeastern China. Hence, the strongest frequency of fog occurs in March-April in the South China Sea region, it is shifted to May in the East China region and finally occurs in June-July in the Yellow Sea region.

As can be inferred from observations, the formation of fog also shows a typical diurnal variation. Fog occurs predominantly during the night and continues markedly until early morning (normally 05.00–07.00 h), but then dissipates with increased warming of the surface layers. Obviously, the diurnal variation of temperature and its effect upon the saturation pressure of water vapour decisively affects the formation of fog.

7 Surface Wind

As far as the *direction* of surface wind is concerned, comments on it were already given as part of the observations on the circulation pattern and air masses (see in particular Sect. 3.2). This section, therefore, directs attention to the velocity of surface wind and to local wind systems (Sects. 7.1 and 7.2).

7.1 Mean and Extreme Wind Velocities

Monthly means of wind velocity on ground are contained in the climate tables of selected stations (see Appendix). According to records, mean wind velocity in China shows a rather homogeneous pattern over space and time, without significant temperature differences. For most stations under consideration, monthly means express comparably low wind velocities mostly between 1 and 4 m s^{-1}. Subsequently, the annual mean is in the same order. Only in rare cases, monthly means <1 respectively >4 m s^{-1} were experienced, for example 0.3 m s^{-1} in December at Lanzhou/247, and 8.1 m s^{-1} in November at Nenjiang/4 which represent the lowest and highest extreme values respectively. The small monthly means of wind velocity usually recorded in China are in the order of the typical mid-latitudinal velocities. Extremely small monthly wind velocities are plausibly connected with very large percentages of calms, as can be seen in the following cases:

Station	Month	Monthly means of	
		Wind velocity (m s^{-1})	Calms (%)
Lanzhou/247	Dec.	0.3	77
	Jan.	0.5	71
	Nov.	0.5	70
Lancang/222	Nov.	0.5	71
	Dec.	0.5	68
Yushu/243	Sept.	0.7	68

In contrast, in the case of higher wind velocities, the percentage of calms decreases, and vice versa.

Wind velocities, recorded on ground, depend on two factors: (1) the speed of the airflow concerned, (2) the effects of landforms upon the airflow. Therefore,

Table 7.1. Wind velocities for different coastal and inland stations; observation period 1961–1970. (After ZHANG Jiacheng and LIN Zhi-quang 1985)

Station[a]	Mean annual wind velocity ($m s^{-1}$)	Maximum velocity ($m s^{-1}$)	No. of days $year^{-1}$ ($\geq 8.0 m s^{-1}$)
A Haiyang	6.1	40	173.1
B Zhuang Hei	2.6	24	19.2
A Dalian	5.3		82.1
B Tianjian	3.0		40.9
A Touji Dao	7.8	40	125.2
B De Zhou	3.5	28	28.2
A Cheng Shan Tou	6.8	40	126.9
B Lai Yang	2.6	22	15.7
A Cheng Si	7.2	40	152.8
B Hang Zhoce	1.9	28	5.9
A Shi Pu	5.5	35	106.4
B Hai Men	2.8	24	4.9
A Nan Bi	7.8	>40	200.5
B Wen Zhou	2.2	25	6.8
A Ping Tan	7.2		107.2
B Yong Tai	1.6		9.2

[a] (A) Coastal station; (B) inland station.

Table 7.2. Wind velocities for different hilltop and valley stations; observation period 1951–1970. (After ZHANG Jiacheng and LIN Zhi-quang 1985)

Station	Elevation (m)	Mean annual wind velocity ($m s^{-1}$)	No. of days $year^{-1}$ ($\geq 17 m^{-1}$)
A Tai Shan	1,534	6.1	127.4
B Tai An	129	2.8	26.3
A Wu Tai Shan	2,896	9.0	175.6
B Yuan Ping	837	2.3	60.5
A Hua Shan	2,065	4.3	171.1
B Xian	397	2.1	10.4
A Huang Shan	1,840	5.7	145.0
B Tuen Xi	147	1.2	9.5
A Nan Ye	1,266	6.2	113.3
B Heng Yang	101	2.2	18.6
A Lou Shan	1,164	5.4	120.8
B Jin Jiang	32	2.7	13.1

[a] (A) Station on top of a hill; (B) station in the valley.
[b] Defined as "strong winds" in China.

it can be deduced that stations on the coast and on top of a hill or a mountain experience the largest wind velocities which express the speed of an unhindered airflow. In comparison, small velocities occur in the case of lee-side mountain stations in valleys and depressions (Tables 7.1 and 7.2).

As can be seen from Table 7.1 and 7.2, wind velocities may vary considerably according to the location of a station and the surrounding landforms. According to Chinese findings (ZHANG Jiacheng and LIN Zhiquang 1985), it must also be taken into consideration that despite small mean wind velocities nevertheless large recorded velocities may be experienced. Expressed by the number of days with "strong winds", defined by a maximum speed ≥ 17 m s^{-1}, their mean number per year demonstrates an extremely large spatial variation between <5 and >100. Although no clear pattern of the regional distribution of strong wind days seems to exist, regions of a maximum frequency are the Qinghai-Xizang Plateau and the northern peripheral areas of Xinjiang and Inner Mongolia. If considering the maximum wind velocity for a 10-min time spell (from 1961–1970), the smallest recorded values in China amounted to around 20 m s^{-1}, while the largest values were >30 m s^{-1}; a clear pattern of regional differences has not been established.

7.2 Local Wind Systems

7.2.1 Mountain and Valley Breezes

With respect to local wind systems, which show a clear diurnal periodicity, mountain and valley breezes are of wide-ranging importance in China, particularly due to the mountainous landforms of the country.

Precise observations on a strongly developed diurnal system of mountain and valley breezes were given for the Tianchi region of the northern slope of Tian Shan. As a result of diurnally alternating and varying temperatures over the slope and valley and thus opposing pressure gradients, also alternating stream flows between the slope and valley were distinctly observed: During the daytime, a persistent flow of valley breezes was recorded which at night changed into mountain breezes. According to the slope and valley direction, the predominant direction of the valley breeze was northwest, of the mountain breeze southeast. As can be deduced from the percentage frequencies of the wind direction over the Tianchi region, given as monthly and annual means from 1961–1970 (Table 7.3), northwesterlies and southeasterlies showed the largest frequencies. As the annual average, strong westerlies were experienced with a 35% frequency, southeasterlies with a 30% frequency. If referring to the whole northwestern and southeastern sector respectively, the valley breeze from northwest occurred with an annual frequency of 44%, while the mountain breeze from southeast occurred with a 38% frequency. Thus, both the valley and mountain breezes together took place with a frequency of 82%, while of the remaining 18% almost 10% were represented by calms. Therefore, it can be inferred that the mountain and valley breezes occurred with great strength.

According to the monthly frequencies of valley and mountain breezes (Table 7.3), it is obvious that a pronounced annual variation between the valley and mountain breezes occurred. The maximum frequencies of the mountain breeze (from Southeast) was experienced in winter, affected by the snow cover of the mountains which caused a stronger pressure and temperature gradient between the upper slopes and the valley. The maximum frequencies of the valley breeze

Table 7.3. Mean monthly and annual percentage frequencies of the wind direction over the Tianchi region, Tian Shan; 1961–1970

	1	2	3	4	5	6	7	8	9	10	11	12	Year
N	3	1	1	2	2	3	3	2	1	1	3	2	2
NNE	0		0		0	0		0		0	0	0	0
NE	0	0	0	0	0	0	1	0	0	0	0	0	0
ENE			0			0	0	0	0		0	0	0
E	1	1	1	0	1	1	2	0	1	1	0	1	1
ESE	3	3	1	1	2	2	2	2	2	2	1	3	2
SE	45	37	29	26	25	23	23	22	25	28	34	43	30
SSE	5	3	3	3	4	6	7	7	7	11	10	8	6
S	1	2	1	1	1	2	2	2	2	4	3	3	2
SSW	0	0	0	0			0		0	0	0	0	0
SW	0	0	0	0	0	1	1	0	0	0	0	0	0
WSW	0	0		0		0	0	0	0	0		0	0
W	1	1	1	1	2	2	3	4	2	1	1	1	2
WNW	1	1	2	3	5	6	4	6	4	2	2	1	3
NW	18	34	40	49	49	40	38	40	38	33	25	17	35
NNW	4	4	6	7	5	7	7	7	7	8	7	3	6
C	17	12	13	7	5	6	7	7	9	9	13	17	10

(from northwest), however, were recorded in April and May, due to the green vegetation cover in spring which leads to smaller differences in temperature between the slopes and the free atmosphere.

The considerably large intensity of the valley and mountain breezes was also expressed by the frequencies of the wind direction in the Tianchi region at 11.00 and 05.00 h respectively (Table 7.4); both are local times. The 11.00-h frequencies, which can be taken as a measure of the valley breezes, clearly expressed prevailing northwesterlies; conversely, the 05.00-h frequencies, which show the mountain breezes, demonstrate governing southeasterlies. In both cases, a frequency of about two-thirds was observed.

Furthermore, for the same region under investigation, the wind velocity of the mountain and valley breezes was studied, by considering 8-day mean values for July 26 to August 2, 1982 (Fig. 7.1). For the mountain breezes, occurring at night, a maximum velocity was experienced between 03.00 and 08.00 h, showing the absolutely highest value at 05.00 h/4.4 m s^{-1}. For the valley breezes, which occurred during day time, a maximum velocity was recorded from 12.00 until 18.00 h, with the absolutely highest speed at 14.00 h (4.0 m s^{-1}). Thus, the moun-

Table 7.4. Percentage frequencies of wind direction at 11.00 and 05.00 h over Tianchi region, Tian Shan, from 1971–1980. Valley breezes are represented by winds from the northwest, mountain breezes from the southeast

N	NNE	NE	ENE	E	ESE	SE	SSW	SW	WSW	W	WNW	NW	NNW	C
Day breezes, 11.00 h														
3.5	0.2	0.3	0	0.4	0.7	4.8	0.8	1.0	0.1	0.1	3.0	64.8	12.2	6.8
Night breezes, 05.00 h														
1.1	0	0.2	0	0.8	3.4	62.0	11.5	2.9	0.2	0.3	0.8	6.3	1.2	8.8

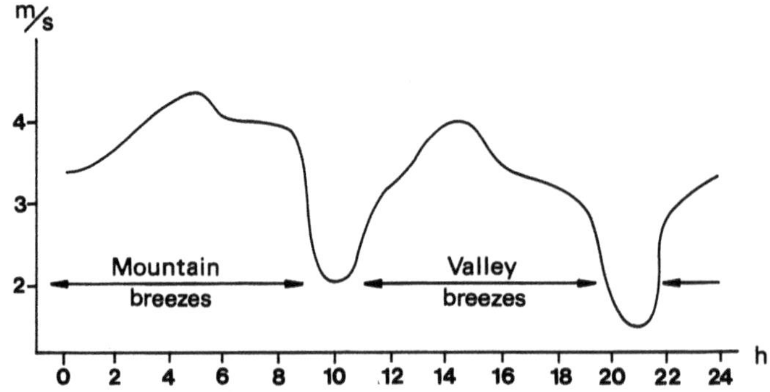

Fig. 7.1. Mean diurnal variation of wind velocity of valley and mountain breezes in Tianchi region, Tian Shan, for July 26 to August 2, 1982. (After ZHANG Jiacheng and LIN Zhi-quang 1985)

tain breezes occurred with a slightly higher velocity than the valley breezes. It can also be seen that the reversion from the mountain to the valley breezes occurred at about 10.00 h, the opposite reversion form the valley to the mountain breezes at about 21.00 h.

The strongly established system of mountain and valley breezes was also expressed by the monthly and annual wind velocities for the mountain breezes (at 05.00 h) and the valley breezes (at 10.00 h), observed for the Tianchi region, Tian Shan (Table 7.5). With the exception of May, for all months the mountain breezes showed a greater velocity than the valley breezes. For the year as a whole, the mountain breezes are 10% stronger. Both the mountain and valley breezes showed a fair annual variation which was underlined for the number of days with mountain and valley breezes (Table 7.5). It is important to note that on 250 days of a year the phenomenon of local breezes was recorded.

Table 7.5. Monthly and annual velocities of mountain and valley winds and number of days of local breezes in Tianchi region, Tian Shan; 1971–1980

	J	F	M	A	M	J	J	A	S	O	N	D
Mountain breeze, 05.00 h	3.1	3.1	3.3	3.9	3.8	4.0	4.4	4.2	3.6	3.5	3.2	3.2
Valley breeze, 11.00 h	2.6	2.8	3.2	3.5	3.9	3.9	3.9	3.8	3.4	3.2	2.6	2.5
Difference	0.5	0.3	0.1	0.4	0.1	0.1	0.5	0.4	0.2	0.3	0.6	0.7
Days with mountain and valley breezes	16.4	20.0	21.4	21.7	23.0	22.4	23.6	25.9	22.3	22.8	16.9	18.6
Mean annual velocity of:												
Mountain breeze, 05.00 h							3.6					
Valley breeze, 11.00 h							3.3					
Difference							0.3					
Annual No. of days with mountain and valley breeze							250.0					

Table 7.6. Annual variation of days with land and lake breezes at Yueyang, Dongtinghu Lake, 1953–1971

	J	F	M	A	M	J	J	A	S	O	N	D	Year
No. of days with land/lake breezes	8.4	6.7	10.0	11.6	11.8	10.8	10.0	10.5	6.3	8.7	7.8	7.4	110

7.2.2 Land and Sea Breezes, Lake Breeze

As can be assumed from the long coastline of China, land and sea breezes can be reasonably expected as a pronounced diurnal wind system which, however, so far has not been investigated by specific studies.

Instead, particular attention was paid to a well-established, diurnal system of a land and lake breeze which has been determined for the Dongtinghu (lake), by analyzing wind observations at Yueyang. The station, located on the northeastern coast of the lake, showed a significant frequency of days with an alternating lake breeze (with governing southwesterlies) during daytime and a land breeze (mainly from southeast) during night. Referring to observations from 1953–1971, land and lake breezes were well established on a total of 110 days year^{-1}, which showed a distinct annual variation (Table 7.6). The maximum frequency could be observed from March to August, with the absolutely highest frequency during April and May, while the minimum was observed in February and September. As can be inferred from the large amount of incoming solar radiation in summer, the larger temperature and pressure gradient between land and water (resp. lake) also leads to a more frequent and also stronger developed lake breeze during daytime and, conversely, a land breeze during night when the temperature and pressure gradient has been reversed.

According to observations, the change from land to lake breezes took place between 08.00 and 11.00 h, while the change from lake to land breezes occurred between 16.00 and 18.00 h. It was also observed that the time of transformation from land to lake breeze and vice versa varied seasonally, showing an earlier time in summer and a later one in winter.

7.2.3 Plateau Monsoon

Quite different to the local wind systems of a diurnal wind change (see Sects. 7.2.1 and 7.2.2), the so-called plateau monsoon represents a seasonal wind, defined by Chinese climatologists as a wind between the Qinghai-Xizang Plateau and the surrounding free atmosphere which is derived from the specific pressure pattern of the Plateau, thus causing particular temperature conditions in summer and winter (see also TAO Shiyan 1984).

As a commonly known fact, the Tibetan Plateau, because of its high elevation, acts as a heat source in summer, while it is a source of coldness in winter. Therefore, a heat Low dominates the planetary boundary layer over the Plateau in summer and, conversely, a cold High in winter. The cyclone over the Plateau

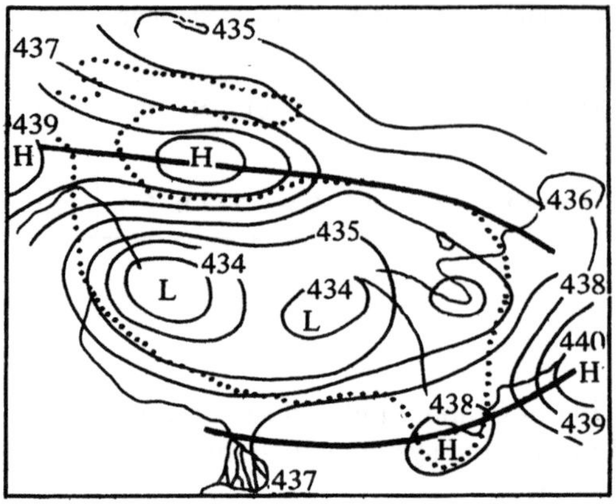

Fig. 7.2. Pressure field of the plateau monsoon in July, given by the mean 600-mb contours, expressed as decameters for 1,200 GMT. Dotted lines indicate the extent of the Qinghai-Xizang (Tibetan) Plateau. (After YED 1977, in TAO Shiyan 1984)

in summer is established between the 500–400 mb level, the anticyclone in winter between the 600–500 mb level. Since the heat Low (in summer) and cold High (in winter) over the Plateau are surrounded by an anticyclonic respectively cyclonic belt, the thus resulting atmospheric pressure gradient leads to strongly developed surface winds on the Plateau which strengthens the prevailing direction of the winter and summer monsoons which are mainly northeasterlies in winter and westerlies in summer.

The strength of both monsoons is closely related with the cold High in winter, and the heat Low in summer on the Tibetan Plateau; stronger force of both monsoons occurs with increasing pressure gradients on the Plateau.

The pressure conditions of the well-established plateau monsoon in summer can best be seen from the mean 600-mb contours in July (Fig. 7.2).

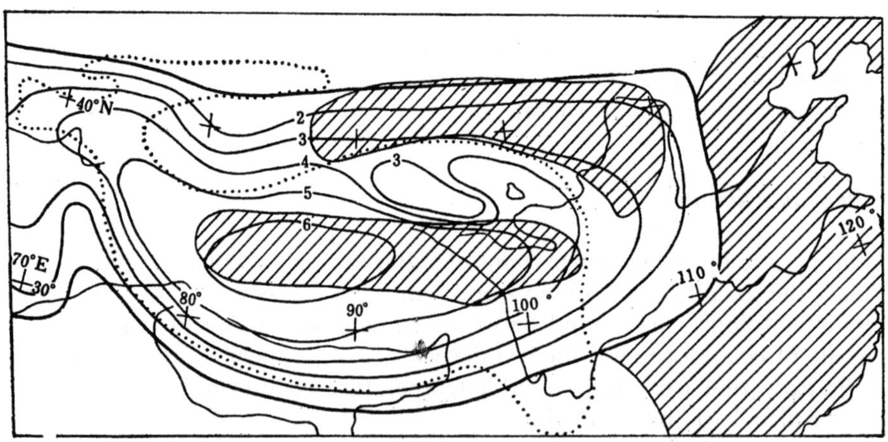

Fig. 7.3. The area and the heights (in km) of the plateau monsoon over the Qinghai-Xizang Plateau and surrounding areas. (After Chinese Geographical Society 1984). The dotted lines indicate the extent of the Qinghai-Xizang (Tibetan) Plateau; 1 . . . 6 = elevation of the Tibetan Plateau a.s.l. (in km). The matched regions are those under the plateau monsoon

The distinct pressure and wind field of the plateau monsoon not only affects the Qinghai-Xizang (Tibetan) Plateau, but also extends northward to the central parts of the Gansu and Xinjiang, southwards to the southern slopes of the Himalayas and eastward to 105°–110° E; the west boundary is uncertain due to the lack of observations (Fig. 7.3).

7.2.4 Local Dry and Hot Winds

Representing an exceptional case of surface winds, the so-called dry and hot winds, nevertheless, can be taken as a characteristic wind which strikes North and Northeast China with far-reaching agroclimatic effects, i.e. serious damage to wheat cultivation.

"Dry and hot winds" are defined by three climate parameters: a daily maximum temperature $\geq 30°$ C, associated with a relative humidity $\leq 30\%$ and a wind velocity > 3 m s^{-1}.

The phenomenon of dry and hot winds, continuously occurring for a series of days at the end of the growing period of winter and spring wheat, appreciably weakens the growing conditions. Badly stricken by dry, hot winds are the following regions of North China: (1) Hua Bei (Shanxi, Hebei, Shandong Provinces) and Guangzhong (Shaansi) from mid-May until early June; (2) Yin Chuan (west of Xian) from mid- to late June; (3) Hei Xi (Yellow River) during late June. Due to dry, hot winds, the harvest of winter and spring wheat may be reduced up to 20%.

Although from a climatological viewpoint, dry and hot winds occur practically every year in the regions mentioned above, the number of winds with a serious agroclimatic affect is much less, showing one case only in 2 to 4 years.

The reasons for the formation of dry, hot winds are not yet verified. It is, however, considered as an important factor that the polar front is shifted far northward, i.e. to 50° N, so that dry winds of a southern origin (and thus hot nature) can penetrate into North China.

8 Climate Classification and Division of China

8.1 General Objectives and Fundamentals of Climate Regionalization

As far as a climate classification and division is concerned, China represents a great complexity and diversity due to the country's vast territory and extremely complex landforms. In detail, three major climate-governing factors must be taken into consideration for the climate of China as such and for its division over space: (1) the geographical latitude, (2) the elevation above sea level and (3) the distance from the Pacific Ocean. Of particular importance for a climate regionalization of China is the country's location on the southeastern corner of the Eurasian continent against the Pacific Ocean, which means that the nature of the surrounding regions varies from open oceans to compact land masses. As can be inferred from the complex geographical setting of China itself, on the one hand, and from the macro-locational context, on the other, the climate division of China plausibly results from interacting earth-atmosphere processes, both within China and from outside.

Any climate division of China obviously depends on the specific reason for such a classification. Thus, from all global climate classifications, China represents a country of an extremely diversified climate division, in both a horizontal and vertical direction. All climate classifications presented so far on a global scale demonstrate an exceptionally large range of climate types and regions valid for China. However, the extremely great vertical extension of China does not appear to be reflected appropriately in any global climate division. Hence, it must be assumed that necessarily all climate divisions must contain uncertainties and even inaccuracies which are particularly true for parts with rare climate data, such as the major parts of the Tibetan Plateau.

Consideration of climate classification with respect to China will be given in this chapter with particular reference to the major principles of regionalization and with particular concern to a hierarchical system of climate division.

In spite of a great number of climate classifications, it is understandable that in the following chapter only a selection can be briefly presented in order to describe the climate division of China. An attempt was made to deal with those global classifications which are most commonly referred to by Chinese scientists. The classifications are distinguished in two ways with two categories each: on the one hand, classifications which are, by nature, either empiric or genetic; on the other hand, classifications developed by either Chinese scientists or foreigners. The scale of climate classifications may refer to a global scale or deal with China alone.

For a climate classification of China, the commonly used guidelines of a climate regionalization must be accepted and should therefore be briefly discussed[24]. As the objective of any climate classification scheme, it must allow the identification of major, or at least significant, differences in climate. Such a scheme of a climate division must aim to simplify and clarify the variations over space in order to enhance comprehension and understanding. A climate regionalization scheme automatically leads to the creation of a series of climate types and may develop a climate hierarchy which, through mapping and identification in spatial terms, leads to various climate regions.

Since a climate classification logically represents a regionalization scheme of the climate conditions, the definition of "climate" is the superior problem in developing a climate classification and also in the user's choice of an appropriate classification.

Climate, as a whole, involves many elements and their interaction which produces a certain type of a climate at a certain location. If only one climate element is considered, it hardly qualifies as a "climate" classification. If, in contrast, an attempt to use all elements is made, the thus following complexity defeats the purpose of classification, i.e. simple understanding and clear identification. Hence, usually two or three climate elements are used which are chosen because they are considered important for the specific use to which the climate classification is to be placed.

Climate regionalization schemes correspondingly try to serve as a source of information for those referring to climate as a resource for a particular purpose. Climate classifications can provide information on human activities and natural processes on earth as well. Thus, a climate regionalization represents, at the same time, an estimate of the spatially variable climate resources of an area.

Once the aims of a climate classification scheme have been fixed, the next step in developing a classification is to identify appropriate threshold values which specify major differences in climate over space. Thereafter, different climate types are automatically derived which subsequently can be transferred onto a map containing climate regions. Their boundaries are established by interpolation only, the accuracy of which depends primarily upon the density of observation stations, and secondly upon the landforms. With increasing density of a network, the accuracy of climate boundaries also increases, and vice versa. However, considering the climate as a spatially continuous variable, no climate boundary can be regarded as an abrupt boundary between different climate regions, rather it must always be interpreted as a transitional zone.

Climate boundaries show not only weaknesses over space, but also over time which means that boundaries between two climate regions obviously cannot express the year to year fluctuations of the threshold values. Instead, climate boundaries are understood as boundaries related to the climatic "normal" conditions which usually represent the arithmetic mean of the threshold values under consideration over a representative observation time series. In this respect, a normal period is commonly taken as a so-called meteorological standard period, covering 30 years, e.g. from 1931–1960, 1941–1970 or from 1951–1980.

---

[24] The objectives of a climate regionalization and regional climates, in both an empiric and genetic approach, have recently been thoroughly described by R. HENDERSON-SELLERS 1986.

Resulting from the great number of choices for a climate classification scheme, numerous classifications have been developed so far, which can be divided into two types:

1. *Empiric,* relating to the observation of climate features (emphasizing the use of observed climate "normals");
2. *Genetic,* relating to the origin of climate features (emphasizing atmospheric dynamics and circulations).

An *empiric* climate classification aims to produce climate types and climate regions based on quantitatively defined threshold values, without regard to the causes of the climate. The threshold parameters most frequently used are temperature and rainfall values, which were chosen in various methods of expression and for various averaging periods. The results are strictly defined climate types and reasonably clear climate boundaries established for certain climate regions. In an empiric approach of a climate regionalization, climate is considered as an analytical variety of mostly two or three elements, while climate in its role as a combined interaction of individual weather events is not considered. Numerous empiric classifications have been developed taking different parameters and threshold values.

Genetic schemes for climate regionalization emphasize the climate-causing factors, and thus refer to the activity and effects of the general circulation of the atmosphere and its dynamics for demonstrating their role in creating climate and its different expressions over space. The basis for this type of climate classification are atmospheric processes and the distribution of distinct atmospheric parameters on earth, e.g. winds, air masses and pressure systems. Because of the dynamic basis and nature of genetic classification schemes, precise quantitative information about certain climate elements cannot be given. Furthermore, accurate boundaries between different climate regions, based on certain threshold values, cannot be established for genetic classifications.

For an appropriate consideration of the extremely complex nature of the climate in any country or region on earth, it seems beneficial to consider empiric and genetic aspects as complementary in order to arrive at a quantitative as well as qualitative description of the climate of a region under consideration.

8.2 China Within Global Climate Classifications

With particular concern towards an understanding of the climate regionalization of China, some of the convenient empiric and genetic climate classifications on a global scale will be briefly introduced and particularly considered for China. In presenting these classification schemes, the authors, however, do not want to be overly concerned with rigorous definitions of either the climate classification or the accurate boundaries of the resultant climate regions.

The Classification of KÖPPEN (1923, 1931)

The climate classification by KÖPPEN, introduced in its first version as early as in 1884, was firmly established in 1923 and – as reprints – in 1931 and 1944 and later modified several times by his scholar GEIGER, in 1954, 1961 and 1968, represents the most widely known climate classification; it has also found much attention by Chinese climatologists. Originally derived from the characteristic distribution of the vegetation, KÖPPEN defined certain mean monthly and annual values of temperature and rainfall as threshold values for different climate types and regions; these values were defined as expressions for specific vegetation boundaries.

As the principal rule of the KÖPPEN classification, the climate regionalization of the earth follows a hierarchial system which consists of the following levels: Five major climate types and regions, symbolized by the letters A to E, represent the highest level of regionalization. Their definition is given by particular temperature criteria, only in the case of the B climates is a specific temperature-rainfall index valid.

The major climate types were termed as follows:

A. Tropical-rainy climates
B. Dry climate;
C. Warm-temperate rainy climate;
D. Subarctic (boreal) climate;
E. Ice-snow climate.

The major climate types and regions are divided into 13 subtypes and subregions, by applying specific temperature and rainfall threshold values which mainly refer to the variation of rainfall and temperature over the year. Each climate type and region is categorized, in symbolic form, by a double-letter formula. A further subdivision of climate types and regions is arrived at by more specific climate criteria which describe typical climate events over space and time. The criteria for the major types and regions and their subtypes and subregions respectively, are given in Table 8.1.

Applying the KÖPPEN classification for China[25], it can be easily seen that China participates in all the five major climate types and regions A to E (Fig. 8.1). This observation already clearly indicates the extremely diversified climate conditions in China. The distribution of the major climate types and regions over space demonstrates a pronounced climate regionalization which expresses a strong latitudinal as well as altitudinal impact.

The major climate types and regions show the following distribution over space: The tropical-rainy climate (A) is located in Hainan Island and in parts of Taiwan Island as well as in the islands of the South China Sea. The dry climate (B) extends in a broad belt through the northern and northwestern parts of China.

[25] It must be pointed out, however, that the boundaries between different climate types and regions, after the KÖPPEN classification, may vary according to the different versions of the classification with variable threshold values and depending upon the network of climate data available. In this respect, compare CHU Co-ching (1929) and his climate map of China given according to KÖPPEN in Fig. 8.1.

Table 8.1. Criteria for classification of major climate types and regions as well as their subtypes and subregions respectively, in the modified KÖPPEN system

Letter symbol 1st	2nd	Explanation
A		Average temperature of coolest month 18° C or higher
	f	Precipitation in driest month at least 60 mm
	m	Precipitation in driest month less than 60 mm, but equal to or greater than (100 − r)/25 [a]
	w	Precipitation in driest month less than (100 − r)/25
B		70% or more annual precipitation falls in the warmer 6 months (April through September in the northern hemisphere) and r/10 less than 2 t + 28 [a]
		70% or more annual precipitation falls in the cooler 6 months (October through March in the northern hemisphere) and r/10 less than 2 t
		Neither half of the year with more than 70% annual precipitation and r/10 less than 2 t + 14
	W	r less than one-half of the upper limit of applicable requirement for B
	S	r less than upper limit for B, but more than one-half of that amount
C		Average temperature of warmest month greater than 10° C and of coldest month between 18° and 0° C
	s	Precipitation in driest month of the summer half of the year less than one-tenth of the amount in wettest summer month
	w	Precipitation in wettest month of summer half of the year more than 10 times of driest month of winter half
	f	Precipitation does not meet conditions of either s or w
D		Average temperature of warmest month greater than 10° C and of coldest month 0° C or below
	s	Same as under C
	w	Same as under C
	f	Same as under C
E		Average temperature of warmest month below 10° C
	T	Average temperature of warmest month between 10° and 0° C
	F	Average temperature of warmest month 0° C or below

[a] In the formula, t is the average annual temperature in °C, r the average annual precipitation in millimetres.

The warm-temperate rainy climate (C) is located south of the Huang Ho and east of the 1,000-m contour line and includes major parts of Taiwan Island. The subarctic (boreal) climate (D) covers all of Northeast China, higher parts of North and Northwest China and Central China. The ice-snow climate (E) occurs in the highest parts which occupy the Tibetan Plateau including Chinghi, northwestern Changtu, and the Tian Shan and Kunlun Shan, as well as the Himalayas.

The subdivision of the climate types and regions is included in Fig. 8.1. The second letter symbol of the subtypes and subregions is explained in Table 8.1. The letter symbol k denotes "cold" which expresses a mean annual temperature below 18° C.

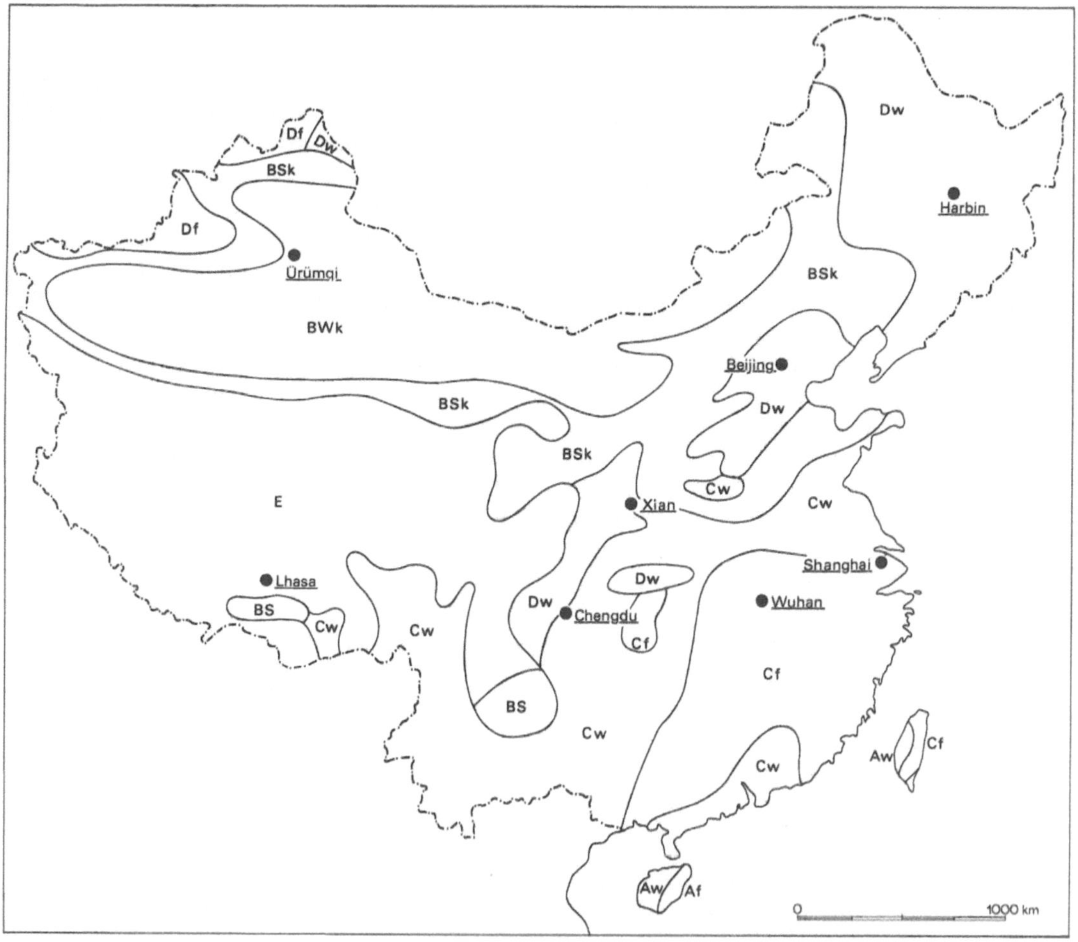

Fig. 8.1. The climate division of China according to the KÖPPEN classification (1923, 1931, MÜLLER 1983)

Despite its weaknesses, which result from the purely statistical and analytical approach of a climate regionalization, KÖPPEN's classification nevertheless enables a global comparison of climates. Criticism of the KÖPPEN classification was particularly concerned with the mountain climates which, according to KÖPPEN, correspond to the climate change in a latitudinal direction. In other words, the climate change with increasing latitude corresponds to the climate change with increasing altitude above sealevel.

The Classification of TROLL and PAFFEN (1964)

Also representing an empiric approach to a regionalization of the earth's climate, the classification scheme of C. TROLL and KH. PAFFEN (1964) is at present a widely applied example. Called "seasonal climates of the earth", the leading concept of the classification is the seasonal rhythm of the climate elements, of which solar radiation, temperature and precipitation represent the basis of climate regionalization. Based on TROLL's preliminary concept of a climate classification scheme, which was first developed in 1955, the final version was established together with KH. PAFFEN in 1963/64, published as a coloured map of the seasonal climates of the earth (in H. E. LANDSBERG et al. 1963, resp. "Erdkunde" 1964).

On the basis of the seasonal variation of the climate elements, TROLL and PAFFEN have established five major climate regions, called 'zones', on earth; the definition of each was based on specific selected threshold values, which refer to the seasonal rhythm of climate. Among the three climate elements selected, temperature plays the leading role, a fact which is underlined by the principal consideration of the "seasonal climates" as temperature-caused climates. Of secondary importance only is the seasonal distribution of precipitation, which was expressed by the duration of alternating humid and arid periods, given by the number of humid or arid months. In this respect, humidity and aridity were defined by the "aridity index" of DE MARTONNE/LAUER (1926/1952) in which the relationship between precipitation and temperature was empirically determined. The definition of climate types and thus the boundaries drawn for the climate zones result from specific threshold values, which aim to express the characteristic annual variation of temperature.

The five major climate types and zones were defined as follows; they are each identified by a roman numeral:

I. polar and subpolar zones;
II. cold-temperate boreal zones;
III. cool-temperate zones;
IV. warm-temperate subtropical zones;
V. tropical zone.

The major climate types and zones (I to V) were divided into a large number of seasonal climate sub-types and sub-zones, expressed by arabic numbers. For this, specific temperature and precipitation threshold values were selected, which refer to the seasonal variation of climate. The threshold values were derived from the plant and vegetation rhythm over the year. It is worth noting that the climate characteristics are only of limited validity for the oceans, which are considered as maritime variations of the corresponding continental climate types. Furthermore, the climate of mountains should be interpreted as altitudinal variations of the lowland climate zone concerned.

According to the climate classification scheme of TROLL and PAFFEN, China participates in the following climate sub-types and sub-zones (Fig. 8.2):

II.3: Highly continental boreal climates (annual fluctuation >40° C) with permanently frozen soils; very long, extremely cold and dry winters (coldest month below −25° C); short, but sufficient warming-up in summertime (warmest month 10° to 20° C) and deep thawing soils; highly continental, dry, coniferous woods.

III.6: Highly continental climates with cold and dry winters, annual fluctuation generally >40° C; coldest month between −10° C and −30° C; short, warm and humid summers, warmest month above 20° C; highly conti-

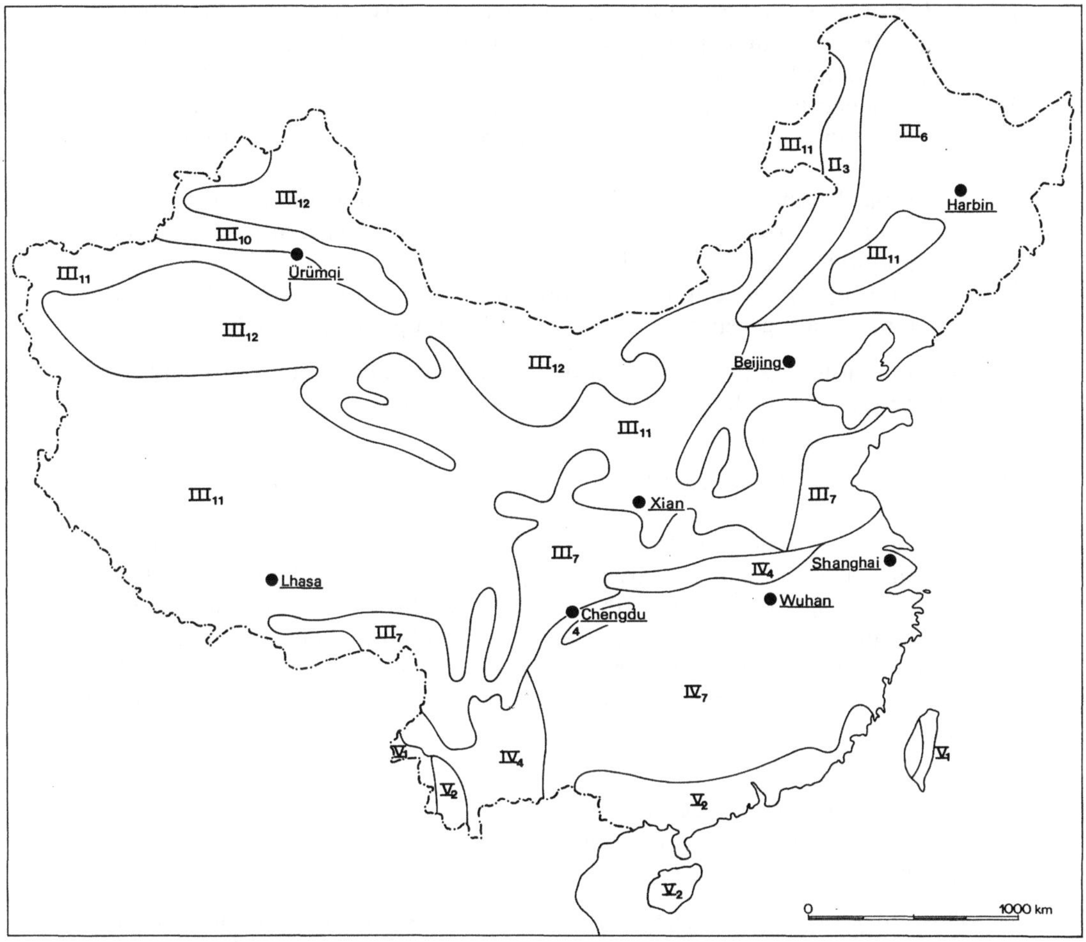

Fig. 8.2. The climate division of China according to the classification of TROLL and PAFFEN 1963 (MÜLLER 1983)

nental, deciduous, broad-leafed and mixed woods as well as wooded steppe.

III.7: Humid and warm summer climates, annual fluctuation 25° to 35° C with moderately cold, but dry winters, coldest month between 0° and −8° C, warmest month between 20° and 26° C; deciduous broad-leafed and mixed wood and wooded steppe favoured by warmth, but withstanding cold and aridity in winter.

III.10: Steppe climates with cold winters, arid summers and less than 6 months of humidity, coldest month below 0° C; short grass, dwarf shrub or thorn steppe.

III.11: Humid-summer steppe climates with cold and dry winters, coldest month below 0° C; Central and East Asian grass and dwarf shrub steppe.

III.12: Semi-desert and desert climates with cold winters, coldest month below 0° C; semi-desert and desert with cold winters.

IV.4: Dry-winter climates with long summer humidity, generally 6 to 9 humid months; subtropical steppe with short grass, hard-leafed monsoon wood and wooded steppe.

IV.7: Permanently humid climates with hot summers and a maximum of precipitation in summer; subtropical humid forests (laurel and coniferous forests).

V.1: Tropical rainy climates with or without short interruptions of the rainy season, 12 to 9.5 humid months; evergreen tropical rain forest and half deciduous transition wood.

V.2: Tropical, humid summer climates with 9.5 to 7 humid and 2.5 to 5 arid months; rain-green humid forest and humid grass savannah.

The terminology and description of the various climate sub-types and subzones were given according to TROLL and PAFFEN 1964.

The Classification of FLOHN (1950)

The genetic basis of the climate classification of FLOHN (1950) is the atmospheric circulation which expresses the large-scale patterns of wind and pressure distribution in the atmosphere, existing either persistently throughout the year or recurring seasonally. Resulting from the imbalance of radiation between low and high altitudes and from significant energy transfers in the atmosphere, FLOHN has established four distinct global wind belts which govern the macro-scale pattern of wind on earth:

1. the polar easterlies;
2. the mid-latitude westerlies;
3. the tropical easterlies resp. trade winds;
4. the equatorial westerlies.

This order indicates the distribution of the four wind belts in a north to south direction in the northern hemisphere and, conversely, the same is valid, in a south to north direction in the southern hemisphere. Three of the four wind belts, namely the polar easterlies, mid-latitude westerlies and trade winds, constitute a circum-global belt in each hemisphere, while the equatorial westerlies occur as one wind belt only which occupies the equatorial low latitudes. As can be inferred from the seasonal imbalance of radiation, all four global wind belts are unstable in their distribution throughout the year, and over land and ocean, resulting in a poleward movement during summer and an equatorial retreat during winter, both associated with a considerable variation of the seasonal extent of the global wind belts over space.

As a result of the four global wind belts and their seasonally varying distribution over space, FLOHN arrived at a total of seven climate types (in each hemisphere), of which four types belong to the so-called *homogeneous* climate types, and the three remaining types to the *heterogeneous* types. The first category includes those climate types, which experience the same global wind belt persistent throughout the year, whereas the latter category includes the types which are under the seasonally alternating influence of two adjoining wind belts.

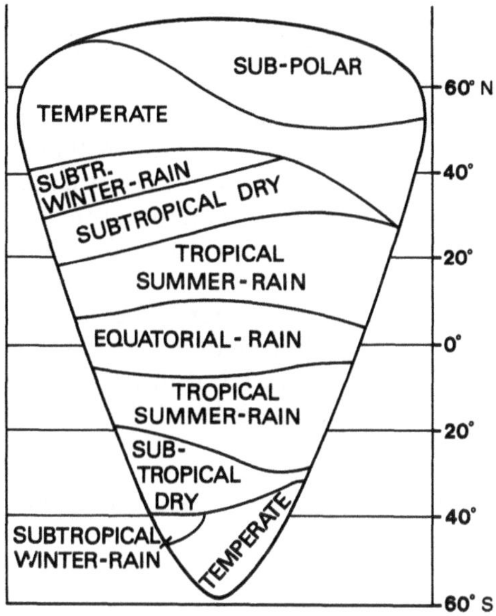

Fig. 8.3. The distribution of FLOHN's climate types and regions on a hypothetical continent of low and uniform elevation

The seven genetic climate types and climate regions of FLOHN and their prevailing wind belt(s) are as follows ("heterogeneous" climate types and regions are printed in *italics*):

Type/region	Summer	Winter
Polar	PE	PE
Subpolar	PE	W
Temperate	W	W
Subtropical winter rain	TW	W
Subtropical dry	TW	TW
Tropical summer rain	EW	TW
Equatorial rain	EW	EW

PE: polar easterlies; W: mid-latitude westerlies; TW: trade winds; EW: equatorial westerlies.

The climate types and regions are given for the northern hemisphere, but they are valid with an opposing seasonal occurrence in the southern hemisphere. The genetic climate types and regions, according to the FLOHN classification, are named after the precipitation characteristics as they result from the acting wind belt. It is noted that temperature does not appear explicitly in the scheme. The global distribution of FLOHN's different climate types and regions has not been compiled on a world map, but only on a hypothetical continent (Fig. 8.3). The boundaries between the different climate regions are of a widely transitional nature and thus represent belts rather than distinct limits.

According to FLOHN's genetic classification scheme of the climates, China comes mostly under the temperate climate which is governed by the mid-latitude westerlies throughout the year. A small part is under the subtropical dry climate, while the marginal northern and southernmost parts are of a subpolar respectively tropical summer rain climate. Although showing the climate belts which govern the climate of China, the FLOHN classification does not describe in detail the climate characteristics in terms of their temperature and precipitation conditions. Particularly due to the extreme contrast between the huge landmasses (including their high-altitude mountains) and the wide oceans, in China the genetic climate types and regions appear to be considerably modified in comparison with the "normal" distribution of wind belts. In this connection, it seems noteworthy to refer to the asymmetrical distribution of the climate types and regions on the western and eastern sides of the continents (see Fig. 8.3).

8.3 National Climate Classifications of China

Climate classifications, compiled for China alone, have been worked out by numerous Chinese scientists. As a common characteristic, all national climate regionalization schemes are empiric by nature, yet the threshold values defined for climate division mostly seem to have been chosen as relevant agroclimatic values.

With regard to the climate regionalization of China, it is worth noting the classification of the climate of Eurasia by VON WISSMANN (1939) with particular reference to China. Based on extensive field studies and research in China, von WISSMANN has worked on a climate classification of Eurasia since 1932. Originally intended as a revised version of KÖPPEN's climate classification (1923, 1931), the final climate division of Eurasia (VON WISSMANN 1939) represented a completely new classification in which the climatic threshold values were defined with particular reference to the vegetation and soil conditions.

Mainly defined by specific temperature values, VON WISSMANN (1939) established six major climate types and associated regions (identified by roman numerals):

I. tropical;
II. warm-temperate;
III. cool-temperate;
IV. boreal (cold-temperate);
V. subarctic (tundra);
VI. arctic.

According to seasonal characteristics in precipitation and temperature, the major climate types and regions were divided into subtypes and subregions; additionally, the steppe and desert climates as well as the mountain climates were defined separately as particular modifications of the prevailing major climate type and associated region. All 21 climate types and subtypes as well as the climate regions and associated subregions were each identified by a specific index, which refers to the thermal-hygric conditions defining the climate regionalization. Without going into a detailed decription of the climate criteria for all 21 climate types

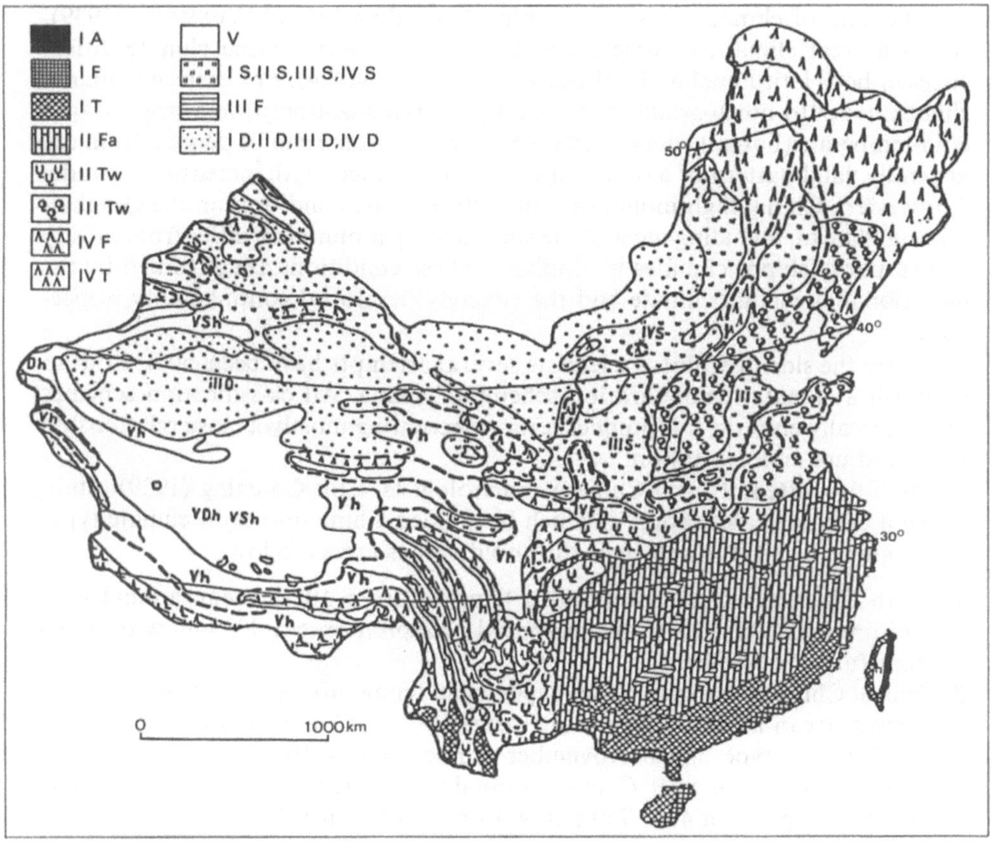

Fig. 8.4. The climate division of China according to Von WISSMANN (1939)

I A – Tropical; equatorial	III F – Cool-temperate; wet
I F – Tropical; with a short dry season	III Tw – Cool-temperate; winter-dry
I T – Tropical; with a long dry season	IV F – boreal; wet
II Fa – Warm-temperate; wet, summer-hot	IV T – boreal; winter-dry
II Tw – Warm-temperate; winter-dry	V – sub-arctic

IS, IIS, IIIS, IVS – Steppe climate: tropical/warm temperate, cool-temperate, boreal.
ID, IID, IIID IVD – Desert climate: tropical/warm temperate, cool-temperate, boreal.
The lowland climates are modified in the mountain climates which are identified by *h*

and subtypes, only the threshold values for the six major climate types and regions shall be given:

1. boundary between tropical/warm-temperate: mean temperature of the coldest month $\geq 13°C$;
2. boundary between warm/cool-temperate: mean temperature of the coldest month $\geq 2°C$;
3. boundary between cool-temperate/boreal: mean annual temperature $\geq 4°C$;
4. boundary between boreal/subarctic: mean temperature of the warmest month $\geq 10°C$;
5. boundary between subarctic/arctic: mean temperature of the warmest month $\geq 0°C$.

The map of climate division of China, according to VON WISSMANN (1939), shows a great climatic complexity which expresses pronounced climate differences in both latitudinal and altitudinal respects (Fig. 8.4). For eastern China, a slightly south to north gradient of a climate regionalization from a tropical to a cold-temperate (boreal) climate region can be seen which can be derived from the generally valid pattern of a decreasing radiation budget with increasing latitude. For western China, high-mountain, subarctic climate conditions in the Qinghai-Xizang Plateau, including most of the surrounding mountains, differ from mostly desert, partly steppe climates in Xinjiang. These conditions are governed by the elevational effects on climate and the strongly developed continentality respectively.

From the side of Chinese scientists several attempts have been carried out to establish a climate regionalization scheme of China. All classification schemes were basically empiric, adopting criteria representing mostly a close relationship with land use and vegetation.

As a first attempt from Chinese climatologists, CHU Co-ching (1929) established a climate classification in which he divided China into eight climate types and regions. The types and their threshold values are as follows:

1. South China type: coldest month's temperature >10° C, mean annual temperature range 12°–20° C, mean annual precipitation >1,000 mm with maximum June–September;
2. Central China or Yangtze Valley type: mean temperatures <10° C in 4 months at most, mean annual temperature range 18°–25° C, precipitation >750 mm;
3. North China type: mean November temperature <10° C but >0° C, mean annual temperature >10° C, mean annual temperature range 25°–35° C, mean annual precipitation 400–750 mm with maximum in July;
4. Manchurian type: mean annual temperature <10° C, mean temperatures <0° C in 5 or more months, mean annual precipitation 400–600 mm;
5. Yunnan Plateau type: mean annual temperature 14°–18° C, mean annual temperature range 12°–15° C, mean annual precipitation mainly >750 mm;
6. Steppe type: mean annual temperature 5°–10° C, mean annual precipitation 200–400 mm;
7. Tibetan type;
8. Mongolian type.

The climate classification of CHU Co-ching created a broad and rough regionalization only, without giving a subdivision of the climate regions; the boundaries between the different regions were drawn rather unspecifically, due to the paucity of climate stations and observations at that time (Fig. 8.5).

CHU's classification, nevertheless, can be seen as a basic climate division of China, stimulating further research of Chinese climatologists towards a more precise climate regionalization of the country. TU (1936) presented a modified classification in which he arrived at some major corrections of CHU's classification. For example, TU divided the climate region of the Tibetan Plateau into two regions, namely Tibet and Southeast Tibet; the division corresponds with the two different precipitation regimes in both parts of the Tibetan Plateau. Furthermore, TU merged the steppe and Mongolian regions into one climate region only,

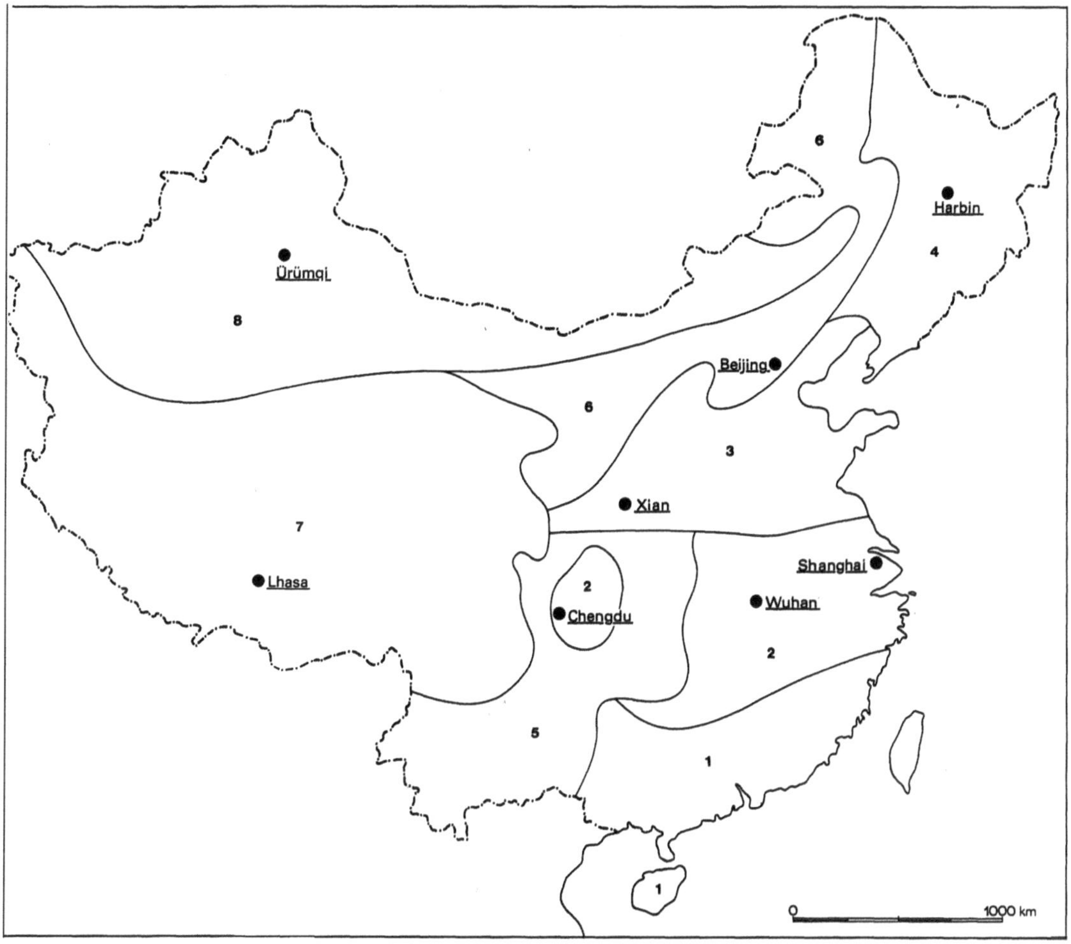

Fig. 8.5. The climate division of China according to CHU Co-ching 1929 (WATTS I.E.M. 1969)

which stands as Mongolia. TU's climate classification successfully created a more detailed regionalization of the climate of China. Major steps towards a further climate regionalization were taken by WU (1945) and TAO (1949) who adopted THORNTHWAITE's indices of both precipitation (1933) and potential evapotranspiration (1948) respectively.

As another empiric approach, LU (1949) developed a climate classification scheme of China by considering four climate criteria: (1) the $-6°$ C isotherm of January which separates winter wheat and spring wheat cultivation; (2) the 6° C isotherm in January which separates the distribution of an annual one-season rice cropping from a two-season cropping pattern; (3) the 750-mm isohyet of annual precipitation which represents the northern limit of rice cultivation; (4) the 1,250-mm isohyet of annual precipitation which is the southern limit of wheat cultivation; this isohyet roughly corresponds with the 50-mm isohyet in January

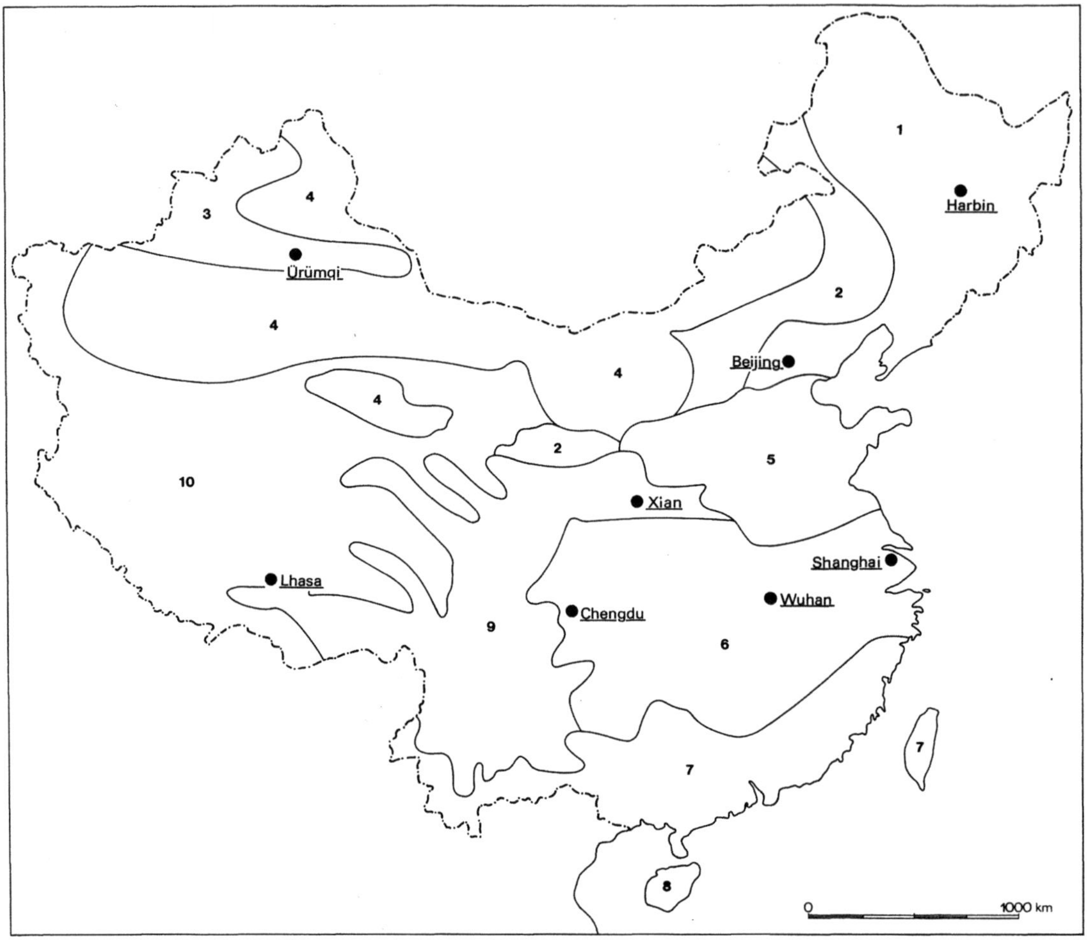

Fig. 8.6. The climate division of China according tu Lu 1949 in a generalized way, showing the major climate types and regions (Watts I.E.M. 1969)

and with the demarcation line between maximum precipitation in June and July. As can be seen from the criteria chosen, the climate division of China by Lu has drawn particular emphasis to land use conditions. On the basis of these criteria, Lu established 10 climate types and divided China accordingly (Fig. 8.6):

1. Northeast type;
2. Northern steppes type;
3. Northern mountain type;
4. Northwest desert type;
5. North China type;
6. Central China type;
7. South China type;
8. Hainan type;
9. West China type;
10. Tibet type.

Each of the ten climate types and regions was divided into several subtypes and subregions, the names and distribution of which correspond to major orographical units. Due to the paucity of climate data at that time, the boundaries of LU's (1949) climate regions and subregions could not be compiled with sound statistical accuracy.

A further empiric approach towards a climate classification of China was given by YAO (1951) who selected certain temperature criteria of the annual variation for establishing three major types of the climate of China: (1) monsoon climate, (2) temperate inland climate, (3) temperate plateau climate. YAO divided the monsoon climate into five subtypes and regions: (a) tropical monsoon region, (b) oceanic monsoon region, (c) warm-temperate monsoon region, (d) cold-temperate monsoon region, (e) continental monsoon region. Although based on temperature only, this climate division nevertheless appears to present a realistic approach towards the climate regionalization of China on a broad scale.

Comprehensive research on a commonly valid climate classification of China started in 1956 with the establishment of the Working Committee of Natural Regionalization of China under the Chinese Academy of Science[26]. As the prime objective of the climate regionalization group of the Committee, a climate classification of China was worked out (1959) which resulted from a close cooperation with the physio-geographical regionalization group. The resultant climate regionalization of China is commonly called the ZHANG classification of 1959, according to the head of the climate regionalization group, ZHANG Pao-kun (cf. footnote[26]).

The climate classification of ZHANG was based on two climate criteria: (1) the accumulated air temperatures, (2) an index of aridity. With secondary importance, other temperature and precipitation values were considered, which express a certain agroclimatic relationship with major cultivated crops.

Accumulated temperatures are taken as the annual total of all daily mean temperature values $>10°C$. The aridity index K, defined by Г. Селяниноб denotes the quotient between the potential evapotranspiration (E) and the amount of precipitation during the period with temperatures $>10°C$ (r). The potential evapotranspiration (E) coincides with the product of accumulated temperatures (Σt) and a coefficient 0.16[27]. The aridity index K thus reads as follows:

$$K = \frac{E}{r} = \frac{0.16\,\Sigma t}{r}. \qquad \text{Г. Селяниноб}$$

[26] Chairman of the Committee was CHU Co-ching, Vice-Chairman HUANG Bing-wei who, at the same time, supervised the physio-geographical regionalization group of the Committee. Head of the climate regionalization group was ZHANG Pao-kun. However, the work of this group was strongly supported by CHU Co-ching and HUANG Bing-wei and, therefore, the findings of the groups can be regarded as a joint success of prominent scientists in the field of meteorology, climatology and physical geography.

[27] The formula is based on the assumption that the annual total of the accumulated temperature is, in general, proportional to that of the net radiation, which is the energy for evapotranspiration. The potential evapotranspiration is obtained by multiplying the annual total of accumulated temperature by a certain coefficient. Since the Qinling-Huaihe line is the most pronounced geographical boundary in China, which is recognized both by Chinese and foreign scientists, this is obviously also a pronounced climate boundary. In various respects, this boundary is accurate, for example the most popular means of transportation is the boat to the south of this boundary, while it is the horse to the north. Furthermore, paddy rice is the main crop in the south, whereas it is winter wheat in the north. The coefficient 0.16 was reached by assuming $K=1$ on this line.

According to the different values computed for the aridity index (K), aridity resp. humidity were divided into the following five classes:

"K" value	Degree of humidity/aridity	Natural vegetation
≤ 0.49	Very humid	Forest
0.50–0.99	Humid	Forest
1.00–1.49	Semi-humid	Forest-steppe
1.50–1.99	Semi-arid	Grassland; arid steppe
2.00–3.99	Semi-arid	Desert steppe
≥ 4.00	Arid	Desert

Based on the criteria described, the climate classification of ZHANG (1959) represents a thermal-hygric regionalization scheme of the climate of China, in which seven climate types were established. They were defined according to specific climate criteria which express distinct climate requirements for crop cultivation; the climate types were identified by the roman numerals I to VI and the letter H.

I. Equatorial climate. Accumulated temperatures exceed 9,000° C, the mean annual temperature is >26° C and the annual rainfall total >1,000 mm. Climate conditions are characterized as most suitable to a wide variety of tropical plants.

II. Tropical climate. Accumulated temperatures are around 8,000° C, air temperature of the coldest 5-day periods is >15° C. Mean minimum temperature is >0° C so that no frosts occur. Climate conditions are therefore abundant for tropical crops, and even three crops of rice can be grown.

III. Subtropical climate. Accumulated temperatures are between 4,500°–8,000° C. The air temperature of the coldest 5-day period is between 0° C and 15° C. The climate conditions lead to the establishment of a subtropical, seasonal rain forest and a monsoonal, evergreen, broad-leafed forest. Two crops of rice can be achieved and wheat can already be planted.

IV. Warm-temperate climate. Accumulated temperatures vary between 3,400° and 4,500° C. Air temperature of the coldest 5-day period falls between 0° and 10° C. Climate conditions are ideal for cotton and a large variety of subtropical fruit trees. Because of the low winter temperatures, no evergreen and broad-leafed trees can survive.

V. Temperate climate. Accumulated temperatures range between 1,600° and 3,400° C. Air temperature of the coldest 5-day period drops between −10° and −30° C. Seasonal crops, such as rice, corn, wheat and soybean, can be grown successfully during summer. In terms of natural vegetation, a wide range is adoptable from coniferous and deciduous trees to steppes and even deserts.

VI. Cold-temperate climate. Accumulated temperatures are <1,600° C. Minimum temperature drops below −30° C. Natural vegetation consists mainly of larches, while land use is concerned with wheat, potatoes, buckwheat and millet.

H. Plateau climate. This type is mainly found on the Tibetan Plateau with an elevation >3,000 m above sea level. Commonly, accumulated temperatures are

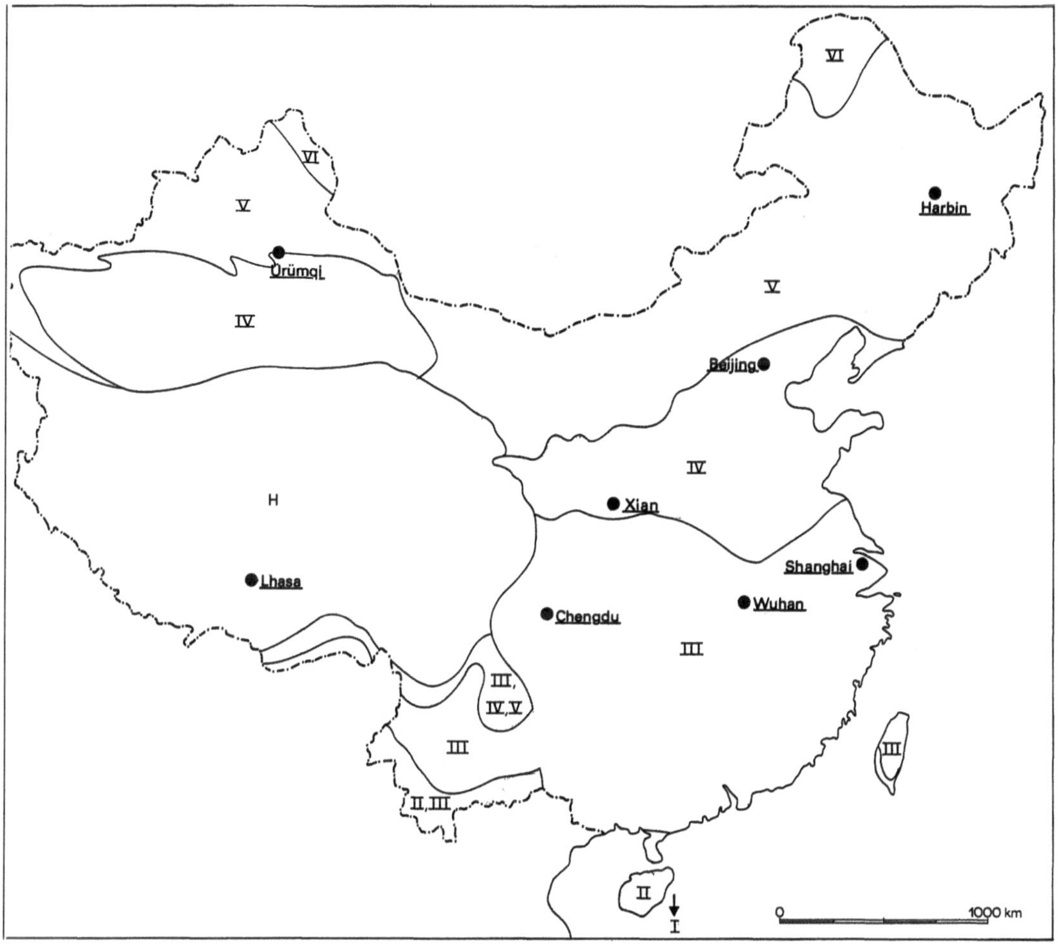

Fig. 8.7. The climate division of China according to ZHANG (1959), based on the major climate types and the associated climate regions

<2,000° and daily mean temperature <10° C. For some parts, the temperature of the warmest 5-day period even drops below 5° or even 0° C. The variation of temperature over a day is large as the variation over the year is rather small. Natural vegetation and land use are distinctly limited by the rigorous climate.

The seven climate types correspond with a total of eight major climate regions which were subdivided into 32 climate provinces, which were again divided into 68 climate areas.

The climate regions of China, according to ZHANG, are as follows (the number of climate provinces and climate areas as well as the types of climate, cf. Fig. 8.7, are given in parentheses):

1. Northeast China (4, 6; V, VI);
2. Inner Mongolia (2, 10; V, VI);
3. Gansu-Xinjiang (4, 10; IV, V, VI);
4. North China (4, 9; IV);
5. Mid-China (6, 9; III);
6. South China (3, 9; I, II, III);
7. Sichuan-Yunnan (3, 6; H)[28];
8. Qinghai-Xizang (Tibetan Plateau).

[28] The climate regions Nos. 7 and 8 are characterized by the same climate type, namely the Plateau climate H which was only modified into a somewhat weaker and stronger type respectively.

The climate division of China (1959), worked out by the Climate Regionalization Group under ZHANG (as a part of the Working Committee of Natural Regionalization of China), served as a sound basis for intense research on a climate regionalization of China. Revisions and corrections of ZHANG's classification were mainly undertaken in two directions, a meteorological and a geographical. From the meteorological approach towards a climate regionalization of China, the climate division of CHEN et al. (1982) represents the most advanced one, which will thus be given particular attention.

The prevailing geographical approach to a climate regionalization is expressed by the climate division of China by HUANG Bing-wei (see Sect. 8.4).

For the climate classification scheme of CHEN et al. (1982), the major criteria of regionalization were the differences in temperature which occur over space, mostly due to the extremely vast territorial extent and the extremely diversified landforms. As expressions of the spatial differences of temperature, CHEN et al. 1982 have chosen the accumulated air temperatures (expressed by the number of days $\geq 10°$ C and the annual totals of degrees $\geq 10°$ C) and the mean air temperature in January. For the Tibetan Plateau, the criteria applied were modified in such that the number of days $\geq 10°$ C and the mean air temperature of the warmest month were considered. The threshold values of the variables, applied for a classification into major climate types and regions, were defined according to the requirements of the leading crops. CHEN et al. 1982, in total, determined 14 major climate types and regions, of which only five belong to the Tibetan Plateau, while the remaining nine climate types constitute the rest of China. The various climate types and regions were identified by the roman numerals I to X and by the symbols HI to HV (for the Tibetan Plateau types); their names and criteria of classification are given in Table 8.2.

The major climate types and regions were divided further into subtypes and -regions. These were identified by the available water budget, expressed by a coefficient of aridity respectively humidity which was calculated as the ratio between the annual amount of potential evapotranspiration and the annual precipitation. Potential evapotranspiration was computed according to the formula of PENMAN (1956). In total, CHEN et al. 1982 have defined five categories of humidity/aridity, identified by the letters A to E, with the following coefficients:

Category of humidity/aridity	coefficient	Natural vegetation
A. Humid	<1.0	Forest
B. Subhumid	1.0– 1.6	Forest-steppe
C. Subarid	1.6– 3.5	Grassland
D. Arid	3.5–16.0	Semi-desert
E. Extremely arid	>16.0	Desert

Applying the humidity/aridity coefficients, CHEN et al. 1982 differentiated a total of 31 climate types and regions for China, of which almost 13 fall into the

Table 8.2. Major climate types and zones of China and their threshold values of accumulated air temperatures and mean temperature in January according to CHEN et al. (1982). (For the Tibetan Plateau, the mean temperature of the warmest month was considered)

Climate type/zone	Days ≧10° C	Accumulated temperature ≧10° C	Mean air temperature in January	Remarks
I. Cold temperate	< 100	< 1,600°	<−30° C	
II. Middle temperate	100−171	1,600°−3,400°	−30° to −12°−−6° C	
III. Warm temperate	171−218	3,200°−3,400° to 4,500°−4,800°	−12° to −6°−0° C	
IV. North subtropical	218−239	4,500°−4,800° to 5,100°−5,300°	0°−4° C	
		3,500°−4,000°	3° to 5°−6° C	Yunnan
V. Middle subtropical	239−285	5,100°−5,300° to 6,400°−6,500°	4°−10° C	
		4,000°−5,000°	5°−6° to 9°−10° C	Yunnan
VI. South subtropical	285−365	6,400°−6,500° to 8,000°	10°−15° C	
		5,000°−7,500°	9°−10° to 13°−15° C	Yunnan
VII. Peripheral tropical	365	8,000°−9,000°	15°−20° C	Yunnan
		7,500°−8,000°	> 13°−15° C	
VIII. Middle tropical	365	9,000°−10,000°	20°−26° C	
IX. Equatorial tropical	365	> 10,000°	> 26° C	

	Days ≧10° C	Mean air temperature of the warmest month
H I Cold plateau	0	< 6° C
H II Sub-cold plateau	< 50	6°−11° C
H III Temperate plateau	50−180	12°−17° C
H IV Subtropical mountain plateau	180−350	18°−24° C
H V Northern tropical mountain plateau	> 350	> 24° C

Tibetan Plateau (Fig. 8.8). Each climate type and region was identified by a specific symbol expressing the temperature conditions by the first letter, while the second one shows the water budget, given as the degree of humidity respectively aridity.

A further subdivision of the climate regionalization scheme by CHEN et al. 1982 was based on the mean air temperature conditions in July which finally resulted in 49 subtypes and -regions for China. This subdivision is not entered on to the map of the climate division by CHEN et al. (Fig. 8.8).

Continued p. 252

Table 8.3. Characteristics of the major climate types and regions according to the climate classification by CHEN et al. (1982)

Formula	Climate type/region	Growing period (months)	Mean air temperature in January (°C)	Mean air temperature in July (°C)	Annual range of air temperature (°C)	Extreme minimum of air temperature (°C)	Annual precipitation (mm)	Humidity type	Remarks
I (A)	Cold temperate	3.0	−30	16−18	5		400−500	Humid	
II (A, B, C, D, E)	Middle temperate	3.5−5.5	−20−−12	20−26	25−40	<−30	500−800 50− 60	Subhumid Over-arid Humid	
III (A, B, D, E)	Warm temperate	5.5−7.5		16−20 19−22	20−30	−20−−30 −5−−10	500−900 50− 60 1,000−1,200	Subhumid Over-arid Humid	
IV (A)	North subtropical	7.5−8.0			15 25−30	−5−−10 −10−−20	900−1,600	Humid	East, West
V (A)	Middle subtropical	8.0−9.5			20−25 12	−5−−10 0−− 5	1,400−1,800 1,000	Humid	
VI (A, B)	South subtropical	9.5−12.0			15−20 10	0−−5	1,600−2,000 1,000−1,500	Humid Subhumid (part)	
VII (A, B)	Peripheral tropical	12.0			8−12	0−−5 0−−2	1,200−2,400 <1,000	Humid Subhumid	
VIII (A)	Middle tropical	12.0			6	15	1,500	Humid	
IX (A)	Equatorial tropical	12.0	26		2	20	1,500−2,000	Humid	

Table 8.3 (continued)

		Growing period (months)	Days with mean daily air temperature ≧10° C		Annual precipitation (mm)	Elevation above sea level (m)
H I (D)	Cold plateau	0	0		100	4,800–5,100
H II (A, B, C)	Subcold plateau	0	< 50		100–300 (west) 400–700 (middle) 600–800 (east)	3,400–4,900
H III (A, B, C, D, E)	Temperate plateau	1 crop a year 3 crops in 2 years	50 100 150–180		< 50 50– 100 400– 600 500–1,000	
H IV (A)	Subtropical mountain plateau	2 crops a year	180–350		1,000	< 2,500
H V (A)	Northern tropical mountain plateau	3 crops a year	> 350		2,500–4,000	100–1,000

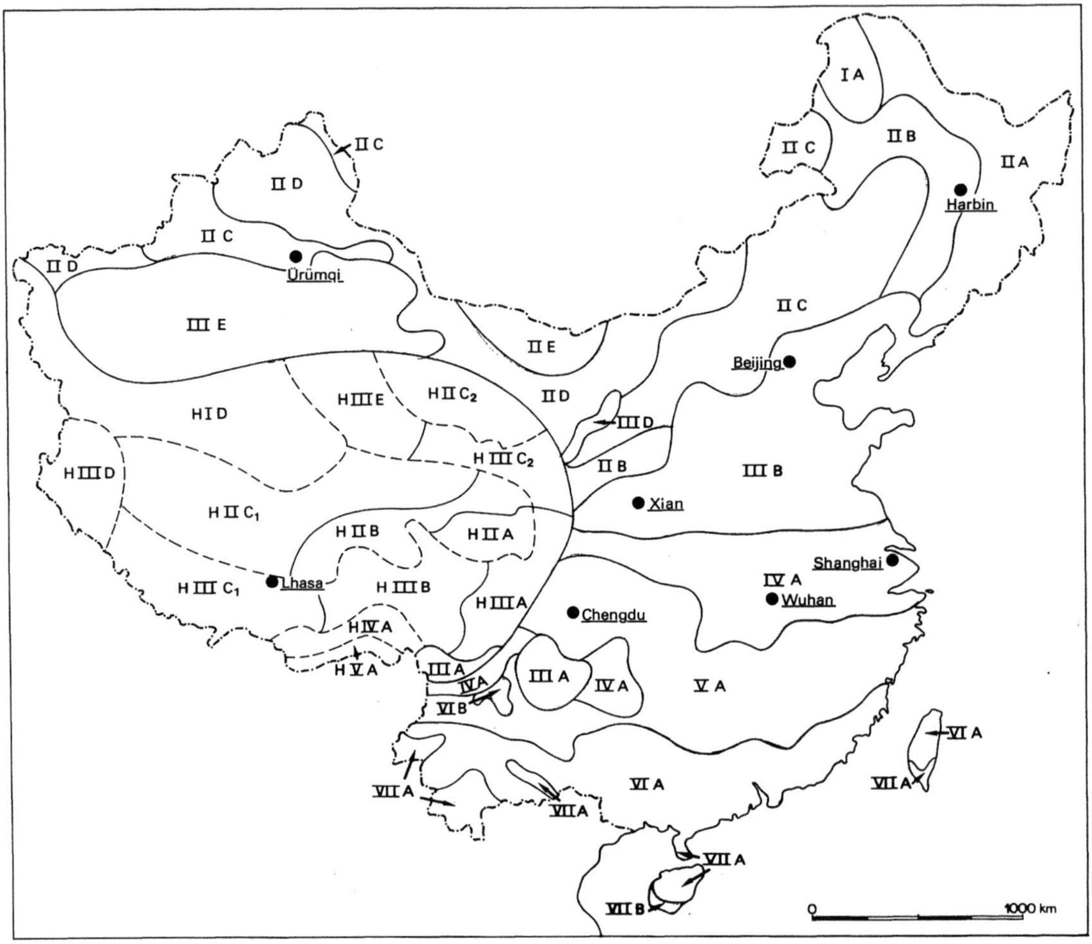

Fig. 8.8. The climate division of China according to CHENG et al. (1981)

The climate conditions of the 14 major climate types and regions of China were carefully described by CHEN et al. 1982, giving particular emphasis to the length of the cropping period (Table 8.3, p. 250 f.).

The climate classification scheme of CHEN et al. (1982) appears to represent a useful climate division, which may particularly help one to classify the climatic land use conditions over space. The criteria applied have particular significance for practical purposes in agricultural land use. No information, however, can be given on the formation conditions of the climate regionalization.

8.4 Climate Division of China According to HUANG Bing-wei (1986)

The most widely quoted and most often applied empiric approach towards a climate regionalization scheme of China was established by HUANG Bing-wei (1986)[29]. This classification summarized the scientific findings of comprehensive climate research work, carried out by HUANG Bing-wei (formerly HUANG Ping-wei) since the 1950s, after the Working Committee of Natural Regionalization of China under the Chinese Academy of Sciences was established with the major objective of developing a climate division of the country.

With particular attention given to crop and animal production, forestry and other land uses, HUANG's classification is substantially based on certain temperature and precipitation criteria which were partly adopted, partly modified in comparison with the original classification of ZHANG (1959). For the boundaries of climate regions, particular emphasis was placed on the physico-geographical conditions as a whole so that the climate division of China should also reliably express the geographical regionalization in total.

For a climate regionalization scheme, temperature criteria were given primary importance, while precipitation criteria played a secondary role only. This order resulted in a hierarchial climate regionalization scheme by HUANG. It consists of the following four levels with different climate characteristics each, thus creating a climate division over space (the names of the different levels of climate regionalization were given by HUANG): climate *realms*, climate *belts*, climate *regions* and climate *provinces*. The hierarchy of these four levels and their partition is given in Fig. 8.9. It shows four climate realms which are formed by 12 climate belts; these were subdivided into 21 climate regions, which were finally divided into a total of 45 climate provinces. For the various levels of climate regionalization, different criteria were selected: temperature criteria for climate realms and belts, precipitation criteria (expressed by different grades of humidity respectively aridity) for climate regions, whereas for climate provinces the various types of relief and landforms, such as mountains, hills, basins and plains were considered.

The boundaries between the four *climate realms*, which are called temperate, subtropical, tropical and plateau climates, can be considered as macro-climate boundaries. While the plateau climate was defined by the 3,000-m contour line (and thus covering all regions above this elevation, namely the Qinghai-Xizang Plateau, Qaidam Basin, Altin Tagh, Nan Shan and Kunlun Shan), the three remaining climate realms were strictly established by thermal criteria: The boundary between the temperate and subtropical realms was represented by the 0° C-January isotherm, whereas the boundary between the tropical and subtropical realms was determined by the isoline of 365 frost-free days. These two boundaries have a far-reaching agroclimatological impact. The polar boundary of tropical China coincides with the northern frontier of perennial crops, while the polar boundary of subtropical China represents the boundary between year-round and only seasonal vegetable cultivation, where the growing season is limited to the warm period only.

---

[29] In the final version presented by the author during the German-Chinese Workshop "The Climate of China" at the Institute of Geography, Mainz University, August 25–September 5, 1986.

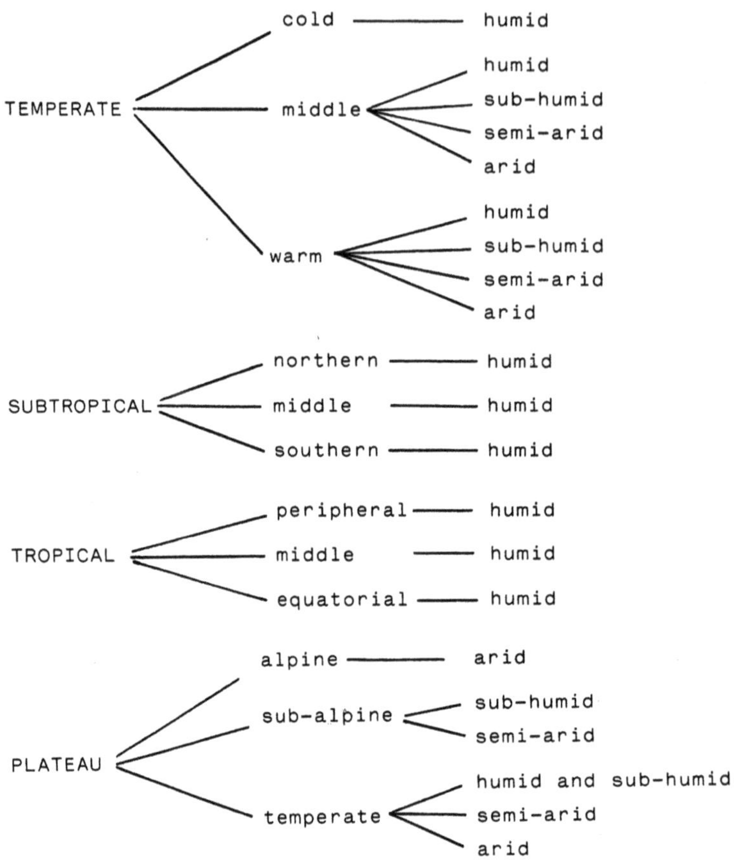

Fig. 8.9. Hierarchy of climate regionalization of China according to HUANG Bing-wei (1986)

For a subdivision of the climate realms into 12 *climate belts*, selected temperature criteria (with a close relationship to agricultural and other land use practices) were applied, among them specific values for accumulated temperatures, distinct numbers of frost-free days, certain mean annual temperatures or mean annual minimum temperatures. Only in the case of the three plateau climates, which were called alpine, subalpine and temperate, the criteria of classification were derived from the distribution of agricultural land use patterns.

With particular emphasis, HUANG considered the division into climate belts as the principal level of the climate regionalization of China. By amalgamating three plateau climates into one climate belt, HUANG arrived at a total of ten climate belts, a fact which leads to the prevailing climate division of China in a mostly zonal direction.

A further division of the climate belts into a total of 21 *climate regions* was carried out according to four grades of humidity resp. aridity: humid, subhumid, semi-arid and arid. Humidity and aridity, in this respect, were determined by the character of the vegetation and land use pattern as a whole. Due to the climate criteria applied, the climate regions may be considered as thermal-hygric regions.

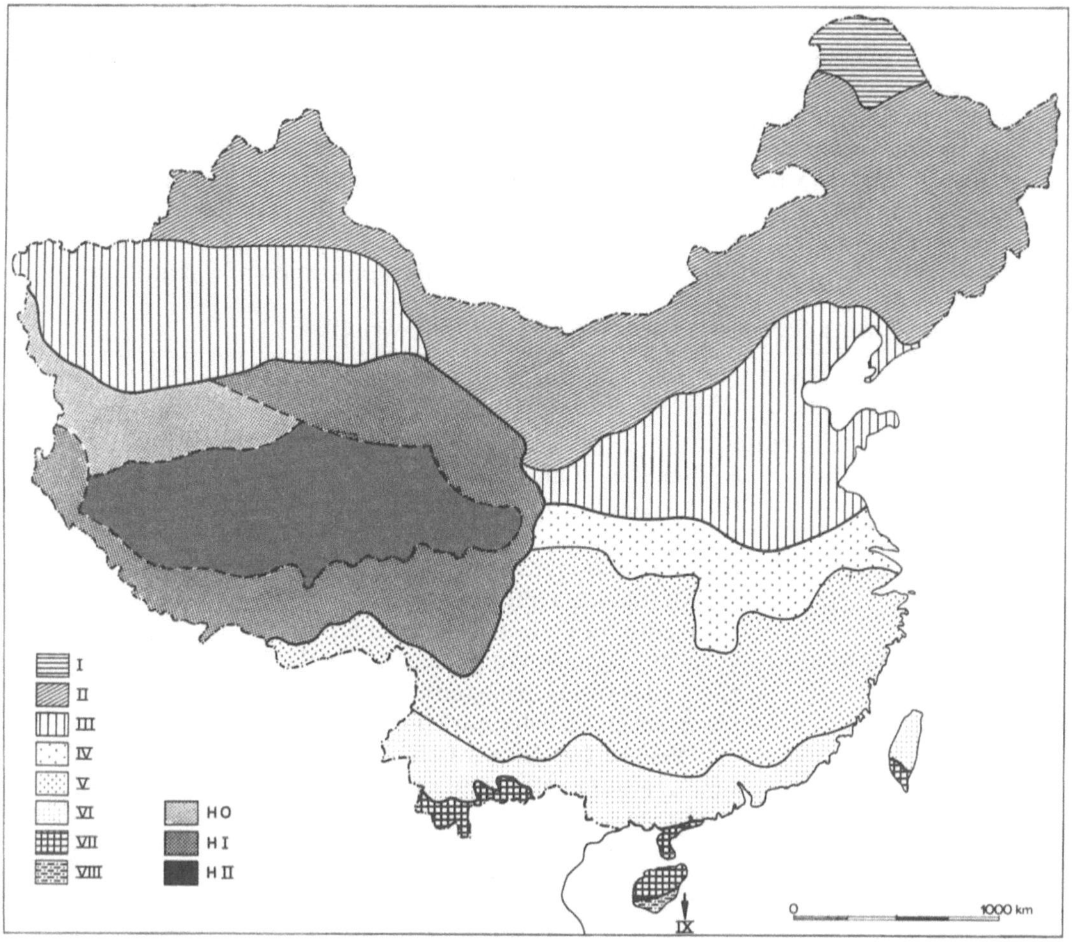

Fig. 8.10. The climate classification of China according to HUANG Bing-wei (1986). Abbreviations in the legend are explained in the text

Their final subdivision according to the topography has led to 45 *climate provinces,* each with specific thermal-hygric and topographical conditions. The climate belts, regions and provinces were identified by particular indices: The climate belts were described by roman numerals (I–IX) and by H0, HI and HII for the plateau climates. For the climate regions, the letters A, B, C and D were added in case of prevailing humid (A), subhumid (B), semi-arid (C) or arid (D) conditions. For the climate provinces, an arabic number was added indicating a distinct regional distribution. Each climate province, therefore, was identified by a combined climate "formula" consisting of three indices. It is noteworthy that the same arabic numerals and Latin characters (in capitals) represent climate similarities between the climate provinces.

As a result of the huge extent of China, the climate map by HUANG (Fig. 8.10) expresses a prevailingly zonal regionalization, complemented by exceptional climate differences dependent on altitude and changes in landforms. Both the zonal classification of climate and the separately established plateau climates represent a mostly temperature-induced climate division, whereas hygric differences rank second only.

HUANG's climate classification of China (1986) reveals an extremely large range of climate valid for China. It is also evident that prevailing zonal and altitudinal elements govern the climate regionalization of the country. HUANG's comprehensive classification also considered the climate potential for various land use types and thus successfully contributes to the application of climatology.

Climate Division Scheme of China (by HUANG Bing-wei 1986)

I. Cold temperate		
	A. Humid	(1) Northern Da Hinggan
II. Middle temperate		
	A. Humid	(1) Sanjiang Plain
		(2) Mountains of Eastern-Northeast China
		(3) Piedmont Plain of Eastern-Northeast China
	B. Subhumid	(1) Central Songhuajiang-Liaohe Plain
		(2) Middle Da Hinggan
		(3) Piedmont Plain and Hills of Sanhe
	C. Semi-arid	(1) Southwestern Songhuajiang-Liaohe Plain
		(2) Southern Da Hinggan
		(3) Eastern Nei-Mongol High Plain
	D. Arid	(1) Western Nei-Mongol High Plain
		(2) Area of Lanzhou and eastern Hexi (Gansu Corridor)
		(3) Junggar Basin
		(4) Altay Mountains, Tacheng Basin and Ertix Valley
		(5) Ili Basin
III. Warm temperate		
	A. Humid	(1) Hills and mountains of Liaodong and eastern Shandong
	B. Subhumid	(1) Hills and mountains of Central Shandong
		(2) North China Plain
		(3) Mountains and hills of North China
		(4) Plains of southern Shanxi and Weihe Valley
	C. Semi-arid	(1) Loess Highlands of Central Shanxi, northern Shanxi
	D. Arid	(1) Tarim Basin and Turpan Basin
IV. Northern subtropical		
	A. Humid	(1) Huainan and Low Reaches of Changjiang (Yangtze)
		(2) Hanzhong Basin
V. Middle subtropical		
	A. Humid	(1) Hills and mountains on Nanling
		(2) Guizhou Plateau
		(3) Sichuan Basin
		(4) Yunnan Plateau
		(5) Southern slope of eastern Himalaya

VI. Southern subtropical
- A. Humid
  - (1) Central and northern Taiwan
  - (2) Hills and plains of Guangdong, Guangxi and Fujian
  - (3) Mountains and hills of Yunnan between Wenschan and Tengchong

VII. Peripheral tropical
- A. Humid
  - (1) Lowlands of southern Taiwan
  - (2) Central and northern Hainan and Leizhou Peninsula
  - (3) Valleys of southernmost Yunnan

VIII. Middle tropical
- A. Humid
  - (1) Southern Hainan and Dongsha, Xisha and Zhongsha Islands

IX. Equatorial tropical
- A. Humid
  - (1) Nansha Islands

H0. Plateau alpine
- D. Arid
  - (1) Kunlun Mountains

HI. Plateau subalpine
- B. Subhumid
  - (1) Aba-Nagqu Area
- C. Semi-arid
  - (1) Southern Qinghai and Qiangtang Plateau

HII. Plateau temperate
- AB. Humid and subhumid
  - (1) High mountains and gorges of western Sichuan and eastern Xizang
- C. Semi-arid
  - (1) Plateaus and Mountains of eastern Qinghai
  - (2) Mountains of southern Xizang
- D. Arid
  - (1) Qaidam Basin
  - (2) Ngari Mountains

9 Climate Zones of China

Note: The climate zones respectively climate types, according to the climate regionalization scheme of HUANG Bing-wei (1986), were described with particular reference to the major climate characteristics and their variation over space and time. Each climate zone also contains a climate map of China, showing distinctly the spatial distribution of the climate zone concerned. A climate diagram for a representative station was added which expresses the annual variation of precipitation and temperature, based on long-term monthly means. Precipitation and temperature were correlated according to the "aridity index" of WALTER/LIETH (1967) which reads as:

$$n = 2t;$$

n represents the mean monthly precipitation total (in mm), t the monthly temperature mean (°C). Thus, precipitation and temperature on the two vertical axes of the climate diagram are in the ratio of 2 to 1. Humid and arid conditions can be easily deduced from the diagrams in the following way: If the precipitation curve (broad line) is above the temperature curve (thin line), humid conditions occur, while conversely arid conditions are found.

The diagrams also contain, besides the name of the reference station(s) of each climate zone, the elevation as well as latitude and longitude of the station concerned. The mean annual precipitation total (R) and the mean annual temperature (T) were also entered.

9.1 Cold Temperate Zone (I)

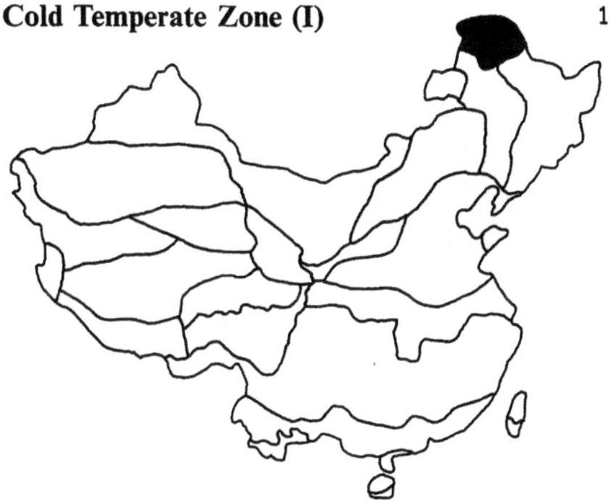

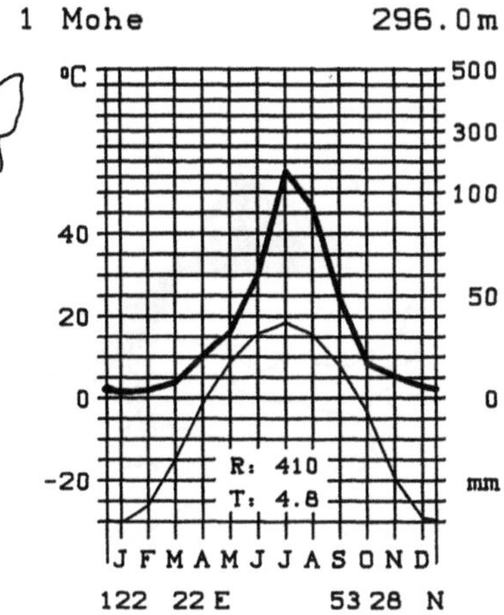

Fig. 9.1. Climate map and reference diagram of the cold temperate zone.

Located in the extreme northeast of China, the comparably only small Cold Temperate Zone (Fig. 9.1) occupies the most northern region in all of China (up to 53°28′N). Temperatures are therefore effectively controlled by the marked seasonal differences in solar radiation. As a further climate constraint, the landforms have to be taken into consideration. They characterize the area as a rather mountainous one, which is mostly composed of the North Daxinganling Mountain, extending up to 1,400 m.

The most significant characteristic of the Cold Temperate Zone is the extremely large mean annual temperature variation which, at the most, reaches up to 50° C, representing the maximum range of temperature over the year in China. The recorded temperatures, however, would result in a much larger annual range, for example at Huma/2: recorded maximum 38.0° C, minimum −48.2° C.

Mean temperatures in summer increase to around 20° C (with July as the warmest month), while they decrease in winter to between −25° and −30° C, thus establishing a pronounced thermal contrast between 45° and 50° C. Since winter is extraordinarily cold and long as well, covering a 9-month period from September through May, frost and snow must be considered among the most serious climate hazards with a far-reaching agroclimatic impact. The larger number of frost and snow days clearly underlines the serious winter conditions. In contrast, summer covers only a 3-month period from June to August, which corresponds to the rainy season. Nearly two-thirds of the annual total of precipitation fall from June to August, thus indicating a sharp contrast in precipitation between summer and winter. Since winter still receives some precipitation (mostly snow) and evaporation can be less expected, this zone is considered as "humid".

From the marked annual range of temperature and precipitation, the Cold Temperate Zone is characterized by a sharp seasonal contrast in both thermal and hygric aspects. A persistent cold and relatively dry winter compared to a short and wet summer form a distinct seasonal climate pattern. It is also worth noting prevailing northerlies throughout the year, however, with an unusually high percentage of calms (Huma/2: 32%), which creates rare conditions in China.

9.2 Middle Temperate Zone (II)

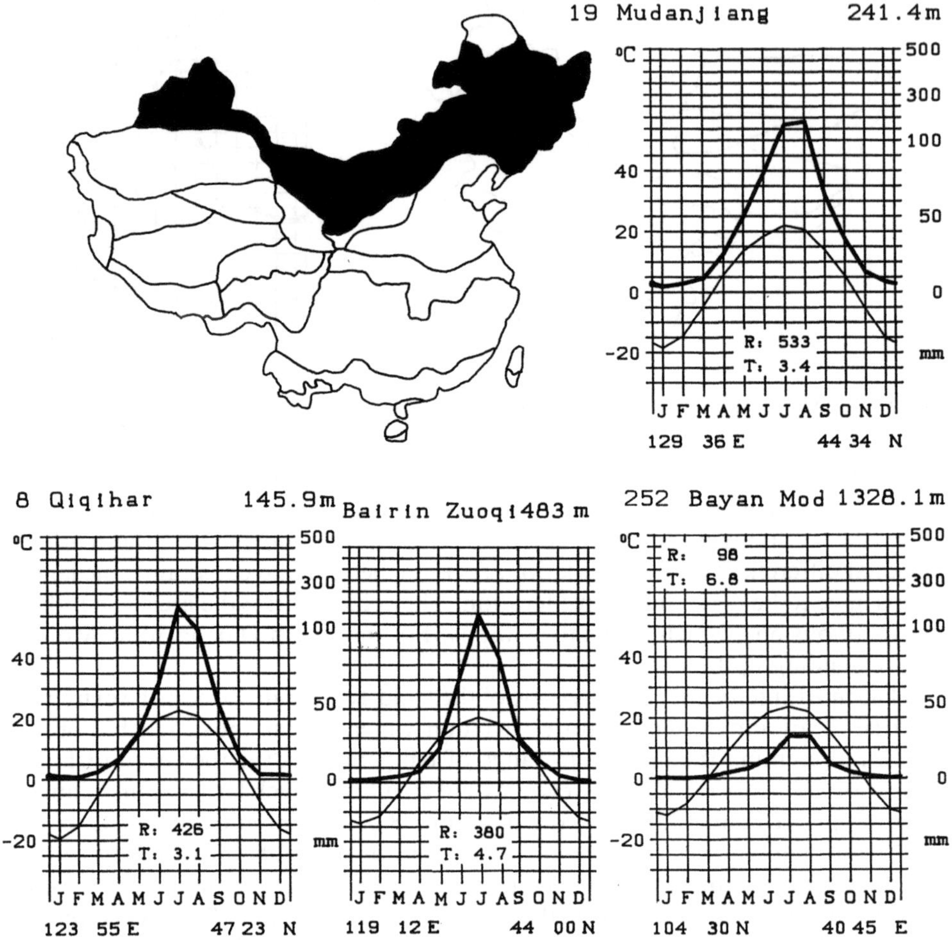

Fig. 9.2. Climate map/diagrams for the Middle Temperate Zone. Reference stations: Mudanjiang/19, Qiqihar/8, Bairin Zuoqi/23, Bayan Mod/252

The Middle Temperate Zone (II) represents the largest climate zone in China and occupies the main parts of Northeast and Northwest China as well as Inner Mongolia (Fig. 9.2). Due to the considerably large spatial variation of precipitation and temperature, this zone is divided into four types: humid (II A), subhumid (IIB), semi-arid (II C) and arid (II D), sometimes an extreme arid type is distinguished. The landforms and the nature of the underlying ground are extremely varying, both from mountains to lowlands and from the fertile Northeast China Plain to the steppes and deserts mostly of Inner Mongolia and Dzungaria.

According to different hygric conditions, which are expressed by the four types II A to II D, the range of precipitation in the Middle Temperate Zone is large, but still the major climate characteristic is considered to be similar due to the fair homogeneity of the mean annual variation of temperature for all four types. Correspondingly, they show a strong seasonal contrast in temperature between a severe winter and a warm summer. Frost and snow play a major role from September to May, representing far-reaching climate hazards. The thermal conditions are, however, markedly strengthened by the altitude above sea level, which may cause severe modifications in temperature. For all four types II A to II D, precipitation synonymously shows a distinct annual variation with a short summer peak (however, of a variable amount in the different types), and a long winter deficit. There is thus an extremely uneven distribution of precipitation over the year.

The humid and subhumid types (II A, II B) of the Middle Temperate Zone, which experience a similar climate, are only divided into two types because of the slight difference in the summer and thus also annual precipitation total which varies between 600–800 mm. With July as the wettest month, a marked summer peak of precipitation is expressed by a normally 50–60% portion of the annual total which falls in July up to August. Since June and September may also record a fair amount of precipitation, a moist period may cover almost 5 months. Summer can really be called a wet season. During the long winter, some precipitation can always be expected, i.e. most often snow, particularly from December to February.

With regards to temperature, the humid and subhumid types are characterized by a long and cold winter and a comparably short, warm summer. January as the coldest and July as the warmest months are distinct, showing a monthly mean of about $-20°$ C, on the one hand, and $>20°$ C, on the other, if downland stations are considered. Additionally, also the diurnal temperature variation may remarkably increase, leading to even hot values of maximum temperature conditions in summer, while in winter temperatures may drop to seriously low values. Thus, frost and snow are of a wide-ranging frequency and intensity.

The semi-arid type (II C) of the Middle Temperate Zone receives much less precipitation than both the humid and subhumid types, ranging normally between 200–400 mm due to the distinctly lower summer totals. However, correspondingly summer has maximum precipitation, which is even more distinctly established (from June to August around 65 to 75% of the annual total), but summer can no longer be regarded as a wet season. Also, in winter some precipitation normally occurs as snow.

The seasonal contrast in precipitation is also valid for temperature. Winter experiences less severe conditions, leading to slightly higher temperatures and a shorter length of winter which, however, still remains as a strongly cold, long season. Although distinctly pronounced, the annual variation of temperature is normally less than for the humid and subhumid types. The temperature mean for January is around $-10°$ to $-12°$ C, for July between $20°-24°$ C.

A strong seasonal contrast in the thermal and hygric conditions is therefore the main characteristic of the semi-arid type. This is also characterized by clear skies which lead to a comparably longer duration of sunshine.

The arid type (II D) of the Middle Temperate Zone records much less summer and thus annual totals of precipitation which normally drop to between 50 and

250 mm. The annual variation of precipitation shows a clear summer maximum which, however, becomes weaker or is nearly extinguished in Northwest China. Evidence is given by comparing the annual variation of precipitation for Bayan Mod/252 with Ürümqi/269 and Altay/270, and by considering the portions of precipitation for winter and summer:

	Bayan Mod	Ürümqi	Altay
Winter (%) (December–February)	1.8	12.2	24.0
Summer (%) (June–August)	69.9	30.4	29.2
Annual total (mm)	99	278	181

Temperatures of the arid type (II D) show in accordance with the other types a significant annual variation in a similarly large range as in the semi-arid type II C. Also in absolute terms, temperatures of the coldest and warmest months fairly correspond with the semi-arid type, thus showing a January mean around −10° to −12° C, and a July mean between 20° to 24° C. Plausibly, temperatures in the mountains gradually decrease. Winter and summer experience severe seasonal contrasts in temperature which establish winter as a long and severe season, whereas summer is a comparably warm, but short period. As an expression of the severe temperature conditions in winter, the remarkably large frequency and intensity of frost and snow events are typical characteristics of the climate.

The extremely strong seasonal contrast of temperature and precipitation, thus establishing a long, severe and dry winter as compared to a short, but relatively warm summer must be considered as the main characteristic of the Middle Temperate Zone, expressed for all four types. Further strengthening the climate extremities, a rapid change of climate takes place in the mountains. Besides the well-established seasons, extreme climate events, such as the severity of frost and snow and the long duration of sunshine, create a particular climate.

9.3 Warm Temperate Zone (III)

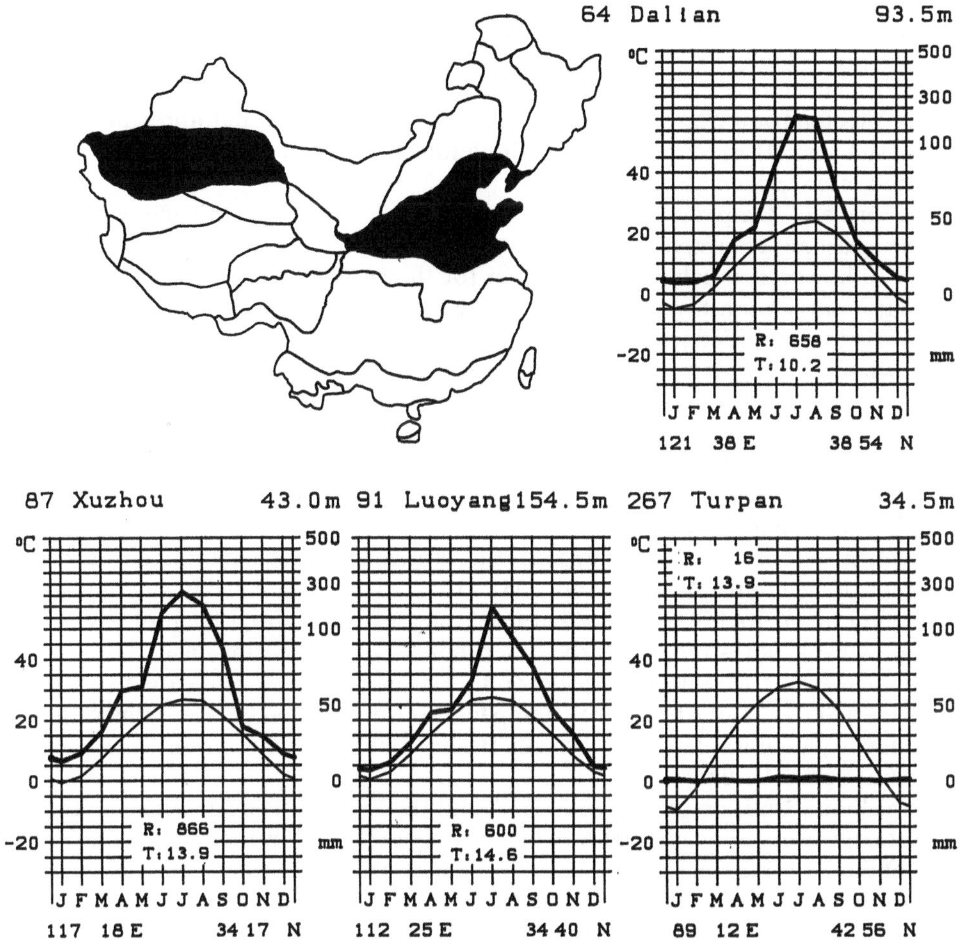

Fig. 9.3. Climate map and reference diagrams of the Warm Temperate Zone. Reference stations: Dalian/64, Xuzhou/87, Luoyang/91, Turpan/267

Divided into four types, i.e. humid (III A), subhumid (III B), semi-arid (III C) and aird (III D), the Warm Temperate Zone is, next to the Middle Temperate Zone, the most widely distributed over space and, at the same time, characterized by a wide range of physio-geographically contrasting regions (Fig. 9.3). They include the most fertile Shaanxi and Shanxi Loess Highlands as well as the North China Plain, but also occupy the large areas of deserts, composed of the Tarim and Turpan Basins.

The extremely variable nature of the Warm Temperate Zone is expressed by the four types III A to III D, thus showing a wide range of moisture conditions. Although, in total, summer and winter represent a distinct wet and dry season, precipitation conditions may considerably vary in each and for the different types. Most spectacularly, the driest location known in China (which is Turpan/267, recording a mean annual total of only 16.4 mm) belongs to the Warm Temperate Zone, although it falls under the arid type. In contrast, mean annual precipitation may increase up to 800–900 mm which is valid in some parts of the North China Plain. However, precipitation is unevenly distributed over the year, showing a 60% portion of the annual total during summer (June–August), while winter precipitation decreases from a weak total (in the humid and subhumid types III A and III B) to nearly diminishing totals (in type III D). Winter precipitation is often represented by snow.

In terms of temperature, the Warm Temperate Zone nevertheless shows a distinct seasonal contrast between a warm, partly hot summer and a cold winter which is, however, much less severe than for the Middle Temperate Zone. Over the major parts, monthly mean temperatures remain >0° C, although due to the extremely low recorded minimum temperatures and thus still large frequencies of frost, winter creates unsuitable growing conditions for temperate crops. As winter temperatures decrease from the humid and semi-humid types (III A, B) to the arid type (III D), summer temperatures increase. Accordingly, the length of winter becomes shorter, yet summer longer.

The hottest location known so far in all of China, which is Turpan/267 reaching in July a mean temperature of 32.7° C (mean maximum 39.9° C), belongs to the Warm Temperate Zone. Turpan also experienced the maximum recorded temperature in China (47.6° C, from 1951–1980). Turpan, located in a depression extending down to 154 m below sea level, creates through its location its own climate which even strengthens the desert climate of the Tarim Basin.

Similar to the Middle Temperate Zone, extreme climate events are a striking characteristic also of the Warm Temperate Zone. Topo-climatic constraints, prominently expressed by the Turpan Basin, may distinctly modify the warm temperate climatic unity.

The humid and subhumid types (III A and III B) show little differences in their climate inventory. For type III A, which is only exhibited in the mostly hilly and mountainous Liaodang and Shandong Peninsulas, major topo-climatic corrections must be considered. The annual variation of precipitation is of the same remarkably uneven nature as for type III B, although a wet season can be recorded from mostly May until September. Strikingly established is an outstanding maximum in July and August, in which heavy downpours of rain, often associated with thunderstorms, result in strong rainfall intensities as well as totals. Evidence is given by the precipitation totals in July and August and their percentages compared to the annual total at selected stations:

	Annual total	July		July–August	
		Total	Percent	Total	Percent
Beijing/61	644	212[a]	32.9	405	62.8
Tianjin/62	570	190	33.3	352	61.8
Xuzhou/87	867	248	28.6	404	46.6
Jinan/77	672	214	31.8	362	53.9

[a] This figure denotes August which records the highest mean monthly rainfall at Beijing.

In comparison with summer, there is less precipitation in winter; it is represented to a small part by snow, thus expressing in total less serious snow conditions in the North China Plain.

Temperatures in both the humid and subhumid types of the Warm Temperate Zone clearly mark a strong seasonal contrast which results from relatively warm, sometimes even hot summer temperatures and (only) normally cold winter temperatures. Mean values for the warmest month (July) are around 25° to 28° C, for the coldest month (January) around −1° to −5° C. However, frost still represents a major climate characteristic; frosts may occur between October/November and March/April.

Worth to mention are the quite hazardous events of duststorms which, in spring (April/May), may affect the northern parts of the subhumid, warm temperate Great Plain, including Beijing. Although of a rare frequency only (about 1 case every two years) and experiencing a duration of less than one day, duststorms result from cold and strong Northwesterlies which blow up the dried-up dust from the Loess Plateaux. At Beijing, the last serious duststorm occurred on April 11, 1988.

The climate characteristics of the semi-arid type (III C) of the Warm Temperate Zone, which is spatially restricted to the famous Loess Highlands of Central Shanxi and North Shaanxi, are modified compared to the humid and subhumid types (III A and III B) in thermal and hygric respects. Precipitation throughout the year decreases, although the annual variation remains same. The same observation is true for temperatures which are in summer and winter lower, thus leading to a slightly more pronounced winter and showing more frequent frost and snow events.

The arid type (III D) particularly differs in its moisture conditions from the three other types of the Warm Temperate Zone. Occupying the Taklamakan and Gashun deserts, the area as a whole receives the lowest precipitation in all of China which mostly drops to an annual total between 15 and 50 mm, showing slightly higher amounts on the foothills towards the Tian Shan as the northern boundary and the Kunlun Shan and Altun Shan, being the southern boundary of the region. Due to the extremely low totals of precipitation, usually all months are dry; evidence for this is also given by the very small number of rainy days which as a monthly figure reaches at the most 4, and varies for the whole year normally between 15 and 40 only. Among the driest stations, Ruoqiang/264 (12.6 rainy days/year), Turpan/267 (15.0) and Hotan/276 (17.4) express arid conditions. Mostly for statistical reasons, the annual variation of precipitation may be considered which shows a rather slight, but still variable amount throughout the year. The typical contrast in precipitation between summer and winter is still valid which can be demonstrated by the percentages of precipitation in winter (December–February) and summer (June–August) at selected stations:

	Annual total (mm)	Winter (mm)	Winter (%)	Summer (mm)	Summer (%)
Turpan/267	16.4	3.2	19.5	9.6	58.5
Ruoqiang/264	17.4	2.5	14.4	10.8	62.1
Hotan/176	33.4	5.0	14.9	15.2	45.5

9.4 Northern Subtropical Zone (IV)

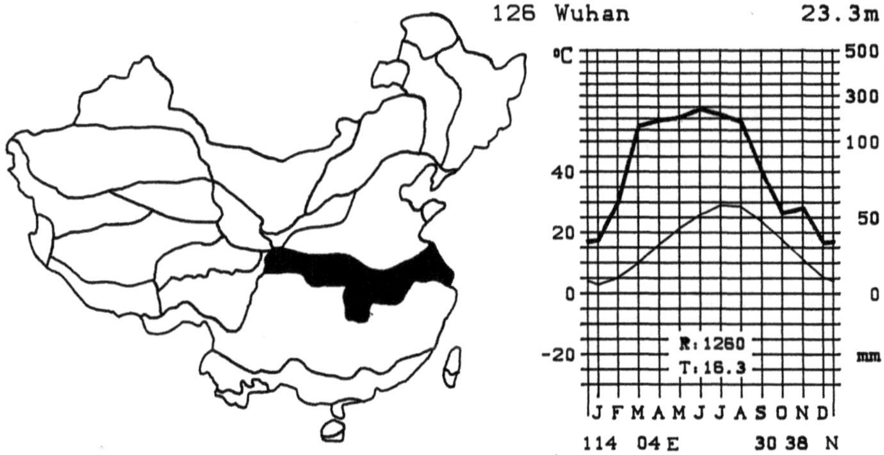

Fig. 9.4. Climate map and reference diagram for the Northern Subtropical Zone. Reference station: Wuhan/126

Occupying the middle and lower reaches of the Yangtze Valley and widened to the north and south over some adjoining regions, such as Hanshui River, North Yunnan and mid-Guizhou, the Northern Subtropical Zone (IV) represents a west to east belt extending from the Chinghai foothills to the East China Sea (Fig. 9.4).

The climate, regarded as humid, shows with regard to rainfall a fair homogeneity over space. A relatively long wet season is predominant, which lasts from March/April until August/September and thus covers a 6- to 7-month period during which ample rains can be expected. Of the annual total which mostly ranges from 1,000 to >1,400 mm (showing an increase west to east), between 75 and 90% can be expected during the long wet season in which May, June and July are the peak months. The ample rains in spring and summer are shown by records for selected stations which give the amount and percentage of rain from March through September:

	Annual total (mm)	March–September amount (mm)	March–September percentage
Shanghai/119	1,128	882	78.1
Hangzhou/123	1,401	1,075	76.7
Wuhan/126	1,262	1,260	81.0
Yichang/129	1,200	1,000	88.6

The heavy rains are in the peak months of June and July of a particularly strong intensity, which is due to the frequently occurring rainstorms, associated with downpours for a few days. On the other hand, the remarkably large early summer and spring rains result from the then typical "mei-yu" depressions which hit the middle and lower reaches of the Yangtze Valley.

As this zone is considered to be subtropical, the particular temperature conditions are expressed by positive monthly means year-round, as long as no major altitudinal decrease in temperature occurs. The coldest month (January) in most cases records a mean of 3°–5° C. Since summer temperatures are not only high, whereby July and August may even be described as hot months (recording means around 27° to 29° C), temperature conditions in all are on a rather high level. Frost temperatures, however, typically occur from December until February; the same is true for snow.

9.5 Middle Subtropical Zone (V)

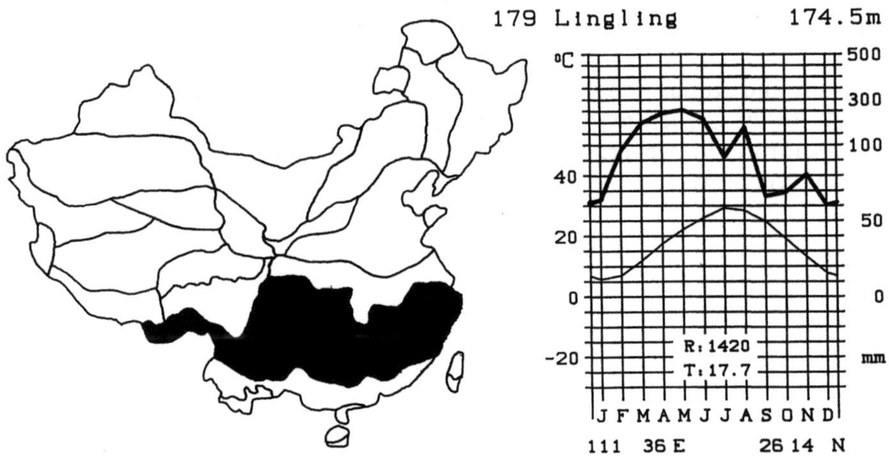

Fig. 9.5. Climate map and reference diagram for the Middle Subtropical Zone. Reference station: Lingling/179

Representing one of the largest climate zones in China, rather variable climate conditions in the Middle Subtropical Zone (Fig. 9.5) result from controversial physio-geographical regions which form five major climate provinces: (1) the hills and mountains of Nanling, (2) the Guizhou plateau, (3) the Sichuan basin, (4) the Yunnan plateau, (5) the southern slope of the eastern Himalaya. Thus, these regions extend from southeastern Xizang to the East China Sea, divided into regions of complex landforms and contrasting altitudinal variations.

The climate conditions, therefore, may vary considerably, but still the "subtropicality" of the whole region is valid. As such, the year-round, positive temperature mean is the main characteristic, which at the same time expresses a remarkably weakened coldness in winter. This is also valid for the mountainous areas, for example, the eastern Himalayas and the Yunnan Plateau, where winter temperatures even at high altitudes drop only slightly below zero (see Deqen/230, 3,589 m above sea level, mean January temperature −2.9° C). Mostly because of the hilly and mountainous as well as plateau character of the region as a whole, summer temperatures are lower than for the Northern Subtropical Zone (IV); the annual variation of temperature drops to 15°–20° C which underlines the less distinct thermal differences between summer and winter.

Considered as an expression of the subtropical temperature conditions, often the moderate spring temperatures are also cited. In this direction, Kunming/225 (located at 1,891 m above sea level) is often taken as a reference location, being called the "spring city" of China, because of moderate spring temperatures (April mean: 16.7° C, May: 19.3° C.

Mostly dependent on the altitude above sea level, the occurrences of frost of a mostly weak intensity can only be regarded as a characteristic from December to February.

From the overall moisture conditions, the whole region is considered as humid. It is, however, interesting to note a remarkably large range of the annual total of precipitation (Wenzhou/154: 1,698 mm, Deqen/230: 676 mm). Throughout the year, a long wet season from about April until September or October (with maximum rains in July and August) alternates with a winter period of meager precipitation, which is partly even snow.

All five climate provinces under the Middle Subtropical Zone (V) have developed their particular climate depending upon the specific landforms and their climate-modifying effects.

Taking the Sichuan Basin as one of the climate provinces, the climate modifications by a basin are expressed in the following way (with reference to Chengdu/234). Cloudiness amounts to between 7.5/10 and 9/10 (in other words, between 75 and 90%, resp.), taken as monthly means, which are the highest records in China. Due to the surrounding mountains, winds enter the basin with an extremely reduced speed, thus leading to an extraordinarily high percentage of calms. Chengdu/134 records an annual average of 42% calms, with only a weak annual variation (May 35%, December 50%). Fog represents the most crucial characteristic of the climate, which, due to recent investigations, is even worsened by the so-called acid fog; this results from the interconnection of fog and acid gas pollution of the atmosphere. Since the sky is overcast, sunshine duration at Chengdu/134 records a minimum for whole China (annual average 3.4 h day^{-1}). The climate calamities of the Sichuan Basin are finally expressed by unusually high records of relative air humidity, which, as monthly means, vary between 77 and 86% and thus reach the same high level as in the tropical zone [cf. Dongfang/210, located in the Peripheral Tropical Zone (VII)].

9.6 Southern Subtropical Zone (VI)

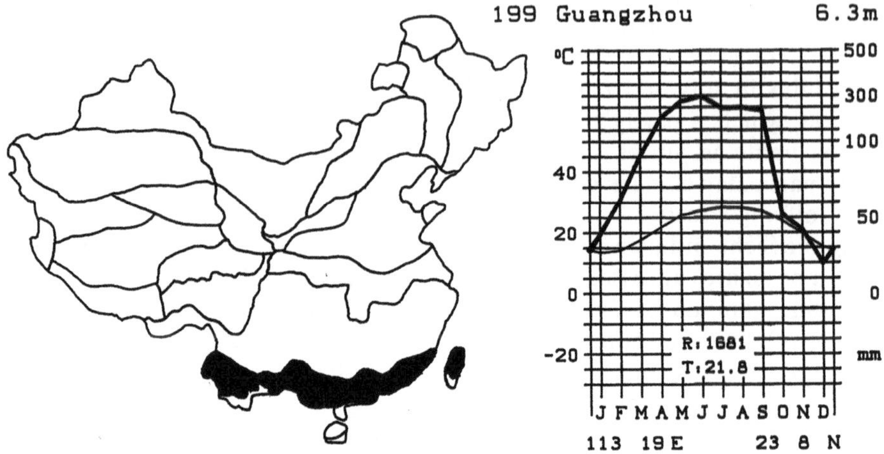

Fig. 9.6. Climate map and reference diagram of the Southern Subtropical Zone. Reference station: Guangzhou/199

Established as a more or less west to east belt extending over southern China, the Southern Subtropical Zone (VI) comprises the southern Yunnan mountains and hills, the Guangxi, Guangdong and Fujian hills and downlands as well as northern and central Taiwan (Fig. 9.6). Due to the differences in altitude and landforms, particular attention should be given to topo-climatological modifications.

The subtropical conditions of this zone are related to mild winter temperatures. When considering lowland stations, such as Guangzhou/199 and Nanning/215, mild temperatures during winter are expressed by monthly means between 13°–15°C. Only in rare cases, daily minimum temperatures may drop slightly below zero, but snow has never been recorded so far (e.g. at locations like Guangzhou and Nanning). Summer is pleasantly warm (around 26°–28°C), but rather humid showing monthly mean of relative humidity between 80–85%. Thus, sultriness may become a major climate constraint in summer.

The obvious evidence of sultriness can be demonstrated by "sultriness diagrams" for Guangzhou, Hongkong and Taibei, constructed after the method of SCHARLAU (1950)[30], see Fig. 9.7. Thus, sultriness is defined by a distinct relationship between a certain amount of relative humidity and temperature, on a monthly basis. Considering the diagrams in all three cases, the sultriness period characterizes the period from April to October, showing the most pronounced conditions from June through August. Distinctly expressed, the reason for sultriness is the high monthly mean of temperature (between 25° and 30°C), while air humidity remains at the same high level throughout the year.

[30] SCHARLAU (1950) has defined sultriness by the revised LANCASTER-CASTENS formula, expressing the threshold values of sultriness by the vapour pressure of 14.08 Hg, which again means a certain relationship between air humidity and temperature.

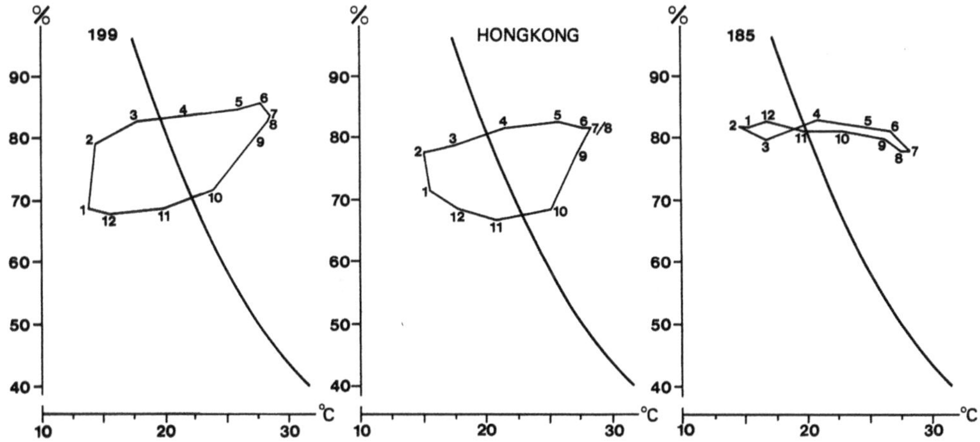

Fig. 9.7. Sultriness diagrams at Guangzhou/199, Hongkong and Taibei/185. (According to SCHARLAU 1950)

The humid character of the Southern Subtropical Zone (VI) is expressed by ample rains which are only weaker from October/November until January. The annual rainfall total mostly varies between 1,200–1,700 mm, showing a distinct peak period from May through August. June experiences the heaviest rains associated with the onset of the strongly developed monsoon and partly resulting from fierce rainstorms and typhoons which both represent in May and June typical phenomena of the Southern Subtropical Zone. The torrential rains often represent climate desasters with far-reaching consequences, such as floods which severely hit the farmland, roads and houses in the densely populated lowlands. The death toll of people, caused by floods, can also reach large numbers, for example in Guangdong Province alone there were 114 deaths during one rainstorm from May 19 to 23, 1987 (according to a press report in CHINA DAILY, on May 30, 1987).

9.7 Peripheral Tropical Zone (VII)

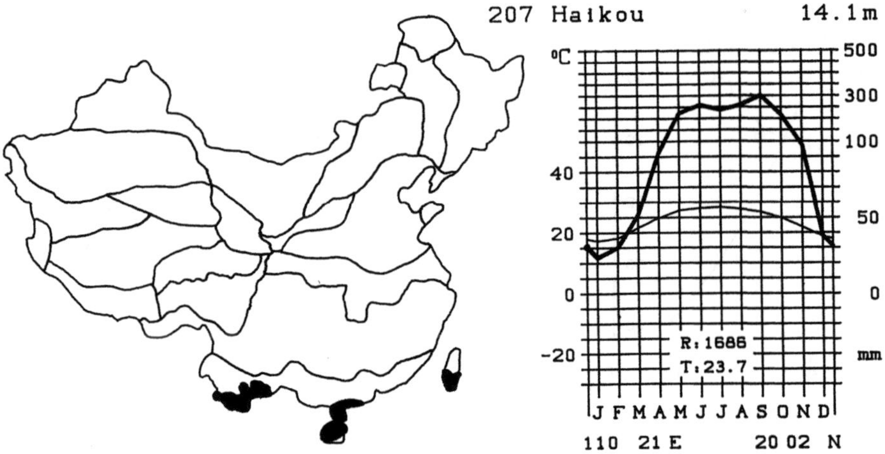

Fig. 9.8. Climate map and reference diagram of the Peripheral Tropical Zone. Reference station: Haikou/207

Among the three tropical zones (VII–IX), located on the southern fringe of China, the Peripheral Tropical Zone (VII) occupies three separate regions: (1) the lowlands of southernmost Yunnan, (2) Leizhou Peninsula and the northern and central Hainan Island, (3) the lowlands of southern Taiwan Island (Fig. 9.8).

Tropicality of this zone is represented by completely frost-free conditions year-round. The abound warmth is also shown by comparably high winter temperatures between 18° and 20° C, given as monthly means. The term "winter", therefore, becomes questionable, but it is still commonly used. This may be particularly due to the rather low daily minimum temperatures during winter which in connection with the high daily maximum clearly mark a pronounced diurnal range of temperature which can also be taken as a typical tropical characteristic. The annual temperature range, expressed by the difference between the warmest and coldest month, is fairly low (about 10° C only).

Characterized as a humid region, the Peripheral Tropical Zone (VII) receives ample rains, yet shows a pronounced dry period from December through March. Between the western, central and eastern parts of this Zone, annual totals usually vary between 1,200–1,500, 1,000–1,200 and 1,400–1,800 mm respectively. The dry period often causes water shortages, affecting the cultivation of perennial crops, mostly rubber and tea which, however, in the mountainous parts are also cold-stricken by low temperatures.

Since the main parts of this Zone are surrounded by the Nan Hai (South China Sea), local climatic effects of the sea-land interaction must be taken into consideration, such as land and sea breezes and frequent fog events.

As air humidity reaches relatively high monthly means, sultriness may affect during the warm summer to a large extent human comfort. Coastal areas, however, benefit from the effects of the sea breezes which diminish the human-climate

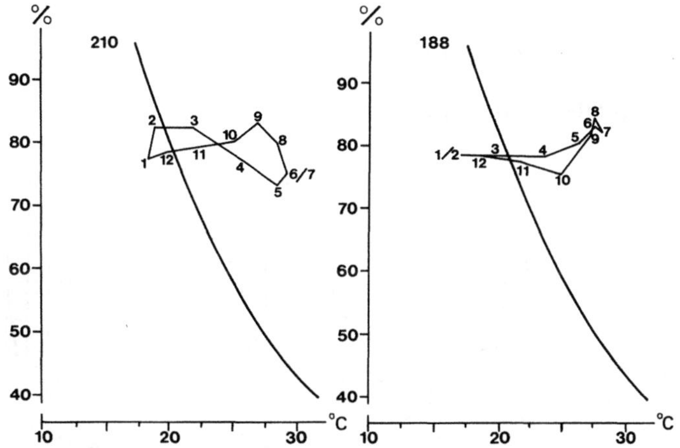

Fig. 9.9. Sultriness diagrams at Dongfang/210 and Tainan/188. (According to SCHARLAU 1950)

constraints of sultriness. As can be seen from the sultriness diagrams at Dongfang and Tainan (Fig. 9.9, according to SCHARLAU 1950), sultriness affects nearly all months of the year, although it is strongest during summer. The reason for this is high temperature associated with high air humidity.

9.8 Middle Tropical Zone (VIII)

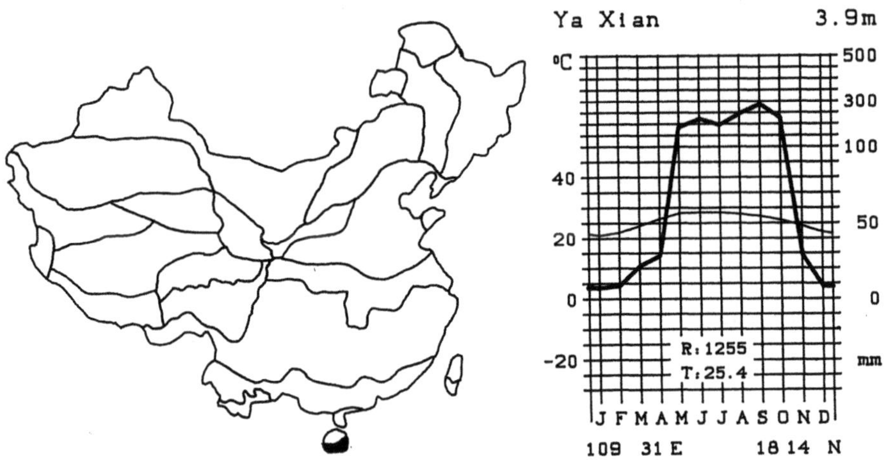

Fig. 9.10. Climate map and reference diagram of the Middle Tropical Zone. Reference station: Ya Xian

Located to the south of the Peripheral Tropical Zone (VII), this zone (VIII) belongs even more to the marginal southern parts of China. It covers southern Hainan Island, as well as Xisha Islands (about 300 km off the southeastern coast of Hainan) and Dongsha Islands which are more than 200 km off Hongkong (Fig. 9.10).

The tropical characteristics of temperature are even more distinctly established than in the Peripheral Tropical Zone (VII) which include a comparably warmer "winter" (>20° C), and a low annual range of temperature (about 6° C), due to the maritime effects on the more balanced annual temperature variation. However, the diurnal range of temperature becomes more pronouncedly established, thus indicating a stronger tropicality in thermal respects.

Although regarded as humid, expressed by a larger annual total of rainfall (1,000–1,500 mm), yet an annual variation is still significantly established. Hence, a distinct wet period from May to November is recorded compared to a dry period from December to March which often results in shortages of rain water for the cultivation of perennial crops.

Since the sea and land interaction can be regarded as fully developed, the thus resulting local climate effects are also well established, particularly the usual diurnal occurrences of land and sea breezes. Human discomfort due to sultriness is effectively counteracted by sea breezes.

9.9 Southern Tropical Zone (IX)

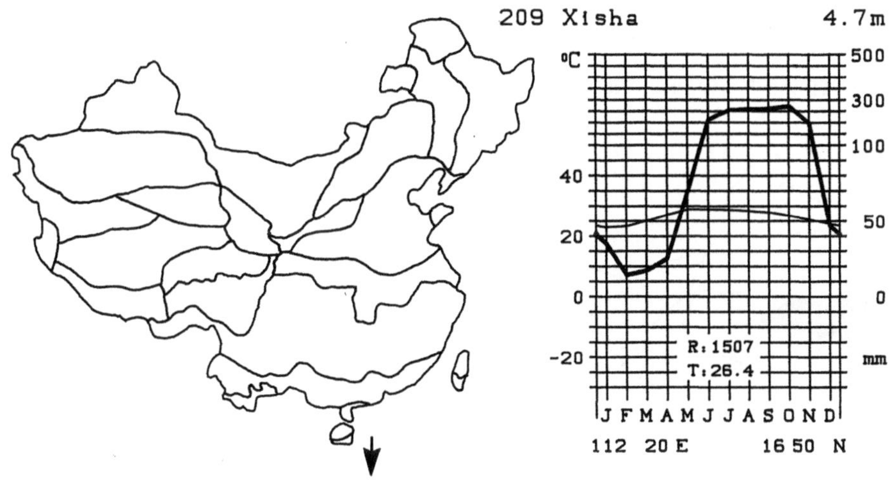

Fig. 9.11. Climate map and reference diagram of the Southern Tropical Zone. Reference station: Xisha/209

This zone (Fig. 9.11) only occupies the very small, extremely southernmost Nan Sha Islands, which experience fully developed equatorial-tropical temperatures and rainfall conditions. As can be seen from Xisha/209, mean monthly temperatures vary between 23° and 29° C and only slightly indicate a thermal seasonality. In terms of temperature, year-round summer conditions prevail; as the most significant characteristic, however, the larger diurnal temperature range (around 10° C) has to be taken into consideration.

Rainfall conditions are characterized by a marked annual variation, dividing the year into a well-established, long rainy season from June to November (which receives almost >80% of the annual total), and a dry season from December to May. The annual total of rainfall reaches a rather high amount (>1,500 mm).

9.10 Alpine Plateau Zone (H 0)

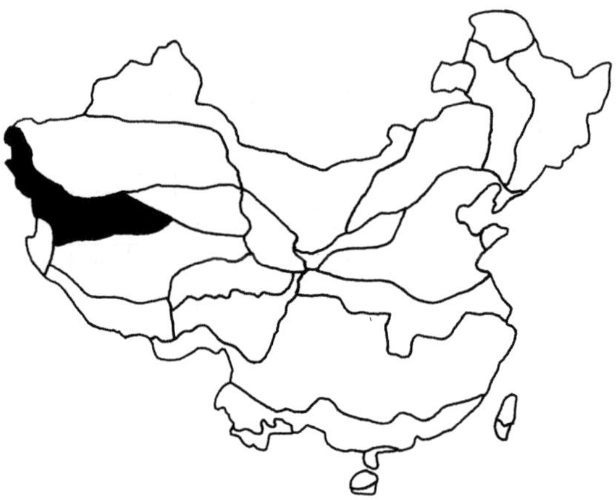

Fig. 9.12. Climate map of the Alpine Plateau Climate

This climatic zone (Fig. 9.12) occupies the western Kunlun Shan, up to about 90° E. It is one of the most spectacular and remote high mountains of China, being glaciered at higher elevation, i.e. above 4,500 m above sea level (Sect. 5.11). Climate data for this area were completely lacking so that not even a climate diagram could be constructed. The Kunlun Shan extends up to 7,719 m above sea level in its most western part which already represents the transitional part to the Russian Pamir. Most parts of the Kunlun Shan consist of mountains above 4,000 m. Therefore, temperatures can be expected at an extremely low level, both the mean monthly temperatures and the annual mean.

As a thermal characteristic, the daily minimum temperature year-round is below 0° C: for the high mountain areas even the monthly means of temperature remain below 0° C in all months. Only the lower and mid-mountains record positive monthly means in summer.

In terms of precipitation, the Alpine Plateau Zone (H 0) has mostly snow, but in all the total of precipitation remains very low so that the climate is called "arid" from the viewpoint of precipitation. The annual total may drop <100 mm, which mostly occurs in winter. The then occurring westerlies were often reported to have strong velocities.

Due to the extreme mountainous conditions of the Kunlun Shan, its climate is considered as one of the most hazardous in whole China.

9.11 Subalpine Plateau Zone (H I)

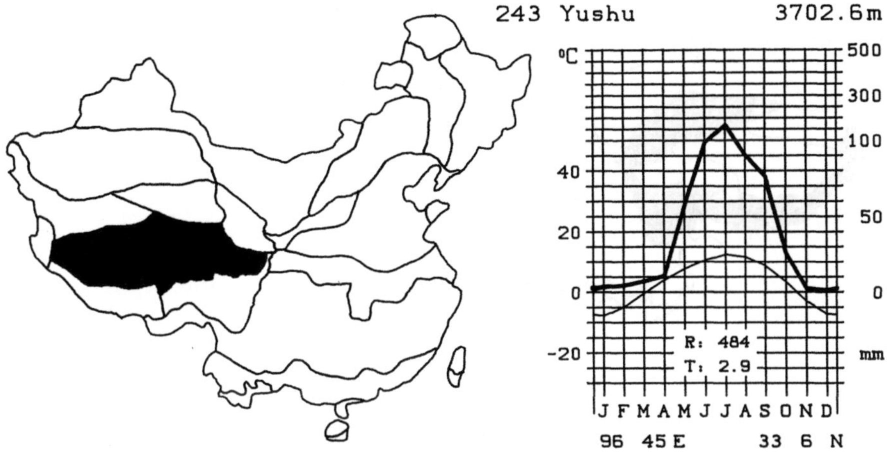

Fig. 9.13. Climate map and reference diagram of the Subalpine Plateau Zone. Reference station: Yushu/243

This climate zone (Fig. 9.13) is composed of most parts of the Qinghai-Xizang Plateau, being divided into a subhumid eastern part (representing the region of Aba-Nagqu) and the semi-arid western part, representing the southern Qinghai Plateau and the Qingtang Plateau. The division into two subtypes is due to different precipitation conditions. Since the elevation of the area extends from 3,500 m in the east to 4,800 m above sea level in the west, subsequently the temperatures will also decrease in the same direction. However, for the western semi-arid part, climate stations were lacking so that the climate description is insufficient.

For the eastern part, Yushu/243 is a representative station, the climate diagram of which shows a pronounced seasonal contrast between summer and winter both in terms of precipitation and temperature. With regard to precipitation, Yushu demonstrates a summer maximum during which >90% of the annual total (484 mm) are recorded. The annual temperature variation marks a fair summer (July: 12.5° C) compared to a more pronounced winter (January: −7.8° C). The number of snow days is, however, rather low, because of the rather dry conditions in winter.

Due to the sheltering effect of the surrounding mountains, calms may be recorded to an extremely large extent, e.g. Yushu/243 where, as an annual average, 60% of all cases are calms which is the highest recorded value for whole China (with a maximum of 86% in September and a minimum of 49% in March, both as monthly means).

9.12 Temperate Plateau Zone (H II)

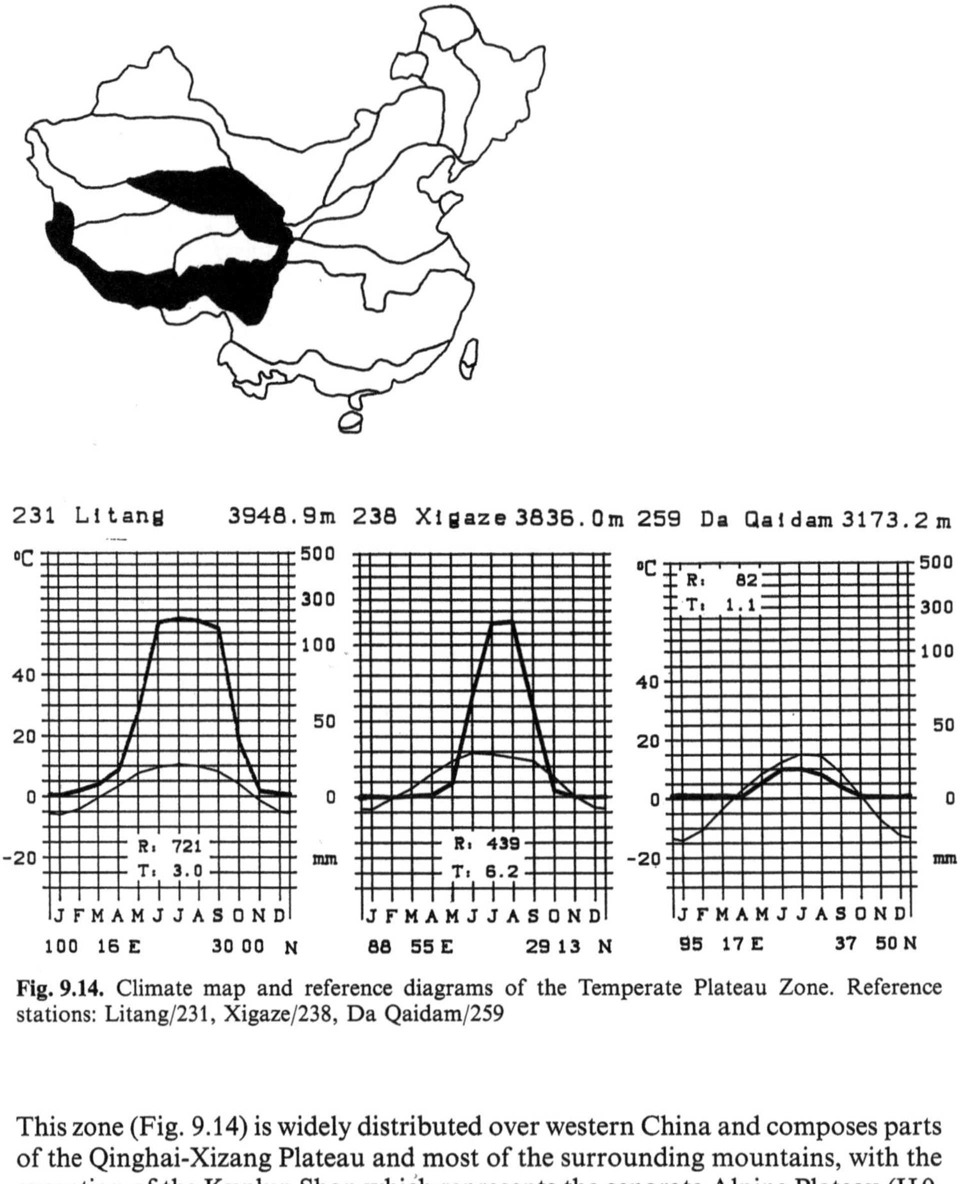

Fig. 9.14. Climate map and reference diagrams of the Temperate Plateau Zone. Reference stations: Litang/231, Xigaze/238, Da Qaidam/259

This zone (Fig. 9.14) is widely distributed over western China and composes parts of the Qinghai-Xizang Plateau and most of the surrounding mountains, with the exception of the Kunlun Shan which represents the separate Alpine Plateau (H 0, see Sect. 9.10). According to the complex nature of the landforms and mountains, in particular, and their location, the Temperate Plateau Zone (H II) is also divided into three types, according to different moisture conditions, namely a humid and subhumid type (in the high mountains and gorges of western Sichuan and eastern Xizang), the semi-arid type (in the plateaus and mountains of eastern Qinghai and southern Xizang) and the arid type which is found in the Qaidam Basin and Ngari Mountains. The three subtypes are each identified by a climate diagram, given for Litang/231, Xigaze/238 and Da Qaidam/259 (Fig. 9.14). At a first glance, the diagrams differ considerably from each other due to both the precipitation and, to a lesser extent, temperature distribution over the year.

Since the whole region of the Temperate Plateau Zone (H II) comes under an extremely high elevation (mostly between 3,000–4,000 m a.s.l.), all three climate subtypes are characterized by a very low atmospheric pressure. Lhasa/239 (3,658 m a.s.l.), which is the favourite tourist destination on the Tibetan Plateau in recent years, shows mean monthly values around 650 mb only. The lack of oxygen at this elevation is the most hazardous climatic feature, and newcomers to Tibet usually experience breathing difficulty and others.

The humid and subhumid types of the Temperate Plateau Zone (H II) show, despite its large elevation (mostly between 3,000–4,000 m above sea level), fair annual precipitation totals between 600–800 mm. The annual variation of precipitation is, however, very large, expressing a pronounced wet season from May to September, alternating with a distinct dry season from November through March. July is the wettest month, December and January are the driest months. During the wet summer season with a duration of 5 to 6 months, up to 90% of the annual total may be experienced. In winter, most precipitation occurs as snow. With increasing elevation above sea level, however, precipitation year-round may turn into snow. Accordingly, the number of snow days and the length of the snow-cover period may be extended from winter to year-round.

Due to the rapid change in landforms and elevation, the temperature conditions may differ to a great extent. In thermal respects, winter and summer are clearly established seasons. The annual variation of temperature, expressed by monthly means, obviously varies according to the elevation of a location. For stations between 3,000–4,000 m above sea level, summer temperatures are still fair (at the most rising to a mean monthly value of 10°–15° C), while winter temperatures at this elevation decrease, at the most, to a monthly (January) mean between −3° and −7° C. The daily range of temperature may reach rather high values, often indicating a negative daily minimum temperature for most months of the year. Subsequently, the number of frost days may reach an extremely high level, for example 147.5 frost days per year at Garze/223; the annual variation clearly marks a maximum during winter and a minimum in summer.

Due to the complex nature of landforms, winds are affected in their direction and velocity by mountains and valleys so that they may either be channelled or sheltered. In this respect Garze/233 shows an extremely large percentage of calms, with an annual average of 45% which is one of the highest values recorded for all of China. The monthly means of calms vary between 59% (in December) and 34% (in May).

The semi-arid type of the Temperate Plateau Zone (H II) is divided into two separate regions, i.e. the plateaux and mountains of eastern Qinghai and the mountains of southern Xizang. Xigaze/238, belonging to the region of the semi-arid type, expresses a clearly established seasonal variation of both precipitation and temperature. The annual total of precipitation is lower than in the humid and subhumid type, and the wet season is shorter, mainly concentrated in the summer months from June to September. Around 90–95% of the annual totals are recorded during these 4 months only, at Xigaze/238 even 96%. Hence, a short wet summer and a long, extremely dry winter alternate throughout the year. Precipitation in winter mostly occurs as snow which is, however, much less, due to the minimal amount of precipitation in winter.

Temperatures in the semi-arid Temperate Plateau Zone (H II) also show a pronounced seasonal contrast between a fair summer and a somewhat mild winter (Xigaze/238: June 14.7° C, January −4.1° C, both monthly means). For southern Xizang, however, to which Xigaze also belongs, higher summer temperatures have to be considered because of the effect of the Tibetan Plateau as an elevated heating surface. The diurnal range of temperature is well marked and daily minimum values drop below zero in most months of the year. Because of the relatively mild winter temperatures, the number of frost days is not as high as for the humid and subhumid temperate plateau types.

The arid Temperate Plateau Zone mostly occupies the Qaidam Basin and the Ngari Mountains on the eastern side of the Karakorum. Surrounded by the Kunlun Shan in the south and the Altun Shan (which both rise to nearly 6,000 m above sea level), the Qaidam Basin is mostly located between 2,700–3,100 m above sea level. Due to the surrounding high mountains and the extremely strong continentality, the Basin is one of the driest parts of China. The annual total of precipitation amounts to 83 mm at Da Qaidam/259, but it drops to <25 mm in the centre of the Basin (ZHANG Jiazhen and LIU En-bao 1985). Due to the extremely dry resp. arid conditions, the annual variation of precipitation is of a statistical value only; however, it clearly shows a maximum from May to August/September, while a period from October to April is practically completely dry. Due to the high elevation of the Basin, the low precipitation totals are to a certain extent represented by snow. The number of snow days, however, is very low (12.8 snow days at Da Qaidam/259).

In thermal respects, the Basin shows pronounced seasons. Despite the high altitude of the Basin, its summer temperatures are fair, while the winter temperatures may drop considerably (Da Qaidam/259: July 15.1° C, January − 14.3° C, both monthly means). Thus, the annual range of temperature varies to a considerably large extent.

Due to the strong continentality, the annual sunshine duration in the Basin is very high, showing at Da Qaidam/259 an annual mean of 8.9 sunshine hours per day which is among the highest values for China. It is also worth noting a large percentage of calms which is recorded as the annual mean at Da Qaidam (36%).

Appendix: Climate Tables

Climate tables are given for 68 stations, containing monthly and annual means for 28 variables, based on the 30-year observation period (=oby) from 1951–1980 or a part of it, depending upon the records available. The climate variables (abbreviations in parentheses) are as follows:

1. Atmospheric pressure (A. press.)
2. Mean temperature (T. mean)
3. Highest recorded maximum temperature (T.a.mx.)
4. Lowest recorded minimum temperature (T.a.mn.)
5. Mean maximum temperature (T.m.mx.)
6. Mean minimum temperature (T.m.mn.)
7. Number of days <0° C (Tmn <0)
8. Number of days <–10° C (Tmn <–10)
9. First and last date showing a maximum temperature ≥0° C (pTmn >0)
10. First and last date showing a maximum temperature ≥10° C (pTmn >10)
11. Number of frost days (Frost d)
12. Relative humidity (r. hum.)
13. Cloudiness (cloud.)
14. Precipitation amount (R. total)
15. Number of days >0.1 mm (RD >0.1)
16. Number of days >5 mm (RD >5)
17. Number of days >10 mm (RD >10)
18. Number of days >25 mm (RD >25)
19. Evaporation (Evap.)
20. Snow depth (Snow h.)
21. Number of snowfall days (Snow d.)
22. Length of the snow cover period (Snow co.)
23. Sunshine duration (total) (SD hrs.)
24. Sunshine duration (percentage of the maximum duration) (SD pc.)
25. Wind velocity (w. vel.)
26. Prevailing wind direction (W. mx, dir)
27. Percentage of prevailing wind direction (W. mx, pc)
28. Percentage of calms (calms)

The stations for which climate tables are given, are as follows:

Huma/2
Nenjiang/4
Hailar/6
Arxan/7
Qiqihar/8
Harbin/15
Jixi/17
Mudanjiang/19
Dong Ujmqin Qi/21
Changchun/27
Yanji/31
Shenyang/38
Chengde/51
Bailingmiao/56
Datong/59
Beijing/61
Tianjin/62
Yulin/68
Taiyuan/69
Huimin/72
Jinan/77
Yan'an/81
Xuzhou/87
Zhengzhou/90
Xi'an/95
Wudu/101
Hefei/114
Nanjing/116
Shanghai/119
Hangzhou/123
Wuhan/126
Yichang/129
Chengdu/134
Xichang/138
Chongqing/142
Changsha/147
Wenzhou/154
Guiyang/173
Guilin/177
Fuzhou/183
Guangzhou/199
Dongfang/210
Nanning/215
Lancang/222
Kunming/225
Deqen/230
Garze/233
Xigaze/238
Lhasa/239
Nagqu/240
Yushu/243
Lanzhou/247
Yinchuan/250
Bayan Mod/252
Xining/256
Golmud/258
Da Qaidam/259
Lenghu/260
Jiuquan/261
Ruoqiang/264
Hami/265
Turpan/267
Ürümqi/269
Altay/270
Tinghe/273
Hotan/276
Taibei/285
Hongkong

2 Huma	51° 43' N 126° 39' E 177.4 m													2
	JAN	FEB	MAR	APR	MAY	JUN	JUL	AUG	SEP	OCT	NOV	DEC	YEAR	oby
A.press.	999	999	995	989	985	984	983	986	990	995	997	999	992	27
T. mean	-27.8	-22.5	-11.5	1.2	10.6	17.4	20.2	17.6	10.2	0.1	-14.6	-25.3	-2.0	27
T.a.mx.	-6.6	2.5	11.6	25.7	34.6	37.4	38.0	36.2	30.4	24.4	10.2	1.9	38.0	27
T.a.mn.	-48.2	-45.2	-40.7	-21.7	-8.8	-1.1	4.6	1.6	-8.7	-22.2	-38.5	-46.3	-48.2	27
T.m.mx.	-20.8	-13.9	-3.6	8.0	17.9	24.5	26.5	24.0	17.3	6.9	-8.2	-19.2	4.9	27
T.m.mn.	-33.5	-29.6	-19.1	-5.2	3.0	10.2	14.3	12.0	4.4	-5.7	-20.2	-30.5	-8.3	27
Tmn < 0														
Tmn < -10														
pTmn > 0														
pTmn > 10														
Frost d	26.1	22.3	21.1	15.1	6.0	0.2	0.0	0.0	7.6	20.0	24.3	26.2	169.0	27
r.hum.	71	68	62	54	52	67	76	79	73	63	70	74	67	27
cloud.	3.4	3.2	4.3	5.5	6.4	6.5	6.6	6.2	5.8	4.8	4.1	4.2	5.1	27
R.total	4	4	8	23	36	70	110	100	63	23	12	8	461	27
RD > .1	6.6	4.6	5.6	7.4	10.6	13.9	14.4	14.0	12.8	7.0	7.7	9.7	114.4	27
RD > 5														
RD > 10														
RD > 25														
Evap.														
Snow h.	35	42	42	39	18	0	0	0	9	26	32	32	42	27
Snow d.	6.4	4.6	5.5	4.8	1.1	0.0	0.0	0.0	0.4	4.6	7.7	9.7	44.9	27
Snow co														
SD hrs.	5.1	7.0	7.8	7.8	8.6	9.2	8.5	8.0	6.8	6.4	5.3	3.9	7.0	27
SD pc.	61	70	66	56	55	56	53	55	53	60	60	50	57	27
W.vel.	1.6	2.0	2.9	3.7	3.6	2.6	2.2	2.3	2.6	2.9	2.2	1.7	2.5	27
W.mx,dir	N	N	N	N	N	N	N	N	N	N	N	N	N	27
W.mx,pc	15	16	19	15	15	12	12	13	14	12	12	13	14	27
Calms	53	44	30	17	16	22	28	28	28	27	38	50	32	27

4 Nenjiang 49° 10' N 125° 13' E 222.3 m **4**

	JAN	FEB	MAR	APR	MAY	JUN	JUL	AUG	SEP	OCT	NOV	DEC	YEAR	oby
A.press.	992	991	987	988	977	975	976	979	982	988	990	991	984	30
T. mean	-25.5	-21.0	-9.5	2.8	11.7	18.1	20.6	18.4	11.3	1.7	-11.4	-22.1	-0.4	30
T.a.mx.	0.4	4.5	15.8	28.6	34.2	35.5	37.4	35.6	29.1	25.4	12.6	4.4	37.4	30
T.a.mn.	-47.3	-44.7	-38.2	-17.9	-8.3	-0.4	4.3	1.1	-9.2	-23.0	-33.6	-41.7	-47.3	30
T.m.mx.	-17.7	-12.1	-1.9	10.1	19.2	25.0	26.5	24.7	18.5	9.1	-4.5	-15.2	6.8	30
T.m.mn.	-32.1	-28.8	-17.3	-4.4	3.4	10.9	14.8	12.5	5.1	-4.6	-17.5	-28.1	-7.2	30
Tmn < 0														
Tmn < -10														
pTmn > 0														
pTmn > 10														
Frost d	28.2	25.0	24.1	15.4	7.0	0.2	0.0	0.0	7.3	21.2	26.7	28.7	183.8	30
r.hum.	75	72	63	52	52	67	78	79	74	63	70	75	68	30
cloud.	2.9	3.2	4.0	5.6	6.3	6.4	6.7	6.1	5.6	4.7	3.8	3.4	4.9	30
R.total	4	4	6	18	36	72	130	112	68	18	5	6	479	30
RD > .1	5.4	5.4	4.9	6.4	9.1	13.3	15.2	13.6	12.1	6.2	4.9	7.2	103.7	30
RD > 5														
RD > 10														
RD > 25														
Evap.														
Snow h.	20	21	31	18	8	0	0	0	4	16	19	17	31	30
Snow d.	5.3	5.5	4.6	3.8	0.4	0.0	0.0	0.0	0.3	3.1	4.7	7.2	34.8	30
Snow co														
SD hrs.	5.6	7.0	8.2	7.9	8.6	9.4	8.2	8.2	7.3	6.7	6.0	4.9	7.3	30
SD pc.	64	69	69	58	56	59	53	57	58	63	66	59	60	30
W.vel.	2.2	3.1	4.1	5.5	6.4	4.5	4.0	3.4	4.0	4.5	8.1	2.6	4.0	8
W.mx,dir	SSW	SSW	N	N	S	N	S	N	N	SSW	SW	SSW	S	30
W.mx,pc	8	9	9	10	10	9	8	9	9	11	11	10	8	30
Calms	41	33	19			14	17	19	19	19	23	35	21	30

6 Hailar 49° 13' N 119° 45' E 612.9 m														6
	JAN	FEB	MAR	APR	MAY	JUN	JUL	AUG	SEP	OCT	NOV	DEC	YEAR	oby
A.press.	948	947	944	939	936	935	934	938	941	945	946	947	942	30
T. mean	-26.8	-23.7	-12.7	0.9	10.1	17.1	19.6	17.1	9.8	0.1	-12.8	-23.2	-2.1	30
T.a.mx.	-3.2	2.1	13.2	29.4	32.2	36.5	36.7	34.0	28.8	26.9	11.8	0.2	36.7	30
T.a.mn.	-48.5	-43.6	-38.4	-19.7	-11.2	-0.4	4.1	1.0	-8.2	-24.6	-37.7	-43.0	-48.5	30
T.m.mx.	-20.3	-16.1	-5.6	7.5	17.3	24.0	25.6	23.4	16.8	7.6	-6.2	-17.1	4.8	30
T.m.mn.	-32.4	-30.0	-19.5	-5.5	2.5	9.6	13.5	11.0	3.6	-5.8	-18.4	-28.5	-8.3	30
Tmn < 0														
Tmn < -10														
pTmn > 0														
pTmn > 10														
Frost d	29.8	26.9	26.2	16.4	6.4	0.2	0.0	0.0	7.3	21.0	26.0	29.7	190.1	30
r.hum.	78	77	71	56	48	60	71	74	69	63	72	78	68	30
cloud.	3.2	3.5	4.1	5.5	6.1	6.3	6.6	5.9	5.4	4.5	4.0	3.8	4.8	30
R.total	3	3	5	13	28	55	95	82	39	12	4	4	343	30
RD > .1	7.3	6.0	5.7	7.0	7.8	11.7	15.3	13.5	10.6	5.5	5.5	9.0	104.9	30
RD > 5														
RD > 10														
RD > 25														
Evap.														
Snow h.	24	39	20	18	15	0	0	0	5	20	17	18	39	30
Snow d.	7.3	6.0	5.5	5.0	1.7	0.0	0.0	0.0	0.6	3.5	5.3	8.4	43.3	28
Snow co														
SD hrs.	5.9	7.3	8.3	8.4	9.5	10.1	9.1	8.7	10.7	7.1	6.1	5.1	10.5	30
SD pc.	68	71	70	61	63	63	58	61	59	66	66	62	63	30
W.vel.	2.4	2.7	3.6	4.7	4.6	3.4	3.1	3.0	3.3	3.6	3.2	2.6	3.3	30
W.mx,dir	S	SSW	S	NNW	NW	E	E	E	NW	W	S	S	S	30
W.mx,pc	16	10	13	12	10	9	11	9	9	10	13	15	10	30
Calms	25	20				11	12	15	14	13	16	22	15	30

7 Arxan 47° 10' N 119° 57' E 1026.5 m														7
	JAN	FEB	MAR	APR	MAY	JUN	JUL	AUG	SEP	OCT	NOV	DEC	YEAR	oby
A.press.	898	898	896	893	891	891	890	894	896	899	899	898	895	27
T. mean	-25.6	-23.0	-13.6	-1.0	7.8	13.8	16.6	14.4	7.5	-1.1	-12.7	-21.8	-3.2	29
T.a.mx.	2.6	3.2	15.3	24.8	31.7	32.9	34.1	32.0	28.8	24.6	12.5	4.0	34.1	29
T.a.mn.	-45.7	-43.4	-40.8	-26.2	-14.3	-7.4	-0.6	-3.4	-13.8	-25.8	-38.4	-44.5	-45.7	29
T.m.mx.	-18.5	-15.0	-6.0	5.6	15.4	21.0	22.9	21.2	15.5	6.8	-5.6	-15.4	4.0	28
T.m.mn.	-32.0	-29.8	-21.1	-8.2	-0.9	5.6	10.1	7.9	0.5	-7.9	-19.2	-28.1	-10.3	28
Tmn < 0														
Tmn < -10														
pTmn > 0														
pTmn > 10														
Frost d	25.8	22.6	24.0	16.7	10.8	2.5	0.0	1.4	12.6	21.9	23.9	24.7	186.9	25
r.hum.	79	76	71	57	53	68	79	80	72	65	73	78	71	28
cloud.	4.4	4.2	4.5	5.4	6.0	6.5	6.7	6.0	5.2	4.5	4.7	5.0	5.2	28
R.total	7	5	10	21	40	80	116	85	47	20	9	8	448	28
RD > .1	16.1	11.8	11.0	9.6	10.5	15.1	17.8	15.2	12.5	9.1	11.8	17.3	157.8	28
RD > 5														
RD > 10														
RD > 25														
Evap.														
Snow h.	34	36	45	32	25	0	0	0	19	16	20	35	45	29
Snow d.	15.8	11.3	11.1	8.2	4.7	0.1	0.0	0.0	1.8	6.9	11.6	16.4	87.9	27
Snow co														
SD hrs.	5.5	6.8	8.0	8.2	8.6	8.4	7.6	7.8	7.0	6.5	6.2	4.7	7.1	24
SD pc.	62	68	68	60	57	53	49	55	56	60	59	56	58	24
W.vel.	1.5	2.0	2.8	4.0	3.9	2.6	2.3	2.3	2.9	3.1	2.7	1.9	2.7	28
W.mx,dir	NW	NW	NW	NW	NW	SE	SE	SE	NW	NW	NW	NW	SE	24
W.mx,pc	7	10	14	17	15	12	15	13	13	14	13	11	11	24
Calms	51	42	37	25	22	32	33	36	31	31	33	44	35	24

8 Qiqihar 47° 23' N 123° 55' E 145.9 m														8
	JAN	FEB	MAR	APR	MAY	JUN	JUL	AUG	SEP	OCT	NOV	DEC	YEAR	oby
A.press.	1005	1005	1000	993	989	987	986	990	994	1000	1003	1004	996	30
T. mean	-19.5	-15.4	-5.3	5.4	14.1	20.2	22.8	20.9	14.0	4.7	-7.2	-16.5	3.2	30
T.a.mx.	2.4	12.8	23.0	30.3	35.9	40.1	39.9	37.5	31.2	26.9	14.5	6.9	40.1	30
T.a.mn.	-39.5	-34.5	-29.4	-14.0	-7.4	2.8	10.4	7.2	-3.5	-16.0	-27.9	-34.8	-39.5	30
T.m.mx.	-12.9	-8.2	1.7	12.3	21.0	26.2	28.0	26.4	20.2	11.1	-1.2	-10.5	9.5	30
T.m.mn.	-25.0	-21.6	-12.0	-1.7	6.8	14.0	17.8	15.9	8.5	-0.9	-12.2	-21.5	-2.7	30
Tmn < 0														
Tmn < -10														
pTmn > 0														
pTmn > 10														
Frost d	29.0	23.4	16.1	7.7	2.0	0.0	0.0	0.0	2.9	16.5	24.8	28.2	150.5	30
r.hum.	71	65	52	46	48	63	73	73	68	60	64	69	63	30
cloud.	2.9	3.2	4.1	5.5	6.0	6.3	6.5	5.8	4.9	4.3	3.4	3.1	4.7	30
R.total	2	2	6	13	31	64	127	99	49	16	4	3	416	30
RD > .1	3.6	3.0	3.0	5.2	7.7	11.2	14.3	11.1	10.4	4.7	3.0	3.4	80.6	30
RD > 5														
RD > 10														
RD > 25														
Evap.														
Snow h.	11	13	17	17	11	0	0	0	0	4	24	11	24	30
Snow d.	3.5	3.1	2.7	2.5	0.3	0.0	0.0	0.0	0.0	1.5	2.9	3.4	19.8	30
Snow co														
SD hrs.	6.2	7.4	8.3	8.3	9.1	9.7	8.8	8.8	8.1	7.3	6.5	5.7	7.9	30
SD pc.	70	72	70	61	61	61	57	62	65	67	70	67	64	30
W.vel.	2.6	3.1	3.9	4.5	4.6	3.5	3.0	3.0	3.4	3.7	3.3	2.6	3.5	17
W.mx,dir	NW	NW	NW	NW	S	N	S	N	N	S	NW	NW	NW	30
W.mx,pc	17	16	17	14	10	11	[illegible]	12	10	12	13	15	11	30
Calms														

15 Harbin 45° 41' N 126° 37' E 171.7 m **15**

	JAN	FEB	MAR	APR	MAY	JUN	JUL	AUG	SEP	OCT	NOV	DEC	YEAR	oby
A.press.	1002	1001	997	991	987	985	984	987	992	998	1001	1001	994	30
T. mean	-19.4	-15.4	-4.8	6.0	14.3	20.0	22.8	21.1	14.4	5.6	-5.7	-15.6	3.6	30
T.a.mx.	4.2	11.0	20.7	27.9	35.6	36.4	36.4	35.8	31.0	26.6	17.2	6.2	36.4	30
T.a.mn.	-38.1	-33.0	-29.0	-12.8	-3.8	4.9	10.4	6.6	-2.2	-15.6	-26.1	-35.7	-38.1	30
T.m.mx.	-13.2	-8.6	1.5	12.7	21.1	25.9	28.0	26.4	20.7	11.8	-0.1	-9.9	9.7	30
T.m.mn.	-24.8	-21.5	-10.9	-0.3	7.4	14.1	18.1	16.2	8.9	0.3	-10.5	-20.5	-2.2	30
Tmn < 0													182.6	30
Tmn < -10													120.7	30
pTmn > 0				2.						29.			211.0	30
pTmn > 10					7.				28.				145.0	30
Frost d	28.8	24.8	21.0	10.5	1.4	0.0	0.0	0.0	2.3	15.5	24.7	27.8	156.8	30
r.hum.	74	70	58	51	51	66	77	78	71	65	67	73	67	30
cloud.	3.1	3.5	4.1	5.4	6.2	6.5	6.7	5.9	4.8	4.4	3.8	3.4	4.8	30
R.total	4	5	11	24	38	78	161	97	66	28	7	6	525	30
RD > .1	6.2	5.6	5.7	6.9	10.4	12.9	15.6	12.6	11.2	7.6	5.5	6.3	106.5	30
RD > 5													29.6	30
RD > 10													15.4	30
RD > 25													3.9	30
Evap.													1508	30
Snow h.	14	17	18	14	0	0	0	0	0	21	19	41	41	30
Snow d.	6.2	5.8	5.2	2.6	0.2	0.0	0.0	0.0	0.1	2.3	4.5	6.2	33.1	30
Snow co				19.						15.			187.6	30
SD hrs.	5.8	7.0	6.6	7.8	8.5	8.7	8.1	7.9	7.6	6.6	5.8	5.1	7.2	30
SD pc.	64	67	66	58	57	56	53	56	61	60	62	58	60	30
W.vel.	3.6	3.8	4.5	5.4	5.0	3.9	3.4	3.2	3.7	4.2	4.4	3.9	4.1	30
W.mx,dir	S	SSW	W	S	SSW	S	S	S	S	S	S	SSW	S	29
W.mx,pc	14	12	11	11	12	12	14	12	15	13	13	15	12	29
Calms														

17 Jixi 45° 17' N 130° 57' E 233.1 m														17
	JAN	FEB	MAR	APR	MAY	JUN	JUL	AUG	SEP	OCT	NOV	DEC	YEAR	oby
A.press.	992	992	988	984	980	979	978	981	985	990	992	992	986	30
T. mean	-17.3	-13.8	-4.6	5.5	13.3	18.2	21.7	20.4	13.8	5.5	-5.4	-14.4	3.6	30
T.a.mx.	4.2	11.0	20.0	29.4	34.6	35.7	37.1	35.0	32.9	27.8	18.1	9.6	37.1	30
T.a.mn.	-35.1	-30.5	-26.1	-13.0	-4.9	2.6	5.5	5.5	-3.9	-12.9	-28.6	-33.3	-35.1	30
T.m.mx.	-11.4	-7.4	1.5	12.2	20.2	24.3	27.3	26.0	20.5	12.2	0.5	-8.8	9.8	30
T.m.mn.	-22.4	-19.6	-10.5	-1.1	6.4	12.4	16.5	15.4	7.9	-0.4	-10.6	-19.3	-2.1	30
Tmn < 0														
Tmn < -10														
pTmn > 0														
pTmn > 10														
Frost d	26.1	21.2	19.9	11.0	2.0	0.0	0.0	0.0	2.3	16.0	22.6	25.7	146.8	30
r.hum.	67	62	57	52	55	71	77	79	72	63	62	66	65	30
cloud.	3.8	4.1	4.9	6.0	6.7	7.0	7.3	6.6	5.2	4.6	4.0	3.9	5.3	30
R.total	5	7	8	23	52	80	114	118	66	38	14	8	533	30
RD > .1	6.0	5.6	6.3	8.7	13.1	16.7	14.8	14.1	11.9	8.8	6.3	6.4	118.7	30
RD > 5														
RD > 10														
RD > 25														
Evap.														
Snow h.	60	60	23	28	1	0	0	0	0	28	40	29	60	30
Snow d.	6.1	5.5	6.4	3.7	0.5	0.0	0.0	0.0	0.0	2.4	5.2	6.4	35.8	30
Snow co														
SD hrs.	6.3	7.6	8.3	8.2	8.6	8.5	8.2	7.9	7.9	7.0	6.3	5.6	7.5	30
SD pc.	69	72	70	61	58	55	54	56	63	64	67	65	62	30
W.vel.	3.7	3.8	4.0	4.1	3.6	2.5	2.2	2.1	2.7	3.1	3.6	3.7	3.3	30
W.mx,dir	W	W	W	W	W	W	W	W	W	W	W	W	W	30
W.mx,pc	35	31	25	20	15	11	9	12	17	21	28	34	21	30
Calms						19	22	23	20					30

19 Mudanjiang 44° 34' N 129° 36' E 241.4 m **19**

	JAN	FEB	MAR	APR	MAY	JUN	JUL	AUG	SEP	OCT	NOV	DEC	YEAR	oby
A.press.	992	992	988	984	980	978	978	980	985	990	992	992	986	30
T. mean	-18.5	-14.5	-4.5	5.8	13.7	18.4	22.0	20.6	13.8	5.4	-5.5	-15.1	3.5	30
T.a.mx.	4.6	11.8	19.7	28.7	34.5	36.3	36.5	35.6	31.9	29.1	19.1	9.4	36.5	30
T.a.mn.	-38.3	-33.5	-30.7	-16.4	-5.8	2.9	8.0	6.2	-5.1	-14.2	-27.3	-35.6	-38.3	30
T.m.mx.	-11.3	-7.0	2.2	12.9	21.0	24.8	27.8	26.4	20.9	12.7	1.0	-8.6	10.2	30
T.m.mn.	-24.6	-21.3	-10.9	-0.8	6.6	12.8	17.0	15.8	7.8	-0.6	-10.9	-20.6	-2.5	30
Tmn < 0														
Tmn < -10														
pTmn > 0														
pTmn > 10														
Frost d	28.0	24.3	22.6	12.6	2.5	0.0	0.0	0.0	3.6	18.1	24.7	28.2	164.5	30
r.hum.	71	66	59	53	55	70	76	79	74	67	67	71	67	30
cloud.	3.8	4.1	4.9	6.0	6.6	7.1	7.3	6.8	5.5	4.7	4.2	4.0	5.4	30
R.total	4	6	10	26	50	79	108	128	65	35	14	8	533	30
RD > .1	5.0	5.1	5.5	8.4	12.1	15.6	15.8	14.1	12.6	9.0	6.4	5.7	115.3	30
RD > 5														
RD > 10														
RD > 25														
Evap.														
Snow h.	22	25	20	17	3	0	0	0	0	39	34	24	39	30
Snow d.	5.2	5.1	4.9	3.1	0.5	0.0	0.0	0.0	0.1	1.9	5.1	5.6	31.5	30
Snow co														
SD hrs.	5.8	7.2	8.0	7.8	8.1	7.8	7.6	7.1	7.2	6.5	5.9	5.2	4.3	30
SD pc.	63	68	67	58	55	51	51	51	58	60	62	59	58	30
W.vel.	2.2	2.5	3.1	3.4	3.2	2.3	2.0	1.9	2.1	2.3	2.6	2.2	2.5	30
W.mx,dir	SW	SW	W	SW	SW	SW	SW	SW	SW	SW	SW	SW	SW	30
W.mx,pc	15	14	15	16	17	16	17	14	15	14	15	15	15	30
Calms	31	26	18			21	23	26	28	28	27	30	24	30

21 Dong Ujmqin Qi 45° 31' N 116° 58' E 839.1 m														21
	JAN	FEB	MAR	APR	MAY	JUN	JUL	AUG	SEP	OCT	NOV	DEC	YEAR	oby
A.press.	923	922	919	915	913	911	909	913	917	922	922	922	917	15
T. mean	-21.8	-19.1	-9.2	3.3	11.9	17.8	20.3	18.4	10.8	2.3	-10.2	-18.7	0.5	15
T.a.mx.	0.5	9.7	23.5	28.5	34.3	37.7	39.0	38.9	31.0	26.0	13.2	4.9	39.0	15
T.a.mn.	-40.5	-39.5	-36.8	-23.8	-10.2	-0.2	5.0	2.2	-8.8	-17.8	-33.9	-37.7	-40.5	15
T.m.mx.	-14.2	-10.3	-0.6	11.2	19.7	24.9	27.2	25.3	19.0	10.4	-2.1	-11.7	8.2	25
T.m.mn.	-27.3	-25.7	-16.2	-4.5	3.3	10.4	14.4	12.1	4.1	-4.7	-15.6	-24.2	-6.2	25
Tmn < 0														
Tmn < -10														
pTmn > 0														
pTmn > 10														
Frost d	28.5	25.0	19.0	5.8	1.6	0.0	0.0	0.0	4.8	12.9	22.7	28.1	148.6	25
r.hum.	72	69	58	41	39	53	65	68	60	52	62	72	59	15
cloud.	3.0	3.1	3.7	5.0	5.4	5.9	6.0	5.3	4.3	3.5	3.4	3.0	4.3	25
R.total	2	2	4	10	19	50	79	65	26	7	3	2	269	15
RD > .1	4.9	3.2	4.1	4.3	6.3	12.3	14.1	12.8	7.9	3.1	4.0	3.5	80.4	15
RD > 5														
RD > 10														
RD > 25														
Evap.														
Snow h.	11	18	17	5	0	0	0	0	10	6	5	4	18	15
Snow d.	3.8	3.4	3.7	2.5	0.5	0.0	0.0	0.0	0.3	2.2	3.8	3.7	24.0	25
Snow co														
SD hrs.	6.6	7.9	8.8	9.2	9.8	9.6	8.8	8.4	8.8	7.9	6.9	6.2	8.2	15
SD pc.	72	76	75	67	65	61	58	60	70	73	73	71	68	15
W.vel.	3.5	3.2	3.9	5.0	4.9	3.7	3.1	2.8	3.4	3.6	3.6	3.3	3.7	15
W.mx,dir	SW	N	N	NW	NW	N	N	N	NW	SW	SW	SW	SW	15
W.mx,pc	17	13	14	13	13	10	[illegible]	9	12	14	18	18	12	15
Calms	26	29	22	16	15	19	23	25	24	23	26	29	23	15

27 C h a n g c h u n 43° 54' N 125° 13' E 236.8 m														**27**
	JAN	FEB	MAR	APR	MAY	JUN	JUL	AUG	SEP	OCT	NOV	DEC	YEAR	oby
A.press.	994	994	990	984	980	977	976	980	985	991	994	994	987	30
T. mean	-16.4	-12.7	-3.5	6.7	15.0	20.1	23.0	21.3	15.0	6.8	-3.8	-12.8	4.9	30
T.a.mx.	5.6	12.1	19.5	27.7	35.2	36.4	38.0	35.6	30.1	27.0	20.7	9.3	38.0	30
T.a.mn.	-36.5	-31.9	-28.0	-15.9	-3.1	4.8	10.9	6.3	-2.4	-13.4	-24.8	-33.2	-36.5	30
T.m.mx.	-10.4	-6.3	2.7	13.5	21.7	26.0	27.9	26.4	21.3	13.0	1.8	-7.2	10.9	30
T.m.mn.	-21.6	-18.5	-9.2	0.4	8.3	14.6	18.6	16.8	9.4	1.5	-8.8	-17.7	-0.5	30
Tmn < 0													174.8	30
Tmn < -10													111.1	30
pTmn > 0			29.								2.		219.5	30
pTmn > 10					3.					1.				
Frost d	27.2	21.3	17.8	8.2	0.9	0.0	0.0	0.0	1.8	14.8	22.7	26.7	141.4	30
r.hum.	68	63	55	51	51	67	78	80	71	64	66	68	65	30
cloud.	3.2	3.5	4.3	5.4	6.0	6.6	7.0	6.1	4.7	4.1	3.8	3.3	4.8	30
R.total	4	5	9	22	42	91	184	128	61	34	12	4	596	30
RD > .1	5.4	4.4	5.4	7.3	9.8	14.2	16.2	12.9	9.4	7.6	5.5	4.8	102.9	30
RD > 5													30.6	30
RD > 10													17.7	30
RD > 25													5.1	30
Evap.													1719	30
Snow h.	13	15	14	10	0	0	0	0	0	22	18	18	22	30
Snow d.	5.4	4.4	4.5	2.0	0.1	0.0	0.0	0.0	0.0	1.7	4.3	4.6	27.1	30
Snow co				23.						14.			191.5	30
SD hrs.	6.3	7.2	7.9	8.8	8.1	7.6	7.6	7.6	6.9	5.6	5.6	3.8	7.2	30
SD pc.	68	68	66	60	58	55	49	55	64	61	61	61	60	30
W.vel.	4.1	4.3	5.0	5.8	5.6	4.1	3.5	3.0	3.5	4.2	4.5	4.3	4.3	30
W.mx,dir	SW	SW	SW	SW	SW	SW	SSW	SW	SW	SW	SW	SW	SW	30
W.mx,pc	21	18	15	17	18	16	16	13	14	16	18	21	17	30
Calms														

31 Y a n j i 42° 53' N 129° 28' E 176.8 m **31**

	JAN	FEB	MAR	APR	MAY	JUN	JUL	AUG	SEP	OCT	NOV	DEC	YEAR	oby
A.press.	1001	1000	996	992	988	986	985	988	993	998	1001	1001	994	26
T. mean	-14.4	-10.8	-2.4	6.6	13.8	17.7	21.3	21.1	14.7	6.7	-3.2	-11.7	5.0	28
T.a.mx.	8.6	15.1	21.7	31.2	35.5	37.2	37.6	36.2	31.6	30.2	21.1	10.1	37.6	28
T.a.mn.	-31.7	-32.7	-25.7	-11.7	-5.2	3.9	9.1	6.5	-4.0	-13.6	-27.5	-32.2	-32.7	28
T.m.mx.	-7.4	-3.4	4.7	14.6	21.7	24.3	26.9	26.8	22.1	14.8	3.7	-5.2	12.0	28
T.m.mn.	-20.1	-17.4	-8.9	-0.7	6.1	12.5	17.1	16.7	8.9	0.1	-8.8	-17.1	-0.9	28
Tmn < 0														
Tmn < -10														
pTmn > 0														
pTmn > 10														
Frost d	23.1	19.0	19.0	10.6	2.5	0.0	0.0	0.0	3.0	17.4	22.3	23.9	140.9	28
r.hum.	60	57	53	54	58	74	80	80	76	67	64	63	66	28
cloud.	3.4	3.9	4.6	5.8	6.4	7.3	7.6	7.0	5.4	4.4	4.2	3.7	5.3	28
R.total	4	5	8	25	50	85	96	129	55	26	15	6	504	28
RD > .1	3.4	4.3	5.0	7.8	12.0	15.4	15.7	13.6	11.2	7.6	5.9	3.8	105.5	28
RD > 5														
RD > 10														
RD > 25														
Evap.														
Snow h.	27	24	25	13	10	0	0	0	0	11	58	27	58	28
Snow d.	3.3	4.3	4.2	2.0	0.1	0.0	0.0	0.0	0.0	1.0	3.9	3.7	22.6	28
Snow co														
SD hrs.	5.7	6.8	7.6	7.4	7.6	6.7	5.9	5.9	6.7	6.5	5.6	5.0	6.4	27
SD pc.	60	64	64	55	52	44	39	43	54	59	58	55	53	27
W.vel.	3.0	3.0	3.5	3.5	3.2	2.6	2.5	1.9	1.8	2.1	2.5	2.8	2.7	28
W.mx,dir	W	W	W	W	ENE	ENE	ENE	ENE	W	W	W	W	W	28
W.mx,pc	21	21	19	16	13	22	26	15	11	15	17	20	14	28
Calms	38	34	29	26	25	26		35	42	43	43	42	34	28

38 S h e n y a n g 41° 46' N 123° 26' E 41.6 m **38**

	JAN	FEB	MAR	APR	MAY	JUN	JUL	AUG	SEP	OCT	NOV	DEC	YEAR	oby
A.press.	1021	1020	1016	1010	1004	1001	999	1003	1009	1015	1019	1021	1011	30
T. mean	-12.0	-8.4	0.1	9.3	16.9	21.5	24.6	23.5	17.2	9.4	0.0	-8.5	7.8	30
T.a.mx.	8.6	14.9	19.8	29.3	34.3	36.2	38.3	35.7	31.3	29.2	21.7	13.4	38.3	30
T.a.mn.	-30.6	-30.1	-25.0	-12.5	0.2	3.6	13.9	8.0	1.0	-8.3	-22.9	-30.5	-30.6	30
T.m.mx.	-5.6	-2.1	5.8	15.7	23.3	27.0	29.2	28.3	23.6	15.9	5.6	-2.6	13.7	30
T.m.mn.	-17.3	-13.8	-5.0	3.2	10.6	16.3	20.5	19.2	11.8	4.0	-4.6	-13.3	2.6	30
Tmn < 0													153.0	30
Tmn < -10													81.8	30
pTmn > 0			19.								12.		299.3	30
pTmn > 10				23.						9.			170.0	30
Frost d	27.0	20.9	20.2	9.6	1.0	0.0	0.0	0.0	0.8	12.8	21.3	26.4	140.0	30
r.hum.	64	58	54	53	55	68	78	79	72	68	65	64	65	30
cloud.	2.7	3.1	4.0	5.0	5.5	6.4	6.9	6.1	4.4	3.9	3.7	2.8	4.6	30
R.total	7	8	13	40	56	89	196	169	82	45	20	11	736	30
RD > .1	4.3	4.0	5.0	6.9	9.4	12.0	14.7	11.8	8.7	7.1	5.2	3.7	92.8	30
RD > 5													34.5	30
RD > 10													20.5	30
RD > 25													7.4	30
Evap.													1431	30
Snow h.	18	20	14	20	0	0	0	0	0	4	19	20	20	30
Snow d.	4.3	3.9	3.5	1.5	0.0	0.0	0.0	0.0	0.0	0.6	3.2	3.5	20.5	30
Snow co				14.						31.			165.6	30
SD hrs.	4.5	6.6	7.5	8.1	8.7	8.2	7.1	7.3	8.0	6.9	5.7	5.0	7.1	30
SD pc.	58	62	63	61	64	54	48	53	65	63	58	55	58	30
W.vel.	3.0	3.3	3.9	4.4	4.1	3.2	2.8	2.6	2.7	3.1	3.3	2.9	3.3	30
W.mx,dir	N	N	N	SSW	SSW	S	S	S	S	N	N	N	S	30
W.mx,pc	13	14	12	14	17	18	19	14	12	11	13	13	12	30
Calms														

51 Chengde 40° 58' N 117° 50' E 375.2 m													51
	JAN	FEB	MAR	APR	MAY	JUN	JUL	AUG	SEP	OCT	NOV	DEC	YEAR	oby
A.press.	981	979	975	970	966	962	961	965	971	977	979	981	972	27
T. mean	-9.3	-5.8	2.0	11.3	18.4	22.2	24.4	22.8	17.2	10.0	0.6	-7.4	8.9	30
T.a.mx.	8.6	18.1	23.9	34.2	36.9	41.3	41.5	36.1	33.2	29.3	22.3	12.3	41.5	30
T.a.mn.	-22.9	-23.3	-20.0	-8.7	0.8	7.6	13.8	6.4	0.8	-7.1	-18.1	-22.7	-23.3	30
T.m.mx.	-2.5	1.3	8.9	18.1	25.4	28.7	29.9	28.5	24.1	17.2	7.3	-0.8	15.5	30
T.m.mn.	-14.6	-11.5	-4.1	4.5	11.3	16.2	19.6	17.9	11.1	4.1	-4.4	-12.3	3.2	30
Tmn < 0														
Tmn < -10														
pTmn > 0														
pTmn > 10														
Frost d	17.5	11.9	11.7	4.7	0.5	0.0	0.0	0.0	1.4	14.8	22.3	21.8	106.6	30
r.hum.	46	45	43	39	44	59	72	74	65	59	54	54	54	30
cloud.	2.5	3.1	4.2	5.2	5.5	6.2	6.7	6.1	4.3	3.5	3.1	2.4	4.4	30
R.total	2	6	8	18	49	94	154	143	55	23	6	1	559	30
RD > .1	1.5	2.7	3.9	5.1	8.1	12.8	15.2	13.4	8.0	5.3	2.4	1.4	79.7	30
RD > 5														
RD > 10														
RD > 25														
Evap.														
Snow h.	20	18	14	15	0	0	0	0	0	0	27	14	27	30
Snow d.	1.3	2.6	2.9	0.8	0.0	0.0	0.0	0.0	0.0	0.1	1.5	1.3	10.5	28
Snow co														
SD hrs.	6.7	7.7	8.2	8.7	8.9	9.1	7.9	7.9	8.5	7.8	6.7	6.3	7.9	29
SD pc.	70	72	69	65	66	60	54	57	68	70	68	68	65	29
W.vel.	1.5	1.5	1.7	2.1	1.9	1.4	1.0	0.8	0.9	1.0	1.2	1.2	1.4	27
W.mx,dir	NW	NW	NW	S	S	S	SE	NW	NW	NW	NW	NW	NW	27
W.mx,pc	12	10	10	9	10	8	7	5	5	6	10	11	7	27
Calms	54	51	44	37	36	43	53	58	64	59	60	61	51	27

56 Bailingmiao 41° 42' N 110° 26' E 1375.9 m **56**

	JAN	FEB	MAR	APR	MAY	JUN	JUL	AUG	SEP	OCT	NOV	DEC	YEAR	oby
A.press.	865	864	862	860	859	857	856	859	863	866	867	866	862	27
T. mean	-15.9	-12.4	-3.7	5.6	13.0	18.4	20.5	18.4	12.1	4.3	-5.5	-13.7	3.4	27
T.a.mx.	7.5	12.4	21.7	29.0	32.1	36.1	36.6	35.3	29.2	24.7	18.7	10.0	36.6	27
T.a.mn.	-41.0	-37.8	-34.1	-17.1	-8.8	-0.1	5.9	0.5	-6.7	-17.0	-29.2	-39.4	-41.0	27
T.m.mx.	-7.9	-3.6	4.4	13.5	20.7	25.6	27.3	25.1	19.7	12.5	2.2	-5.9	11.1	27
T.m.mn.	-22.3	-19.5	-10.6	-1.9	4.9	10.5	14.0	12.3	5.4	-2.3	-11.4	-19.8	-3.4	27
Tmn < 0														
Tmn < -10														
pTmn > 0														
pTmn > 10														
Frost d	17.1	13.4	7.5	2.3	0.7	0.0	0.0	0.0	2.1	8.6	11.9	17.9	81.6	27
r.hum.	58	54	43	34	34	41	55	61	52	50	54	60	50	27
cloud.	2.8	3.1	4.3	5.0	5.2	5.4	5.8	5.2	4.1	3.0	3.1	2.7	4.1	27
R.total	3	3	5	10	15	29	71	72	32	12	5	2	259	27
RD > .1	3.7	3.7	4.5	4.2	5.9	8.8	13.2	11.7	7.1	4.1	3.6	3.1	73.6	27
RD > 5														
RD > 10														
RD > 25														
Evap.														
Snow h.	11	10	10	6	4	0	0	0	0	21	16	13	21	27
Snow d.	3.9	3.7	4.3	2.2	0.8	0.0	0.0	0.0	0.1	1.6	3.4	3.2	23.1	25
Snow co														
SD hrs.	7.3	8.3	8.7	9.4	10.1	10.2	9.2	8.9	9.0	8.6	7.5	7.0	8.7	27
SD pc.	77	78	74	70	70	67	62	64	73	78	76	77	72	27
W.vel.	4.0	4.2	4.7	5.7	5.5	4.7	3.8	3.6	3.8	4.0	4.4	4.2	4.4	26
W.mx,dir	SE	SE	SE	SW	SW	SW	SW	SE	SE	SW	SW	SE	SW	27
W.mx,pc	20	18	12	14	15	13	13	14	13	15	18	17	14	27
Calms							15	16	15					27

59 Datong 40° 06' N 113° 20' E 1067.6 m														59
	JAN	FEB	MAR	APR	MAY	JUN	JUL	AUG	SEP	OCT	NOV	DEC	YEAR	oby
A.press.	900	898	896	893	891	888	887	891	896	899	901	900	895	26
T. mean	-11.3	-7.7	-0.1	8.3	15.4	19.9	21.8	20.1	14.3	7.5	-1.4	-8.9	6.5	26
T.a.mx.	11.2	16.9	23.5	30.5	35.4	37.7	37.7	35.9	31.6	26.2	20.9	9.6	31.7	26
T.a.mn.	-29.1	-27.6	-20.9	-15.6	-2.9	4.0	8.8	6.2	-3.4	-10.4	-21.3	-27.6	-29.1	26
T.m.mx.	-4.3	-0.3	7.2	16.0	23.1	27.0	28.1	26.3	21.3	14.8	5.4	-2.1	13.6	26
T.m.mn.	-17.0	-13.9	-6.2	1.1	7.7	12.8	16.1	14.8	8.2	1.6	-6.7	-14.4	0.3	26
Tmn < 0														
Tmn < -10														
pTmn > 0														
pTmn > 10														
Frost d	12.4	11.8	8.6	3.6	1.0	0.0	0.0	0.0	1.3	9.7	15.6	14.5	78.5	26
r.hum.	50	49	46	42	42	52	66	70	62	57	53	51	53	26
cloud.	2.8	3.5	4.6	5.2	5.4	5.6	6.0	5.6	4.7	3.6	3.1	2.8	4.4	30
R.total	3	4	10	19	29	45	95	94	51	24	8	2	384	26
RD > .1	2.5	3.0	4.3	5.3	7.1	10.8	15.2	12.2	9.3	5.6	3.1	1.9	80.2	26
RD > 5														
RD > 10														
RD > 25														
Evap.														
Snow h.	9	15	22	10	5	0	0	0	0	4	9	5	22	27
Snow d.	2.4	3.0	3.3	1.3	0.1	0.0	0.0	0.0	0.0	0.5	2.0	1.9	14.4	26
Snow co														
SD hrs.	6.4	7.3	7.6	8.1	9.2	9.3	8.3	8.7	8.1	7.7	6.7	6.0	7.8	26
SD pc.	67	68	64	62	64	63	57	59	65	69	67	65	64	26
W.vel.	3.0	3.1	3.3	3.8	3.5	2.8	2.2	2.1	2.2	2.6	2.9	2.9	2.9	26
W.mx,dir	W	N	N	N	N	N	N	N	N	N	N	N	N	26
W.mx,pc	18	18	17	14	13	12	10	12	15	17	16	19	15	26
Calms	20		18	15	18	21	28	28	28	23	21	20	21	26

61 Beijing 39° 48' N 116° 28' E 31.2 m													61
	JAN	FEB	MAR	APR	MAY	JUN	JUL	AUG	SEP	OCT	NOV	DEC	YEAR	oby
A.press.	1021	1019	1015	1008	1003	998	997	1001	1007	1014	1019	1021	1010	30
T. mean	-4.6	-2.2	4.5	13.1	19.8	24.0	25.8	24.4	19.4	12.4	4.1	-2.7	11.5	30
T.a.mx.	12.9	18.5	24.4	31.1	38.3	40.6	39.6	38.3	32.3	29.8	23.3	13.9	40.6	30
T.a.mn.	-22.8	-27.4	-15.4	-3.2	2.1	10.0	15.3	11.4	3.7	-3.2	-12.3	-18.3	-27.4	30
T.m.mx.	1.4	3.9	10.7	19.6	26.4	30.2	30.8	29.4	25.7	18.9	9.9	2.9	17.5	30
T.m.mn.	-9.9	-7.4	-1.0	6.6	12.7	17.9	21.5	20.2	13.8	6.9	-0.6	-7.3	6.1	30
Tmn < 0													127.9	30
Tmn < -10													31.0	30
pTmn > 0			3.								26.		269.2	30
pTmn > 10				7.						23.			199.4	30
Frost d	13.4	13.0	12.9	2.6	0.3	0.0	0.0	0.0	0.1	7.2	18.2	20.1	87.8	30
r.hum.	45	49	52	48	52	62	78	80	71	66	60	51	60	30
cloud.	3.0	3.8	4.8	5.3	5.5	6.0	7.0	6.3	4.8	4.0	3.7	3.0	4.8	30
R.total	3	7	9	19	33	78	113	212	57	24	7	8	570	30
RD > .1	3.0	3.1	4.1	4.6	5.9	9.7	14.1	13.2	6.8	5.4	3.7	1.6	73.9	30
RD > 5													29.8	30
RD > 10													15.9	30
RD > 25													7.5	30
Evap.													1842	30
Snow h.	21	24	15	6	0	0	0	0	0	0	11	13	24	30
Snow d.	1.8	3.0	2.3	0.2	0.0	0.0	0.0	0.0	0.0	0.0	1.0	1.3	9.5	30
Snow co			19.								26.		114.3	30
SD hrs.	6.6	7.1	7.7	8.4	9.4	9.3	7.4	7.4	8.2	7.4	6.4	6.2	7.6	30
SD pc.	68	66	64	63	65	62	51	55	66	67	65	66	63	30
W.vel.	2.9	2.9	3.1	3.4	2.9	2.4	1.8	1.5	1.8	2.1	2.5	2.7	2.5	30
W.mx,dir	NNW	N	NNW	SSW	SSW	S	S	N	N	N	N	N	N	30
W.mx,pc	14	12	11	12	15	9	9	10	11	11	13	14	10	30
Calms	18	17	14			17	25	30	26	26	22	23	20	30

62 T i a n j i n 39° 06' N 117° 10' E 3.3 m														**62**
	JAN	FEB	MAR	APR	MAY	JUN	JUL	AUG	SEP	OCT	NOV	DEC	YEAR	oby
A.press.	1027	1026	1021	1015	1009	1005	1003	1007	1014	1021	1025	1027	1017	26
T. mean	-4.0	-1.6	5.0	13.2	20.0	24.1	26.4	25.5	26.8	13.6	5.2	-11.6	12.2	26
T.a.mx.	12.7	18.8	25.1	32.7	37.4	39.6	39.7	37.0	33.4	30.1	23.1	13.5	39.7	26
T.a.mn.	-18.0	-22.9	-17.7	-2.8	4.5	11.2	16.2	13.7	6.2	-0.3	-11.4	-14.7	-22.9	26
T.m.mx.	1.3	3.9	10.6	19.3	26.2	29.7	30.7	29.8	25.9	19.2	10.3	3.3	17.5	26
T.m.mn.	-8.2	-5.7	0.5	8.0	14.4	19.4	22.8	22.0	16.5	9.3	1.4	-5.2	7.9	26
Tmn < 0													112.8	30
Tmn < -10													16.1	30
pTmn > 0			1.									1.	275.8	30
pTmn > 10				7.						23.			204.1	30
Frost d	17.7	16.1	10.4	1.0	0.0	0.0	0.0	0.0	0.0	2.1	13.3	19.9	80.5	30
r.hum.	53	56	56	53	54	64	78	78	69	66	63	59	63	26
cloud.	3.1	4.0	4.8	5.3	5.2	5.7	6.6	6.1	4.6	3.8	3.8	3.2	4.7	30
R.total	3	6	6	21	31	69	190	162	43	25	9	4	569	30
RD > .1	1.8	2.6	3.4	4.7	5.5	9.1	13.2	11.6	6.4	4.2	3.5	2.1	68.1	30
RD > 5													24.8	30
RD > 10													15.9	30
RD > 25													7.5	30
Evap.													1779	30
Snow h.	18	20	20	4	0	0	0	0	0	0	10	7	20	26
Snow d.	1.7	1.4	1.9	0.3	0.0	0.0	0.0	0.0	0.0	0.0	0.8	1.4	8.4	30
Snow co			18.									1.	108.5	30
SD hrs.	6.2	6.7	6.9	9.2	9.5	7.7	7.9	7.9	7.9	7.4	6.2	5.6	7.5	30
SD pc.	63	62	59	60	67	63	55	58	64	66	62	62	61	30
W.vel.	3.1	3.2	3.5	3.9	3.6	3.1	2.5	2.2	2.4	2.7	2.8	2.9	3.0	26
W.mx,dir	NNW	NNW	NNW	SSW	SSW	SE	SE	SE	SSW	SSW	NNW	NNW	SSW	26
W.mx,pc	14	12	9	13	14	13	11	9	9	10	12	13	8	26
Calms								15	11	11	15	15	10	26

68 Yulin 38° 14' N 109° 42' E 1057.5 m														**68**
	JAN	FEB	MAR	APR	MAY	JUN	JUL	AUG	SEP	OCT	NOV	DEC	YEAR	oby
A.press.	903	900	898	895	893	890	888	892	897	901	903	903	897	29
T. mean	-10.0	-5.5	2.5	10.2	16.7	21.4	23.4	21.6	15.5	8.7	0.1	-8.0	8.1	30
T.a.mx.	10.4	17.2	24.7	31.1	36.2	38.5	38.6	36.9	32.4	27.8	22.4	10.1	38.6	30
T.a.mn.	-30.9	-28.5	-19.5	-10.5	-2.8	4.7	10.3	5.4	-3.2	-9.7	-21.1	-32.7	-32.7	30
T.m.mx.	-1.9	2.4	9.8	17.9	24.0	28.6	29.9	27.8	22.5	16.1	7.3	-0.3	15.3	30
T.m.mn.	-16.1	-11.8	-3.6	3.3	9.1	13.6	17.3	16.1	9.7	2.8	-5.3	-13.7	1.8	30
Tmn < 0														
Tmn < -10														
pTmn > 0														
pTmn > 10														
Frost d	18.3	14.0	11.8	3.8	0.9	0.0	0.0	0.0	1.3	12.7	20.9	22.0	105.6	29
r.hum.	58	55	50	45	44	49	62	68	66	63	62	62	57	29
cloud.	2.8	3.8	4.9	5.2	5.1	5.0	5.5	5.2	4.7	3.7	3.3	2.7	4.3	29
R.total	3	5	10	24	26	34	97	116	59	27	11	3	415	30
RD > .1	2.1	3.2	4.1	5.3	6.0	7.7	11.4	11.8	9.6	5.7	3.4	2.0	72.5	30
RD > 5														
RD > 10														
RD > 25														
Evap.														
Snow h.	13	14	8	2	0	0	0	0	0	1	8	15	15	30
Snow d.	2.2	2.7	2.5	0.5	0.0	0.0	0.0	0.0	0.0	0.4	1.8	2.0	12.1	30
Snow co														
SD hrs.	7.0	7.5	7.8	8.3	9.2	9.7	9.0	8.4	7.8	7.6	7.2	6.7	8.0	26
SD pc.	71	69	65	64	65	66	63	62	63	68	71	71	66	26
W.vel.	1.7	2.1	2.6	3.2	3.0	2.6	2.5	2.3	2.1	2.0	1.9	1.7	2.3	27
W.mx,dir	NNW	NNW	NNW	SSE	SSE	SSE	SSE	SSE	SSE	SSE	NNW	NNW	SSE	28
W.mx,pc	14	13	11	13	13	12	16	15	12	11	12	14	11	28
Calms	39	34	27	23	23	27	25	27	35	39	40	41	32	28

69 Taiyuan 37° 47' N 112° 33' E 777.9 m														69
	JAN	FEB	MAR	APR	MAY	JUN	JUL	AUG	SEP	OCT	NOV	DEC	YEAR	oby
A.press.	933	931	929	925	922	919	918	921	927	931	934	934	927	30
T. mean	-6.6	-3.1	3.7	11.4	17.7	21.7	23.5	21.8	16.1	9.9	2.1	-4.9	9.5	30
T.a.mx.	14.3	19.1	26.3	32.2	36.5	38.4	39.4	36.6	31.8	28.5	23.2	11.3	39.4	30
T.a.mn.	-25.5	-24.6	-18.0	-9.7	-0.5	4.4	7.2	7.4	-2.0	-7.4	-21.2	-21.9	-25.5	30
T.m.mx.	1.1	4.6	11.2	19.0	25.2	28.8	29.5	27.8	23.2	17.6	9.1	2.3	16.6	30
T.m.mn.	-13.0	-9.4	-2.7	4.2	10.0	14.6	18.2	17.0	10.2	3.6	-3.3	-10.5	3.2	30
Tmn < 0													149.1	30
Tmn < -10													55.5	30
pTmn > 0			6.								20.		259.9	30
pTmn > 10				18.						12.			178.0	30
Frost d	14.8	11.0	11.1	2.4	0.2	0.0	0.0	0.0	0.6	9.0	19.3	20.4	88.8	30
r.hum.	51	50	52	48	49	57	72	76	72	67	64	57	60	30
cloud.	3.0	4.0	5.0	5.5	5.4	5.5	6.2	6.0	5.2	4.0	3.6	2.9	4.7	30
R.total	3	6	10	24	30	53	118	104	64	31	13	3	459	30
RD > .1	2.3	3.2	4.3	5.2	6.1	10.1	14.0	12.8	9.4	6.1	4.2	1.9	79.9	30
RD > 5													25.0	30
RD > 10													13.8	30
RD > 25													3.5	30
Evap.													1798	30
Snow h.	11	15	10	16	0	0	0	0	0	2	13	15	16	30
Snow d.	2.3	2.8	2.4	0.3	0.0	0.0	0.0	0.0	0.0	0.1	1.7	1.8	11.4	30
Snow co			26.								27.		120.9	30
SD hrs.	6.3	6.2	7.2	7.7	8.8	9.2	6.9	7.4	7.2	7.2	6.2	6.1	6.2	30
SD pc.	65	67	60	59	63	63	54	55	59	65	62	64	60	30
W.vel.	2.5	2.7	3.1	3.3	3.0	2.5	2.0	1.9	1.8	2.0	2.4	2.5	2.5	17
W.mx,dir	NNW	NNW	N	NNW	N	NNW	NNW	NNW	NNW	NNW	NNW	NNW	NNW	17
W.mx,pc	14	14	12	13	10	12	13	15	13	14	13	15	13	17
Calms	24	22	18	16	18	21	29	29	34	31	26	25	24	17

72 Huimin 37° 30' N 117° 32' E 11.3 m														71
	JAN	FEB	MAR	APR	MAY	JUN	JUL	AUG	SEP	OCT	NOV	DEC	YEAR	oby
A.press.	1027	1025	1020	1014	1009	1004	1002	1006	1013	1020	1024	1026	1016	30
T. mean	-4.0	-1.6	5.1	13.2	20.1	24.7	26.3	25.2	20.0	13.6	5.5	-1.4	12.2	30
T.a.mx.	14.7	23.0	26.2	35.0	38.5	41.7	42.2	38.8	33.7	31.0	24.9	16.9	42.2	30
T.a.mn.	-22.4	-21.4	-16.6	-4.5	1.2	8.8	12.5	11.0	2.9	-3.2	-12.0	-20.9	-22.4	30
T.m.mx.	2.4	4.9	12.0	20.4	27.2	31.4	31.5	30.1	26.3	20.4	11.8	4.5	18.6	30
T.m.mn.	-9.2	-6.7	-0.6	6.6	13.1	18.3	21.8	21.0	14.5	7.8	0.4	-6.1	6.7	30
Tmn < 0														
Tmn < -10														
pTmn > 0														
pTmn > 10														
Frost d	22.0	17.5	11.5	1.9	0.0	0.0	0.0	0.0	0.0	5.6	17.4	24.7	100.9	30
r.hum.	60	61	58	54	54	61	79	82	74	69	68	65	66	30
cloud.	3.6	4.3	5.0	5.4	5.4	5.5	6.5	5.9	4.8	3.9	4.0	3.5	4.8	30
R.total	4	7	8	32	29	74	191	150	48	36	18	6	603	30
RD > .1	2.7	3.6	3.8	5.7	5.6	8.3	14.1	11.0	7.0	5.0	4.7	3.2	74.1	30
RD > 5														
RD > 10														
RD > 25														
Evap.														
Snow h.	18	16	15	6	0	0	0	0	0	0	6	9	18	30
Snow d.	2.4	3.0	1.3	0.3	0.0	0.0	0.0	0.0	0.0	0.0	0.8	1.8	9.7	30
Snow co														
SD hrs.	5.9	6.4	7.9	7.8	9.1	9.0	7.5	7.4	7.6	7.3	6.2	5.8	7.2	30
SD pc.	59	58	58	59	64	62	52	55	62	65	61	61	60	30
W.vel.	3.1	3.4	3.9	4.3	3.7	3.4	2.5	2.1	2.1	2.6	3.0	2.9	3.1	30
W.mx,dir	SW	SW	SW	SW	SW	SW	SE	NE	SW	SW	SW	SW	SW	27
W.mx,pc	11	10	11	14	17	12	10	10	10	10	11	11	11	27
Calms							13	21	19	14	12			27

77 Jinan 36° 41' N 116° 59' E 51.6 m														77
	JAN	FEB	MAR	APR	MAY	JUN	JUL	AUG	SEP	OCT	NOV	DEC	YEAR	oby
A.press.	1021	1019	1015	1009	1004	999	997	1000	1008	1015	1019	1021	1011	20
T. mean	-1.7	0.9	7.3	15.1	21.8	26.3	27.6	26.3	21.7	15.8	7.8	0.8	14.2	20
T.a.mx.	15.9	23.4	26.2	34.9	38.1	40.7	42.5	40.7	35.9	32.6	24.6	18.7	42.5	20
T.a.mn.	-19.7	-16.5	-11.3	-1.9	4.2	10.9	16.9	14.2	6.4	0.3	-9.1	-16.0	-19.7	20
T.m.mx.	3.4	6.1	13.1	21.1	27.7	32.1	32.1	30.7	26.9	21.2	12.8	5.7	19.4	30
T.m.mn.	-5.4	-3.1	2.8	10.0	16.4	21.0	23.2	22.3	17.1	11.1	3.8	-2.8	9.7	30
Tmn < 0													90.1	30
Tmn < -10													6.8	30
pTmn > 0		22.										12.	293.4	30
pTmn > 10				3.							4.		216.0	30
Frost d	14.8	11.0	4.4	0.4	0.0	0.0	0.0	0.0	0.0	0.9	7.3	14.9	53.7	30
r.hum.	54	56	52	49	49	53	73	76	65	59	61	58	59	20
cloud.	3.7	4.4	5.2	5.6	5.4	5.6	6.6	6.0	5.1	4.1	4.1	3.7	5.0	30
R.total	6	10	16	36	37	74	214	148	61	33	29	8	672	20
RD > .1	2.7	4.0	4.9	6.1	5.9	7.7	14.6	11.6	7.0	5.1	5.5	3.4	78.5	20
RD > 5													30.7	30
RD > 10													19.3	30
RD > 25													7.5	30
Evap.													2417	30
Snow h.	10	15	19	5	0	0	0	0	0	0	5	9	19	20
Snow d.	2.0	3.0	1.4	0.3	0.0	0.0	0.0	0.0	0.0	0.0	0.6	1.8	9.3	30
Snow co			22.								30.		111.6	30
SD hrs.	6.4	6.8	7.3	8.3	9.2	9.7	8.2	7.8	7.8	7.5	6.2	6.1	7.6	20
SD pc.	64	62	61	63	65	67	58	58	63	67	61	63	63	20
W.vel.	3.1	3.7	4.0	4.3	3.7	3.3	2.6	2.3	2.5	2.9	3.3	3.0	3.2	20
W.mx,dir	NE	NE	SSW	SSW	SSW	SSW	SSW	NE	SSW	SSW	SSW	SSW	SSW	19
W.mx,pc	13	15	16	18	21	17	14	13	12	17	17	14	15	19
Calms	18	16					17	23	20	19		17	16	19

81 Yan'an 36° 36' N 109° 30' E 957.6 m **81**

	JAN	FEB	MAR	APR	MAY	JUN	JUL	AUG	SEP	OCT	NOV	DEC	YEAR	oby
A.press.	914	912	909	906	904	900	899	902	908	912	914	914	908	26
T. mean	-6.4	-2.5	4.5	11.4	16.9	21.1	22.9	21.5	15.7	9.7	2.5	-4.5	9.4	30
T.a.mx.	16.6	21.7	29.4	33.8	36.5	38.0	39.7	36.9	32.6	29.7	25.5	15.8	39.7	30
T.a.mn.	-25.4	-20.8	-18.4	-8.8	-3.4	4.8	8.8	5.3	-3.0	-8.5	-15.6	-24.1	-25.4	30
T.m.mx.	2.0	5.5	12.4	19.7	25.0	29.3	29.8	28.4	23.0	17.7	10.1	3.3	17.2	30
T.m.mn.	-12.3	-8.3	-1.4	4.6	9.6	13.7	17.4	16.5	10.6	4.0	-2.6	-9.8	3.5	30
Tmn < 0														
Tmn < -10														
pTmn > 0														
pTmn > 10														
Frost d	15.8	11.7	10.7	1.8	0.2	0.0	0.0	0.0	0.2	7.5	18.8	21.9	88.7	30
r.hum.	54	54	54	52	54	58	72	76	76	70	65	60	62	30
cloud.	3.6	4.7	5.6	5.8	5.9	5.7	6.2	6.0	6.0	4.8	4.2	3.3	5.2	30
R.total	3	6	15	32	45	62	123	119	84	40	16	4	549	29
RD > .1	2.4	3.8	5.3	7.0	8.0	9.6	13.8	13.3	11.1	7.7	4.8	2.3	89.1	29
RD > 5														
RD > 10														
RD > 25														
Evap.														
Snow h.	11	10	7	16	0	0	0	0	0	0	10	17	17	30
Snow d.	2.4	3.4	2.8	0.5	0.1	0.0	0.0	0.0	0.0	0.4	1.8	2.3	13.7	30
Snow co														
SD hrs.	6.4	6.4	6.3	6.6	7.7	8.4	7.3	7.0	5.9	6.1	6.1	6.1	6.7	26
SD pc.	65	59	53	51	55	58	51	52	48	55	60	64	55	26
W.vel.	2.2	2.1	2.0	2.1	2.0	1.9	1.5	1.4	1.5	1.7	1.9	2.1	1.9	29
W.mx,dir	SW	SW	SW	SW	SW	SW	SW	SW	SW	SW	SW	SW	SW	29
W.mx,pc		21	19	22	20	22	17	14	17	20	21	22	20	29
Calms	23	23	25	24	23	23	34	36	32	29	25	22	26	29

87 Xuzhou 34° 17' N 117° 18' E 43.0 m **87**

	JAN	FEB	MAR	APR	MAY	JUN	JUL	AUG	SEP	OCT	NOV	DEC	YEAR	oby
A.press.	1023	1021	1017	1011	1006	1001	999	1002	1010	1017	1021	1023	1013	29
T. mean	-0.7	1.6	7.3	14.2	20.0	25.0	26.9	26.3	21.3	15.4	8.3	1.9	14.0	30
T.a.mx.	19.7	23.6	30.1	34.3	36.7	40.1	39.5	38.3	35.1	32.1	28.9	20.7	40.1	30
T.a.mn.	-18.9	-23.3	-10.5	-3.5	1.8	9.9	15.2	13.9	5.1	-1.7	-8.5	-13.8	-23.3	30
T.m.mx.	5.2	7.5	13.8	20.5	26.4	31.1	31.6	31.1	26.5	21.5	14.2	7.4	19.7	21
T.m.mn.	-4.1	-2.6	2.7	9.2	14.5	19.9	23.3	22.8	17.1	10.7	4.0	-2.1	9.6	21
Tmn < 0														
Tmn < -10														
pTmn > 0														
pTmn > 10														
Frost d	19.5	14.9	7.7	1.1	0.0	0.0	0.0	0.0	0.0	1.8	10.6	19.4	75.1	30
r.hum.	70	69	66	67	67	67	83	83	78	72	72	71	72	30
cloud.	5.0	5.9	6.3	6.5	6.5	6.3	7.4	6.4	6.0	5.0	4.9	4.7	5.9	30
R.total	13	19	33	60	62	107	248	156	87	36	29	18	868	30
RD > .1	3.9	5.6	6.6	8.5	7.9	8.7	14.8	10.9	8.4	6.1	5.6	4.7	91.8	30
RD > 5														
RD > 10														
RD > 25														
Evap.														
Snow h.	20	22	9	0	0	0	0	0	0	0	3	10	22	30
Snow d.	2.5	3.0	1.2	0.1	0.0	0.0	0.0	0.0	0.0	0.0	0.4	1.9	9.0	30
Snow co														
SD hrs.	5.5	5.7	6.4	6.9	7.8	8.6	6.9	7.5	6.8	6.7	5.9	5.6	6.7	30
SD pc.	54	52	53	53	56	60	49	56	55	60	57	57	55	30
W.vel.	3.0	3.4	3.8	3.9	3.4	3.4	2.9	2.6	2.5	2.6	2.9	2.9	3.1	30
W.mx,dir	ENE	ENE	ENE	E	E	E	E	E	ENE	E	E	ENE	E	30
W.mx,pc	11	14	13	13	12	12	14	16	15	13	12	13	13	30
Calms	16								16	16	15	16		30

90 Z h e n g z h o u 34° 43' N 113° 39' E 110.4 m														**90**
	JAN	FEB	MAR	APR	MAY	JUN	JUL	AUG	SEP	OCT	NOV	DEC	YEAR	oby
A.press.	1014	1012	1007	1002	997	992	990	993	1001	1008	1012	1013	1003	17
T. mean	-0.3	2.1	7.7	14.8	21.1	26.3	27.5	25.9	21.0	15.1	7.8	1.6	14.2	20
T.a.mx.	18.8	25.3	31.8	34.8	40.8	42.3	43.0	40.6	35.9	33.3	25.1	20.6	43.0	20
T.a.mn.	-17.9	-14.7	-13.7	-4.9	3.9	12.1	16.4	13.4	5.0	-0.2	-8.3	-14.5	-17.9	20
T.m.mx.	5.3	7.9	14.1	21.2	27.4	32.4	32.4	30.8	26.8	21.6	13.9	7.5	20.1	30
T.m.mn.	-4.7	-2.6	2.4	8.8	14.3	19.8	22.9	21.7	15.9	9.7	3.0	-2.9	9.0	30
Tmn < 0													92.8	30
Tmn < -10													3.3	30
pTmn > 0		15.										18.	306.7	30
pTmn > 10				2.							4.		217.7	30
Frost d	18.1	14.7	6.4	0.4	0.0	0.0	0.0	0.0	0.0	1.1	9.2	18.6	68.6	30
r.hum.	60	64	63	62	60	56	75	79	72	69	69	63	66	20
cloud.	4.8	5.9	6.5	6.6	6.5	6.2	7.0	6.4	6.2	5.2	5.1	4.8	5.9	30
R.total	9	13	29	51	46	68	135	135	67	41	34	8	636	20
RD > .1	3.1	4.4	6.3	7.4	7.5	7.3	12.3	10.8	8.8	6.9	6.1	3.1	83.7	20
RD > 5													29.5	30
RD > 10													18.1	30
RD > 25													7.0	30
Evap.													2013	30
Snow h.	18	19	15	0	0	0	0	0	0	0	23	18	23	19
Snow d.	2.6	3.2	1.7	0.2	0.0	0.0	0.0	0.0	0.0	0.0	1.2	2.1	10.9	30
Snow co			15.									1.	105.1	30
SD hrs.	5.7	6.0	6.0	6.5	8.0	8.7	7.6	7.5	6.4	6.4	5.7	5.6	6.7	16
SD pc.	57	54	50	50	57	61	54	57	52	57	55	57	55	16
W.vel.	3.8	3.4	3.5	3.8	3.3	3.3	2.7	2.3	2.5	2.7	3.3	3.5	3.2	20
W.mx,dir	WNW	NE	NE	S	S	S	S	NE	NE	NE	NE	NE	NE	20
W.mx,pc	14	16	16	15	14	14	13	13	11	10	14	13	12	20
Calms								15	15	16		14		20

95 X i ' a n 34° 18' N 108° 56' E 396.9 m														95
	JAN	FEB	MAR	APR	MAY	JUN	JUL	AUG	SEP	OCT	NOV	DEC	YEAR	oby
A.press.	980	977	972	968	964	960	957	961	968	974	978	980	970	30
T. mean	-1.0	2.1	8.1	14.1	19.1	25.2	26.6	25.5	19.4	13.7	6.6	0.7	13.3	30
T.a.mx.	17.0	24.1	27.9	33.6	37.5	41.7	41.0	39.7	34.5	33.7	23.9	17.5	41.7	30
T.a.mn.	-20.6	-18.7	-7.6	-4.0	3.5	9.2	15.1	12.1	4.8	-1.9	-16.8	-19.3	-20.6	30
T.m.mx.	4.5	7.7	14.3	20.6	25.6	32.1	32.4	31.2	24.8	19.4	12.0	6.0	19.2	30
T.m.mn.	-5.0	-2.1	3.1	8.5	13.2	18.6	21.7	20.8	15.2	9.3	2.6	-3.1	8.6	30
Tmn < 0													93.7	30
Tmn < -10													2.6	30
pTmn > 0		9.										15.	310.4	30
pTmn > 10				3.						28.			209.2	30
Frost d	22.9	15.7	6.4	0.9	0.0	0.0	0.0	0.0	0.0	1.8	12.7	23.6	83.9	30
r.hum.	67	67	67	70	70	59	72	74	80	78	77	71	71	30
cloud.	5.1	6.2	6.7	6.7	6.8	6.4	6.5	5.8	6.7	6.2	5.6	5.0	6.1	30
R.total	8	11	25	52	63	52	99	72	98	62	32	7	581	30
RD > .1	4.3	5.5	6.9	8.9	9.6	8.4	10.9	8.9	12.0	9.9	7.2	4.1	96.6	30
RD > 5													32.7	30
RD > 10													18.5	30
RD > 25													4.5	30
Evap.													1546	30
Snow h.	22	11	11	1	0	0	0	0	0	0	17	15	22	30
Snow d.	4.3	3.7	1.3	0.2	0.0	0.0	0.0	0.0	0.0	0.0	1.3	3.1	13.9	30
Snow co			14.									28.	106.8	30
SD hrs.	4.6	4.5	4.8	5.6	6.4	7.6	7.3	7.5	5.1	4.9	4.3	4.3	5.6	30
SD pc.	45	41	41	43	46	53	51	56	41	44	42	44	46	30
W.vel.	1.7	2.0	2.4	2.4	2.1	2.3	2.2	2.2	1.7	1.7	1.8	1.6	2.0	26
W.mx,dir	NE	NE	NE	NE	NE	NE	NE	NE	NE	NE	NE	NE	NE	30
W.mx,pc	11	17	17	16	12	12	17	19	15	12	10	11	14	30
Calms	34	29	24	23	26	22	25	26	35	38	34	35	29	30

101 Wudu 33° 23' N 104° 41' E 1079.1 m														101
	JAN	FEB	MAR	APR	MAY	JUN	JUL	AUG	SEP	OCT	NOV	DEC	YEAR	oby
A.press.	897	895	893	890	889	886	884	887	892	897	899	898	892	26
T. mean	2.8	5.7	10.9	16.0	19.6	22.7	24.8	24.2	19.4	14.9	9.3	4.2	14.5	27
T.a.mx.	15.3	21.5	28.4	33.2	34.5	37.6	36.1	36.3	34.0	29.2	23.0	17.0	37.6	27
T.a.mn.	-7.5	-7.8	-2.7	-2.1	6.2	10.9	14.8	12.3	8.4	0.9	-4.8	-8.1	-8.1	27
T.m.mx.	7.6	10.9	16.5	22.0	25.3	28.6	30.2	29.7	24.3	19.8	14.1	9.2	19.9	27
T.m.mn.	-1.0	1.7	6.6	11.2	14.7	18.0	20.4	19.9	15.8	11.3	5.5	0.2	10.4	27
Tmn < 0														
Tmn < -10														
pTmn > 0														
pTmn > 10														
Frost d	14.0	6.1	1.1	0.1	0.0	0.0	0.0	0.0	0.0	0.2	7.6	16.8	46.0	29
r.hum.	56	53	55	56	59	60	67	66	72	69	63	59	61	28
cloud.	5.5	6.8	7.5	7.4	7.8	7.6	7.1	6.7	7.8	7.3	5.9	4.8	6.8	26
R.total	2	3	15	38	57	61	93	81	75	39	9	.9	474	30
RD > .1	2.0	2.4	7.3	11.0	13.0	12.6	14.0	11.8	13.2	11.8	5.0	1.3	105.5	29
RD > 5														
RD > 10														
RD > 25														
Evap.														
Snow h.	7	4	2	0	0	0	0	0	0	0	1	2	7	30
Snow d.	1.5	0.8	0.4	0.1	0.0	0.0	0.0	0.0	0.0	0.0	0.2	0.3	3.4	27
Snow co														
SD hrs.	4.9	5.0	5.0	5.7	5.8	6.0	6.1	6.4	4.3	4.2	4.5	5.0	5.2	27
SD pc.	48	45	42	44	42	42	44	48	35	37	43	50	43	27
W.vel.	1.0	1.5	1.8	2.0	2.0	1.8	1.9	1.9	1.3	1.2	1.0	0.9	1.5	25
W.mx,dir	SE	SE	SE	SE	SE	SE	SE	SE	SE	SE	SE	SSE	SSE	29
W.mx,pc	7	11	12	14	14	14	14	16	13	10	8	6	12	29
Calms	47	37	34	33	32	33	36	36	46	46	50	51	40	29

114 Hefei 31° 51' N 117° 17' E 23.6 m													114
	JAN	FEB	MAR	APR	MAY	JUN	JUL	AUG	SEP	OCT	NOV	DEC	YEAR	oby
A.press.	1023	1021	1017	1011	1006	1002	999	1002	1009	1016	1020	1023	1012	18
T. mean	1.9	4.2	9.2	15.3	20.7	25.1	28.5	28.2	23.0	17.0	10.6	4.6	15.7	18
T.a.mx.	18.4	23.7	28.1	32.5	35.7	37.1	39.7	41.0	37.6	33.0	25.9	22.5	41.0	18
T.a.mn.	-20.6	-14.1	-7.3	-0.4	6.2	12.7	17.9	17.2	11.0	1.5	-5.0	-13.1	-20.6	18
T.m.mx.	6.5	8.7	13.7	20.2	25.2	29.3	32.4	32.5	27.3	22.2	15.6	9.3	20.2	29
T.m.mn.	-1.2	0.6	5.5	11.5	16.6	21.3	24.9	24.4	19.3	12.9	6.6	1.0	12.0	29
Tmn < 0													48.8	30
Tmn < -10													1.0	30
pTmn > 0	1/31												339.4	30
pTmn > 10			29.								14.		230.6	30
Frost d	16.7	11.9	4.1	0.5	0.0	0.0	0.0	0.0	0.0	0.7	5.9	15.5	55.2	29
r.hum.	74	74	75	77	76	76	81	79	78	74	76	75	76	18
cloud.	5.8	6.6	7.0	7.0	7.4	7.3	7.1	6.1	6.5	5.6	5.5	5.4	6.4	29
R.total	34	50	75	106	106	96	182	114	80	43	53	32	971	18
RD > .1	7.9	8.8	11.9	12.3	11.9	10.3	12.2	10.1	10.2	7.6	8.9	7.4	119.6	18
RD > 5													47.3	30
RD > 10													30.1	30
RD > 25													9.8	30
Evap.													1488	30
Snow h.	39	24	13	0	0	0	0	0	0	0	3	45	45	19
Snow d.	3.6	3.2	1.2	0.1	0.0	0.0	0.0	0.0	0.0	0.0	0.3	1.9	10.3	29
Snow co			12.									10.	91.9	30
SD hrs.	5.1	5.3	5.3	5.9	6.6	7.6	8.2	8.7	6.2	6.5	5.5	5.0	6.4	18
SD pc.	50	47	44	46	48	54	59	66	50	57	52	50	52	18
W.vel.	2.6	2.7	3.1	3.0	2.8	2.7	2.7	2.2	2.2	2.1	2.4	2.4	2.6	18
W.mx,dir	ENE	ENE	ENE	ESE	ESE	S	S	E	ENE	ENE	NW	NW	ENE	18
W.mx,pc	9	9	12	10	10	14	20	9	12	12	9	9	9	18
Calms	18	16		11	13			19	20	21	19	19	16	18

116 N a n j i n g 32° 00' N 118° 48' E 8.9 m **116**

	JAN	FEB	MAR	APR	MAY	JUN	JUL	AUG	SEP	OCT	NOV	DEC	YEAR	oby
A.press.	1026	1024	1020	1014	1010	1005	1003	1005	1012	1019	1023	1026	1016	30
T. mean	2.0	3.8	8.4	14.8	19.9	24.5	28.0	27.8	22.7	16.9	10.5	4.4	15.3	30
T.a.mx.	21.0	23.3	29.5	34.0	36.0	38.1	39.7	40.7	39.0	33.1	28.6	23.1	40.7	30
T.a.mn.	-14.0	-13.0	-7.1	-0.2	5.0	11.8	16.8	17.4	7.7	0.2	-6.3	-11.8	-14.0	30
T.m.mx.	6.8	8.6	13.4	20.0	25.0	29.2	32.2	32.4	27.3	22.3	15.9	9.7	20.2	30
T.m.mn.	-1.6	0.1	4.4	10.4	15.6	20.5	24.7	24.3	19.1	12.5	6.3	0.4	11.4	30
Tmn < 0													56.2	30
Tmn < -10													0.8	30
pTmn > 0	2.	3.											335.7	30
pTmn > 10				1.							11.		252.2	30
Frost d	17.9	13.2	5.9	0.7	0.0	0.0	0.0	0.0	0.0	0.7	6.7	16.7	61.9	30
r.hum.	73	75	74	75	75	77	81	80	80	76	76	75	77	30
cloud.	5.5	6.5	6.8	6.9	7.2	7.2	7.1	5.9	6.3	5.4	5.3	5.1	6.3	30
R.total	31	50	73	94	100	167	184	113	96	46	48	29	1031	30
RD > .1	7.7	9.6	10.8	12.0	11.2	11.0	12.4	11.2	10.3	7.7	8.2	6.8	118.8	30
RD > 5													48.5	30
RD > 10													30.5	30
RD > 25													10.1	30
Evap.													1585	30
Snow h.	51	14	18	0	0	0	0	0	0	0	7	50	51	30
Snow d.	3.2	2.9	1.2	0.0	0.0	0.0	0.0	0.0	0.0	0.0	0.1	1.0	8.4	30
Snow co			10.									14.	87.4	30
SD hrs.	4.8	4.7	4.9	5.7	6.3	6.7	7.3	7.9	5.9	5.9	5.3	5.1	5.9	30
SD pc.	46	42	41	44	46	48	53	59	48	55	51	50	49	30
W.vel.	2.7	2.8	3.2	3.1	2.8	2.7	2.6	2.6	2.4	2.3	2.4	2.4	2.7	25
W.mx,dir	NE	NE	E	SE	SE	SE	SE	SE	NE	NE	NE	NE	NE	25
W.mx,pc	11	11	12	12	12	15	12	12	16	12	9	9	9	25
Calms	25	21	17	17	17	16	19	19	24	28	29	29	22	25

119 Shanghai 31° 10' N 121° 26' E 4.5 m **119**

	JAN	FEB	MAR	APR	MAY	JUN	JUL	AUG	SEP	OCT	NOV	DEC	YEAR	oby
A.press.	1026	1024	1021	1016	1011	1006	1004	1006	1012	1019	1023	1027	1016	20
T. mean	3.3	4.6	8.3	13.8	18.8	23.2	27.9	27.8	23.8	17.9	12.5	6.2	15.7	20
T.a.mx.	19.8	23.6	27.6	33.3	35.5	36.9	38.3	38.9	37.3	31.3	28.5	23.3	38.9	20
T.a.mn.	-9.4	-7.9	-5.4	0.0	6.9	12.3	18.9	19.2	12.4	1.7	-3.8	-8.5	-9.4	20
T.m.mx.	7.6	8.7	12.6	18.5	23.2	27.3	31.8	31.6	27.4	22.4	16.8	10.7	19.9	30
T.m.mn.	0.3	1.4	4.9	10.4	15.3	20.1	24.7	24.7	20.5	14.3	8.6	2.7	12.3	30
Tmn < 0													39.7	30
Tmn < -10													0.0	30
pTmn > 0	8/29												347.7	30
pTmn > 10				2.							20.		232.5	30
Frost d	15.7	11.2	5.7	0.7	0.0	0.0	0.0	0.0	0.0	0.2	4.9	14.1	52.5	30
r.hum.	74	78	78	80	82	84	83	82	81	77	78	77	80	20
cloud.	6.0	6.8	7.1	7.4	7.9	8.1	7.2	6.1	6.7	5.9	5.8	5.6	6.7	30
R.total	44	63	81	111	129	157	142	116	146	47	53	39	1128	20
RD > .1	9.0	10.2	13.1	13.5	15.0	13.1	11.4	10.0	11.6	8.4	9.1	8.6	132.6	20
RD > 5													55.8	30
RD > 10													35.8	30
RD > 25													11.1	30
Evap.													1427	30
Snow h.	6	14	12	0	0	0	0	0	0	0	0	1	14	20
Snow d.	1.7	2.7	0.8	0.1	0.0	0.0	0.0	0.0	0.0	0.0	0.0	0.3	5.5	31
Snow co	5.		11.										66.4	30
SD hrs.	4.7	4.4	4.5	5.1	5.2	5.7	7.9	8.3	5.9	5.7	5.0	4.6	5.6	20
SD pc.	45	39	38	39	38	40	57	63	48	50	47	45	46	20
W.vel.	3.4	3.4	3.6	3.6	3.3	3.2	3.3	3.3	3.1	2.8	3.0	3.1	3.3	20
W.mx,dir	NW	NW	NE	SE	SE	SE	SSE	SE	E	E	NW	NW	ESE	20
W.mx,pc	15	11	10	14	16	18	19	16	12	12	15	17	10	20
Calms														

123 H a n g z h o u 30° 14' N 120° 10' E 41.7 m **123**

	JAN	FEB	MAR	APR	MAY	JUN	JUL	AUG	SEP	OCT	NOV	DEC	YEAR	oby
A.press.	1026	1024	1020	1015	1010	1006	1003	1005	1011	1019	1023	1026	1016	20
T. mean	3.6	5.0	9.2	15.1	20.3	24.3	28.7	28.2	23.5	17.4	12.1	6.1	16.1	20
T.a.mx.	23.9	28.3	29.5	33.9	36.5	39.7	38.9	39.6	36.3	32.4	31.2	26.5	39.7	20
T.a.mn.	-8.0	-9.6	-3.5	0.7	7.3	14.1	19.4	18.2	12.0	1.0	-3.6	-6.5	-9.6	20
T.m.mx.	8.8	9.3	13.7	20.2	24.7	28.6	33.3	32.7	27.6	22.5	16.9	10.8	20.7	30
T.m.mn.	0.7	1.9	5.9	11.6	16.6	21.0	25.0	24.6	20.3	14.0	8.4	2.9	12.7	30
Tmn < 0													33.8	30
Tmn < -10													0.0	30
pTmn > 0	14/30												354.8	30
pTmn > 10			31.								19.		234.2	30
Frost d	12.1	8.1	2.7	0.1	0.0	0.0	0.0	0.0	0.0	0.1	3.6	10.7	37.4	30
r.hum.	78	81	82	82	82	82	83	80	82	85	81	82	81	20
cloud.	6.1	6.8	7.3	7.5	8.0	8.1	6.9	5.8	6.7	6.0	5.8	5.7	6.7	27
R.total	64	84	117	130	186	192	132	136	183	67	61	49	1401	20
RD > .1	10.6	12.8	15.4	15.6	16.9	14.4	11.8	12.5	13.3	9.8	10.5	10.1	153.7	20
RD > 5													69.8	30
RD > 10													45.5	30
RD > 25													14.4	30
Evap.													1309	30
Snow h.	14	16	15	0	0	0	0	0	0	0	0	4	16	20
Snow d.	3.8	3.5	1.1	0.1	0.0	0.0	0.0	0.0	0.0	0.0	0.1	1.3	9.8	30
Snow co			11.									20.	82.0	30
SD hrs.	4.4	4.1	4.2	4.5	4.9	5.3	7.8	8.2	5.8	5.8	5.0	4.5	5.4	18
SD pc.	42	37	35	35	36	38	57	62	47	51	48	45	45	18
W.vel.	2.2	2.2	2.4	2.1	2.0	1.8	1.9	1.9	1.9	1.8	2.0	2.0	2.0	20
W.mx,dir	NNW	N	E	E	E	E	E	E	NNE	N	N	NNW	E	20
W.mx,pc	11	10	10	11	11	11	8	10	9	9	10	11	8	20
Calms	26	23	22	25	26	29	26	27	26	27	27	26	26	20

126 Wuhan	30° 38' N	114° 04' E	23.3 m											126
	JAN	FEB	MAR	APR	MAY	JUN	JUL	AUG	SEP	OCT	NOV	DEC	YEAR	oby
A.press.	1024	1022	1017	1012	1007	1002	1000	1002	1009	1017	1021	1024	1013	20
T. mean	2.8	5.0	10.0	16.0	21.3	25.8	29.0	28.5	23.6	17.5	11.2	5.3	16.3	20
T.a.mx.	20.9	24.9	29.6	33.1	36.0	37.8	38.2	39.4	36.8	34.4	28.5	23.3	39.4	20
T.a.mn.	-17.3	-14.8	-5.0	-0.3	7.2	14.2	17.3	18.0	10.1	1.5	-5.1	-8.7	-17.2	20
T.m.mx.	7.8	9.7	14.6	20.8	25.8	30.1	33.0	23.9	28.2	22.9	16.4	10.5	21.1	30
T.m.mn.	-0.9	1.3	6.2	12.1	17.5	22.0	25.4	24.7	19.5	13.3	7.1	1.4	12.5	30
Tmn < 0													43.8	30
Tmn < -10													0.7	30
pTmn > 0	8/27												351.3	30
pTmn > 10			26.								16.		236.3	30
Frost d	15.8	10.1	3.0	0.1	0.0	0.0	0.0	0.0	0.0	0.3	4.9	14.2	48.4	30
r.hum.	76	78	81	81	80	78	79	78	77	77	79	77	79	20
cloud.	6.3	7.1	7.4	7.4	7.5	7.3	6.7	5.7	6.0	6.1	6.0	5.8	6.6	30
R.total	36	61	105	144	161	218	179	133	81	53	57	34	1262	20
RD > .1	8.1	10.0	13.6	14.7	14.1	11.7	9.6	8.5	8.1	8.8	10.3	8.1	125.6	20
RD > 5													52.6	30
RD > 10													33.8	30
RD > 25													12.5	30
Evap.													1147	30
Snow h.	32	12	11	0	0	0	0	0	0	0	0	30	32	20
Snow d.	2.9	3.1	0.8	0.1	0.0	0.0	0.0	0.0	0.0	0.0	0.4	2.0	9.2	30
Snow co			4.									6.	89.7	30
SD hrs.	4.1	4.0	3.9	4.8	5.7	7.1	8.4	8.8	6.7	5.6	4.8	4.3	5.7	20
SD pc.	40	36	33	38	42	51	61	67	55	49	45	43	47	20
W.vel.	2.7	3.0	3.1	2.9	2.6	2.6	2.9	2.6	2.8	2.5	2.8	2.7	2.4	20
W.mx,dir	NNE	NNE	NNE	NNE	SE	SE	SW	NNE	NNE	NNE	NNE	NNE	NNE	20
W.mx,pc	18	18	15	11	10	11	11	12	20	15	21	18	14	20
Calms					11									

129 Y i c h a n g 30° 42' N 111° 05' E 131.1 m														129
	JAN	FEB	MAR	APR	MAY	JUN	JUL	AUG	SEP	OCT	NOV	DEC	YEAR	oby
A.press.	1019	1016	1011	1006	1002	997	994	997	1005	1012	1016	1018	1008	17
T. mean	4.7	6.4	11.1	16.7	21.5	25.7	28.3	27.6	23.6	18.0	12.3	6.6	16.9	20
T.a.mx.	21.6	24.6	29.5	35.5	36.8	39.7	40.7	41.4	36.6	35.7	29.3	23.6	41.4	20
T.a.mn.	-8.9	-4.4	-1.3	2.1	8.8	14.9	19.0	17.2	11.4	6.3	-0.8	-5.4	-8.9	20
T.m.mx.	9.0	10.9	15.7	21.7	26.0	30.5	33.1	32.7	28.2	23.0	16.8	10.8	21.5	29
T.m.mn.	1.4	3.0	7.5	13.1	17.6	21.6	24.4	23.9	19.7	14.5	8.9	3.5	13.2	29
Tmn < 0														
Tmn < -10														
pTmn > 0														
pTmn > 10														
Frost d	12.2	6.7	1.0	0.0	0.0	0.0	0.0	0.0	0.0	0.0	1.5	8.4	29.9	30
r.hum.	74	73	77	78	78	77	81	79	75	76	78	77	77	19
cloud.	6.8	7.4	7.6	7.5	7.7	7.5	7.0	6.4	7.0	6.8	6.7	6.7	7.1	29
R.total	22	28	65	106	140	158	236	195	101	74	49	26	1200	19
RD > .1	6.7	8.6	12.8	13.2	15.1	13.2	14.9	13.9	11.6	11.0	10.6	8.4	140.1	19
RD > 5														
RD > 10														
RD > 25														
Evap.														
Snow h.	14	8	3	0	0	0	0	0	0	0	0	20	20	19
Snow d.	2.5	2.5	0.7	0.1	0.0	0.0	0.0	0.0	0.0	0.0	0.3	2.3	8.3	30
Snow co														
SD hrs.	3.4	3.7	3.5	4.2	4.8	6.1	6.8	7.2	5.3	4.3	3.7	3.2	4.7	17
SD pc.	33	32	29	34	35	44	51	56	44	40	35	31	39	18
W.vel.	1.1	1.2	1.2	1.2	1.1	1.1	1.2	1.1	1.1	0.9	1.0	1.0	1.1	18
W.mx,dir	SE	SE	SE	SE	SE	SE	SE	SE	SE	SE	SE	SE	SE	19
W.mx,pc	16	17	17	13	12	14	13	11	11	11	13	15	14	19
Calms	46	44	43	45	47	45	43	47	46	53	48	49	46	19

134 Chengdu 30° 40' N 104° 01' E 505.9 m														134
	JAN	FEB	MAR	APR	MAY	JUN	JUL	AUG	SEP	OCT	NOV	DEC	YEAR	oby
A.press.	964	961	958	954	952	949	946	949	955	961	964	965	956	30
T. mean	5.5	7.5	12.1	17.0	20.9	23.7	25.6	25.1	21.2	16.8	11.9	7.3	16.2	30
T.a.mx.	17.7	22.5	29.2	32.8	36.1	35.3	36.2	37.3	34.8	29.8	24.5	20.3	37.3	30
T.a.mn.	-4.6	-3.5	-1.0	2.1	8.6	13.2	17.0	15.7	11.6	5.3	-0.1	-5.9	-5.9	30
T.m.mx.	9.5	11.5	16.7	22.0	25.7	28.0	30.0	29.9	25.4	20.5	15.6	11.1	20.5	30
T.m.mn.	2.4	4.4	8.6	13.1	17.3	20.3	22.3	21.7	18.5	14.3	9.3	4.5	13.0	30
Tmn < 0													12.9	30
Tmn < -10													0.0	30
pTmn > 0													365.2	30
pTmn > 10			16.								22.		252.6	30
Frost d	9.7	4.0	0.4	0.0	0.0	0.0	0.0	0.0	0.0	0.0	1.0	6.0	21.2	30
r.hum.	80	80	78	78	77	81	85	85	85	86	84	83	82	30
cloud.	8.0	8.6	8.5	8.4	8.8	8.8	8.0	7.5	8.7	9.0	8.6	8.2	8.4	30
R.total	6	11	21	51	89	111	236	234	118	46	18	6	947	30
RD > .1	6.0	7.7	10.9	13.8	16.1	15.9	17.4	15.5	16.2	15.2	8.3	5.7	148.6	30
RD > 5													91.9	30
RD > 10													24.3	30
RD > 25													9.5	30
Evap.													1021	30
Snow h.	4	4	0	0	0	0	0	0	0	0	0	5	5	30
Snow d.	1.0	0.8	0.0	0.0	0.0	0.0	0.0	0.0	0.0	0.0	0.1	0.4	2.4	28
Snow co		6.										5.	28.5	30
SD hrs.	2.3	2.3	3.1	3.8	4.1	4.4	5.3	5.7	2.9	2.1	2.1	2.1	3.4	30
SD pc.	22	21	26	29	30	31	38	43	24	19	20	21	28	30
W.vel.	1.4	1.6	1.5	1.7	1.7	1.7	1.5	1.6	1.6	1.5	1.4	1.5	1.6	6
W.mx,dir	NNE	NNE	NNE	NNE	NNE	NNE	NNE	N	NNE	NNE	NNE	NNE	NNE	24
W.mx,pc	14	12	15	12	9	7	[illegible]	9	12	10	12	11	11	24
Calms	45	43	35	36	35	40	41	44	42	49	47	50	42	24

138 Xichang 27° 53' N 102° 18' E 1590.7 m														138
	JAN	FEB	MAR	APR	MAY	JUN	JUL	AUG	SEP	OCT	NOV	DEC	YEAR	oby
A.press.	838	837	836	835	835	835	834	836	839	842	842	840	837	17
T. mean	9.5	11.7	16.4	19.8	21.5	21.0	22.7	22.2	20.0	16.7	13.1	10.2	17.1	20
T.a.mx.	24.1	26.3	30.9	35.2	36.5	36.2	33.8	34.9	33.3	29.4	27.6	23.1	36.5	20
T.a.mn.	-3.4	-3.2	0.7	3.1	7.8	10.7	13.6	13.4	9.3	4.8	1.5	-2.1	-3.4	20
T.m.mx.	16.5	18.9	23.9	27.0	27.7	26.3	27.7	27.7	25.1	21.9	19.1	16.7	23.2	30
T.m.mn.	3.6	5.7	9.8	12.9	16.0	17.3	18.9	18.2	16.3	13.1	8.4	4.7	12.1	30
Tmn < 0														
Tmn < -10														
pTmn > 0														
pTmn > 10														
Frost d	14.5	7.5	1.4	0.0	0.0	0.0	0.0	0.0	0.0	0.0	1.4	13.9	38.7	30
r.hum.	50	46	41	44	55	73	76	75	75	74	65	59	61	20
cloud.	2.7	3.3	3.7	5.1	6.9	8.7	8.6	8.0	8.1	7.2	4.8	3.3	5.9	30
R.total	5	8	9	25	80	215	223	191	174	88	21	4	1043	20
RD > .1	1.9	3.9	3.9	7.0	13.0	20.8	18.9	18.0	16.9	14.4	5.1	2.3	126.0	20
RD > 5														
RD > 10														
RD > 25														
Evap.														
Snow h.	9	9	0	0	0	0	0	0	0	0	0	0	9	20
Snow d.	0.7	1.1	0.1	0.0	0.0	0.0	0.0	0.0	0.0	0.0	0.0	0.3	2.3	30
Snow co														
SD hrs.	7.4	7.8	8.4	8.4	7.0	4.5	5.7	6.2	5.1	5.0	7.0	7.2	6.6	19
SD pc.	70	69	70	66	52	32	42	58	42	44	65	70	55	19
W.vel.	1.5	1.8	2.1	1.8	1.6	0.9	0.8	0.8	0.8	0.8	1.0	1.1	1.2	16
W.mx,dir	S	S	S	S	S	SE	SSW	N	S	N	N	S	S	16
W.mx,pc	9	12	15	13	10	5	5	5	5	6	8	7	8	16
Calms	43	35	31	35	41	57	55	56	54	58	52	53	47	16

142 Chongqing 29° 35' N 106° 28' E 260.6 m														142
	JAN	FEB	MAR	APR	MAY	JUN	JUL	AUG	SEP	OCT	NOV	DEC	YEAR	oby
A.press.	992	989	985	981	978	975	971	974	981	988	991	993	983	30
T. mean	7.5	9.5	14.1	18.8	22.2	25.2	28.6	28.5	23.8	18.6	13.9	9.5	18.3	30
T.a.mx.	18.8	23.7	34.8	36.4	38.9	39.7	40.4	42.2	40.4	33.4	28.7	21.5	42.2	30
T.a.mn.	-1.8	-0.8	2.8	2.8	12.0	15.5	19.4	17.8	14.3	8.0	0.7	-1.7	-1.8	30
T.m.mx.	9.5	11.8	17.1	22.4	25.7	28.2	32.5	32.9	26.7	21.6	15.8	11.1	21.3	16
T.m.mn.	5.6	6.9	11.0	15.1	18.9	21.4	24.6	24.5	20.4	16.3	11.4	7.4	15.3	16
Tmn < 0														
Tmn < -10														
pTmn > 0														
pTmn > 10														
Frost d	2.6	1.0	0.1	0.0	0.0	0.0	0.0	0.0	0.0	0.0	0.0	0.6	4.3	30
r.hum.	82	78	75	76	79	79	75	73	79	84	84	84	79	30
cloud.	8.4	8.4	7.9	7.6	8.1	7.9	6.8	5.9	7.4	8.4	8.5	8.5	7.8	30
R.total	20	20	39	90	158	166	142	138	136	97	48	25	1079	30
RD > .1	9.5	9.3	11.2	13.4	17.7	15.5	11.1	10.8	13.9	16.4	13.5	10.1	152.4	30
RD > 5														
RD > 10														
RD > 25														
Evap.														
Snow h.	3	2	0	0	0	0	0	0	0	0	0	0	3	30
Snow d.	0.5	0.2	0.0	0.0	0.0	0.0	0.0	0.0	0.0	0.0	0.0	0.1	0.9	28
Snow co														
SD hrs.	1.3	1.7	2.9	4.0	3.6	4.2	6.9	7.3	4.0	2.1	1.7	1.1	3.4	29
SD pc.	13	15	24	31	27	30	50	55	32	19	16	11	28	29
W.vel.	1.1	1.3	1.6	1.6	1.4	1.3	1.5	1.5	1.3	1.0	1.1	1.1	1.3	30
W.mx,dir	N	N	N	N	N	N	N	N	N	N	N	N	N	30
W.mx,pc	11	13	14	12	10	7	7	7	9	9	12	10	10	30
Calms	46	39	32	35	37	41	37	39	43	51	47	47	41	30

147 Changsha 28° 12' N 113° 04' E 44.5 m **147**

	JAN	FEB	MAR	APR	MAY	JUN	JUL	AUG	SEP	OCT	NOV	DEC	YEAR	oby
A.press.	1016	1014	1009	1005	1000	995	993	995	1002	1009	1013	1016	1006	20
T. mean	4.6	6.2	10.9	16.7	21.7	26.0	29.5	28.9	24.5	18.5	12.5	7.0	17.2	20
T.a.mx.	24.7	28.3	31.2	34.2	35.5	38.2	39.4	40.6	38.2	34.8	29.7	24.9	40.6	20
T.a.mn.	-9.5	-8.4	-2.3	2.5	9.1	16.5	21.0	19.6	12.3	6.1	-1.5	-5.0	-9.5	20
T.m.mx.	8.6	9.8	14.8	21.1	25.7	30.1	34.0	33.5	28.9	23.2	16.9	11.2	21.5	30
T.m.mn.	1.6	3.4	7.8	13.6	18.4	22.5	25.6	25.1	20.9	15.0	9.1	3.9	13.9	30
Tmn < 0													11.9	30
Tmn < -10													0.0	30
pTmn > 0													358.6	30
pTmn > 10			23.								21.		243.3	30
Frost d	9.2	4.0	1.0	0.0	0.0	0.0	0.0	0.0	0.0	0.1	1.4	6.4	22.0	30
r.hum.	79	82	83	83	82	80	74	75	77	78	80	80	80	20
cloud.	7.1	7.9	8.0	7.8	8.0	7.8	6.6	6.2	6.2	6.5	6.5	6.6	7.1	30
R.total	53	87	152	199	245	185	123	106	69	85	70	48	1422	20
RD > .1	11.5	14.4	17.4	17.5	17.4	12.0	8.9	10.3	7.8	11.6	10.9	10.8	150.2	20
RD > 5													67.1	30
RD > 10													43.1	30
RD > 25													14.5	30
Evap.													1383	30
Snow h.	9	10	2	0	0	0	0	0	0	0	0	4	10	20
Snow d.	4.3	2.9	0.4	0.0	0.0	0.0	0.0	0.0	0.0	0.0	0.1	1.1	8.8	30
Snow co		28.										20.	71.0	30
SD hrs.	3.0	2.5	2.8	3.7	3.9	5.5	8.7	8.0	6.5	4.7	4.0	3.2	4.7	17
SD pc.	29	22	24	29	29	29	63	61	52	41	37	31	39	17
W.vel.	2.8	2.8	2.8	2.8	2.4	2.2	2.8	2.5	2.9	2.5	2.8	2.7	2.7	20
W.mx,dir	NW	NW	NW	NW	NW	S	S	NW	NW	NW	NW	NW	NW	20
W.mx,pc	33	31	28	23	23	13	23	15	33	33	35	35	26	20
Calms						20		18						20

154 W e n z h o u 28° 01' N 120° 40' E 6.0 m **154**

	JAN	FEB	MAR	APR	MAY	JUN	JUL	AUG	SEP	OCT	NOV	DEC	YEAR	oby
A.press.	1024	1023	1019	1015	1010	1006	1005	1005	1010	1017	1021	1024	1015	20
T. mean	7.5	8.1	11.2	16.0	20.5	24.2	27.9	28.0	25.1	20.0	15.7	10.4	17.9	20
T.a.mx.	24.4	25.2	29.5	31.0	35.7	36.8	39.3	38.1	35.4	33.4	28.5	25.8	39.3	20
T.a.mn.	-4.5	-3.9	-0.1	4.1	10.5	15.9	20.9	19.7	13.7	5.7	1.3	-3.3	-4.5	20
T.m.mx.	12.0	12.1	15.3	20.3	24.0	27.8	31.9	31.7	28.8	24.6	19.8	14.8	21.9	30
T.m.mn.	4.4	5.2	8.2	13.1	17.8	21.8	25.0	24.9	22.0	16.7	12.1	7.0	14.9	30
Tmn < 0														
Tmn < -10														
pTmn > 0														
pTmn > 10														
Frost d	8.4	4.5	1.2	0.1	0.0	0.0	0.0	0.0	0.0	0.0	0.8	5.2	20.4	30
r.hum.	74	80	84	85	86	89	86	83	83	77	76	75	81	20
cloud.	6.3	7.2	7.8	8.1	8.7	8.8	7.3	6.5	6.5	6.0	6.1	6.0	7.1	27
R.total	50	87	135	152	211	247	146	224	255	75	65	50	1697	20
RD > .1	11.2	14.8	18.5	17.9	20.3	18.4	12.1	15.2	14.1	9.2	10.0	10.9	172.2	20
RD > 5														
RD > 10														
RD > 25														
Evap.														
Snow h.	4	10	0	0	0	0	0	0	0	0	0	1	10	20
Snow d.	1.2	1.8	0.6	0.0	0.0	0.0	0.0	0.0	0.0	0.0	0.0	0.3	3.9	30
Snow co														
SD hrs.	4.3	3.7	3.5	4.1	3.7	4.2	7.9	8.0	6.5	5.7	4.7	4.2	5.1	19
SD pc.	41	32	30	32	27	30	58	61	53	50	44	41	42	19
W.vel.	2.3	2.3	2.1	2.1	2.0	1.8	2.2	2.2	2.1	1.9	2.1	2.1	2.1	15
W.mx,dir	NW	NW	ESE	ESE	ESE	ESE	E	E	ESE	NW	NW	NW	ESE	15
W.mx,pc	23	18	22	26	25	22	21	16	14	16	20	17	17	15
Calms		22	25	29	33	38	33	32	28	24	22	27	27	15

173 G u i y a n g 26° 35' N 106° 43' E 1071.2 m **173**

	JAN	FEB	MAR	APR	MAY	JUN	JUL	AUG	SEP	OCT	NOV	DEC	YEAR	oby
A.press.	898	896	894	892	890	888	887	889	893	897	898	899	893	20
T. mean	4.9	6.4	11.5	16.3	19.6	21.9	24.0	23.5	20.8	15.9	11.6	7.0	15.3	20
T.a.mx.	25.8	28.4	31.7	35.3	34.6	35.6	37.5	35.9	34.4	32.1	27.5	26.1	37.5	20
T.a.mn.	-7.8	-6.0	-3.1	0.1	6.3	10.4	16.4	14.3	8.1	3.3	-1.4	-6.1	-7.8	20
T.m.mx.	9.1	11.1	16.9	21.8	24.2	26.4	28.7	28.5	25.6	20.5	15.8	11.4	20.0	30
T.m.mn.	2.2	3.6	7.9	12.5	16.1	18.6	20.7	19.8	17.1	13.0	8.5	4.2	12.0	30
Tmn < 0													19.2	30
Tmn < -10													0.0	30
pTmn > 0													357.5	30
pTmn > 10			26.								16.		235.3	30
Frost d	4.1	1.3	0.8	0.1	0.0	0.0	0.0	0.0	0.0	0.3	1.6	3.8	11.9	30
r.hum.	78	78	75	74	77	78	78	78	76	79	79	79	77	20
cloud.	8.5	8.5	8.0	8.0	8.6	8.6	8.0	7.4	7.4	8.1	8.1	8.2	8.1	30
R.total	20	22	33	108	192	213	179	142	83	89	56	26	1163	20
RD > .1	13.6	13.5	14.4	15.0	19.4	16.9	16.0	15.3	11.3	17.1	14.0	12.1	178.4	20
RD > 5													56.1	30
RD > 10													34.6	30
RD > 25													12.0	30
Evap.													1312	30
Snow h.	1	7	8	0	0	0	0	0	0	0	0	7	8	20
Snow d.	2.8	2.3	0.4	0.0	0.0	0.0	0.0	0.0	0.0	0.0	0.2	0.9	6.5	30
Snow co		19.										10.	71.9	30
SD hrs.	1.8	2.1	3.4	4.5	4.5	4.4	6.3	6.4	5.2	2.7	2.8	2.1	3.9	20
SD pc.	17	19	29	36	33	32	46	49	42	24	26	21	32	20
W.vel.	2.2	2.5	2.7	2.6	2.2	1.9	2.3	1.7	1.9	1.9	2.1	2.1	2.2	20
W.mx,dir	NE	NE	NE	NE	NE	S	S	S	NE	NE	NE	NE	NE	20
W.mx,pc	20	22	19	16	12	13	24	14	12	17	19	20	15	20
Calms				18	21	29	25	34	27	23	22	22	23	20

177 G u i l i n 25° 20' N 110° 18' E 166.7 m **177**

	JAN	FEB	MAR	APR	MAY	JUN	JUL	AUG	SEP	OCT	NOV	DEC	YEAR	oby
A.press.														
T. mean	8.0	9.0	13.1	18.4	23.1	26.2	28.3	27.8	25.8	20.7	15.2	10.1	18.8	20
T.a.mx.	27.6	28.8	31.9	35.6	35.0	37.4	38.2	39.4	38.5	35.2	30.4	27.6	39.4	20
T.a.mn.	-4.9	-3.6	0.0	4.0	11.2	13.0	21.3	18.3	12.9	8.0	2.7	-1.8	-4.9	20
T.m.mx.	12.0	12.9	16.9	22.4	27.1	30.4	32.9	32.9	30.9	25.8	20.1	14.7	23.3	30
T.m.mn.	5.0	6.5	10.5	15.5	20.1	23.3	25.0	24.3	22.0	17.1	11.8	7.0	15.7	30
Tmn < 0														
Tmn < -10														
pTmn > 0														
pTmn > 10														
Frost d	3.4	1.3	0.2	0.0	0.0	0.0	0.0	0.0	0.0	0.0	0.3	2.7	7.8	30
r.hum.	70	76	80	81	81	80	78	79	71	69	72	70	76	20
cloud.	7.8	8.5	8.7	8.7	8.6	8.6	7.8	7.3	6.0	6.5	6.5	6.7	7.7	30
R.total	56	76	134	280	319	316	226	167	66	97	83	56	1876	20
RD > .1	12.7	14.7	19.5	20.1	20.7	17.7	15.8	14.6	8.6	9.3	10.6	11.4	175.4	20
RD > 5														
RD > 10														
RD > 25														
Evap.														
Snow h.	1	1	0	0	0	0	0	0	0	0	0	0	1	20
Snow d.	0.8	0.7	0.0	0.0	0.0	0.0	0.0	0.0	0.0	0.0	0.0	0.5	2.0	30
Snow co														
SD hrs.	2.8	2.2	2.2	3.0	3.8	5.1	7.4	7.6	7.4	5.7	4.9	3.9	4.7	18
SD pc.	27	20	18	23	28	37	55	58	61	49	45	37	39	18
W.vel.	3.5	3.5	2.9	2.5	2.2	1.7	1.6	1.5	2.6	3.1	3.1	3.2	2.6	16
W.mx,dir	NNE	NNE	NNE	NNE	NNE	NNE	S	NNE	NNE	NNE	NNE	NNE	NNE	18
W.mx,pc	51	49	43	34	33	18	14	14	37	50	50	51	36	18
Calms						31	34	38						

183 F u z h o u 26° 05' N 119° 17' E 84.0 m **183**

	JAN	FEB	MAR	APR	MAY	JUN	JUL	AUG	SEP	OCT	NOV	DEC	YEAR	oby
A.press.	1013	1012	1009	1005	1001	997	996	996	1000	1007	1011	1013	1005	20
T. mean	10.4	10.6	13.4	18.1	22.2	25.3	28.7	28.3	26.0	21.6	17.8	13.1	19.6	20
T.a.mx.	27.0	29.5	32.1	33.0	35.7	38.0	39.3	39.3	38.2	35.6	31.5	29.8	39.3	20
T.a.mn.	-1.2	-0.8	2.5	5.2	11.1	16.4	19.0	21.0	15.0	9.8	4.2	1.5	-1.2	20
T.m.mx.	15.0	15.1	18.2	23.0	26.5	29.9	34.0	33.1	30.5	26.2	21.8	17.6	24.3	30
T.m.mn.	7.6	7.8	10.3	14.9	19.1	22.6	25.2	24.8	23.0	18.7	14.6	10.0	16.5	30
Tmn < 0													0.4	30
Tmn < -10													0.0	30
pTmn > 0													365.3	30
pTmn > 10			6.									29.	299.0	30
Frost d	2.8	1.3	0.4	0.0	0.0	0.0	0.0	0.0	0.0	0.0	0.1	1.8	6.3	30
r.hum.	73	78	81	80	82	84	78	78	77	71	71	73	77	20
cloud.	6.9	7.5	7.8	7.8	8.4	8.5	6.7	6.7	6.9	6.4	6.8	6.6	7.2	30
R.total	53	80	121	136	210	224	119	142	155	31	29	29	1329	20
RD > .1	11.4	13.9	16.7	16.1	18.3	17.0	9.9	12.5	12.7	6.6	7.5	8.5	150.9	20
RD > 5													62.7	30
RD > 10													41.0	30
RD > 25													14.7	30
Evap.													1455	30
Snow h.	0	0	0	0	0	0	0	0	0	0	0	0	0	30
Snow d.	0.2	0.5	0.0	0.0	0.0	0.0	0.0	0.0	0.0	0.0	0.0	0.1	0.8	30
Snow co													5.4	30
SD hrs.	4.1	3.4	3.6	4.6	4.5	5.1	8.7	7.8	6.2	5.4	4.4	4.1	5.2	20
SD pc.	39	30	30	36	34	37	64	60	51	47	41	39	43	20
W.vel.	2.8	2.6	2.6	2.6	2.6	2.5	3.3	3.1	3.1	3.1	3.2	3.0	2.9	20
W.mx,dir	NW	N	SE	SE	SE	SE	SE	SE	N	N	N	NW	SE	17
W.mx,pc	14	11	15	22	23	23	34	22	13	17	17	15	15	17
Calms	16	19	23	25		27			15				19	17

199 Guangzhou 23° 08' N 113° 19' E 6.3 m														199
	JAN	FEB	MAR	APR	MAY	JUN	JUL	AUG	SEP	OCT	NOV	DEC	YEAR	oby
A.press.	1020	1018	1016	1012	1008	1005	1004	1004	1007	1014	1017	1020	1013	20
T. mean	13.4	14.2	17.7	21.8	25.7	27.2	28.3	28.2	27.0	23.8	19.7	15.2	21.8	20
T.a.mx.	27.9	28.6	31.0	33.0	36.0	36.6	37.5	38.7	37.6	33.6	32.4	29.6	38.7	20
T.a.mn.	0.1	0.0	3.3	7.7	14.7	18.9	21.6	21.1	15.5	11.3	5.3	1.8	0.0	20
T.m.mx.	18.3	18.6	21.7	25.6	29.5	31.1	32.6	32.4	31.3	28.5	24.5	20.5	26.2	30
T.m.mn.	9.7	11.2	15.0	19.0	22.8	24.4	25.2	25.0	23.8	20.1	15.5	11.3	18.6	30
Tmn < 0													0.0	30
Tmn < -10													0.0	30
pTmn > 0													365.3	30
pTmn > 10	9.	10.											337.8	30
Frost d	1.6	0.4	0.0	0.0	0.0	0.0	0.0	0.0	0.0	0.0	0.1	0.7	2.8	30
r.hum.	69	78	83	84	85	86	84	83	80	72	69	68	78	20
cloud.	6.4	7.6	8.3	8.4	8.3	8.5	7.3	7.4	6.5	5.3	5.1	5.4	7.1	30
R.total	39	63	92	159	267	299	220	225	204	52	42	20	1682	20
RD > .1	7.8	11.4	14.7	15.1	17.8	20.3	16.6	16.2	13.1	6.0	5.6	6.2	150.5	20
RD > 5													67.5	30
RD > 10													46.8	30
RD > 25													20.2	30
Evap.													1629	30
Snow h.	0.0	0.0	0.0	0.0	0.0	0.0	0.0	0.0	0.0	0.0	0.0	0.0	0.0	20
Snow d.	0	0	0	0	0	0	0	0	0	0	0	0	0	20
Snow co													0.0	30
SD hrs.	4.5	3.1	2.5	3.2	5.1	5.3	7.5	7.1	6.7	7.0	6.4	5.4	5.3	20
SD pc.	42	27	21	25	38	40	56	55	54	61	59	50	44	20
W.vel.	2.3	2.2	2.1	2.0	2.0	1.8	1.9	1.8	2.0	2.0	2.3	2.3	2.1	20
W.mx,dir	N	N	N	SE	SE	SE	SE	E	N	N	N	N	N	20
W.mx,pc	28	26	17	16	17	15	17	12	14	24	31	31	16	20
Calms			23	22	22	24	24	30	29	33			26	20

210 Dongfang 19° 06' N 108° 37' E 8.4 m														210
	JAN	FEB	MAR	APR	MAY	JUN	JUL	AUG	SEP	OCT	NOV	DEC	YEAR	oby
A.press.	1017	1015	1012	1009	1006	1004	1003	1003	1006	1012	1015	1016	1010	16
T. mean	18.3	18.9	22.0	25.7	28.6	29.1	29.1	28.2	27.0	25.2	22.4	19.7	24.5	19
T.a.mx.	33.2	35.1	37.3	38.8	38.7	38.1	36.8	36.5	34.6	34.9	33.8	32.3	38.8	19
T.a.mn.	1.4	6.7	10.2	11.5	19.4	20.2	20.4	20.4	17.7	9.3	6.6	4.8	1.4	19
T.m.mx.	23.0	23.1	26.0	29.6	31.9	32.0	32.1	31.4	30.9	29.4	26.8	24.4	28.4	28
T.m.mn.	15.4	16.3	19.2	22.7	25.6	26.4	26.4	25.8	24.4	22.6	19.6	16.9	21.8	28
Tmn < 0														
Tmn < -10														
pTmn > 0														
pTmn > 10														
Frost d	0.0	0.0	0.0	0.0	0.0	0.0	0.0	0.0	0.0	0.0	0.0	0.0	0.0	19
r.hum.	78	83	83	78	74	76	76	81	84	81	80	79	79	18
cloud.	6.7	7.6	7.3	7.4	7.9	8.8	8.3	8.9	8.2	7.2	6.7	6.8	7.6	18
R.total	9	16	19	29	58	130	131	279	197	82	21	5	976	18
RD > .1	3.6	5.8	5.4	4.6	6.3	8.9	8.1	12.7	14.9	9.1	5.6	3.1	88.1	18
RD > 5														
RD > 10														
RD > 25														
Evap.														
Snow h.	0.0	0.0	0.0	0.0	0.0	0.0	0.0	0.0	0.0	0.0	0.0	0.0	0.0	19
Snow d.	0.0	0.0	0.0	0.0	0.0	0.0	0.0	0.0	0.0	0.0	0.0	0.0	0.0	19
Snow co														19
SD hrs.	6.4	5.8	6.1	7.8	9.4	8.8	9.1	6.9	7.2	8.2	7.4	6.9	7.6	18
SD pc.	59	51	52	61	73	66	69	60	59	69	66	63	62	18
W.vel.														
W.mx,dir														
W.mx,pc.														
Calms														

215 Nanning 22° 49' N 108° 21' E 72.2 m														**215**
	JAN	FEB	MAR	APR	MAY	JUN	JUL	AUG	SEP	OCT	NOV	DEC	YEAR	oby
A.press.	1012	1010	1007	1003	999	996	995	996	1000	1007	1010	1012	1004	20
T. mean	12.9	13.9	17.3	21.9	26.0	27.4	28.2	27.9	26.7	23.3	18.9	14.8	21.6	20
T.a.mx.	32.6	32.6	35.0	36.8	40.4	37.9	39.0	39.1	38.0	35.2	32.4	30.5	40.4	20
T.a.mn.	-2.1	1.4	4.1	9.8	14.6	18.6	21.8	20.8	15.4	12.4	4.6	2.1	-2.1	20
T.m.mx.	17.3	18.3	21.5	26.3	30.7	31.9	33.0	32.5	31.6	28.6	24.1	19.9	26.3	30
T.m.mn.	9.6	11.2	14.8	19.0	22.7	24.4	25.1	24.7	23.3	19.5	14.8	11.0	18.3	30
Tmn < 0													0.0	30
Tmn < -10													0.0	30
pTmn > 0													365.3	30
pTmn > 10	5.	13.											330.3	30
Frost d	1.8	0.6	0.0	0.0	0.0	0.0	0.0	0.0	0.0	0.0	0.2	1.6	4.3	30
r.hum.	74	79	83	82	80	82	82	82	78	74	75	74	79	20
cloud.	7.7	8.2	8.4	8.1	8.0	8.6	8.0	8.0	6.7	6.2	6.2	6.5	7.5	30
R.total	40	42	63	84	183	241	180	204	110	67	44	25	1283	20
RD > .1	9.5	11.0	15.1	14.1	16.0	17.8	16.2	16.6	11.8	8.2	8.7	8.4	153.1	20
RD > 5													56.7	30
RD > 10													36.4	30
RD > 25													14.7	30
Evap.													1693	30
Snow h.	0	0	0	0	0	0	0	0	0	0	0	0	0	30
Snow d.	0.0	0.0	0.0	0.0	0.0	0.0	0.0	0.0	0.0	0.0	0.0	0.1	0.1	30
Snow co													0.1	30
SD hrs.	3.2	2.5	2.3	3.7	5.6	5.8	7.2	7.1	7.1	6.5	5.5	4.2	5.1	20
SD pc.	30	22	19	29	43	43	54	55	58	56	50	40	42	20
W.vel.	1.9	2.0	2.0	2.2	2.2	1.9	2.1	1.8	1.7	1.6	1.6	1.7	1.9	19
W.mx,dir	ENE	ENE	E	E	SE	SE	E	E	ENE	ENE	ENE	ENE	E	19
W.mx,pc	17	17	16	15	14	14	16	15	12	12	14	16	14	19
Calms	22	20	20	18	16	20		24	19	19	18	25	23	19

222 Lancang 22° 34' N 99° 06' E 1054.0 m													222
	JAN	FEB	MAR	APR	MAY	JUN	JUL	AUG	SEP	OCT	NOV	DEC	YEAR	oby
A.press.	898	897	895	894	892	891	890	891	893	897	899	899	895	16
T. mean	12.5	14.1	17.4	20.5	22.7	23.0	22.6	22.5	22.0	20.1	16.3	13.2	18.9	17
T.a.mx.	28.2	29.9	32.8	36.1	37.1	37.2	33.5	32.7	33.3	32.0	29.3	27.2	37.2	17
T.a.mn.	-0.3	0.4	2.5	7.6	11.4	14.5	17.5	14.5	12.6	7.3	4.7	0.9	-0.3	17
T.m.mx.	23.4	22.2	29.0	30.9	30.4	28.6	27.6	28.0	28.5	26.9	24.8	23.1	27.3	27
T.m.mn.	6.0	6.2	8.8	12.7	17.3	19.9	20.2	19.8	18.8	16.7	12.1	8.0	13.9	27
Tmn < 0														
Tmn < -10														
pTmn > 0														
pTmn > 10														
Frost d	1.5	1.4	0.0	0.0	0.0	0.0	0.0	0.0	0.0	0.0	0.0	0.5	3.4	28
r.hum.	78	71	64	65	74	84	88	88	85	86	84	82	79	17
cloud.	4.6	3.6	2.9	4.4	7.0	9.2	9.4	9.1	8.2	7.7	6.3	5.4	6.5	27
R.total	30	10	14	32	166	255	368	341	169	170	63	34	1652	17
RD > .1	9.6	5.1	4.8	8.3	17.9	25.4	27.9	26.6	21.3	20.0	13.8	11.2	191.8	17
RD > 5														
RD > 10														
RD > 25														
Evap.														
Snow h.														
Snow d.														
Snow co														
SD hrs.	6.4	7.8	8.4	7.8	6.6	4.1	3.2	3.9	5.4	4.9	5.8	6.0	5.8	17
SD pc.	59	68	70	61	50	31	24	30	44	43	52	56	48	17
W.vel.	0.7	1.1	1.3	1.3	1.1	0.8	0.7	0.7	0.7	0.6	0.5	0.5	0.8	17
W.mx,dir	SSW	SW	W	W	WNW	SSW	SSW	SSW	E	SSW	SSW	SW	SSW	17
W.mx,pc	4	5	7	6	6	4	5	4	4	3	4	3	4	17
Calms	66	61	59	56	58	62	61	62	61	64	71	68	62	17

225 Kunming 25° 01' N 102° 41' E 1891.4 m														225
	JAN	FEB	MAR	APR	MAY	JUN	JUL	AUG	SEP	OCT	NOV	DEC	YEAR	oby
A.press.	811	810	810	809	808	808	807	809	811	814	814	813	810	20
T. mean	7.8	9.8	13.2	16.7	19.3	19.5	19.9	19.2	17.6	15.0	11.5	8.3	14.8	20
T.a.mx.	22.0	24.5	27.5	30.4	31.5	31.3	28.8	29.7	28.4	26.3	25.0	21.1	31.5	20
T.a.mn.	-5.4	-1.9	-2.8	0.5	6.0	9.2	12.0	8.8	6.3	2.0	0.6	-4.6	-5.4	20
T.m.mx.	15.3	17.2	20.8	24.0	24.9	23.8	24.0	23.9	22.6	20.1	17.5	15.2	20.8	30
T.m.mn.	1.4	2.9	5.7	9.2	13.8	16.1	16.8	15.9	14.1	11.3	6.6	2.6	9.7	30
Tmn < 0													16.8	30
Tmn < -10													0.0	30
pTmn > 0													365.3	30
pTmn > 10			3.								19.		262.7	30
Frost d	23.0	17.2	11.9	1.2	0.0	0.0	0.0	0.0	0.0	0.3	7.1	20.4	81.1	30
r.hum.	68	62	58	56	64	78	83	84	82	82	76	73	72	20
cloud.	3.4	3.4	3.3	4.2	6.4	8.7	8.8	8.2	7.6	7.1	5.0	3.8	5.8	30
R.total	10	10	14	20	78	182	216	195	123	95	34	16	993	20
RD > .1	4.1	3.9	5.0	5.5	11.4	18.9	21.3	21.1	15.2	16.1	6.9	4.6	133.6	20
RD > 5													50.7	30
RD > 10													31.4	30
RD > 25													9.6	30
Evap.													1871	30
Snow h.	6	6	0	0	0	0	0	0	0	0	0	3	6	17
Snow d.	0.8	0.5	0.3	0.0	0.0	0.0	0.0	0.0	0.0	0.0	0.0	0.6	2.2	30
Snow co	29.											30.	27.5	30
SD hrs.	7.8	8.4	9.0	7.6	7.8	5.1	4.9	5.9	5.7	5.0	7.1	7.2	6.9	17
SD pc.	73	74	76	73	59	37	36	45	47	44	65	68	57	17
W.vel.	2.4	2.8	2.8	2.8	2.6	2.1	1.9	1.4	1.4	1.7	1.9	2.0	2.1	20
W.mx,dir	SW	SW	SW	SW	SW	SW	SW	S	S	SW	SW	SW	SW	20
W.mx,pc	24	27	24	25	24	18	20	10	12	13	21	23	19	20
Calms	33	30	29	27		24	28	39	40	35	36	35	31	20

230 Deqen	28° 39' N	99° 10' E	3588.6 m										230
	JAN	FEB	MAR	APR	MAY	JUN	JUL	AUG	SEP	OCT	NOV	DEC	YEAR	oby
A.press.	657	657	659	661	661	660	661	662	663	663	662	660	660	14
T. mean	-2.9	-2.6	0.4	3.8	7.5	10.7	11.9	11.3	9.9	5.8	1.3	-1.2	4.6	14
T.a.mx.	11.9	12.7	15.6	19.3	21.4	22.1	22.7	22.0	21.9	18.9	15.4	12.6	22.7	14
T.a.mn.	-11.7	-11.6	-9.0	-7.9	-1.9	0.7	3.8	4.7	1.2	-4.8	-10.2	-13.1	-13.1	14
T.m.mx.	3.2	3.5	6.6	9.9	13.8	16.2	17.2	16.9	15.6	12.2	8.0	5.3	10.7	24
T.m.mn.	-6.8	-5.9	-3.1	0.0	3.7	7.2	8.6	8.2	6.6	2.2	-2.5	-5.3	1.1	24
Tmn < 0														
Tmn < -10														
pTmn > 0														
pTmn > 10														
Frost d	24.3	20.0	22.3	18.1	5.4	0.3	0.0	0.0	0.5	16.1	25.9	26.5	159.4	24
r.hum.	56	66	66	70	74	80	84	85	83	73	61	52	71	14
cloud.	3.6	5.3	5.3	6.5	7.1	8.4	9.1	8.8	8.4	5.6	3.3	2.5	6.2	24
R.total	12	20	30	55	69	70	145	141	56	59	11	8	676	14
RD > .1	7.2	13.3	13.0	12.8	15.1	17.9	21.3	22.4	16.1	10.6	5.3	3.9	158.9	14
RD > 5														
RD > 10														
RD > 25														
Evap.														
Snow h.	17	23	65	70	13	0	0	0	0	23	14	32	70	18
Snow d.	7.3	11.7	13.1	10.1	2.0	0.0	0.0	0.0	0.0	2.4	5.6	4.0	56.4	24
Snow co														
SD hrs.	6.2	5.3	5.8	5.8	5.5	4.7	4.0	4.2	4.7	5.8	6.7	7.0	5.5	14
SD pc.	59	47	48	45	40	34	29	32	38	50	63	69	45	14
W.vel.	2.3	2.3	2.3	2.1	2.3	2.1	1.7	1.7	1.8	1.8	2.0	2.0	2.1	14
W.mx,dir	S	SSW	SSW	SSW	SW	SSW	SSW	SSW	SSW	SSW	SSW	SSW	SSW	14
W.mx,pc	14	16	16	15	17	20	16	17	17	13	14	14	15	14
Calms	32	29	30	34	27	32	36	37	32	36	35	35	33	14

233 Garze 31° 38' N 99° 59' E 3393.5 m														233
	JAN	FEB	MAR	APR	MAY	JUN	JUL	AUG	SEP	OCT	NOV	DEC	YEAR	oby
A.press.	671	670	672	673	674	674	675	676	677	677	676	674	674	29
T. mean	-4.4	-1.5	2.6	6.6	10.5	12.5	14.0	13.4	11.0	6.5	0.5	-4.1	5.6	30
T.a.mx.	15.4	18.5	21.4	26.6	28.7	31.7	28.8	29.8	28.1	25.4	21.3	17.0	31.7	30
T.a.mn.	-22.5	-23.6	-16.4	-9.6	-5.3	-2.4	0.6	-1.1	-4.4	-9.5	-15.7	-28.7	-28.7	30
T.m.mx.	5.3	7.5	11.4	15.3	18.4	19.5	21.2	21.0	18.8	14.6	10.0	5.7	14.1	30
T.m.mn.	-11.9	-8.6	-4.6	-0.7	3.8	7.0	8.6	7.7	5.7	0.8	-6.4	-11.2	-0.8	30
Tmn < 0														
Tmn < -10														
pTmn > 0														
pTmn > 10														
Frost d	21.3	16.4	18.1	12.2	4.5	1.5	0.7	1.3	4.0	14.5	26.3	26.5	147.5	30
r.hum.	43	45	46	50	56	67	71	71	71	65	51	47	57	29
cloud.	4.4	5.9	6.4	7.1	7.3	8.0	8.0	7.5	7.1	5.7	3.7	3.5	6.2	29
R.total	4	9	15	32	72	128	110	95	113	47	7	4	636	30
RD > .1	3.4	5.4	7.6	11.2	16.8	22.2	22.2	19.7	19.4	12.6	3.7	2.8	146.9	30
RD > 5														
RD > 10														
RD > 25														
Evap.														
Snow h.	13	11	13	18	4	0	0	0	4	9	12	16	18	30
Snow d.	3.5	5.4	7.5	5.5	1.4	0.1	0.0	0.0	0.2	3.2	3.6	2.8	33.4	28
Snow co														
SD hrs.	7.0	6.8	7.2	7.5	7.6	6.8	7.0	7.0	6.9	7.3	7.9	7.5	7.2	27
SD pc.	68	61	60	58	56	48	50	53	56	64	75	74	60	27
W.vel.	1.6	2.0	2.5	2.5	2.3	2.0	1.6	1.5	1.6	1.5	1.3	1.2	1.8	30
W.mx,dir	W	W	W	W	W	W	E	W	W	W	W	W	W	30
W.mx,pc	6	10	11	11	10	9	7	6	7	7	6	5	8	30
Calms	51	42	35	36	34	41	46	47	46	48	53	59	45	30

238 X i g a z e 29° 13' N 88° 55' E 3836.0 m **238**

	JAN	FEB	MAR	APR	MAY	JUN	JUL	AUG	SEP	OCT	NOV	DEC	YEAR	oby
A.press.	636	634	636	638	637	637	638	640	640	640	639	638	638	15
T. mean	-4.1	-0.5	2.9	7.7	11.9	14.7	14.2	13.0	11.8	6.5	0.4	-3.5	6.3	15
T.a.mx.	14.6	19.6	21.5	23.7	25.9	27.5	27.3	25.2	24.1	21.2	17.9	14.8	27.5	15
T.a.mn.	-25.1	-18.6	-16.2	-10.0	-4.8	0.5	2.0	0.3	-0.6	-11.5	-15.4	-20.2	-25.1	15
T.m.mx.	5.7	8.0	11.0	15.5	19.5	22.1	21.0	19.6	18.7	15.2	10.7	7.0	14.5	26
T.m.mn.	-13.1	-9.5	-5.4	-0.7	3.4	7.6	8.7	8.1	5.8	-1.2	-8.2	-12.3	-1.4	26
Tmn < 0														
Tmn < -10														
pTmn > 0														
pTmn > 10														
Frost d	10.9	5.4	6.8	4.8	3.1	0.3	0.2	0.1	3.4	21.6	21.8	20.4	98.6	26
r.hum.	28	19	23	25	33	50	66	71	63	43	32	31	40	15
cloud.	2.0	2.8	4.3	4.9	4.8	6.3	8.0	8.3	6.1	2.5	1.4	1.6	4.4	26
R.total	.6	0	1	2	9	66	144	154	58	4	.2	0	439	15
RD > .1	0.5	0.0	0.9	1.6	4.3	12.1	19.9	23.3	12.9	2.2	0.5	0.0	78.3	15
RD > 5														
RD > 10														
RD > 25														
Evap.														
Snow h.	4	0	6	0	2	0	0	0	0	4	1	0	6	15
Snow d.	0.4	0.2	0.9	1.4	1.8	0.0	0.0	0.0	0.1	0.8	0.6	0.1	6.2	26
Snow co														
SD hrs.	8.6	8.9	8.6	9.1	10.4	9.8	7.4	7.1	8.5	9.7	9.4	8.9	9.1	15
SD pc.	82	79	72	71	77	70	54	54	69	85	88	86	73	15
W.vel.	1.7	2.6	2.6	2.8	2.6	2.1	1.4	1.1	1.2	1.1	1.2	1.0	1.8	15
W.mx,dir	SW	WSW	WSW	SW	SE	SE	N	N	N	SE	SW	SW	SW	15
W.mx,pc	9	17	16	16	11	18	11	11	10	6	7	6	7	15
Calms	52	35	34	29	26	31	47	53	51	56	61	66	45	15

239 L h a s a 29° 42' N 91° 08' E 3658.0 m **239**

	JAN	FEB	MAR	APR	MAY	JUN	JUL	AUG	SEP	OCT	NOV	DEC	YEAR	oby
A.press.	650	649	651	652	651	651	652	654	655	654	654	652	652	15
T. mean	-2.3	0.8	4.3	8.3	12.6	15.5	14.9	14.1	12.8	8.1	1.9	-1.9	7.5	20
T.a.mx.	15.8	20.7	23.6	23.6	28.7	29.4	26.7	25.6	24.9	23.2	19.3	16.7	29.4	20
T.a.mn.	-16.5	-15.4	-13.6	-8.1	-2.7	2.3	4.7	3.7	0.6	-7.2	-11.2	-16.1	-16.5	20
T.m.mx.	6.8	9.2	12.0	15.7	19.7	22.5	21.7	20.7	19.6	16.4	11.6	7.7	15.3	26
T.m.mn.	-10.2	-6.9	-3.2	0.9	5.1	9.2	9.9	9.4	7.6	1.4	-5.0	-9.1	0.8	26
Tmn < 0													169.8	30
Tmn < -10													36.7	30
pTmn > 0		19.									25.		280.4	30
pTmn > 10					11.					6.			148.9	30
Frost d	19.8	14.1	11.3	7.9	3.7	0.1	0.0	0.0	1.5	19.6	25.7	26.1	129.9	30
r.hum.	28	27	30	35	40	53	67	70	65	48	37	35	45	16
cloud.	2.8	3.9	5.4	6.1	6.0	7.1	8.1	8.3	6.8	3.2	1.9	2.2	5.1	30
R.total	.2	.1	2	4	21	73	142	149	57	5	.8	.3	454	16
RD > .1	0.5	0.5	1.8	3.7	7.7	14.4	20.4	21.2	13.8	2.9	0.6	0.2	87.7	16
RD > 5													30.9	30
RD > 10													15.0	30
RD > 25													1.1	30
Evap.													2206	30
Snow h.	4	3	10	3	4	0	0	0	0	0	4	2	10	20
Snow d.	0.5	0.5	1.8	2.2	1.3	0.0	0.0	0.0	0.1	0.9	0.7	0.4	8.3	30
Snow co					13.					23.			203.8	30
SD hrs.	8.1	8.1	7.6	8.0	9.4	8.6	7.1	7.0	8.0	9.5	9.0	8.4	8.2	16
SD pc.	78	72	63	62	69	62	51	53	66	83	85	82	68	16
W.vel.	2.2	2.4	2.6	2.6	2.5	2.2	1.7	1.6	1.7	1.8	1.9	1.9	2.1	16
W.mx,dir	E	E	E	E	E	W	ESE	ESE	ESE	ESE	E	E	E	16
W.mx,pc	18	15	13	13	13	15	15	14	14	15	18	20	14	16
Calms	21	19	17	16	18	21	28	29	27	28	26	27	23	16

240 N a g q u 31° 29' N 92° 03' E 4507.0 m **240**

	JAN	FEB	MAR	APR	MAY	JUN	JUL	AUG	SEP	OCT	NOV	DEC	YEAR	oby
A.press.	584	583	585	587	587	588	589	590	591	590	588	586	587	15
T. mean	-14.4	-11.1	-6.5	-1.4	3.3	7.3	8.8	8.1	5.2	-1.3	-9.0	-13.7	-2.1	17
T.a.mx.	5.4	9.2	11.3	15.5	20.4	21.2	22.0	20.6	18.4	15.7	10.8	8.8	22.0	17
T.a.mn.	-41.2	-34.5	-26.2	-22.1	-15.9	-8.0	-6.8	-7.0	-12.4	-23.6	-32.5	-35.3	-41.2	17
T.m.mx.	-3.3	-1.0	2.8	7.1	11.1	14.4	15.5	14.6	12.7	7.1	1.6	-2.3	6.7	27
T.m.mn.	-23.2	-20.0	-15.2	-9.7	-4.2	0.8	2.9	2.7	-0.6	-7.0	-16.4	-22.1	-9.3	27
Tmn < 0														
Tmn < -10														
pTmn > 0														
pTmn > 10														
Frost d														
r.hum.	39	29	32	36	48	64	72	73	71	59	44	41	51	16
cloud.														
R.total	2	1	4	6	24	81	100	101	63	18	3	3	405	16
RD > .1	2.4	1.8	4.8	4.4	10.6	19.6	20.7	20.2	17.2	7.4	2.7	2.7	114.4	16
RD > 5														
RD > 10														
RD > 25														
Evap.														
Snow h.	8	4	5	11	10	3	0	0	16	17	8	8	17	17
Snow d.														
Snow co														
SD hrs.	7.7	7.7	7.5	8.0	8.9	7.8	7.1	7.0	7.5	8.2	8.5	7.9	7.8	16
SD pc.	75	69	62	62	65	55	51	53	61	73	81	78	64	16
W.vel.	2.9	3.6	3.6	3.5	3.2	2.8	2.2	1.9	2.1	2.4	2.5	2.4	2.8	16
W.mx,dir	W	W	W	WSW	WSW	NE	NE	NE	NE	NE	NE	NE	NE	16
W.mx,pc	11	16	13	10	8	9	8	8	9	8	11	10	8	16
Calms	30	22	22	21	23	24	29	31	31	31	33	34	28	16

243 Yushu	33° 06' N	96° 45' E	3702.6 m										243
	JAN	FEB	MAR	APR	MAY	JUN	JUL	AUG	SEP	OCT	NOV	DEC	YEAR	oby
A.press.	646	645	647	649	650	651	651	652	653	653	651	649	650	25
T. mean	-7.8	-5.0	-0.5	4.0	7.7	10.6	12.5	11.6	8.7	3.3	-3.0	-7.2	2.9	30
T.a.mx.	12.5	13.0	17.9	22.9	26.0	28.7	28.3	27.5	26.4	25.1	18.5	14.4	28.7	30
T.a.mn.	-26.1	-25.9	-18.5	-12.8	-7.9	-3.8	-1.9	-2.3	-7.8	-12.3	-20.6	-26.1	-26.1	30
T.m.mx.	1.7	3.6	8.1	12.6	15.6	17.6	19.9	19.5	17.0	12.0	7.0	3.2	11.5	27
T.m.mn.	-15.5	-12.2	-7.6	-3.3	1.2	4.9	6.7	5.7	3.1	-2.5	-10.1	-14.9	-3.7	27
Tmn < 0														
Tmn < -10														
pTmn > 0														
pTmn > 10														
Frost d	14.4	11.8	16.9	16.5	10.7	4.7	3.2	4.7	9.5	20.9	26.0	20.4	159.8	27
r.hum.	43	42	41	47	57	65	69	70	71	64	48	43	55	27
cloud.	5.2	6.5	6.9	7.1	7.3	7.7	7.4	7.0	6.8	5.7	3.9	4.0	6.3	27
R.total	4	5	7	11	53	99	105	90	76	26	3	1	480	27
RD > .1	3.0	3.7	5.9	8.6	16.7	21.9	21.4	18.5	19.0	11.3	2.3	1.7	134.1	27
RD > 5														
RD > 10														
RD > 25														
Evap.														
Snow h.	7	8	13	4	9	4	0	0	0	12	9	4	13	30
Snow d.	3.1	3.5	6.2	8.0	5.0	0.5	0.0	0.0	0.7	6.1	2.3	1.7	37.3	27
Snow co														
SD hrs.	5.9	6.2	6.9	7.2	7.4	6.6	7.0	6.9	6.5	6.7	7.1	6.4	6.7	27
SD pc.	57	56	58	56	53	47	50	52	52	59	68	64	55	27
W.vel.	1.1	1.4	1.6	1.5	1.3	1.1	0.9	0.8	0.7	0.8	1.0	1.0	1.1	26
W.mx,dir	W	W	W	W	W	ENE	ENE	ENE	ENE	W	W	W	W	27
W.mx,pc	10	12	11	9	7	8	7	7	6	5	8	8	7	27
Calms	62	54	49	50	52	57	63	65	68	67	65	65	60	27

247 L a n z h o u 36° 03' N 103° 53' E 1517.2 m **247**

	JAN	FEB	MAR	APR	MAY	JUN	JUL	AUG	SEP	OCT	NOV	DEC	YEAR	oby
A.press.	852	849	848	846	845	843	842	844	849	852	854	853	848	30
T. mean	-6.9	-2.3	5.2	11.8	16.6	20.3	22.2	21.0	15.8	9.4	1.7	-5.5	9.1	30
T.a.mx.	17.1	19.9	27.5	31.8	33.9	36.7	39.1	38.3	33.0	27.6	20.3	16.2	39.1	30
T.a.mn.	-21.7	-17.6	-16.3	-8.6	1.0	4.6	9.7	5.4	0.4	-6.0	-14.9	-19.9	-21.7	30
T.m.mx.	1.0	5.4	12.7	19.3	23.8	27.5	29.2	27.6	22.1	16.5	8.8	2.1	16.3	30
T.m.mn.	-12.6	-8.0	-0.8	5.3	10.2	13.6	16.3	15.3	10.7	4.1	-3.1	-10.4	3.4	30
Tmn < 0													139.0	30
Tmn < -10													50.0	30
pTmn > 0			2.								21.		265.5	30
pTmn > 1				19.						12.			176.3	30
Frost d	22.8	16.8	11.4	3.3	0.5	0.0	0.0	0.0	0.5	12.4	23.0	27.3	118.0	30
r.hum.	58	53	49	48	51	53	61	65	69	69	65	65	59	30
cloud.	4.0	5.1	6.4	6.5	7.0	6.5	6.2	5.8	6.4	5.3	4.2	3.1	5.5	27
R.total	1	2	8	17	36	33	64	85	49	25	5	1	326	30
RD > .1	1.9	2.5	3.9	6.3	8.4	9.2	11.9	11.5	10.9	6.5	2.6	1.5	77.0	30
RD > 5													20.8	30
RD > 10													9.7	30
RD > 25													1.6	30
Evap.													1438	30
Snow h.	8	6	7	10	0	0	0	0	0	5	6	9	10	30
Snow d.	2.0	2.5	2.7	0.9	0.0	0.0	0.0	0.0	0.0	0.8	1.8	1.6	12.2	28
Snow co				9.							1.		160.0	30
SD hrs.	6.1	6.8	6.9	7.6	8.2	8.7	8.2	7.9	6.6	6.6	6.6	5.8	7.1	30
SD pc.	61	62	58	58	58	60	57	59	53	59	61	61	59	30
W.vel.	0.5	0.8	1.2	1.5	1.5	1.4	1.4	1.2	0.9	0.7	0.5	0.3	1.0	30
W.mx,dir	NE	NE	NE	NE	NE	E	E	NE	NE	NE	NE	NE	NE	30
W.mx,pc	3	7	11	12	10	9	9	8	7	6	4	3	7	30
Calms	71	59	47	41	40	42	44	48	57	64	70	77	55	30

250 Yinchuan 38° 29' N 106° 13' E 1111.5 m														**250**
	JAN	FEB	MAR	APR	MAY	JUN	JUL	AUG	SEP	OCT	NOV	DEC	YEAR	oby
A.press.	896	894	891	889	887	884	882	885	891	895	897	897	891	30
T. mean	-9.0	-4.8	2.8	10.6	16.9	21.4	23.4	21.6	16.0	9.1	0.9	-6.7	8.5	30
T.a.mx.	16.7	17.2	24.9	32.2	35.0	37.0	39.3	37.8	32.4	27.6	21.5	15.0	39.3	30
T.a.mn.	-30.6	-25.4	-19.3	-9.6	-2.5	3.9	11.2	6.8	-3.3	-7.9	-15.8	-29.3	-30.6	30
T.m.mx.	-1.5	3.0	10.5	18.5	24.4	28.2	29.6	27.7	22.9	16.6	7.1	-0.3	15.5	30
T.m.mn.	-15.0	-11.3	-3.5	3.2	9.4	14.5	17.7	16.5	10.4	2.9	-3.9	-11.6	2.4	30
Tmn < 0													156.9	30
Tmn < -10													67.1	30
pTmn > 0			10.								17.		253.2	30
pTmn > 10				20.						8.			172.0	30
Frost d	18.9	11.9	9.1	2.6	0.4	0.0	0.0	0.0	0.8	11.7	22.5	25.8	104.1	30
r.hum.	58	52	50	46	47	53	64	70	67	64	66	64	59	30
cloud.	3.0	4.0	5.1	5.6	5.6	5.3	5.5	5.3	4.9	3.7	3.2	2.7	4.5	30
R.total	1	2	6	12	15	20	44	56	27	14	5	.7	203	30
RD > .1	1.2	1.4	1.9	2.9	4.3	5.4	7.4	8.9	6.4	3.9	1.7	0.8	46.2	30
RD > 5													12.0	30
RD > 10													5.9	30
RD > 25													1.0	30
Evap.													1593	30
Snow h.	8	7	17	11	0	0	0	0	0	3	4	13	17	29
Snow d.	1.3	1.3	1.2	0.2	0.0	0.0	0.0	0.0	0.0	0.4	1.0	0.8	6.2	30
Snow co			21.							19.			129.7	30
SD hrs.	7.4	8.0	7.9	8.5	9.4	10.1	9.2	8.7	8.2	7.9	7.4	7.1	8.3	28
SD pc.	76	73	67	65	67	68	63	64	66	72	74	75	69	28
W.vel.	1.6	2.0	2.2	2.5	2.3	2.0	1.7	1.5	1.4	1.4	1.7	1.6	1.8	30
W.mx,dir	N	N	N	S	S	S	S	S	S	N	NNE	N	N	30
W.mx,pc	11	12	11	8	14	12	11	9	9	8	8	11	8	30
Calms	35	27	24	21	24	26	32	36	40	40	37	38	32	30

252 Bayan Mod 40° 45' N 104° 30' E 1328.1 m **252**

	JAN	FEB	MAR	APR	MAY	JUN	JUL	AUG	SEP	OCT	NOV	DEC	YEAR	oby
A.press.	872	870	868	866	864	862	860	863	867	872	873	873	867	23
T. mean	-12.1	-8.1	0.0	8.7	16.3	21.7	23.6	21.9	15.6	7.1	-2.8	-10.2	6.8	24
T.a.mx.	10.9	15.2	22.9	30.6	34.4	38.0	37.9	37.6	31.8	27.8	20.0	12.1	38.0	24
T.a.mn.	-30.7	-28.4	-22.8	-14.5	-4.4	2.5	7.1	6.5	-4.6	-12.8	-25.1	-31.7	-31.7	24
T.m.mx.	-4.5	0.1	8.0	16.5	23.5	28.6	30.3	28.6	22.7	15.0	4.8	-2.9	14.2	23
T.m.mn.	-17.9	-14.5	-6.7	1.1	8.5	14.0	16.9	15.5	9.2	0.8	-8.3	-15.7	0.3	23
Tmn < 0														
Tmn < -10														
pTmn > 0														
pTmn > 10														
Frost d	10.4	4.9	1.5	0.5	0.0	0.0	0.0	0.0	0.0	1.3	3.8	10.2	32.5	24
r.hum.	51	42	30	26	26	30	41	44	37	37	43	52	38	23
cloud.	2.4	3.0	4.6	5.0	5.1	5.2	5.6	4.7	4.1	2.8	2.6	2.3	4.0	23
R.total	.7	.5	2	4	7	13	28	28	10	4	2	.6	100	23
RD > .1	1.3	1.0	1.7	2.4	3.0	4.9	8.4	6.5	4.0	1.7	1.3	1.2	37.3	23
RD > 5														
RD > 10														
RD > 25														
Evap.														
Snow h.	6	4	3	5	0	0	0	0	0	1	12	8	12	24
Snow d.	1.4	1.0	1.5	1.0	0.1	0.0	0.0	0.0	0.0	0.6	1.2	1.4	8.1	22
Snow co														
SD hrs.	7.7	8.5	8.7	9.9	10.4	10.8	9.8	9.6	9.3	8.9	7.9	7.4	9.1	23
SD pc.	81	79	73	72	73	72	67	70	75	81	80	80	74	23
W.vel.	3.7	3.7	4.3	4.7	4.7	4.3	4.0	3.8	3.5	3.4	3.9	3.7	4.0	23
W.mx,dir	W	WNW	WNW	WNW	W	W	W	NE	W	W	W	W	W	23
W.mx,pc	22	17	16	16	13	13	11	10	10	14	24	23	15	23
Calms							13	12	13					23

256 Xining	36° 35' N 101° 55' E 2261.2 m												256
	JAN	FEB	MAR	APR	MAY	JUN	JUL	AUG	SEP	OCT	NOV	DEC	YEAR	oby
A.press.	775	773	773	773	774	774	773	775	777	779	779	777	775	26
T. mean	-8.4	-4.9	1.9	7.9	12.0	15.2	17.2	16.5	12.1	6.4	-0.8	-6.7	5.7	27
T.a.mx.	12.6	16.3	24.3	27.6	28.4	32.4	32.2	33.5	29.9	24.6	19.3	12.2	33.5	27
T.a.mn.	-24.9	-20.7	-16.1	-12.5	-2.0	0.2	4.5	4.0	-0.4	-8.8	-19.0	-26.6	-26.6	27
T.m.mx.	0.4	3.7	10.0	15.9	19.4	22.6	24.4	23.6	18.7	13.7	6.9	1.9	13.5	27
T.m.mn.	-15.1	-11.5	-4.2	1.5	5.9	8.8	11.3	10.9	7.3	1.2	-6.3	-12.9	-0.3	27
Tmn < 0													169.1	30
Tmn < -10													79.8	30
pTmn > 0			16.								8.		237.9	30
pTmn > 10					14.				24.				133.6	30
Frost d	13.9	8.5	6.1	4.3	1.7	0.2	0.0	0.0	1.9	13.3	21.0	22.3	93.1	27
r.hum.	48	46	46	49	55	58	65	67	68	65	58	54	56	27
cloud.	3.8	5.0	6.4	6.5	7.0	6.6	6.5	6.0	6.5	5.1	3.8	3.2	5.5	27
R.total	1	2	5	20	45	49	81	82	55	25	3	.9	369	27
RD > .1	2.6	2.7	3.7	6.4	11.6	12.7	15.1	14.3	13.4	7.2	2.9	2.2	95.1	27
RD > 5													24.6	30
RD > 10													10.6	30
RD > 25													1.1	30
Evap.													1763	30
Snow h.	9	9	7	18	1	0	0	0	0	7	14	5	18	28
Snow d.	2.7	2.7	3.7	2.9	0.6	0.0	0.0	0.0	0.0	1.9	2.9	2.2	19.5	27
Snow co					6.					12.			206.6	30
SD hrs.	7.0	7.5	7.6	8.0	8.1	8.4	8.0	8.0	6.8	7.2	7.2	7.0	7.6	27
SD pc.	70	69	64	61	58	58	56	59	55	64	71	72	62	27
W.vel.	1.5	2.1	2.6	2.6	2.3	2.0	1.9	1.9	1.9	1.7	1.7	1.4	2.0	27
W.mx,dir	SE	SE	SE	SE	SE	SE	SE	SE	SE	SE	SE	SE	SE	27
W.mx,pc	21	28	34	29	25	18	22	26	30	26	24	18	25	27
Calms	46	37			26	27	29	30	35	42	42	49	35	27

258 G o l m u d 36° 12' N 94° 38' E 2807.7 m **258**

	JAN	FEB	MAR	APR	MAY	JUN	JUL	AUG	SEP	OCT	NOV	DEC	YEAR	oby
A.press.	723	722	723	724	724	724	723	725	727	729	728	726	725	25
T. mean	-10.9	-6.7	-0.2	6.5	11.5	15.3	17.6	16.7	11.5	3.8	-4.6	-9.9	4.2	26
T.a.mx.	10.5	10.0	23.6	29.7	31.0	32.7	33.1	32.4	29.7	25.1	15.4	9.8	33.1	26
T.a.mn.	-33.6	-26.6	-26.1	-13.4	-7.8	-5.2	1.9	-1.4	-8.2	-20.2	-25.4	-28.5	-33.6	26
T.m.mx.	-1.8	2.4	8.9	15.0	19.1	22.4	24.9	24.3	19.2	12.7	4.6	-0.6	12.6	25
T.m.mn.	-18.4	-14.7	-8.2	-1.5	3.8	7.9	10.5	9.6	4.7	-3.5	-11.8	-17.2	-3.2	25
Tmn < 0														
Tmn < -10														
pTmn > 0														
pTmn > 10														
Frost d	5.7	1.6	1.0	0.3	0.1	0.0	0.0	0.0	0.2	1.4	3.6	5.0	18.9	26
r.hum.	41	31	26	25	27	32	36	36	34	32	36	40	33	25
cloud.	4.9	6.1	6.6	6.6	6.5	6.6	6.0	5.4	5.4	4.1	3.8	4.0	5.5	25
R.total	.6	.5	.9	1	4	7	9	8	5	1	.9	.3	38	25
RD > .1	1.3	1.0	1.0	1.3	2.8	4.2	5.4	4.0	2.4	1.0	1.0	0.6	25.9	25
RD > 5														
RD > 10														
RD > 25														
Evap.														
Snow h.	3	3	4	2	5	0	0	0	0	4	5	3	5	26
Snow d.	1.2	0.9	1.0	1.1	0.5	0.2	0.0	0.0	0.0	0.7	1.0	0.6	7.2	26
Snow co														
SD hrs.	6.9	7.5	8.1	9.0	9.4	9.3	9.0	9.0	8.6	8.9	8.1	7.3	8.4	25
SD pc.	70	68	68	69	67	64	63	67	69	79	80	76	70	25
W.vel.	2.6	2.9	3.3	3.5	3.9	3.7	3.6	3.3	3.0	2.8	2.6	2.4	3.2	25
W.mx,dir	SW	SW	SW	W	W	W	W	W	W	SW	SW	SW	SW	25
W.mx,pc	19	19	17	18	21	24	24	21	19	19	20	21	17	25
Calms														

259 Da Qaidam 37° 50' N 95° 17' E 3173.2 m														259
	JAN	FEB	MAR	APR	MAY	JUN	JUL	AUG	SEP	OCT	NOV	DEC	YEAR	oby
A.press.	689	688	689	691	692	692	692	693	695	696	695	692	692	23
T. mean	-14.3	-10.6	-3.6	2.8	8.4	12.3	15.1	14.4	8.6	0.8	-7.4	-12.8	1.1	25
T.a.mx.	6.9	11.0	17.5	24.2	26.1	29.6	29.9	29.6	26.1	21.1	13.6	7.5	29.9	25
T.a.mn.	-33.6	-31.7	-26.2	-19.8	-12.4	-5.7	-1.2	-6.6	-10.4	-20.5	-30.6	-31.6	-33.6	25
T.m.mx.	-4.3	-1.4	4.9	11.1	15.5	19.0	21.7	21.3	16.1	9.6	1.8	-2.8	9.4	24
T.m.mn.	-22.4	-18.9	-11.8	-5.7	0.2	4.7	7.8	7.1	1.2	-7.1	-15.0	-20.4	-6.7	24
Tmn < 0														
Tmn < -10														
pTmn > 0														
pTmn > 10														
Frost d	11.6	7.4	2.2	0.8	1.5	0.7	0.3	1.0	2.1	3.3	6.4	10.0	47.2	25
r.hum.	41	37	27	26	30	36	40	38	35	31	35	39	35	24
cloud.	4.4	5.4	6.3	6.3	6.3	6.1	5.8	5.2	4.9	3.7	3.6	3.8	5.2	24
R.total	2	2	2	1	11	20	20	16	7	1	.8	.6	83	24
RD > .1	2.5	2.2	1.4	1.3	3.3	5.0	6.6	5.6	2.8	1.0	1.1	1.1	33.9	24
RD > 5														
RD > 10														
RD > 25														
Evap.														
Snow h.	5	6	10	6	7	1	0	0	0	4	5	5	10	26
Snow d.	2.4	2.1	1.4	1.6	1.4	0.5	0.1	0.0	0.2	0.9	1.1	1.2	12.8	25
Snow co														
SD hrs.	7.4	8.0	8.7	9.6	10.0	9.9	9.6	9.4	9.1	9.1	8.1	7.4	8.9	24
SD pc.	75	74	73	73	71	68	66	71	74	82	81	78	73	24
W.vel.	1.2	1.8	2.6	3.0	2.9	2.8	2.5	2.3	2.1	1.8	1.4	1.1	2.1	24
W.mx,dir	NE	NE	W	W	W	W	W	W	W	W	NE	NE	W	24
W.mx,pc	12	12	14	17	16	15	15	15	13	11	9	10	12	24
Calms	49	37	29	27	27	29	30	31	34	38	46	52	36	24

260 Lenghu 38° 50' N 93° 23' E 2733.0 m **260**

	JAN	FEB	MAR	APR	MAY	JUN	JUL	AUG	SEP	OCT	NOV	DEC	YEAR	oby
A.press.	727	726	727	728	728	728	728	729	731	733	733	730	729	15
T. mean	-13.1	-9.2	-2.5	4.2	10.1	14.5	17.0	16.4	10.4	2.0	-6.6	-11.9	2.6	25
T.a.mx.	5.4	10.7	19.0	24.9	28.7	31.2	34.2	33.0	28.8	21.6	13.2	5.3	34.2	25
T.a.mn.	-34.3	-29.9	-28.4	-20.0	-11.7	-8.8	-2.0	-4.3	-12.8	-21.0	-29.6	-33.2	-34.3	25
T.m.mx.	-3.7	0.3	7.0	13.4	18.4	22.3	24.8	24.5	19.0	11.7	3.1	-2.2	11.6	24
T.m.mn.	-21.5	-18.1	-11.9	-5.3	1.0	5.8	8.9	8.3	1.8	-7.0	-15.2	-20.4	-6.1	24
Tmn < 0													232.2	30
Tmn < -10													148.7	30
pTmn > 0														
pTmn > 10														
Frost d	8.6	4.7	2.6	1.0	0.2	0.1	0.0	0.0	0.4	1.5	6.4	8.5	34.0	25
r.hum.	36	31	26	23	22	27	31	29	26	27	34	37	29	24
cloud.	4.2	4.8	6.0	6.0	5.9	5.5	5.3	4.5	4.2	3.2	3.5	3.7	4.7	24
R.total	.5	.1	.2	.2	1	5	5	5	.2	0	.1	.1	17	24
RD > .1	0.7	0.3	0.2	0.3	0.8	2.2	3.3	2.2	0.6	0.1	0.3	0.3	11.4	24
RD > 5													0.9	30
RD > 10													0.3	30
RD > 25													0.0	30
Evap.													3298	30
Snow h.	3	1	3	1	0	0	0	0	0	0	0	2	3	25
Snow d.	0.7	0.3	0.2	0.2	0.2	0.0	0.1	0.0	0.0	0.1	0.2	0.3	2.4	25
Snow co				24.							5.		171.0	30
SD hrs.	7.9	8.9	9.5	10.1	11.0	11.3	10.7	10.8	10.3	9.8	8.6	7.8	9.7	24
SD pc.	81	82	80	77	77	76	74	79	83	88	86	83	80	24
W.vel.	3.2	3.7	4.2	4.5	4.9	4.9	4.8	4.7	4.1	3.4	2.9	2.5	4.0	24
W.mx,dir	ENE	ENE	W	W	NE	NE	NE	NE	ENE	ENE	ENE	ENE	ENE	23
W.mx,pc	19	13	13	12	15	16	17	18	15	11	13	16	14	23
Calms	25	18	15	13						16	25	29	15	23

261 Jiuquan 39° 46' N 98° 31' E 1477.2 m														261
	JAN	FEB	MAR	APR	MAY	JUN	JUL	AUG	SEP	OCT	NOV	DEC	YEAR	oby
A.press.	857	854	853	851	850	847	846	848	853	857	858	857	853	26
T. mean	-9.7	-5.9	1.8	9.4	15.6	20.1	21.8	20.7	15.0	7.4	-1.2	-7.9	7.3	30
T.a.mx.	10.7	15.2	23.3	30.2	33.6	36.1	38.4	37.5	32.2	29.0	17.9	17.3	38.4	30
T.a.mn.	-28.6	-31.6	-20.7	-10.6	-3.4	2.4	7.7	6.4	-3.6	-9.4	-24.2	-27.4	-31.6	30
T.m.mx.	-2.3	1.6	9.5	17.2	22.9	27.2	28.7	28.0	22.6	15.4	6.0	-0.8	14.7	30
T.m.mn.	-15.6	-12.1	-4.4	2.3	8.2	12.6	14.8	14.0	8.5	1.2	-6.5	-13.3	0.8	30
Tmn < 0														
Tmn < -10														
pTmn > 0														
pTmn > 10														
Frost d	17.2	11.6	5.5	1.6	0.5	0.0	0.0	0.0	1.6	9.8	14.4	18.6	80.8	30
r.hum.	55	49	40	35	35	42	52	50	47	44	49	57	46	29
cloud.	3.8	4.7	5.8	6.1	5.7	5.4	5.5	4.6	4.2	3.1	3.5	3.7	4.7	27
R.total	2	2	3	5	9	11	21	17	11	2	2	1	86	30
RD > .1	1.9	2.8	2.3	2.2	2.9	5.8	8.1	6.3	3.4	1.2	1.9	2.0	40.8	30
RD > 5														
RD > 10														
RD > 25														
Evap.														
Snow h.	9	14	7	4	0	0	0	0	0	5	10	9	14	30
Snow d.	2.0	2.9	2.2	1.0	0.1	0.0	0.0	0.0	0.0	0.6	1.7	2.0	12.5	28
Snow co														
SD hrs.	7.0	7.4	7.6	8.6	9.4	9.7	9.1	9.1	8.9	8.6	7.5	6.7	8.3	30
SD pc.	73	69	64	65	65	65	62	67	72	78	75	72	68	30
W.vel.	2.0	2.3	2.7	3.2	2.7	2.6	2.2	2.1	2.1	2.1	2.3	2.0	2.4	23
W.mx,dir	SW	SW	SW	NW	E	NW	E	E	SW	SW	SW	SW	SW	23
W.mx,pc	13	13	11	11	11	10	9	10	11	14	16	14	11	23
Calms	18	15	13	12	14	15	20	21	18	17		19	16	23

264 Ruoqiang 39° 02' N 88° 10' E 888.3 m **264**

	JAN	FEB	MAR	APR	MAY	JUN	JUL	AUG	SEP	OCT	NOV	DEC	YEAR	oby
A.press.	923	920	915	911	909	906	904	906	912	918	923	924	914	23
T. mean	-8.5	-2.3	7.1	15.4	21.0	25.3	27.4	26.0	20.1	11.2	1.6	-6.3	11.5	28
T.a.mx.	7.8	18.5	29.8	36.2	40.7	42.6	43.6	42.2	37.5	32.0	25.1	13.6	43.6	28
T.a.mn.	-27.2	-21.9	-13.6	-6.8	-0.5	4.9	10.1	8.6	-0.8	-8.2	-16.3	-25.0	-27.2	28
T.m.mx.	-1.5	5.4	15.7	24.1	29.3	33.6	35.6	34.7	29.4	21.1	9.8	0.7	19.8	28
T.m.mn.	-14.1	-8.9	-0.8	6.9	12.0	16.4	19.1	17.3	11.2	2.9	-4.6	-11.5	3.8	28
Tmn < 0														
Tmn < -10														
pTmn > 0														
pTmn > 10														
Frost d	15.3	5.7	0.6	0.0	0.1	0.0	0.0	0.0	0.1	3.1	9.7	15.9	50.5	28
r.hum.	57	45	31	26	27	32	35	33	33	38	47	57	38	27
cloud.	5.1	5.3	6.4	6.7	6.0	5.2	4.8	4.3	4.0	2.9	3.8	4.6	4.9	17
R.total	1	.6	.2	.6	1	3	6	2	.5	.2	.4	.7	17	27
RD > .1	1.7	0.7	0.4	0.6	1.1	2.0	2.8	1.3	0.4	0.2	0.4	1.0	12.6	27
RD > 5														
RD > 10														
RD > 25														
Evap.														
Snow h.	15	12	2	0	0	0	0	0	0	0	6	18	18	28
Snow d.	1.6	0.8	0.1	0.0	0.0	0.0	0.0	0.0	0.0	0.0	0.3	1.3	4.1	28
Snow co														
SD hrs.	6.5	7.2	7.5	7.9	8.9	10.1	9.8	9.8	9.8	9.6	7.6	6.5	8.4	26
SD pc.	67	66	62	60	66	68	67	72	79	83	76	68	69	26
W.vel.	1.8	2.3	3.2	3.9	3.9	3.4	3.0	3.1	2.8	2.2	1.7	1.7	2.7	26
W.mx,dir	SW	NE	NE	NE	NE	NE	NE	NE	NE	NE	NE	SW	NE	26
W.mx,pc	12	13	19	20	21	21	20	23	22	18	12	11	17	26
Calms	34	29	22			24	27	26	30	34	41	38	29	26

265 H a m i 42° 49' N 93° 31' E 737.9 m **265**

	JAN	FEB	MAR	APR	MAY	JUN	JUL	AUG	SEP	OCT	NOV	DEC	YEAR	oby
A.press.	941	937	932	929	926	922	919	922	928	935	939	941	931	27
T. mean	-12.2	-5.8	4.5	13.2	20.2	25.2	27.2	25.9	19.1	9.9	-0.6	-9.0	9.8	30
T.a.mx.	6.0	14.3	26.5	33.4	38.2	42.5	43.9	43.2	37.4	31.7	19.3	7.0	43.9	30
T.a.mn.	-31.9	-28.8	-24.5	-11.7	-1.6	5.6	9.4	6.6	-1.7	-9.6	-27.6	-32.0	-32.0	30
T.m.mx.	-4.7	2.4	12.6	21.0	27.6	32.5	34.5	33.6	27.7	18.7	7.0	-2.3	17.6	30
T.m.mn.	-17.7	-12.2	-2.9	5.1	11.9	17.0	19.4	18.1	11.4	2.9	-6.1	-13.9	2.7	30
Tmn < 0														
Tmn < -10														
pTmn > 0														
pTmn > 10														
Frost d	20.3	10.3	2.0	0.1	0.0	0.0	0.0	0.0	0.1	1.6	7.8	19.3	61.7	30
r.hum.	63	50	34	27	27	33	34	34	36	46	51	62	41	29
cloud.	3.4	4.1	5.0	5.7	5.4	5.5	5.1	4.3	3.5	3.0	3.3	3.6	4.3	29
R.total	2	1	1	3	3	6	6	5	3	2	2	2	36	29
RD > .1	1.8	1.1	0.9	1.3	1.6	3.1	3.7	3.0	1.6	1.0	0.8	2.2	21.9	29
RD > 5														
RD > 10														
RD > 25														
Evap.														
Snow h.	11	14	6	2	0	0	0	0	0	1	16	10	16	30
Snow d.	1.8	1.1	0.5	0.2	0.0	0.0	0.0	0.0	0.0	0.1	0.6	2.1	6.5	30
Snow co														
SD hrs.	6.9	8.1	8.8	9.7	10.9	11.2	10.8	10.7	10.2	9.1	7.5	6.5	9.2	27
SD pc.	73	76	74	72	75	74	73	77	83	83	77	72	76	27
W.vel.	2.3	2.5	3.0	3.7	3.7	3.3	3.1	2.9	2.7	2.5	2.4	2.2	2.8	29
W.mx,dir	NE	NE	NE	NE	NE	NE	NE	NE	NE	NE	ENE	NE	NE	28
W.mx,pc	22	16	13	14	15	16	14	14	15	15	14	16	15	28
Calms								16	18		16	18		28

267 T u r p a n 42° 56' N 89° 12' E 34.5 m **267**

	JAN	FEB	MAR	APR	MAY	JUN	JUL	AUG	SEP	OCT	NOV	DEC	YEAR	oby
A.press.	1031	1024	1016	1009	1004	999	996	999	1007	1017	1026	1030	1013	26
T. mean	-9.5	-2.1	9.3	18.9	25.7	31.0	32.7	30.4	23.3	12.6	1.8	-7.2	13.9	30
T.a.mx.	8.5	19.5	29.7	37.6	43.6	47.5	47.6	46.6	43.4	34.3	23.0	9.0	47.6	30
T.a.mn.	-28.0	-24.5	-10.4	-1.8	4.7	11.5	15.1	11.6	1.3	-5.7	-14.2	-26.1	-28.0	30
T.m.mx.	-3.1	5.1	16.6	26.1	33.1	38.2	39.9	38.2	32.0	21.8	9.4	-1.0	21.4	30
T.m.mn.	-14.5	-8.2	2.2	11.3	17.5	22.9	25.1	22.6	15.5	5.9	-3.5	-11.7	7.1	30
Tmn < 0	31.0	27.1	10.2	0.2	0	0	0	0	0	2.1	23.0	31.0	124.6	30
Tmn < -10	26.5	9.6	0.1	0	0	0	0	0	0	0	2.8	19.9	59.9	30
pTmn > 0		23.									21.		272.2	30
pTmn > 10			23.							22.			213.9	30
Frost d	16.8	6.6	0.8	0.0	0.0	0.0	0.0	0.0	0.0	3.5	14.2	21.5	63.4	30
r.hum.	59	46	33	27	27	29	31	36	41	49	53	62	41	29
cloud.	3.9	4.1	4.9	5.5	5.4	5.6	5.3	4.4	3.5	3.0	3.3	3.7	4.4	27
R.total	2	.3	1	.4	.5	3	2	3	1	1	.4	1	16	29
RD > .1	2.2	0.3	0.4	0.5	0.7	2.2	2.7	2.5	1.2	0.3	0.3	1.7	15.0	29
RD > 5													0.7	30
RD > 10													0.2	30
RD > 25													0.0	30
Evap.													2838	30
Snow h.	17	13	1	0	0	0	0	0	0	0	2	14	17	30
Snow d.	2.1	0.2	0.0	0.0	0.0	0.0	0.0	0.0	0.0	0.0	0.2	1.6	4.5	30
Snow co		4.									24.		30.2	30
SD hrs.	5.8	7.3	7.9	8.7	9.7	10.3	10.3	10.1	9.6	8.5	6.8	5.3	8.4	28
SD pc.	62	68	67	65	67	67	69	73	77	77	70	58	68	28
W.vel.	0.9	1.2	1.8	2.3	2.4	2.5	2.3	2.0	1.7	1.2	1.0	0.8	1.7	25
W.mx,dir	N	N	E	E	E	E	E	E	E	E	N	N	E	25
W.mx,pc	10	12	10	12	11	11	9	9	10	8	8	9	9	25
Calms	49	37	23	22	21	21	23	26	31	39	44	51	32	25

269 Urumqi 43° 54' N 87° 28' E 653.5 m **269**

	JAN	FEB	MAR	APR	MAY	JUN	JUL	AUG	SEP	OCT	NOV	DEC	YEAR	oby
A.press.	921	918	916	913	911	908	905	908	913	918	921	921	914	13
T. mean	-15.4	-12.1	-4.0	9.0	15.9	21.2	23.5	22.0	16.8	7.4	-4.2	-11.6	5.7	30
T.a.mx.	8.8	13.5	19.8	30.0	35.7	36.5	40.5	38.8	34.9	28.1	17.4	8.8	40.5	30
T.a.mn.	-34.1	-41.5	-33.4	-14.9	-1.2	4.9	9.1	6.2	-5.0	-12.4	-36.6	-38.3	-41.5	30
T.m.mx.	-9.6	-6.4	0.7	15.1	22.2	27.0	29.6	28.4	23.4	13.4	0.7	-6.4	11.5	30
T.m.mn.	-20.3	-16.8	-8.1	3.5	9.8	15.3	17.3	16.0	11.0	2.5	-7.9	-16.1	0.5	30
Tmn < 0													163.3	30
Tmn < -10													98.6	30
pTmn > 0			31.								3.		217.5	30
pTmn > 10					2.					3.			154.3	30
Frost d	23.1	19.6	19.6	5.4	0.9	0.0	0.0	0.0	0.4	11.1	19.5	23.4	123.0	14
r.hum.	80	81	77	54	44	45	44	42	45	57	76	82	61	14
cloud.	4.3	5.4	5.9	5.8	5.6	5.9	5.4	4.6	3.7	3.5	5.2	5.4	5.0	14
R.total	9	11	21	34	35	39	22	24	26	24	19	15	279	14
RD > .1	7.6	8.8	7.3	7.7	7.1	8.5	7.9	6.2	5.3	4.8	7.5	9.6	88.3	14
RD > 5													17.5	30
RD > 10													6.7	30
RD > 25													0.9	30
Evap.													1914	30
Snow h.	44	40	48	20	8	0	0	0	3	12	33	33	48	30
Snow d.	7.7	8.6	7.4	2.8	0.9	0.0	0.0	0.0	0.1	2.3	7.1	9.6	46.5	14
Snow co					1.					11.			200.1	30
SD hrs.	4.9	5.6	6.2	8.1	9.4	9.8	10.0	9.8	9.3	7.8	4.9	3.8	7.5	12
SD pc.	53	53	52	60	64	64	67	71	75	72	51	43	61	12
W.vel.	1.7	1.9	2.2	3.2	3.5	3.2	3.1	3.1	3.0	2.6	2.0	1.6	2.6	13
W.mx,dir	S	S	N	N	NW	NW	NW	NW	NW	NNW	NNW	S	NW	13
W.mx,pc	12	12	12	13	15	15	15	16	14	12	9	10	11	13
Calms	30	27	19							17	24	32	17	13

270 A l t a y 47° 44' N 88° 05' E 735.1 m **270**

	JAN	FEB	MAR	APR	MAY	JUN	JUL	AUG	SEP	OCT	NOV	DEC	YEAR	oby
A.press.	943	941	937	933	930	926	923	926	932	938	942	943	935	24
T. mean	-17.0	-15.1	-6.1	7.0	14.9	20.4	22.1	20.5	14.6	5.8	-5.2	-14.1	4.0	27
T.a.mx.	2.3	6.0	18.8	30.7	33.3	36.0	36.7	37.4	35.0	27.7	15.9	5.5	37.6	27
T.a.mn.	-41.2	-41.5	-36.4	-16.3	-4.3	1.9	6.0	0.4	-6.2	-14.8	-37.1	-43.5	-43.5	27
T.m.mx.	-10.9	-8.5	-0.2	13.0	21.1	26.6	28.2	26.9	21.2	12.0	0.3	-8.7	10.1	27
T.m.mn.	-23.2	-21.4	-11.9	1.4	8.0	13.4	15.3	13.6	8.1	0.5	-10.2	-19.9	-2.2	27
Tmn < 0														
Tmn < -10														
pTmn > 0														
pTmn > 10														
Frost d	19.6	16.7	16.4	5.6	0.3	0.0	0.0	0.0	1.4	7.6	15.9	19.9	103.5	27
r.hum.	71	72	69	49	39	44	47	45	45	51	67	73	56	26
cloud.	4.7	5.0	5.6	5.6	5.6	5.7	5.4	4.8	4.5	5.0	5.5	5.3	5.2	26
R.total	12	14	9	14	15	18	20	16	15	14	18	17	182	26
RD > .1	6.9	6.8	5.5	6.2	6.2	7.0	7.4	7.0	5.7	6.0	8.0	9.4	82.1	26
RD > 5														
RD > 10														
RD > 25														
Evap.														
Snow h.	56	73	67	36	11	0	0	0	1	8	30	45	73	27
Snow d.	6.9	6.8	5.1	1.4	0.3	0.0	0.0	0.0	0.2	1.2	6.8	9.3	37.9	27
Snow co														
SD hrs.	5.4	6.8	7.7	9.2	10.4	10.9	10.8	10.3	9.3	7.1	5.1	4.4	8.1	27
SD pc.	61	65	65	67	69	69	70	73	74	66	55	51	67	27
W.vel.	1.3	1.5	2.5	4.0	4.1	3.5	2.9	3.0	3.0	2.9	2.2	1.4	2.7	24
W.mx,dir	NE	NE	NNE	NNE	W	W	W	W	NNE	NNE	NE	NE	NNE	24
W.mx,pc	12	10	11	14	15	18	15	15	13	14	11	11	11	24
Calms	50	47	32				20	19	21	22	34	48	28	24

273 Tinghe 44° 37' N 82° 54' E 320.1 m														273
	JAN	FEB	MAR	APR	MAY	JUN	JUL	AUG	SEP	OCT	NOV	DEC	YEAR	oby
A.press.	996	993	987	981	977	972	969	972	979	986	992	994	983	26
T. mean	-16.6	-12.3	0.2	11.5	18.8	23.5	25.2	23.5	17.7	8.2	-1.5	-11.0	7.3	28
T.a.mx.	9.5	10.8	22.0	33.3	38.6	40.0	40.4	40.8	37.8	28.6	17.2	7.0	40.5	28
T.a.mn.	-36.4	-35.1	-22.7	-10.1	-1.9	5.4	10.3	5.8	-3.3	-7.8	-26.6	-35.6	-36.4	28
T.m.mx.	-10.7	-6.3	5.7	18.5	26.0	30.6	32.3	30.8	25.2	15.0	3.4	-6.5	13.7	28
T.m.mn.	-21.5	-17.6	-4.7	5.0	11.6	16.3	18.1	16.4	10.8	2.5	-5.3	-14.7	1.4	28
Tmn < 0														
Tmn < -10														
pTmn > 0														
pTmn > 10														
Frost d	26.4	22.4	14.3	2.3	0.1	0.0	0.0	0.0	0.7	9.1	19.2	23.0	117.6	27
r.hum.	79	78	69	51	44	47	50	50	51	60	74	81	61	28
cloud.	5.1	5.5	6.8	6.1	5.9	5.7	5.0	4.2	3.8	4.5	5.9	6.5	5.4	28
R.total	3	3	10	11	13	12	12	10	6	5	3	4	92	28
RD > .1	7.0	5.7	5.7	5.2	6.3	7.5	7.6	6.2	3.9	2.7	3.5	7.8	69.2	28
RD > 5														
RD > 10														
RD > 25														
Evap.														
Snow h.	13	13	10	6	0	0	0	0	0	8	6	12	13	28
Snow d.	7.1	5.8	4.2	0.3	0.0	0.0	0.0	0.0	0.0	0.2	2.8	7.7	28.1	28
Snow co														
SD hrs.	4.8	5.8	6.0	7.9	9.4	10.0	10.2	10.2	9.1	7.2	4.4	3.3	7.4	27
SD pc.	53	55	50	59	64	65	68	73	73	66	46	37	60	27
W.vel.	1.2	1.5	2.1	2.9	3.0	2.6	2.2	2.1	2.0	1.7	1.6	1.3	2.0	26
W.mx,dir	S	S	NE	N	N	N	N	N	S	S	S	S	S	26
W.mx,pc	18	17	13	14	14	15	14	15	13	12	12	14	13	26
Calms	41	34	29	18	18	21	22	22	26	34	40	42	29	26

276 Hotan	37° 08' N	79° 56' E	1374.6 m										276
	JAN	FEB	MAR	APR	MAY	JUN	JUL	AUG	SEP	OCT	NOV	DEC	YEAR	oby
A.press.	868	865	862	860	859	857	855	857	862	866	869	869	866	25
T. mean	-5.6	-0.3	9.0	16.5	20.4	23.9	25.5	24.1	19.7	12.4	3.8	-3.2	12.2	28
T.a.mx.	17.0	22.0	30.4	34.3	36.3	39.2	40.6	39.4	35.5	28.2	22.9	21.2	40.6	28
T.a.mn.	-21.6	-18.2	-4.6	-0.2	3.3	8.1	11.4	9.0	4.9	-4.0	-13.3	-18.5	-21.6	28
T.m.mx	0.1	5.4	15.5	23.4	27.1	30.9	32.6	31.1	27.0	20.2	10.6	2.7	18.9	28
T.m.mn.	-10.3	-5.1	3.0	10.1	14.0	17.4	19.1	18.0	13.3	5.9	-1.6	-7.8	6.3	28
Tmn < 0														
Tmn < -10														
ppTmn > 0														
pTmn > 10														
Frost d	14.2	8.6	2.2	0.1	0.0	0.0	0.0	0.0	0.0	5.1	18.2	20.6	68.9	28
r.hum.	53	49	35	29	35	37	40	44	43	40	45	54	42	27
cloud.	5.6	6.5	6.2	6.3	6.1	5.5	5.4	5.4	4.3	3.0	3.6	4.8	5.2	22
R.total	2	3	.8	3	7	7	4	3	3	.6	.4	.7	35	27
RD > .1	1.9	2.3	0.5	1.0	2.3	2.2	2.3	1.8	1.1	0.3	0.6	1.0	17.4	27
RD > 5														
RD > 10														
RD > 25														
Evap.														
Snow h.	14	12	4	0	0	0	0	0	0	0	5	9	14	28
Snow d.	2.0	2.2	0.3	0.0	0.0	0.0	0.0	0.0	0.0	0.1	0.5	1.2	6.3	28
Snow co														
SD hrs.	5.6	5.6	6.2	6.6	7.5	8.6	8.0	7.5	7.9	8.6	7.5	6.1	7.2	26
SD pc.	57	51	52	50	53	59	56	56	64	77	74	64	59	26
W.vel.	1.5	1.8	2.4	2.5	2.6	2.6	2.3	2.1	2.0	1.9	1.8	1.6	2.1	26
W.mx,dir	SW	SW	SW	W	W	SW	W	SW	SW	SSW	SSW	SW	SW	25
W.mx,pc	10	10	11	10	11	12	9	10	12	15	13	10	11	25
Calms	31	25	17	17	15	15	19	20	21	21	23	28	21	25

Taibei 25° 02' N 121° 31' E 8.0 m (after MUELLER)														
	JAN	FEB	MAR	APR	MAY	JUN	JUL	AUG	SEP	OCT	NOV	DEC	YEAR	oby
A.press.														
T. mean	15.2	14.8	16.9	20.6	24.1	26.6	28.2	27.9	26.2	23.0	19.8	16.9	21.7	44
T.a.mx.	28.8	31.2	32.6	34.8	36.5	37.1	38.6	37.7	36.4	36.1	33.6	31.5	38.6	37
T.a.mn.	2.6	-0.2	1.4	7.5	10.0	15.6	19.5	18.9	13.5	10.8	1.1	1.8	-0.2	37
T.m.mx.														
T.m.mn.														
Tmn < 0														
Tmn < -10														
pTmn > 0														
pTmn > 10														
Frost d														
r.hum.	84	84	84	83	82	81	78	78	80	81	81	83	82	
cloud.														
R.total	91	147	164	182	205	322	269	266	189	117	71	77	2100	52
RD > .1	17	17	18	15	16	16	14	15	14	15	15	16	188	
RD > 5														
RD > 10														
RD > 25														
Evap.	32	30	45	81	132	162	181	171	142	101	64	41	1182	
Snow h.														
Snow d.														
Snow co														
SD hrs.	2.8	2.6	2.9	3.7	4.5	5.6	7.3	7.0	6.4	4.6	3.6	3.0	4.5	
SD pc.														
W.vel.														
W.mx,dir	E	E	E	E	E	E	E	E	E	E	E	E	E	
W.mx,pc														
Calms														

H o n g k o n g 22° 18' N 114° 10' E 33 m (after MUELLER)

	JAN	FEB	MAR	APR	MAY	JUN	JUL	AUG	SEP	OCT	NOV	DEC	YEAR	oby
A.press.														
T. mean	15.6	15.0	17.5	21.7	25.6	27.5	28.1	28.1	27.2	25.0	20.9	17.5	22.5	50
T.a.mx.	26.1	26.1	28.3	31.7	32.8	34.4	34.4	36.1	34.4	34.4	30.0	27.8	36.1	50
T.a.mn.	0.0	3.3	7.2	11.1	15.6	19.4	22.2	22.2	18.3	13.9	6.7	5.0	0.0	50
T.m.mx.	17.8	17.2	19.4	23.9	27.8	29.4	30.6	30.6	29.4	27.2	23.3	20.0	25.0	50
T.m.mn.	13.3	12.8	15.6	19.4	23.3	25.6	25.6	25.6	25.0	22.8	18.3	15.0	20.0	50
Tmn < 0														
Tmn < -10														
pTmn > 0														
pTmn > 10														
Frost d														
r.hum.	72	78	79	82	83	82	82	82	78	69	67	50	75	50
cloud.														
R.total	33	46	74	292	394	381	394	361	247	114	43	30	2162	50
RD > .1	4	5	7	8	13	18	17	15	12	6	2	3	110	20
RD > 5														
RD > 10														
RD > 25														
Evap.	32	29	46	81	131	158	166	161	143	112	76	46	1181	56
Snow h.														
Snow d.														
Snow co														
SD hrs.	4.6	3.5	3.1	3.8	5.0	5.4	6.8	6.5	6.6	7.0	6.2	5.5	5.3	60
SD pc.														
W.vel.														
W.mx,dir	E	E	E	E	E	E	SW	W	E	NE	NE	NE	S	16
W.mx,pc														
Calms														

References

Acad Meteorol Science, State Meteorol Administr China (1981) Yearly charts of dryness/ wetness in China for the last 500-year period. Beijing, Map Press (in Chinese)

AKIYAMA T (1973a) The large-scale aspects of the characteristic features of Baiu front. Pap Meteorol Geophys 24:157–188

AKIYAMA T (1973b) Frequent occurrence of heavy rainfall along the north side of the low-level jet stream in the Baiu season. Pap Meteorol Geophys 24:380–388

ALISSOW BP (1954) Die Klimate der Erde. Berlin, pp 277

ARAKAWA H (1969) see WATTS IEM (1969)

ASAI T (1986) (ed) Proceed Internat Conf Monsoons in the Far East. Ocean Res Inst, Univ of Tokyo, 1985, Tokyo

BAO C (1987) Synoptic meteorology in China. Berlin Heidelberg New York, Springer

Central Meteorol Bureau of People's Rep China (1976) Atlas of Aerological Climate in China. Beijing, Geogr Map Press, pp 210

Central Weather Bureau of China, see State Meteorol Administr of China

CHANG CP, KRISHNAMURTHI TN (eds) (1987) Monsoon meteorology. Oxford, Oxford Univ Press

CHANG P (1934) On the duration of the four seasons in China. In: Collected Scient Papers, Meteorol. Acad Sinica 1954:273–323

CHANG P et al. (1956) A preliminary study on dividing Chinese climate. In: A tentative division of natural landscape of China. Science Press, Beijing, pp 37–49 (in Chinese)

CHANG Y (1961) The climate of Taipei. Meteorol Bull (Taipei) 1:15–24 (in Chinese)

CHEN G, CHI S (1981) On the frequency and speed of the "Mei-Yu" front over Southern China and adjacent areas. Pap Meteorol Res 3:31–42

CHEN L et al. (1980) The structure of the Asian monsoon circulation in summer and its relation to seasonal variation of the general circulation. Proc Conf Tropical Weather, Beijing, Science Press, pp 82–92 (in Chinese)

CHEN L, LI W (1981) The heat sources and sinks in the monsoon region of Asia. In: Proc Sympos Summer Monsoon in South East Asia 1981, Kunming, People's Press of Yunnan Province, pp 86–101 (in Chinese)

CHEN L, LUO S (1979) The circulation during the strong and weak ITCZ in the West Pacific in summer. Collect Papers Inst Atmos Phys. Acad Sinica 8, Beijing, Science Press, pp 77–85 (in Chinese)

CHEN L, YANG Y (1982) The long-range influence of the tropical monsoon circulation on the circulation over the Tibetan Plateau and its relation to drought and flood in China in summer. In: Proc Symp Summer Monsoon in South East Asia, Kunming, 1982: Kunming, People's Press of Yunnan Province, pp 158–172 (in Chinese)

CHEN S (1957) The climate of China. Shanghai, Press of New Knowledge (in Chinese)

CHEN X (1982) A new approach to the climatic division of China. Acta Meteorol Sinica 40:35–48 (in Chinese)

Chinese Geograph Soc (1984) Physical Geography of China: Climate. Beijing, Science Press, pp 161 (in Chinese)

CHROMOW SP (1950) Der Monsun als geographische Rarität. Sowjetwissenschaft, Naturwiss Abt 1950:39–62

CHROMOW SP (1957) Die geographische Verbreitung der Monsune. Peterm Geogr Mitt 101:234–237

CHU C (1929) Climatic provinces of China. Mem Inst Meteorol Nanking 1:pp 11

CHU C (1934) Southeast monsoon and rainfall in China. Journal Chin Geogr Soc 1:1–27 (in Chinese)

Chu C (1954) Southeast monsoon and rainfall in China. In: Coll Scientific Pap Meteorol, 1919–1949, Peking. Acad Sinica (in Chinese)
Chu C (1973) A preliminary study on the climate fluctuations during the last 5,000 years in China. Scientia Sinica 16:226–256
Climate atlas of China (1966) see State Meteorol Administr of China
Climate atlas of China (1978) see State Meteorol Administr of China
De Martonne E (1926) Une nouvelle fonction climatologique: L'indice d'aridité. Meteorologie 2:449–459
Ding Y, He S (1984) The mean circulation in the tropics in South Asia and the West Pacific. Science Bull 29:414–416 (in Chinese)
Feng L, Zheng S (1985) An experiment of reconstructing seasonal drought and flood sequences for the last 500 years in the western part of Honan Province. Geogr Res 4:55–61 (in Chinese)
Flohn H (1950) Studien zur allgemeinen Zirkulation der Atmosphäre. In: Berichte Deutscher Wetterdienst US-Zone 18, pp 52
Flohn H (1951) Grundzüge der atmosphärischen Zirkulation und Klimagürtel. In: Wissenschaftl Abh Deutscher Geographentag Frankfurt, pp 105–118
Flohn H (1957a) Zur Kenntnis des „Monsuns" in Ostasien. Stuttgarter Geogr Studien 69:263–275
Flohn H (1957b) Large-scale aspects of the "summer monsoon" in South and East Asia. 75th Anniversary Vol Journal Meteorol Soc Japan, pp 180–186
Flohn H (1957c) Zur Frage der Einteilung der Klimazonen. In: Erdkunde 11:161–175
Flohn H (1960) Monsoon winds and general circulation. Proc Symp Monsoons of the World, New Delhi 1960, pp 65–74
Flohn H (1968) Contributions to a meteorology of the Tibetan Highlands. Dep of Atmosph Science, Colorado State Univ Fort Collins: Atmospheric Science Paper 130, pp 120
Fu C, Fletcher J, Slutz R (1983) The structure of the Asian monsoon surface wind field over the ocean. Journal Climate and Applied Meteorol 52:1242–1252
Gao Y et al. (1962) Problem on monsoons over East Asia. Collect Papers Inst Geophys Meteorol, Acad Sinica 5. Beijing, Science Press, pp 1–106
Gong G, Chen E, Wen H (1979) The climatic fluctuations in the Heilongjiang Province, China. Acta Geogr Sinica 34:129–138 (in Chinese)
Guo Q (1983) The summer monsoon intensity index in East Asia and its variation. Acta Geogr Sinica 38:207–217 (in Chinese)
Guo Q, Wang Y (1986) The snow cover on Tibet Plateau and its effect on the monsoon over East Asia. Plateau Meteorol 5:116–124 (in Chinese)
Guo Q, Ye W (1979) The circulation in Southern and Northern Hemisphere and the monsoon over East Asia. Acta Meteorol Sinica 37:86–95 (in Chinese)
Guo Q et al. (1979) Classification of precipitation distribution during the wet season in China and its seasonal change. Dili-jikan 11:16–39 (in Chinese)
Hanson-Lowe J (1941) Notes on the climate of the South Chinese-Tibetan borderland. Geogr Rev 31:444–453
Hayashi Y (1977) Distribution on rainfall accompanied with the monsoon around the Indo-China Peninsula. Geogr Rev Jpn 50:238–248 (in Japanese)
He D (1980) On the tropical climate of Hainan Island. South China Teacher's College (Natural Science Studies) 2:114–122 (in Chinese)
Henderson-Sellers R (1986) Contemporary climatology. New York
Hou G, Jiang S (1980) To estimate the average temperatures from January to December at the Tibetan Plateau. Acta Geogr Sinica 35:265–269 (in Chinese)
Huang B (1984a) Dr. Zhu Kezhen (Co-ching Chu) and scientific investigations of tropical China (1). The demarcation of tropical and subtropical belts. Geogr Res 3:9–18 (in Chinese)
Huang B (1984b) Dr. Zhu Kezhen (Co-ching Chu) and scientific investigations of tropical China (2). The prospects of the development of Hainan. Geogr Res 3:81–96 (in Chinese)
Huang B (1986) Climatic division and physio-geographic division of China: retrospectives and prospects. Paper presented at the German-Chinese Workshop on The Climate of China. Inst of Geogr, Mainz Univ, FRG 1986 (unpubl)
Huang S (1986) (ed) Heavy rainfall in South China in pre-typhoon season. Guangzhou: Guangdong Press of Technology, pp 362 (in Chinese)

JIANG A (1982) On the variation of temperature with height of mountainous regions in tropical China and its effect on cultivation of rubber tree. Climatol Not (Tsukuba) 29:19–29

JIANG A (1984) Climate and agricultural land use in China. In: YOSHINO MM (ed) Climate and agricultural land use in monsoon Asia. Tokyo, Univ of Tokyo Press, pp 297–316

JIANG A (1986) Climate and agriculture in China. Paper presented at the German-Chinese Workshop on The Climate of China. Inst of Geogr. Mainz Univ, FRG 1986 (unpubl)

JIANG A, HAO Y (1986) A comparative study of agroclimates in Hainan Island and Xishuangbanna. Climatol Not (Tsukuba) 35

KAO Y, TSANG Y (1957) Typhoon tracks and some related statistical analyses. Peking, Science Press, pp 136 (in Chinese)

KÖPPEN W (1884) Die Wärmezonen der Erde. Meteorol Zeitschrift 1:215–226

KÖPPEN W (1923) (1st ed) 1931 (2nd ed) Grundriß der Klimakunde. Berlin, pp 388

KOLB A (1986) Xinjiang als Naturraum und ökologisches Problemgebiet. Geoökodynamik 7:29–40

LANDSBERG HE et al. (1963) Weltkarten zur Klimatologie (eds RODENWALDT E, JUSATZ HJ). Berlin Göttingen Heidelberg

Lanzhou Inst of Glaciology, Cryopedology and Desert Research, Division of Glaciology (1975) Basic features of the glaciers of the Mt. Qomolangma Region, southern part of the Tibet Autonomous Region. Scientia Sinica 18:106–130

LAUER W (1952) Humide und aride Jahreszeiten in Afrika und Südamerika und ihre Beziehung zu den Vegetationsgürteln. Bonner Geogr Abhandlungen 9:15–98

LIN C (1982) The establishment of summer monsoon over the middle and lower reaches of the Yangtze River and the seasonal transition of circulation over East Asia in early summer. Proc Symp Summer Monsoon South East Asia, Kunming, People's Press of Yunnan Province. 21–28 (in Chinese)

LIN Z (1981) A study of the influence of orography on diurnal changes of temperature, pressure, wind velocity, precipitation, relative humidity and duration of sunshine. Acta Geogr Sinica 36:392–403 (in Chinese)

LU A (1937) The cold waves of China. In: Collected Scientific Papers (Meteorology). Nanjing, The National Res Inst Meteorol 10:6–34 (in Chinese)

LU A (1946) Discussion on a climate classification in China. Acta Geogr Sinica 12:1–10 (in Chinese)

LU A (1949) The climatic provinces of China. In: Collect Sci Pap (Meteorology). Acad Sinica Peking 1954, pp 441–466

LU A (1954) The climatic provinces of China. In: Collect Sci Pap Meteorol, Peking, pp 469–474

LU C et al. (1965) A study on the wet and dry periods and regionalization of China according to aridity. Acta Geogr Sinica 31:15–24 (in Chinese)

LU W (1954) General comments of the climate of China. Shanghai, Sangwu Press (in Chinese)

LUI S (1957) Climate of Lhasa. Monthly Weather 10:15–19 (in Chinese)

MECKELEIN W (1986) Zur physischen Geographie und agraren Nutzungsproblemen in den innerasiatischen Wüsten Chinas. Geoökodynamik 7:1–28

MU F (1986) Fireland in China, Turpan Basin: Its drier-hotter climate and utilization. Arid Land Geogr 9:7–12 (in Chinese)

MÜLLER MJ (1983) (3rd ed) Handbuch ausgewählter Klimastationen der Erde. Trier, Universität Trier, Forschungsstelle Bodenerosion

National Physiogeogr Atlas of the People's Rep of China (1965) Inst of Geography, Academy of Science of China, Beijing Map Press (in Chinese)

Natural Division Committee, Academia Sinica (1959) Climatic division of China. In: Natural division of China. Peking, Science Press. pp 311–323 (in Chinese)

PAULHUS JLH (1965) Indian Ocean and Taiwan rainfalls set new records. Monthly Weather Review 93:331–335

PENG G (1965) Synoptic and climatic studies of the genesis and development of the East China Sea and Northeast China cyclones. Proc Inst Geogr, Academia Sinica, No 9

PENG G, ZUO D, CHENG J (1966) Basic characteristics of the climate of the Xizang Plateau, Proceed of complex expeditions of the Academia Sinica

PENG G, LU W, YIN Y (1980) Some questions of the pole displacement and climate, Atmospheric Sciences, Vol. 4, No 44

PENG G, LU W (1982) Development of atmospheric circulation and irregularity of earth position, Scientia Sinica (Ser B), Vol 25, No 5
PENG G, SI Y, LU W (1983) Application of geographical factors to long-range weather forecasts, Kexue Tongbao (Journal Sciences) Vol 28, No 7
PENG G, LU W (1983) Natural factors of fourth kind of the climate. Beijing, Scientific Press
PENG G, DOMRÖS M (1987) Connections of the West Pacific subtropical high and some hydroclimatic regimes in China with Antarctic ice-snow indices. Meteorology and Atmospheric Physics 37:61–71
PENG G, DOMRÖS M (1987) Statistical studies of the atmospheric circulation of the Northern Hemisphere, hydroclimatic regimes in China and Antarctic ice-snow cover. Proceed Vancouver Symp, Aug 1987. IAHS Publ No 166:61–72
PENMAN HL (1956) Evapotranspiration – an introductory survey. Meteorol Journal of Agricult Sciences 4:9–29
QION J, LIN L (1965) A preliminary study on the dry and wet climatic regionalization of China. Acta Geogr Sinica 31:1–14 (in Chinese)
QIU B (1983) Further study on the regionalization of agroclimate of China. Acta Geogr Sinica 38:154–165 (in Chinese)
QIU B, LU Q (1980) A tentative regionalization of agroclimate of China. Acta Geogr Sinica 35:116–125 (in Chinese)
RAMAGE CS (1952) Diurnal variation of summer rainfall over East China, Korea and Japan. Journal Meteorol 9:83–86
SCHARLAU K (1950) Zur Einführung eines Schwülemaßstabes und Abgrenzung von Schwülezonen durch Isohygrothermen. Erdkunde 4:188–201
SHA W et al. (1979) Relationship between the subtropical anticyclone and the Mai-yu rainfall in the lower course of the Yangtze River and SST over the Pacific. Dili-jikan 11:126–137 (in Chinese)
SHENG C et al. (1986) General comments of the climate of China. Beijing Science Press, pp 538 (in Chinese)
SHI Y, LI J (1981) Glaciological research of the Qinghai-Xizang Plateau in China. In: Geological and ecological studies of the Qinghai-Xizang Plateau, Vol II, Environment and Ecology of Qinghai-Xizang Plateau (Proceedings of a Symposium on Qinghai-Xizang (Tibet) Plateau), Beijing, 1980; Beijing, New York, pp 1589–1597
SHI Y, XIE Z, ZHENG B, LI J (1980) Distribution, features and variations of glaciers in China. In: World Glacier Inventory (Proceedings of the Riederalp Workshop, September 1978) IAHS Publ 61, no 126:111–116
SI Y (1984) Summer precipitation characteristics in the North China Plain. Acta Geogr Sinica 39:115–120 (in Chinese)
State Meteorol Administration of China (1966) Climate atlas of China. Beijing, Map Press (in Chinese)
State Meteorol Administration of China (1978) Climate atlas of China. Beijing, Map Press (in Chinese)
State Meteorol Administration of China (1981) Yearly charts of dryness/wetness in China for the last 500 year period. Beijing: Map Press (in Chinese)
SUN A, FAN J (1985) Characteristics of the distribution of temperature variability in China. Acta Geogr Sinica 40:11–19 (in Chinese)
TANG M (1985) The distribution of precipitation in mountain Qilion (Nanshan). Acta Geogr Sinica 40:323–332 (in Chinese)
TANG M, HUANG S (1981) On the advance of the summer monsoon in South East Asia. Proc Symp Summer Monsoon in Southeast Asia 1981, Kunming. People's Press of Yunnan Province: 15–30 (in Chinese)
TANG M, SHEN Z, CHEN Y (1979) On climatic characteristics of the Xizang plateau monsoon. Acta Geogr Sinica 34:33–42 (in Chinese)
TAO S (1949) Analysis of rainfall in China at various places and a new climate classification in China. Acta Meteor Sinica 20
TAO S (1984) Advances in climatology in China. In: WU C, WANG N, LIN C, ZHAO S (eds) Geographical Soc of China, Geography in China. Beijing, Science Press, pp 33–49
TAO S (1984) (ed) The climate of China. In: Physical geography of China. Beijing, Science Press, pp 161 (in Chinese)

TAO S (1986a) The summer monsoon in eastern Asia. Paper presented at the German-Chinese Workshop on The Climate of China, Institute of Geogr, Mainz University, FRG 1986 (unpubl)
TAO S (1986b) The severe rainstorms in China. Paper presented at the German Chinese Workshop on The Climate of China, Inst of Geogr Mainz University, FRG 1986 (unpubl)
TAO S et al. (1980) Severe rainstorms in China. Beijing, Science Press, pp 225 (in Chinese)
TAO S, CHEN L (1957) The structure of the summer circulation over Asia. Acta Meteorol Sinica 28:234–247 (in Chinese)
TAO S, CHEN L (1986) The East Asian summer monsoon. In: Proc Conf Monsoons in the Far East, Tokyo 1985, pp 1–11
TAO S, HE S (1983) An observational study on the onset of the summer monsoon over Eastern Asia in 1979. Scientia Atmosph Sinica 7:348–355 (in Chinese)
TAO S, XU S, GUO Q (1962) The circulations with the persistent flood or drought in the Yangtze Valley in summer. Acta Meteorol Sinica 32:91–103 (in Chinese)
THORNTHWAITE CW (1933) The climate of the earth. Geogr Rev 23:433–440
THORNTHWAITE CW (1948) An approach towards a rational classification of climate. Geogr Rev 38:55–94
TIETZE W, DOMRÖS M (1987) The climate of China. GeoJournal 4:265–266
TROLL C (1955) Der jahreszeitliche Ablauf des Naturgeschehens in den verschiedenen Klimagürteln der Erde. Studium Generale 8
TROLL C, PAFFEN K-H (1964) Karte der Jahreszeitenklimate der Erde. Erdkunde 18:5–28
Tropical Oceanic and Meteorol Inst (1984) Rainstorms in the pretyphoon period in Guangdong. Guangzhou, Guangdong Science Popularization Press (in Chinese)
TU C (1936a) Climatic provinces of China. J Geophys Soc China 3:1–34 (in Chinese)
TU C (1936b) Climate classification in China. J Geophys Soc China 3:495–528 (in Chinese)
TU C, HUANG S (1944a) The onset and retreat of summer monsoon in China. Acta Meteorol Sinica 18:81–92 (in Chinese) (Bull American Meteorol Soc 26:9–22)
TU C, HUANG S (1944b) The advance and retreat of the summer monsoon. Meteorol Map 18:1–20
WALTER H, LIETH H (1967) Klimadiagramm-Weltatlas. Jena
WANG B (1983) Sea fog. Beijing, Ocen Publ, pp 352 (in Chinese)
WANG S (1981) Summer rainfall for last 500 years. GeoJournal 5:117–122
WANG S, TAO S (1984) The structure of the cross-equatorial flow in the South China Sea. Acta Oceanol Sinica 6:160–173 (in Chinese)
WANG S, ZHANG P, ZANG D (1981a) Further studies on the climatic change during historical times in China. Proc Tech Conf Climate Asia and Western Pacific, WMO 578:376–393
WANG S, ZHAO Z, CHENG Z (1981b) Reconstruction of the summer rainfall regime for the last 500 years in China. GeoJournal 5:117–122
WANG S, ZHAO Z (1982) The 36-year wetness oscillation in China and its mechanism. Acta Meteorol Sinica 37:61–72 (in Chinese)
WANG T (1941) Die Dauer der ariden, humiden und nivalen Zeiten des Jahres in China. Tübinger Geogr und Geolog Abhandlungen (2) 7:33
WATTS IEM (1962) The diurnal variation of frequency of precipitation over Southeast Asia. Reg Conf Southeast Asian Geographers, Kuala Lumpur
WATTS IEM (1969) Climates of China and Korea. In: ARAKAWA H (ed): Climates of Northern and Eastern Asia. World Survey of Climatology (LANDSBERG HE, ed) Vol 8. Amsterdam, London, New York, Elsevier Publ Comp, pp 1–74 and climate tables
VON WISSMANN H (1937) Niederschlagskarte von China mit Begleitworten. Zeitschrift Gesellschaft für Erdkunde zu Berlin 1937:38–43
VON WISSMANN H (1939) Die Klima- und Vegetationsgebiete Eurasiens. Zeitschrift Gesellschaft für Erdkunde zu Berlin 1939:1–14
VON WISSMANN H (1959) Die heutige Vergletscherung und Schneegrenze in Hochasien. Verlag der Akademie der Wissenschaften und der Literatur in Mainz
XU S (1982) Summer monsoon activity and the drought and flood in the Yangtze Valley. Geogr Research 1; Beijing, Science Press, pp 58–68 (in Chinese)
XU S, ZHENG S (1979) Thirty years of climatology in China. Acta Geogr Sinica 34:293–304 (in Chinese)

XU X, ZHU M (1984) The vegetational and climatic changes in the Zhenjiang region since 1,500 years B.P. Acta Geogr Sinica 39:277–284 (in Chinese)
YANG C et al. (1960) Meteorology of the Tibetan Plateau. Beijing, Science Press (in Chinese)
YANG Q, SHEN K (1984) On vertical zonation of the Northwestern Yunnan. Acta Geogr Sinica 39:141–147 (in Chinese)
YAO C (1958) The variability of precipitation in eastern China. Acta Meteorol Sinica 29:225–238 (in Chinese)
YAO L, WANG A, WANG Q, LUO S (1984) A study of the mean heat source over Qinghai-Xizang (Tibet) Plateau and its neighbouring areas in Summer. Proc Symp Meteorol Experiment on Qinghai-Xizang Plateau (I). Beijing, Science Press, pp 291–302 (in Chinese)
YE D, FU C, CHAO J, YOSHINO M (1987) The climate of China and global climate. Proc Beijing Intern Symp on Climate, 1984; Beijing: China Ocean Press 1987, Berlin Heidelberg New York, Springer, pp 441
YE D, GAO Y (1979) Meteorology of Qinghai-Xizang (Tibet) Pateau. Beijing, Science Press (in Chinese)
YE D, LUO S, ZHU B (1957) The structure of the circulation and the heat balance in Tibetan Plateau and its neighbourhoods. Acta Meteorol Sinica 28:108–121 (in Chinese)
YEH T, CHAO J, YOSHINO M (1985) An international symposium on climate in Beijing, People's Republic of China. Bull Amer Meteorol Soc 66, No 9:1147–1152
YOSHINO MM (1963) Rainfall, frontal zones and jet streams over East Asia. Bonner Meteorol Abh 3:1–129
YOSHINO MM (1969) Climatological studies on the polar frontal zone and intertropical convergence zones over South, Southeast and East Asia. Climatol Notes (Tsukuba) 1:1–71
YOSHINO MM (1971) Water balance of Monsoon Asia. Univ of Tokyo Press, Tokyo, and Univ of Hawaii Press, Honolulu, pp 308
YOSHINO MM (1984) Climate and agriculture on the Hainan Island, South China: A Preliminary Study. Geogr Rev of Jpn, Ser B 57 (2):166–182
YOSHINO MM (1986) (ed) Climates, geoecology and agriculture in tropical China. Climatol Notes (Tsukuba) 35, pp 244
YOSHINO MM, CHIBA M (1984) Regional division of China by precipitation. Geogr Rev Jpn, Ser A 57:583–590
YOSHINO MM, URUSHIBARA K (1981) Regionality of climatic change over East Asia. Geo Journal 5:123–132
ZHANG B (1959) Climate classification in China. Beijing, Science Press (in Chinese)
ZHANG J (1959) Some problems about the monsoon features in China. Acta Meteorol Sinica 30:350–361 (in Chinese)
ZHANG J (1982) The seasonal variation of the atmospheric circulation and the scientific concept of monsoon. Proc Symp Summer Monsoon South East Asia, Kunming People's Press of Yunnan Province, pp 1–9 (in Chinese)
ZHANG J, LIU E et al. (1985) Hydrological characteristics of streams in Qaidam Basin. Acta Geogr Sinica 40:242–255 (in Chinese)
ZHANG J, LIU J, ZHOU Y (1984) Some patterns of the heat island in Beijing. Acta Geogr Sinica 39:428–435 (in Chinese)
ZHANG J, LIN Z (1985) Climate of China. Shanghai, Science and Technology Press of Shanghai, pp 603 (in Chinese)
ZHANG P, GONG G (1979) Some characteristics of climatic fluctuation in China since 16th century. Acta Geogr Sinica 34:238–247 (in Chinese)
ZHANG X (1981) The analysis of the index of dryness in the last 500 years in eastern China. Proc Conf Fluctuation of the Climate, 1978; Beijing: Science Press (in Chinese)
ZHANG J et al. (1983) The interannual variation of the summer monsoon and its association with floods and droughts in China (unpubl) after TAO S and CHEN L (1986)
ZHAO S (1986) Physical Geography of China. Beijing, Science Press, pp 209
ZHAO W, XU B, HU X (1984) The seasonal variation in circulation in Asia in early summer, 1979. Proc Symp Meteorol Experiment on Qinghai-Xizang (Tibet) Plateau (I). Beijing Science Press, pp 142–153 (in Chinese)
ZHENG S (1983) The patterns of dryness from cold and warm decades in historical times of China. Geogr Res 2:32–40 (in Chinese)

ZHENG S, FENG L (1986) Historical evidence on climatic instability above normal in cool periods in China. Scientia Sinica, Ser B 24:441–448 (in Chinese)
ZHENG Z (1980) A discussion on northern limit of the tropics zone in China. Acta Geogr Sinica 35:87–92 (in Chinese)
ZHOU S (1981) On the climatic features of the Hainan Island. Journal East China Normal Univ, Natural Science Ed 1:61–71 (in Chinese)
ZHU K, see CHU C
ZHU B (1962) The climate of China. Beijing, Science Press, pp 1–363 (in Chinese)

Subject Index

Autumn, see Seasons

Certain limited temperature 101–114
Climate Atlas of China 17–19
Climate classification 230–232
  of China 16, 232–257
  Chen 248–252
  Chu Koching 242–243
  Flohn 238–240
  Huang 253–257
  Köppen 232–235
  Troll and Paffen 235–238
  v. Wissmann 240–242
  Zhang 245–249
Climate, controlling factors 20–29
  map 3
  monographs on China 17–19
  periodicals on China 18–19
  stations 2–14
  zones of China 258–280
    Alpine Plateau 276
    Cold Temperate 259
    Middle Subtropical 268–269
    Middle Temperate 260–262
    Middle Tropical 274
    Northern Subtropical 266–267
    Peripheral Tropical 272–273
    Southern Subtropical 270–271
    Southern Tropical 275
    Sub-Alpine Plateau 277
    Temperate Plateau 278–280
    Warm Temperate 263–265
Climatological front 59–65
Cloudiness 210–212
  annual 210–212
  January 211, 212
  July 212, 213
Cold waves 47–50
Crachin 29

Disturbances 65–76
Diurnal range of temperature
  119–122
Dry and hot wind 229
Dryness/wetness index 196–199
Duration of the monsoon,
  see Monsoon

Empty Mei-Yu 64–65

Fog 220–221
Front 59–65
  Mei-Yu 62–65
Frost 102, 103–107, 133–135
Frost damage 51–52

Heavy rainfall 186–190
Historical change of precipitation
  195–201
Historical change of temperature
  130–138
Hot day 110, 111
Hot wind 229

Index
  dryness/wetness 196–199
  monsoon 41–42
  summer monsoon 57–58

Lake breeze 227
Land and sea breeze 227
Lapse-rate of temperature 114–116

Mei-Yu 29, 58
  empty 62–65
  front 29, 58
  precipitation 157
Monsoon 28–29, 39–58
  angle 41

index 41–42
plateau 227–229
Mountain and valley breeze 224–226

Onset of the monsoon, see Monsoon

Plum rains, plum season 29, 62, 64
Pre-typhoon season 59–62
Precipitation 139–209
annual 139–141
April 151, 152
diurnal variation 190–191
historical change 195–201
inter-annual variability 168–175
January 142, 151
July 151–153
Mei-Yu 157
October 153–155
rainfall intensity 183–186
rainy days 182–183
summer 168–169
types of annual variation 155–160
wet and dry months 160, 165–168

Rainfall, see Precipitation
Rainfall intensity 183–186
Rainy day 182–183
Rainstorms 186–190

Seasons 28–29
Mei-Yu 62–65
pressure distribution 30–33
pre-typhoon 59–62
summer precipitation 168–169
winds and air masses 34–39
Snow 201–209
cover period 201–204
depth 206–207
snowfall day 204–206
snowline 207–209
Solar radiation 218–220
Spring, see Seasons
Summer, see Seasons
index 57–58
monsoon, see Monsoon
Sunshine 212–218
annual 213–215
annual variation 217–218
January 216
July 217

Temperature 77–138
annual 77–80
annual range 86–92
April 82, 83
certain limited 101–114
diurnal range 119–122
global comparison 116–119
historical change 130–138
inter-annual variability 122–124, 130
January 80–83
July 83–85
lapse-rate 114–116
October 85–86
types of annual variation 90–92
vertical distribution 114–116
Typhoon 71–76

Variability of temperature 122–124, 130
Variation of temperature 90–92
of diurnal precipitation 190–191
Vertical distribution of temperature 114–116

Wetness/dryness index 196–199
Waves, cold 47–50
Wind 222–229
direction 34–39
dry and hot 229
velocity 222–224
Winter, see Seasons
Winter monsoon, see Monsoon
active and weak, see Monsoon
Withdrawal of the monsoon, see Monsoon

Index of Climate Tables

Altay/270 346
Arxan/7 286
Bailingmiao/56 296
Bayan Mod/252 336
Beijing/61 298
Changchun/27 292
Changsha/147 318
Chengde/51 295
Chengdu/134 315
Chongqing/142 317
Da Qaidam/259 339
Datong/59 297
Deqen/230 328
Dongfang/210 324
Dong Ujmqin Qi/21 291
Fuzhou/183 322
Garze/233 329
Golmud/258 338
Guangzhou/199 323
Guilin/177 321
Guiyang/173 320
Hailar/6 285
Hami/265 343
Hangzhou/123 312
Harbin/15 288
Hefei/114 309
Hongkong 350
Hotan/276 348
Huimin/72 302
Huma/2 283
Jinan/77 303
Jiuquan/261 341
Jixi/17 289
Kunming/225 327
Lancang/222 326
Lanzhou/247 334
Lenghu/260 340
Lhasa/239 331
Mudanjiang/19 290
Nagqu/240 332
Nanjing/116 310
Nanning/215 325
Nenjiang/4 284
Qiqihar/8 287
Ruoqiang/264 342
Shanghai/119 311
Shenyang/38 294
Taibei/285 349
Taiyuan/69 301
Tianjin/62 299
Tinghe/273 347
Turpan/267 344
Ürümqi/269 345
Wenzhou/154 319
Wudu/101 308
Wuhan/126 313
Xi'an/95 307
Xichang/138 316
Xigaze/238 330
Xining/256 337
Xuzhou/87 305
Yan'an/81 304
Yanji/31 293
Yichang/129 314
Yinchuan/250 335
Yulin/68 300
Yushu/243 333
Zhengzhou/90 306

MIX
Papier aus verantwortungsvollen Quellen
Paper from responsible sources
FSC® C105338

If you have any concerns about our products,
you can contact us on
ProductSafety@springernature.com

In case Publisher is established outside the EU,
the EU authorized representative is:
Springer Nature Customer Service Center GmbH
Europaplatz 3, 69115 Heidelberg, Germany

Printed by Libri Plureos GmbH
in Hamburg, Germany